Electron Transfer in Biology and the Solid State

ADVANCES IN CHEMISTRY SERIES **226**

Electron Transfer in Biology and the Solid State

Inorganic Compounds with Unusual Properties

Michael K. Johnson, EDITOR
R. Bruce King, EDITOR
Donald M. Kurtz, Jr., EDITOR
Charles Kutal, EDITOR
Michael L. Norton, EDITOR
Robert A. Scott, EDITOR

University of Georgia

Developed from a symposium sponsored
by the Division of Inorganic Chemistry
at the Biennial Inorganic Chemistry Symposium
of the American Chemical Society,
Athens, Georgia,
March 1–4, 1989

American Chemical Society, Washington, DC 1990

Library of Congress Cataloging-in-Publication Data

Electron transfer in biology and the solid state: inorganic compounds with unusual properties
  Michael K. Johnson, editor . . . [et al.].
  p. cm.—(Advances in chemistry series, ISSN 0065-2393; 226).

"Developed from a symposium sponsored by the Division of Inorganic Chemistry at the Biennial Inorganic Chemistry Symposium of the American Chemical Society, Athens, Georgia, March 1–4, 1989."

  Includes bibliographical references.

  ISBN 0–8412–1675–4

  1. Metalloproteins—Congresses. 2. Electron transport—Congresses. 3. Solid state chemistry—Congresses.

  I. Johnson, Michael K., 1953– . II. American Chemical Society. Division of Inorganic Chemistry.
  III. Inorganic Chemistry Symposium (1989: Athens, Ga.)
  IV. Series.

QD1.A355 no. 226
[QP552.M46]
540 s—dc20
[574.19'245]   90–30119   CIP

The paper used in this publication meets the minimum requirements of American National Standard for Information Sciences—Permanence of Paper for Printed Library Materials, ANSI Z739.48—1984. ∞

Copyright © 1990

American Chemical Society

All Rights Reserved. The appearance of the code at the bottom of the first page of each chapter in this volume indicates the copyright owner's consent that reprographic copies of the chapter may be made for personal or internal use or for the personal or internal use of specific clients. This consent is given on the condition, however, that the copier pay the stated per-copy fee through the Copyright Clearance Center, Inc., 27 Congress Street, Salem, MA 01970, for copying beyond that permitted by Sections 107 or 108 of the U.S. Copyright Law. This consent does not extend to copying or transmission by any means—graphic or electronic—for any other purpose, such as for general distribution, for advertising or promotional purposes, for creating a new collective work, for resale, or for information storage and retrieval systems. The copying fee for each chapter is indicated in the code at the bottom of the first page of the chapter.

The citation of trade names and/or names of manufacturers in this publication is not to be construed as an endorsement or as approval by ACS of the commercial products or services referenced herein; nor should the mere reference herein to any drawing, specification, chemical process, or other data be regarded as a license or as a conveyance of any right or permission to the holder, reader, or any other person or corporation, to manufacture, reproduce, use, or sell any patented invention or copyrighted work that may in any way be related thereto. Registered names, trademarks, etc., used in this publication, even without specific indication thereof, are not to be considered unprotected by law.

PRINTED IN THE UNITED STATES OF AMERICA

Advances in Chemistry Series

M. Joan Comstock, *Series Editor*

1990 ACS Books Advisory Board

Paul S. Anderson
Merck Sharp & Dohme Research
 Laboratories

V. Dean Adams
Tennessee Technological
 University

Alexis T. Bell
University of California—
 Berkeley

Malcolm H. Chisholm
Indiana University

Natalie Foster
Lehigh University

G. Wayne Ivie
U.S. Department of Agriculture,
 Agricultural Research Service

Mary A. Kaiser
E. I. du Pont de Nemours and
 Company

Michael R. Ladisch
Purdue University

John L. Massingill
Dow Chemical Company

Robert McGorrin
Kraft General Foods

Daniel M. Quinn
University of Iowa

Elsa Reichmanis
AT&T Bell Laboratories

C. M. Roland
U.S. Naval Research Laboratory

Stephen A. Szabo
Conoco Inc.

Wendy A. Warr
Imperial Chemical Industries

Robert A. Weiss
University of Connecticut

FOREWORD

The ADVANCES IN CHEMISTRY SERIES was founded in 1949 by the American Chemical Society as an outlet for symposia and collections of data in special areas of topical interest that could not be accommodated in the Society's journals. It provides a medium for symposia that would otherwise be fragmented because their papers would be distributed among several journals or not published at all. Papers are reviewed critically according to ACS editorial standards and receive the careful attention and processing characteristic of ACS publications. Volumes in the ADVANCES IN CHEMISTRY SERIES maintain the integrity of the symposia on which they are based; however, verbatim reproductions of previously published papers are not accepted. Papers may include reports of research as well as reviews, because symposia may embrace both types of presentation.

ABOUT THE EDITORS

MICHAEL K. JOHNSON is an associate professor of chemistry and biochemistry at the University of Georgia. He received his Ph.D. from the University of East Anglia (England) and his B.A. in natural sciences at Cambridge University (England). He is currently an Alfred P. Sloan fellow. Johnson has authored approximately 60 research papers and book chapters. His research expertise in the field of bioinorganic and biophysical chemistry focuses on the use of spectroscopic techniques for investigating the structural, electronic, and magnetic properties of transition metal centers in metalloproteins.

R. BRUCE KING is the Regents' Professor of Chemistry at the University of Georgia. He received his Ph.D. from Harvard University and his B.A. in chemistry from Oberlin College. Among his numerous awards are the American Chemical Society Award in Pure Chemistry, Georgia Scientist of the Year Award, and a Japan Society for the Promotion of Science Fellowship. King has authored approximately 450 research papers and monograph chapters. His research expertise includes synthetic organometallic and organophosphorus chemistry, as well as the development of new applications of mathematical methods to chemical problems. King has also authored or edited a dozen books, including *Organometallic Syntheses* and symposium volumes in the areas of inorganic compounds with unusual properties, solar energy storage, and chemical applications of topology and graph theory. In addition, King has served as the American regional editor of the *Journal of Organometallic Chemistry* since 1981 and is an original editorial board member of the *Journal of Mathematical Chemistry*, founded in 1987.

DONALD M. KURTZ, JR., is an associate professor of chemistry at the University of Georgia. He graduated with a B.S. degree in chemistry from the University of Akron. He received his Ph.D. degree in chemistry (majoring in physical biochemistry) at Northwestern University under Irving M. Klotz. He was a National Institutes of Health postdoctoral fellow under Richard H. Holm at Stanford University during 1977–79. In 1979 he joined the faculty in the Department of Chemistry at Iowa State University. In 1986 he moved to the University of Georgia. Professor Kurtz is a National Institutes of Health Research Career Development Awardee during 1988–1993. He has coauthored more than 50 research papers dealing with various aspects of inorganic biochemistry. His research interests involve the inorganic chemistry and biochemistry associated with nonheme iron proteins.

CHARLES KUTAL is a professor and associate director in the Department of Chemistry at the University of Georgia. He received his Ph.D. and M.S. from the University of Illinois and his B.S. from Knox College. He is the recipient of the Undergraduate Teaching Award and the Chemist of the Year Award of the Northeast Georgia Section of the American Chemical Society. Kutal has authored over 60 articles and chapters dealing with the photochemistry of inorganic and organometallic compounds, solar energy conversion and storage, and photosensitive polymer materials. He is coeditor of *Solar Energy: Chemical Conversion and Storage*.

MICHAEL L. NORTON is an assistant professor of inorganic chemistry at the University of Georgia. He received his Ph.D. in solid state chemistry from Arizona State University and his B.S. degree in chemistry from Louisiana State University in Shreveport. Norton is a former National Research Council postdoctoral fellow. Norton's research focuses on the design and preparation of natural and artificial superlattices capable of supporting the exciton mechanism of high-temperature superconductivity. After completing graduate work in the area of layered antiferromagnetic materials under the direction of W. S. Glaunsinger, Norton took a position as a research chemist in the Naval Weapons Center in China Lake, California, where his investigations centered on oxide superconductors, including the tungsten bronzes and bismuth bronzes. Norton has authored or coauthored numerous research papers and chapters in the area of high-temperature superconductivity. He has also organized and chaired sections related to superconductivity at several meetings.

ROBERT A. SCOTT is an associate professor of chemistry and biochemistry at the University of Georgia. He received his Ph.D. from the California Institute of Technology and his B.S. from the University of Illinois and was an assistant professor of chemistry at the University of Illinois before moving to Georgia. He was the recipient of a Beckman Research Award at the University of Illinois, a Presidential Young Investigator Award from the National Science Foundation, and an Alfred P. Sloan Research Fellowship. Scott has coauthored more than 50 research papers on various aspects of inorganic biochemistry. Aside from his expertise in the application of X-ray absorption spectroscopy to biological systems, he has an active research program to study the mechanisms of long-range electron transfer in metalloproteins. He organized and chaired a symposium on this subject in 1987 at the American Chemical Society meeting in Denver.

CONTENTS

Preface xv

BIOLOGICAL ELECTRON TRANSFER

1. Overview of Biological Electron Transfer 3
   R. J. P. Williams

THEORETICAL ASPECTS OF BIOLOGICAL ELECTRON TRANSFER

2. Formalism for Electron Transfer and Energy Transfer in Bridged Systems 27
   J. R. Reimers and N. S. Hush

3. Some Aspects of Electron Transfer in Biological Systems 65
   Norman Sutin and Bruce S. Brunschwig

EXPERIMENTAL APPROACHES TO BIOLOGICAL ELECTRON TRANSFER: PEPTIDES AND PROTEINS

4. Directional Electron Transfer in Ruthenium-Modified Cytochrome c 91
   Stephan S. Isied

5. Photoinduced Electron Transfer Across Peptide Spacers 101
   Leonardo A. Cabana and Kirk S. Schanze

6. Energetics and Dynamics of Gated Reactions: Control of Observed Rates by Conformational Interconversion 125
   Brian M. Hoffman, Mark A. Ratner, and Sten A. Wallin

7. Electronic Coupling and Protein Dynamics in Biological Electron-Transfer Reactions 147
   J. S. Bashkin and G. McLendon

8. Electrostatic, Steric, and Reorganizational Control of Electron Self-Exchange in Cytochromes 161
   Dabney White Dixon and Xiaole Hong

9. Electron-Transfer Kinetics of Singly Labeled Ruthenium(II) Polypyridine Cytochrome c Derivatives 181
   Bill Durham, Lian Ping Pan, Seung Hahm, Joan Long, and Francis Millett

EXPERIMENTAL APPROACHES TO BIOLOGICAL ELECTRON TRANSFER: INORGANIC COMPLEXES

10. High-Pressure Studies of Long-Range Electron-Transfer Reactions in Solution 197
    Nita A. Lewis and Daniel V. Taveras

11. Intramolecular Electron Transfer from Photoexcited Ru(II) Diimine Complexes to N,N'-Diquaternarized Bipyridines 211
    Russell H. Schmehl, Chong Kul Ryu, C. Michael Elliott, C. L. E. Headford, and S. Ferrere

12. Bridged Mixed-Valence Systems: How Polarizable Bridging Ligands Can Lead to Interesting Spectroscopic and Conductive Properties 225
    Mary Jo Ondrechen, Saeed Gozashti, Li-Tai Zhang, and Feimeng Zhou

13. Chiral Recognition by Metal–Ion Complexes in Electron-Transfer Reactions 237
    Rosemary A. Marusak, Thomas P. Shields, and A. Graham Lappin

14. The Role of Free Energy in Interligand Electron Transfer 253
    L. K. Orman, D. R. Anderson, T. Yabe, and J. B. Hopkins

THEORETICAL ASPECTS OF SOLID-STATE SYSTEMS

15. Band Orbital Mixing and Electronic Instability of Low-Dimensional Metals 269
    Myung-Hwan Whangbo

16. Ceramic Superconductors: Single-Valent versus Mixed-Valent Oxides 287
    John B. Goodenough

17. Geometrical Control of Superconductivity in Copper Oxide Based Superconductors 323
    Jeremy K. Burdett and Gururaj V. Kulkarni

EXPERIMENTAL ASPECTS OF SOLID-STATE SYSTEMS

18. Organometallic Chemical Vapor Deposition: Strategies and Progress in the Preparation of Thin Films of Superconductors Having High Critical Temperatures 351
    Lauren M. Tonge, Darrin S. Richeson, Tobin J. Marks, Jing Zhao, Jiming Zhang, Bruce W. Wessels, Henry O. Marcy, and Carl R. Kannewurf

19. Centered Cluster Halides for Group-Three and Group-Four Transition Metals: A Versatile Solid-State and Solution Chemistry 369
    Friedhelm Rogel, Jie Zhang, Martin W. Payne, and John D. Corbett

20. Oxidative Intercalation of Graphite by Fluoroanionic Species: Evidence for Thermodynamic Barrier 391
    Neil Bartlett, Fujio Okino, Thomas E. Mallouk, Rika Hagiwara, Michael Lerner, Guy L. Rosenthal, and Kostantinos Kourtakis

21. Intramolecular Electron Transfer and Electron Delocalization in Molybdophosphate Heteropoly Anions 403
    Julie N. Barrows and Michael T. Pope

22. Organometallic Electron-Transfer Salts with Tetracyanoethylene Exhibiting Ferromagnetic Coupling 419
   Joel S. Miller and Arthur J. Epstein

23. Stabilization of Conducting Heteroaromatic Polymers in Large-Pore Zeolite Channels 433
   Thomas Bein, Patricia Enzel, Francois Beuneu, and Libero Zuppiroli

INDEXES

Author Index 452

Affiliation Index 453

Subject Index 454

PREFACE

IN BIOLOGICAL SYSTEMS, FUNDAMENTAL QUESTIONS REMAIN about the mechanism of electron transfer between metalloprotein active sites separated by polypeptide chains and about factors controlling the rate and specificity of biological electron transfer. The work presented herein includes both theoretical and experimental aspects of electron transfer in metalloproteins, as well as in appropriate model systems. In solid-state materials, theoretical interest in electron transfer is currently centered on understanding materials that display electronic phase transitions, particularly the high-transition-temperature oxide superconductors. Experimental work on solid-state materials includes studies of the design, preparation, and characterization of numerous one-, two-, and three-dimensional systems displaying a variety of electronic properties. This volume includes chapters pertaining to electron transfer in both biological systems and solid-state materials.

In March 1989 a symposium on "Inorganic Compounds with Unusual Properties. III. Electron Transfer in Biology and the Solid State" was held at the Georgia Center for Continuing Education of the University of Georgia as one of the biennial symposia of the Division of Inorganic Chemistry of the American Chemical Society. The purpose of this symposium was to stimulate communication among scientists who are studying mechanisms of electron transfer in metalloproteins and those who are studying similar processes in solid state materials. This symposium was a sequel to symposia entitled "Inorganic Compounds with Unusual Properties. I and II." that were held at the University of Georgia in January 1975 and February 1978 and were published as Volumes 150 and 173 of the *Advances in Chemistry Series*.

The program of the 1989 symposium included 27 oral presentations, of which 23 are included in this volume. The speakers at the symposium represented industry, government laboratories, and diverse academic institutions.

Acknowledgments

We acknowledge financial support for the symposium from the Air Force Office of Scientific Research, the Petroleum Research Fund of the American Chemical Society, the University of Georgia Research Foundation, and E. I. du Pont de Nemours and Company.

MICHAEL K. JOHNSON
R. BRUCE KING
DONALD M. KURTZ, JR.

CHARLES KUTAL
MICHAEL L. NORTON
ROBERT A. SCOTT

University of Georgia
Athens, Georgia 30602

October 1989

BIOLOGICAL ELECTRON TRANSFER

Overview of Biological Electron Transfer

R. J. P. Williams

Inorganic Chemistry Laboratory, University of Oxford, South Parks Road, Oxford OX1 3QR, United Kingdom

Electron transfer in biological systems uses metalloproteins to a very large degree. In this chapter, the properties of the metal ion sites are examined first. Subsequently the proteins are analyzed. The discussion is divided into metalloproteins for simple wirelike electronic conductor systems and proteins that couple electron transfer with other movements (e.g., of protons). Stress is placed on both structural and dynamic features of metalloproteins. It is shown that the metal ion and the protein are cooperatively adjusted in their properties, differently in different molecules, so as to optimize function.

THIS SURVEY OF LONG-RANGE OUTER-SPHERE ELECTRON TRANSFER in biology deals first with simple electron-transfer proteins (Table I). The following description of the sites of the metal ions in these proteins is considered to be proven.

1. The ligands around metal ions in simple electron-transfer proteins are selected to aid electron-transfer rates; they reduce the central charge on the metal or induce the low-spin rather than the high-spin state of a metal ion. Examples are sulfur, imidazole, and porphyrin ligands for iron and copper atoms in all the most important simple electron-transfer proteins.

2. The geometry around the metal ion in simple electron-transfer proteins is generated by the protein fold. This fold is so strong that there is minimal ligand rearrangement on change of charge. Examples are the close-to-tetrahedral, entatic, state

0065–2393/90/0226–0003$06.25/0
© 1990 American Chemical Society

Table I. Simple Electron-Transfer Proteins

Protein	Electronic Circuit
Cytochrome c	Mitochondrial periplasmic space
Copper blue proteins	Chloroplast chains
Ferredoxins	All energy capture chains
Cytochrome b_5	Reductases

of the blue copper proteins and the coordination geometry of low-spin iron in cytochrome c.

3. The metal ion sites are buried some 10 Å into the protein so that adventitious reactions of the metal ions with small ligand molecules cannot occur. The coordination sphere is effectively complete and inviolate because the protein fold does not relax readily. The metal ion sites are also removed from the solvent, water.

4. The conditions in items 1, 2, and 3 generate ideal electron-transfer sites. Such sites would provide very fast electron transfer if the metal coordination spheres could come into contact with each other (i.e., as in small-molecule outer-sphere reactions, in which the unimolecular rate constant for electron transfer would be close to 10^{10} s^{-1} if there were no thermodynamic barriers). In the context of the buried sites of the electron-transfer proteins, this means that a factor of up to about 10^5 in unimolecular rate constant has been forfeited in order to gain positional depth in an insulator. The electron-transfer sites are partially or completely hidden centers. The rate constants now achieved, around 10^5 s^{-1}, are adequate for biological systems.

5. Placing the metal ion just below the surface of a protein allows it to be covered by a recognition zone so that the electron-transfer distance is in the range of 10 Å between its coordination sphere and one or perhaps two electron-transfer spheres of other protein centers.

6. The metal ion is oriented toward one quadrant edge of a roughly spherical protein (e.g., the cytochromes and the blue copper proteins), so that electron-transfer distances to all other quadrant edges exceed 15 Å.

7. The distance constraints in item 6 and the recognition surface constraints in item 5 ensure a highly selective route for electrons in and out of the available similarly designed electron-transfer proteins.

8. Thermodynamic control (the redox potential) is generated by a complex set of factors, including the coordination sphere atoms and geometry, the surrounding protein, and the exposure to solvent. These potentials have evolved to control rates and to allow devices such as potential droppers to connect electrons with proton movements.

9. Apart from these simple electron-transfer proteins, which transfer electrons at low driving force, there are two further groups of simple biological electron-transfer centers. In very fast light-induced reactions there is often a considerable reaction driving force, and of course the center is highly energized. This chapter will not consider such excited-state electron-transfer reactions. After the initial light absorption, electron transfer in biological photoreaction centers becomes thermal transfer. The thermodynamic driving force is also important in another group of proteins, oxidases, in which there is a large energy change between the redox couples. One typical case is the reactions of oxocations (such as FeO) in simple oxidases (such as peroxidases) with restricted relaxation of the protein.

This description applies to simple electron-transfer proteins that are involved only in allowing electrons to go from one site to another. The prime examples are cytochrome c, blue copper proteins, and many iron–sulfur proteins. We have called these proteins uncoupled electron-transfer proteins. Another variety of proteins is involved in electron-transfer reactions that are coupled to other events (e.g., opening of clefts, cytochrome P_{450}; proton movement, cytochrome oxidase). In these proteins, which are often based on heme centers and quite different copper centers from the blue copper proteins (*see* item 6), there is evidence for considerable change in protein structure associated with change of oxidation state.

At least two extreme sets of electron-transfer proteins have evolved. Both appear to be designed for the functions in hand. We can combine X-ray crystallographic data and high-resolution NMR spectroscopy to look closely at protein structure and dynamics to evaluate the design. Crystallographic data reveal the outline structure most accurately. These data can assist NMR spectroscopy to provide a wider perspective on the dynamics.

To begin, we must identify the function. For convenience, I separate three types of function of electron-transfer proteins (Figure 1).

1. Simple electron transfer, as in a wire made from hop centers (e.g., many FeS centers in electron-transfer chains (Figure 2).

2. Scavenger simple electron transfer, as when a protein brings

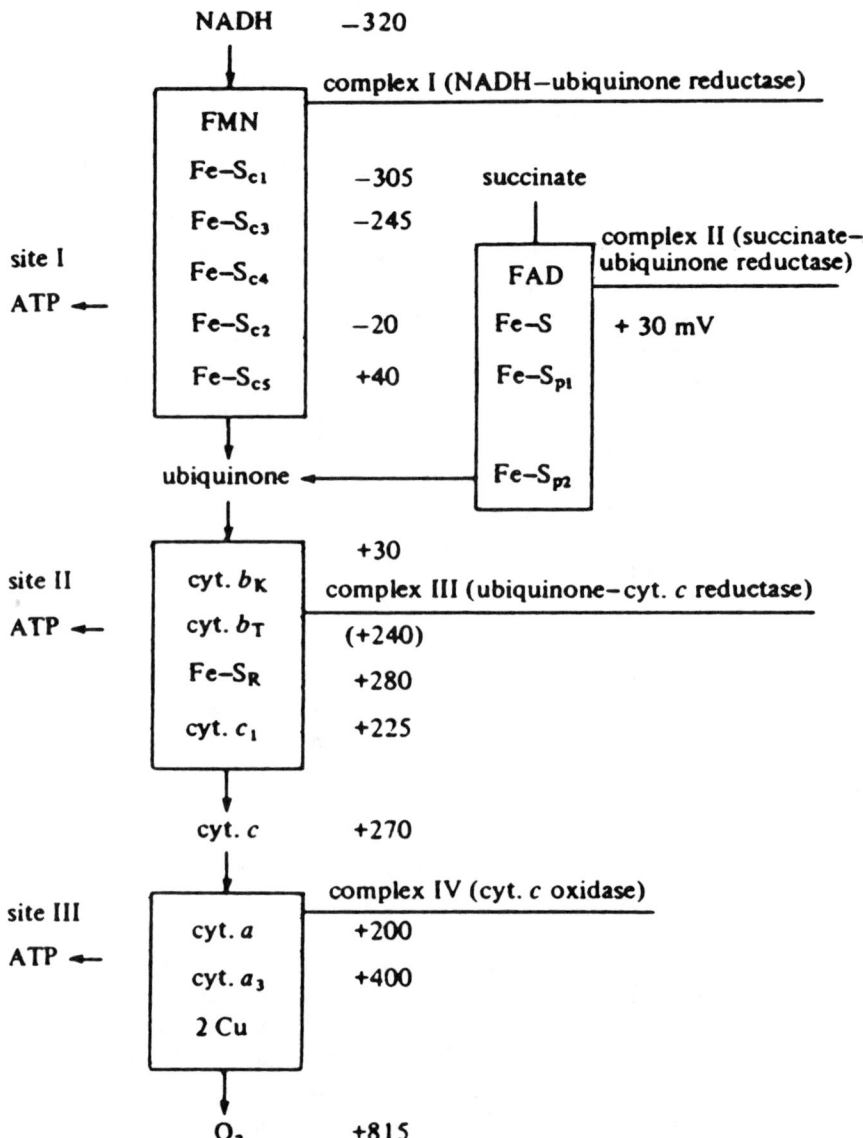

Figure 1. A representation of the electron-transfer chain of the mitochondrial inner membrane. There are three types of electron-transfer protein. Those described in the text as static simple electron-transfer proteins are represented by Fe–S and cytochrome (cyt) c_1. Simple scavenger electron-transfer proteins are represented by cytochrome c. The remaining cytochromes belong to coupled proteins. Some copper proteins belong to each of the three classes. (Subscripts identify different protein centers and are not of consequence in this chapter. Reproduced from ref. 28.)

1. WILLIAMS *Overview of Biological Electron Transfer* 7

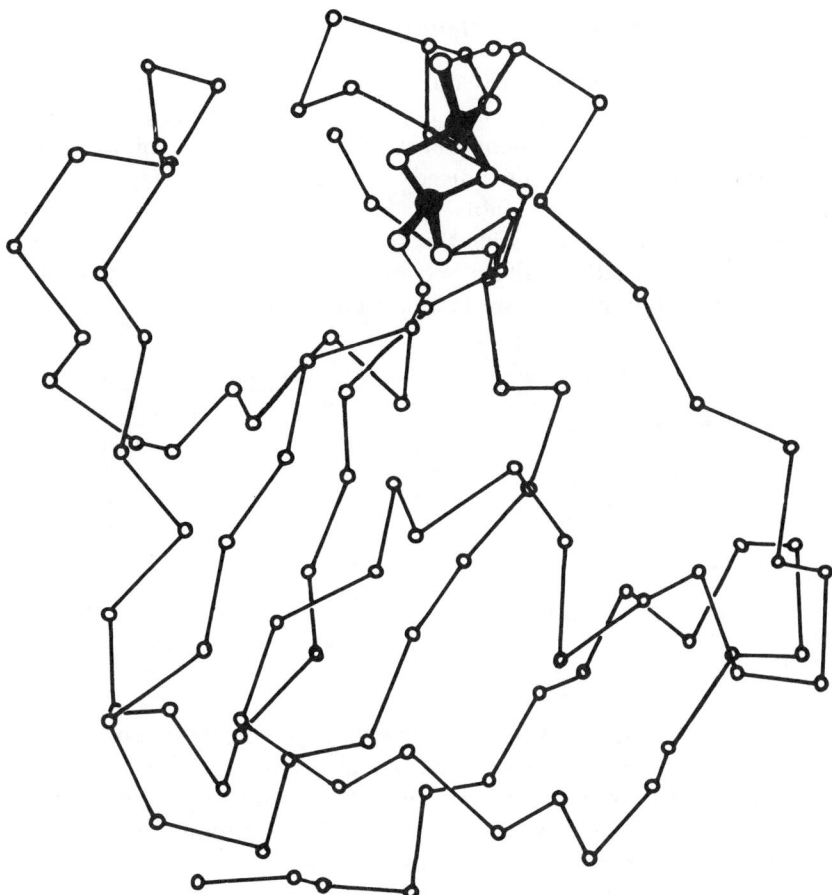

Figure 2. A typical fold of an Fe_2S_2 electron-transfer protein, a ferredoxin, which shows a β-sheet structure of parallel strands cross-linked by H bonds. (Reproduced with permission from reference 29. Copyright 1981 Japanese Biochemical Society.)

    reducing (oxidizing) equivalents from many centers and delivers them to one site by diffusion of the protein in a matrix (e.g. cytochrome c of mitochondrial electron-transfer chains). These proteins are similar to the first group, except for their surfaces.

3. Coupled electron-transfer proteins that, after or before the electron-transfer event, undergo considerable conformational adjustment so as to couple electron transfer with proton movements, for example.

Further understanding of these functions depends on a detailed analysis of protein structure against the background of knowledge from model studies and theory.

Simple Electron-Transfer Proteins

How closely do these proteins resemble solid-state electron-transfer matrices with which we are familiar, such as hop semiconductors? In my own model work (1) on electronic conduction in what is effectively a solid, crystal or glass, matrix, we made various mixed-valent solids from complex ions and analyzed their conductivity (Table II). Some were quite good conductors, but the overwhelming impression this research made upon me was that the cooperative electrostatics of the lattice made it difficult to create rapid charge movement by a hop mechanism in a molecular solid (1).

Table II. Some Examples of Model Semiconductor Crystals

Compound	Log (Resistivity), 300 °C	Activation Energy, ev
$KFeFe(CN)_6$	~8	0.75
$M^+{}_5(CuCl_4)^{3-}(CuCl_4)^{2-}$	~10	0.3
$Tl_3Fe(CN)_6$	6	0.4
$[Fe(o-phen)_3]^{2+}[IrCl_6]^{2-}$	>8	>1.0
Fe_3O_4	−2	0.1
Crocidolite[a]	10	>0.5

NOTE: In these solids, high conductivity and low activation energy were observed only for very short metal-to-metal distances.
[a] Fe(II), Fe (III) asbestos.

I would suggest that those who study simple electron-transfer chains should look at the problems we met. We also investigated many black inorganic ionic lattice solids at room temperature, but failed to find any really good conductors (unpublished observations). I was astonished when others found superconductors in some of these families.

A great advantage of a protein matrix is that the electrostatics are not cooperative, as is obviously the case in an NaCl matrix. Moreover, although crystal lattices are rigid and proteins crystallize in ordered matrices, protein molecules are not crystallites. They more closely resemble glasses or even rubberlike molecules that are easily open to low-energy minor relaxations. Typical examples are the iron–sulfur proteins described in Table III. However, their relaxation is also limited by cross-linking, as in a hard rubber. The Fe_nS_n proteins are cross-linked by the Fe–S links, but they are also built of β sheets cross-linked over remote parts of the sequence. They are

Table III. Structural Features of Fe–S Proteins

Protein	Structural Feature
Ferredoxin (Fe_2S_2)	Antiparallel β sheet
Ferredoxin (Fe_4S_4)	Short antiparallel sheet
HIPIP[a] (Fe_4S_4)	Three-stranded β sheet
Rubredoxin (Fe)	Small sheet segments

[a] HIPIP is high-potential iron protein.

found in fixed positions in biological organization. Finally, the electron-transfer centers are buried away from solvent, so the charge switch does not involve solvent electrostatic energies.

It is worth stressing some other points from Table II. In mixed-valent crystals, we found strong inter-complex-ion charge-transfer spectral absorption bands only at very short distances (e.g., Fe_3O_4 and Prussian blue). At somewhat longer distances (e.g., the Cu(I)–Cu(II) system) the bands were much weaker and were not detected in the $[Fe(o\text{-phen})_3]^{2+}$ $[IrCl_6]^{2-}$ systems, where o-phen is 1,10-orthophenanthroline. They have not been detected in complexes of simple electron-transfer proteins. The overlap integral in these systems must be very small indeed.

Consider a metal center taken from such a protein, but now surrounded by a protein sphere of radius close to the shortest distance to the protein surface, for example, 10 Å. In this position it can make the usual protein–protein contact distance for biological electron transfer in all directions. A crystal from an equimolar mixture of such molecules in its two oxidation states will show an extremely weak charge-transfer band and it will be a very poor conductor. This characteristic is generated not because of the rather low number of potential carriers, nor because of a high activation energy, but because the overlap is so small in the lattice and motion is restricted.

Contrast this situation to a cluster, Fe_nS_n, within which the conductivity will be high, as in Fe_3O_4 (Table II). Biological systems can make Fe_3O_4 or FeO(OH) crystals, but do not use them in electronic devices. Biological electron transfer is slow, even in proteins where some or all of the barriers found in crystals have been removed to gain very selective local circuit pathways.

Analysis. Shortly after our "failures" to develop suitable electronic conductivity devices in mixed-valent crystals, we devoted our attention to the analysis of what simple electron-transfer proteins are really like. We and others now have very detailed NMR spectroscopic studies of two series: cytochromes c and b_5 (2, 3, 15) and blue copper proteins (4, 5). These are all simple electron-transfer proteins.

I shall concentrate on the properties of cytochrome c. We have full (90%) proton NMR spectroscopic assignments for yeast, tuna, and horse cytochrome c in both Fe(II) and Fe(III) states as well as 3 crystal structures for all the proteins (6–8). The NMR spectroscopic work established not just the resemblance of the solution and crystal structures, but the many dynamic features of these proteins. Space does not permit detailed analysis here, and I shall therefore state the conclusions briefly.

There is a general belief that proteins are to be compared with ordered crystal solids. Our NMR spectroscopic data accumulated since 1972 do not match this picture, which we believe to have arisen through a misinterpretation of X-ray diffraction data. The X-ray data, diffraction patterns, reflect the highly populated states of proteins in a regular lattice surrounded by water. This water contributes little to the diffraction pattern because much of it is in a liquid state. Only a very small number of water molecules have residence times in excess of 10^{-9} s.

The outside of proteins (e.g., lysines) must be in rapidly fluctuating states. This rapid fluctuation is confirmed by the narrow line widths of ϵ-CH_2 groups of lysines in NMR spectra. The inside of proteins has been represented as fixed to a fixed secondary structure. The degree to which this structure fluctuates can be determined in part experimentally by using NMR spectroscopic analysis of, for example, aromatic ring flip rates, NH–ND exchange, and relaxation times, especially for isolated ^{13}C nuclei. The results of such studies give a mobility map of a protein (Figure 3) (2).

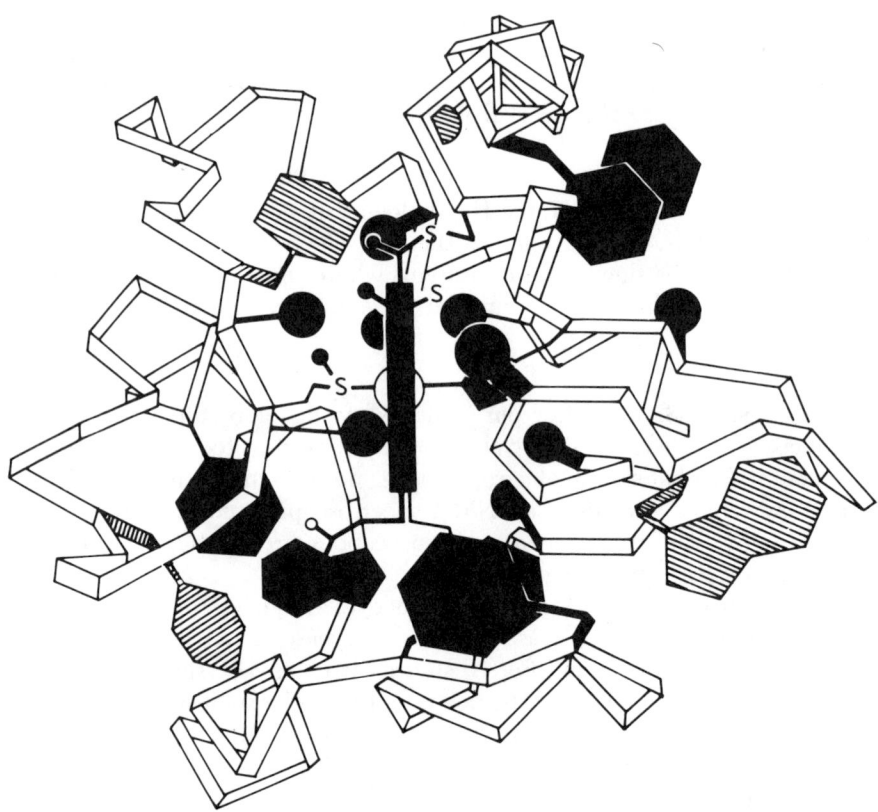

Figure 3. A mobility map imposed on the structure of cytochrome c. Some fast-moving flipping or flapping groups are hatched. Some slow-moving groups are filled in. Helices are illustrated. (Reproduced from ref. 2.)

A more detailed map showing line-width studies throughout a protein would distinguish atoms that have NMR relaxation times equal to the tumbling time of the protein, approximately 10^9 s^{-1} for a 10,000-dalton protein, from those that tumble more rapidly. All parts of the major secondary structure are resistant to very fast independent motion. On a longer time scale, slower motions of considerable amplitude affect the intensities of cross-peaks in two-dimensional NMR spectra. Furthermore, signal averaging is possible when more than one site is occupied by a given atom for appreciable lengths of time. These analyses show that motions of the main chain increase in loops of the structure relative to helices and β sheets. They also show that relative motions of main-chain segments must exceed 1.0 Å on the time scale of 10^5 s^{-1}. This conclusion follows from the observation of aromatic ring flipping with this rate constant inside most proteins, but noticeably not in cytochrome c.

Molecular Dynamics. Since proteins were first described in these terms, there has been a very considerable advance in computational studies of molecular motion in proteins. These molecular dynamic calculations give a picture of a protein as a highly fluctuating assembly. From both this work with the experimental NMR spectroscopic data and today's X-ray diffraction B factors, we can be certain that amino acid side chains such as valines and leucines flip as rapidly as phenylalanines. However, they are asymmetric tops and the motion is hard to detect.

A major concern, then, is not just the structure of a protein and its relationship to function, but the way in which the dynamics within the structure underlies function (9, 10). Electron-transfer proteins are a particularly good case for study. So far electron-transfer proteins have been described mainly in terms of isotropic motion. The larger-scale motions of a helix or of a segment causing a groove-opening reaction are anisotropic, slow ($>10^{-4}$ s), and difficult to detect except by perturbation of the whole structure. However, such slow large-scale motions as these are required to explain some reactions and much NH–ND exchange in proteins. They will be seen to be of great functional consequence later.

Stiffness. The simple electron-transfer protein, cytochrome c, although it is a helical protein, is one of the stiffest proteins we have examined (Figure 3). It is cross-linked. The blue copper proteins are formed from a stiff β-sheet barrel, and cytochrome b$_5$ is supported by a β sheet (*see* Figure 2). β sheets are stiff because they are cross-linked by H bonds. In the blue proteins, the copper changes valence state or can be removed with little effect on the structure. This condition is the basis of the entatic state hypothesis (11). The overall impression of these proteins and of the Fe$_n$S$_n$ proteins (again β sheets, Figure 1) is that the uptake or loss of an electron has little effect on the protein conformation. This lack of effect confirms their description as simple electron-transfer proteins. Their stiffness is not to be confused with the rigidity of crystals. The interior of all these proteins remains quite mobile on the time scale of the electron-transfer reactions.

From other studies we know that other helical proteins that are not cross-linked are readily adjustable by ion binding (calmodulin) (10), change of oxidation state or oxygenation (hemoglobin, hemerythrin, hemocyanin), or by phosphorylation (phosphorylase) (12). We must look at all the electron-transfer proteins in terms of their secondary structure and the cross-linking of that structure. Before turning to more mobile frameworks, we need to inspect in more detail the significance of individual amino acids in the sequences of the simple electron-transfer proteins, both in the interior and on the periphery.

Interior Side Chains and Electron-Transfer Rates. There is the suggestion that, independent of the redox potential of the metals in electron-transfer proteins, aromatic groups in the interior of proteins can assist electron transfer. This suggestion is contrary to our experience in two respects. First, if it were true, electron transfer from the heme of cytochrome c could proceed through tryptophan-59 to the wrong side of the protein. This result is not observed. Second, there is no evidence from models that electron-transfer rates from centers that have redox potentials of around $+0.2$ V are affected by aromatic residues outside the coordination sphere (13). However, in view of the assertion that aromatic residues could have such an effect, we examined the electron-transfer rates in the following two unimolecular reactions.

1. Electron exchange between two cytochromes c, self-exchange, where the two proteins are held together by hexametaphosphate anions (HMP) in a complex [Cyt.c.HMP]$_2$ (14).

2. Electron transfer between cytochrome c and cytochrome b$_5$ in their complex (14).

Both reactions can be examined by NMR spectroscopic measurements of electron exchange seen through line-broadening. The test of the effect of aromatic groups has been made by site-specific mutagenesis where Phe-82 of cytochrome c (*see* Figure 3) is replaced by Tyr-82, Gly-82, or Ser-82 (6). The results are shown in Table IV. No assistance to electron-transfer rates of Phe-82 and Tyr-82 relative to Gly-82 or Ser-82 was found. We maintain the position that aromatic amino acid residues can assist electron transfer only where the redox centers are at $> +0.8$ V, such as in reactions of FeO(IV) states of peroxidases (13, 16).

A second claim (17) has been made for the functioning of Phe-82 in the electron-transfer reactions of cytochrome c with cytochrome b$_5$. According to this claim, Phe-82 is so mobile that it can flap out into the space between

Table IV. Electron-Transfer Rates of Cytochrome c Measured at Equilibrium by NMR Spectroscopy

Cytochrome c Variant	$[Fe^{II}Cc, Fe^{III}Cb_5]^a$ $\downarrow$ $[Fe^{III}Cc, Fe^{II}Cb_5]$, s^{-1}	Self-Exchange Bimolecular[b] $M^{-1}s^{-1}$	Self-Exchange Intracomplex[c] s^{-1}
Horse: wild-type	0.6 ± 0.2	1800 ± 200	400 ± 100
Yeast: C102T mutant[d]	0.3 ± 0.1	300 ± 100	200 ± 100
Yeast: C102T, F82G mutant[e]	0.5 ± 0.2	1200 ± 100	400 ± 100

NOTE: Ref. 14. Cc is cytochrome c, and Cb is cytochrome b.
[a] Determined by saturation transfer to Met-80 methyl resonance of $Fe^{II}Cc$. Intensity decrease was independent of the $Fe^{III}Cc/Fe^{II}Cc$ ratio and total protein concentration (10 mM phosphate, pH 7.1, 30 °C).
[b] Determined by saturation transfer to Met-80 methyl resonance of $Fe^{II}Cc$ (100 mM phosphate, pH 7.0, 30 °C).
[c] Determined from line broadenings: observed in the presence of hexametaphosphate (HMP), a polyphosphate anion known to cause aggregation of cytochrome c (pH 7.0, 30 °C). The intramolecular complex is $[(Cyt\ c)_2(HMP)_2]$.
[d] The yeast C102T, F82Y behaves very like the native protein. The code C102T means that, in this yeast, cytochrome c residue cysteine (C) at position 102 of the sequence has been replaced by a threonine (T) residue. F82G means that phenylalanine (F) 82 in the sequence has been replaced by glycine (G). Y is tyrosine and S is serine.
[e] The yeast C102T,F82S behaves very like F82G.

the two cytochromes in their complexes. Our full assignments of the two cytochromes separately and in their complex allow us to state that no such movement of Phe-82 is detectable. Although it flips rapidly in the free cytochrome and in the complex, there is no evidence for flapping outward at rates around 10^5 s^{-1} (18).

The interior of all these proteins allows quite considerable vibronic–rotational motion. Our evidence indicates that many aromatic rings in most proteins flip rapidly, $>10^4$ s^{-1}. We can observe few other motions in the interior. The cavity size needed to flip Phe and Tyr residues is such that we believe that many Val, Leu, Ile, and Met residues can rotate easily, although through their asymmetry they are seen in one heavily populated state (as if they were rigid). The exterior of the protein and its motions become particularly important when we turn to the scavenger proteins.

Surfaces of Scavenger Proteins. The distinction between a scavenger protein and a protein that is a permanent part of an organization lies not in basic electron-transfer reactions, but in the organization of the proteins. Just as some classes of proteins have very mobile interiors and others have relatively rigid frameworks, so some proteins form relatively rigid assemblies and others do not. Scavenger proteins (e.g., cytochrome c) are often nearly spherical and have virtually no concave surfaces. The scavenger proteins also have highly charged surfaces. The charges on protein surfaces attributable to Glu, Asp, Lys, and Arg side chains are known to be relatively mobile themselves. Furthermore, their long-range electrostatic interactions with other proteins allow considerable movement in the complexes without much loss of energy. Wide areas of the surface of cytochrome c are zones of approximately the same electrostatic potential. We have examined these surfaces by using complex ion probes. We cannot define a binding site for ions such as $[Cr(CN)_6]^{3-}$, but only a binding zone (Figure 4) (19).

The scavenger protein has to be able to find an electron-transfer site, react, and leave that site in $\sim 10^{-3}$ s to maintain electron flow. Its binding has to be effectively diffusion-controlled to allow a binding constant of 10^6. The easiest way to achieve such a reaction between surfaces is to have many electrostatic mobile fingers (lysines) that will bind to many electrostatic mobile anions. The assembly is mobile. We call this hand-in-glove fitting.

The exact structure for the electron-transfer act may be of little consequence. The surface dynamics is of greater importance. In contrast, some proteins, such as many Fe_nS_n electron-transfer proteins, are fixed parts of organizations. We suspect that the interactions they form are based more largely on shorter range hydrophobic forces than on electrostatic forces. Again, these proteins have much more distinctive shapes, insofar as they are known (Figure 2). It may well be that, when bound in an organized system, their surfaces are not mobile.

Simple Oxidases and Their Electron-Transfer Reactions

Some of the major simple oxidases are given in Table V. In their reactions, oxidation of substrate is not coupled to any other reaction. For example,

$$\text{dioxygen + substrate} \rightarrow \text{oxidized product + reduced oxygen}$$

The reaction is written in long-hand because reduced oxygen can appear as O_2^-, H_2O_2, or H_2O. The alternative of O_2-incorporation in an oxidation does not require electron transfer except in inner-sphere activation (e.g., in dioxygenases of soil bacteria).

In the case of plant peroxidases that oxidize phenols and indoles, the reaction is a one-electron oxidation by peroxide. The iron cycles from Fe(III) to a nominal Fe(V) (i.e., Fe(IV) plus a free radical, through FeO(IV)). The substrate goes to a free radical that may polymerize. We studied these electron-transfer proteins by using NMR spectroscopy to determine the structure with substrate bound. We showed that electron transfer took place over a distance of some 10 Å (outer-sphere reaction) (20). Our present interest lies in the structure of these proteins. All the peroxidases in this class of known structure are similar to simple electron-transfer proteins, in that they are cross-linked either chemically (horseradish peroxidases) or by β sheets (cytochrome c peroxidases). The pocket of the high-spin heme is open to small molecules, O_2 or H_2O_2, but not to the larger substrates such as phenols or indoles. Electron transfer is not linked to a conformational change. Other oxidases in this class may well include the copper oxidases such as ascorbic

1. WILLIAMS *Overview of Biological Electron Transfer*   15

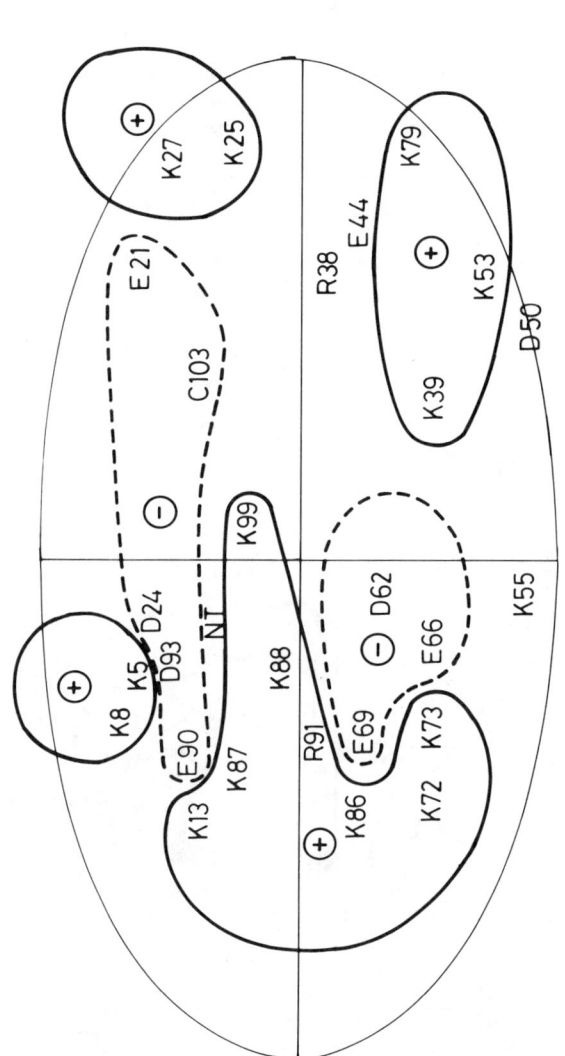

Figure 4. The electrostatic potentials on the surface of cytochrome c, represented as a Mercato projection and estimated from anion- and cation-binding experiments, as well as by calculation.

acid and phenol oxidases, but we do not know the structure of the dioxygen sites involving type 2 Cu and the copper pairs of these proteins. We have

Table V. Long-Range Electron Transfer to Aromatic Compounds

Protein	Group Involved
Cytochrome c peroxidase	Possibly tryptophan
Horseradish peroxidase	Phenols or indoles
Copper oxidases (laccase)	Phenols
Reaction center	Tyrosine(?), phaeophytin
Ribonucleotide reductase	Tyrosine

NOTE: All systems involve high redox potential centers. Quinones and flavins are, of course, very different aromatic electron-transfer centers from metal ions with low-lying molecular orbitals (redox potentials).

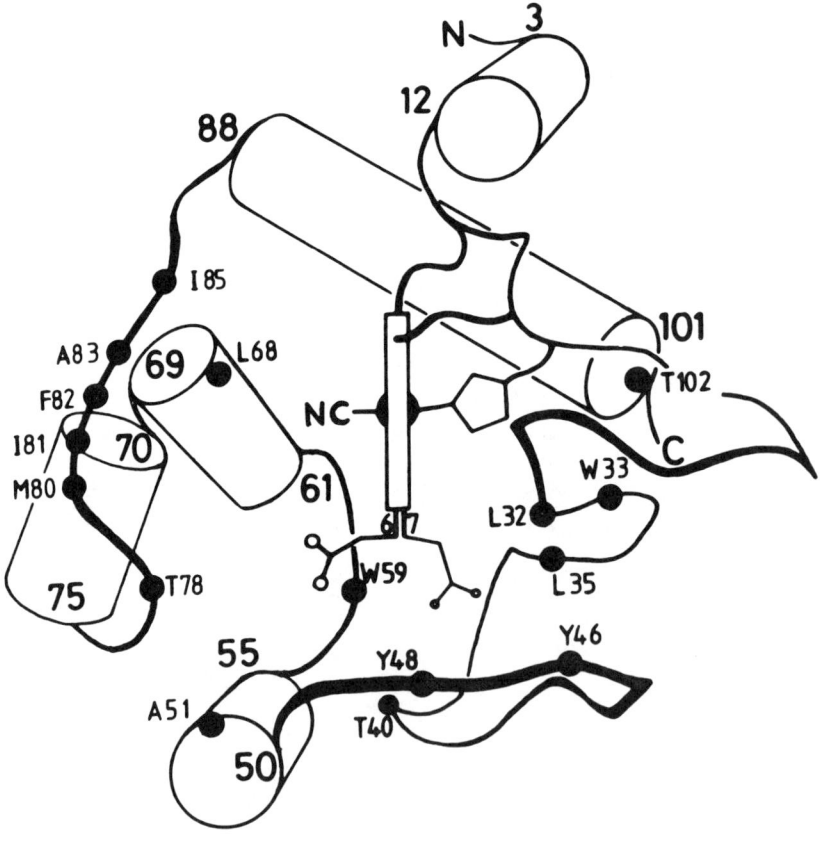

(a)

Figure 5. The change in structure from (a) Fe(III) cytochrome c to (b) Fe(III)–CN⁻ cytochrome c.

NMR spectroscopic evidence that the inside of many of these proteins is as mobile as that of the simple electron-transfer proteins.

Conformational Changes: Groove Openings

In the description of cytochrome c, we have deliberately left aside one much-studied reaction, that with cyanide (21). The reaction is very slow and quite unrelated to any fast, simple, electron-transfer reaction. However, it is suitable as a model for coupled reactions. At temperatures above 60 °C, the electron-transfer reaction of cytochrome c is coupled to a breaking of the Met-80–Fe(III) bond. We shall consider the P_{450} reactions in similar terms.

A detailed NMR spectroscopic study with full assignment of the cyanide Fe(III) state of the protein (21) shows that the groove opening is a leverlike action in which the segment of the cytochrome below the sequential chain 79 to 85, which includes Met-80, moves away from the heme to leave room for the cyanide anion (Figure 5). The reaction shows that relatively small

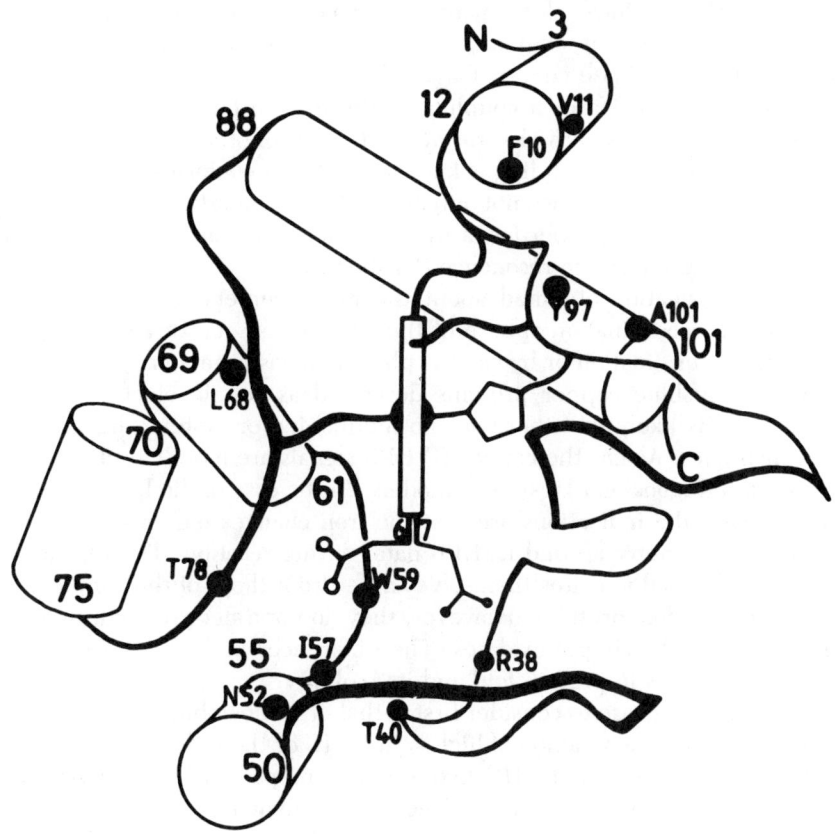

(b)

Figure 5.—Continued.

domains of a protein can move away from the bulk, even on change from low-spin Fe(II) to high-spin Fe(III). The reaction is

low-spin Fe(II) → low-spin Fe(III) → high-spin Fe(III) → cyanide complex

We shall next look for similar changes in other proteins, noting that in cytochrome c the movable part of the protein is not cross-linked.

Coupled Electron Transfer

These are the most difficult proteins to study. Among iron proteins, they include hemoglobin, which is linked to a cytochrome b_5 reduction to protect it; hemerythrin; cytochrome P_{450}, which is linked to either a cytochrome b_5 or an Fe_nS_n protein to ensure the controlled order of its reaction kinetics; and cytochromes b and a of the electron-transfer chains of chloroplasts and mitochondria, in which electron transfer is coupled to proton movement. Among copper proteins, they include the dioxygen carrier hemocyanin and the cytochrome oxidase center, Cu_B.

Coupling here is not a coupling of the electrostatic potentials of the electron-transfer ions with electrostatic potentials of other groups, as it may be in cytochrome c_3. Coupling is known to involve conformational changes (22–25). This result is shown not only through the properties of the proteins, but by following the properties of the metal ion (for example, its spin state or electron paramagnetic resonance (EPR) signals).

The information obtained about the metal centers themselves is revealing. The coordination sphere of the copper, as seen directly by X-ray structure determination or by various physical measurements, is unlike that found in the blue copper proteins described as simple electron-transfer proteins. It is likely that there are no methionine or cysteine ligands, but only histidines. Again, the copper(II) EPR signals are more closely those of distorted tetragonal Cu(II) sites in models. In the case of the heme proteins in this class, the indications are that the iron changes either spin state or the ligand geometry around it (EPR data) during reaction. The indications are that the metal ion sites themselves differ from those of the simple electron-transfer heme proteins. However, they do not differ greatly from some of the sites of the simple oxidases. The differences in function are related to the differences in protein fold and its mobility.

The simplest case to consider first is that of hemoglobin. In hemoglobin the dioxygen uptake reaction of high-spin Fe(II) or the redox switch of high-spin Fe(II) to high-spin Fe(III) hydrate is accompanied by a considerable conformational change (Figure 6). The change involves some movement of the iron atom toward the heme plane, some adjustment of the iron ligands, both the histidine and the porphyrin, and most importantly, the adjustment of helices. The adjustment of helices is, we consider, a common feature of all these coupling proteins. The evidence that coupling is related to helix movements comes from general studies of proteins other than hemoglobin (e.g., the calcium trigger proteins and phosphorylase). Coupling is also very likely in the cooperative action of helical hemerythrin and hemocyanin, where the iron and copper ions undergo redox switches on dioxygen uptake rather than spin-state changes.

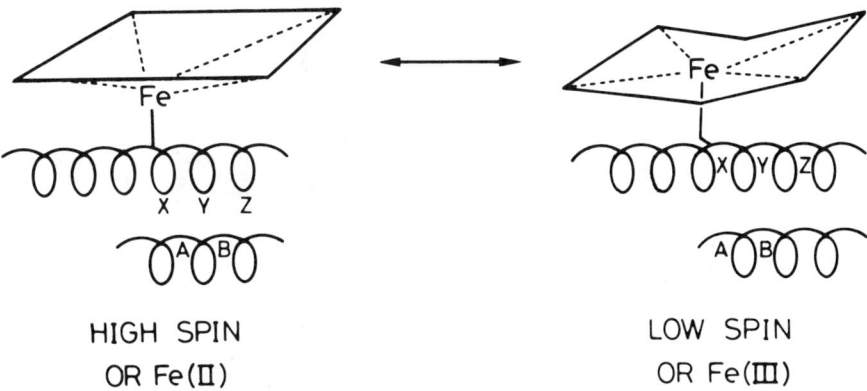

Figure 6. A representation of coupled helix movements with chemical changes at the iron atom of a heme protein.

A revealing example is provided by our study of cytochrome b_{562}. The structure, shown in Figure 7, is a four-helix bundle protein with the iron bound as in cytochrome c (26). The redox potential of the protein is pH-dependent; there is a switch of 260 mV within the pH range 5.0–8.5. This proton-coupled redox potential lies in the physiological pH range. We know that a conformational change accompanies the protonation–redox switch. In this respect, the protein is entirely comparable with hemoglobin, but the reactions are now confined to low-spin states. These reactions are to be contrasted with those of cytochrome c, Fe_nS_n, and the blue copper proteins described in the section on simple electron-transfer proteins.

Given the obvious ease of helix–helix motions; the strong evidence for the highly helical character of the membrane portions of cytochrome oxidase (i.e., containing cytochrome a_3 and Cu_B), of the cytochrome b complex, and of cytochrome P_{450}; and the fact that these heme proteins are not cross-linked by either extended β sheets or chemical cross-links, it appears that coupling in all these cases is a property of helix motion (10). This is not the place to extend the argument, but a whole range of other coupled activities of proteins appears to be associated with helix motion (27). It is immediately clear by comparison with simple electron-transfer proteins that cross-linking separates the protein structures of coupled and uncoupled activities (27).

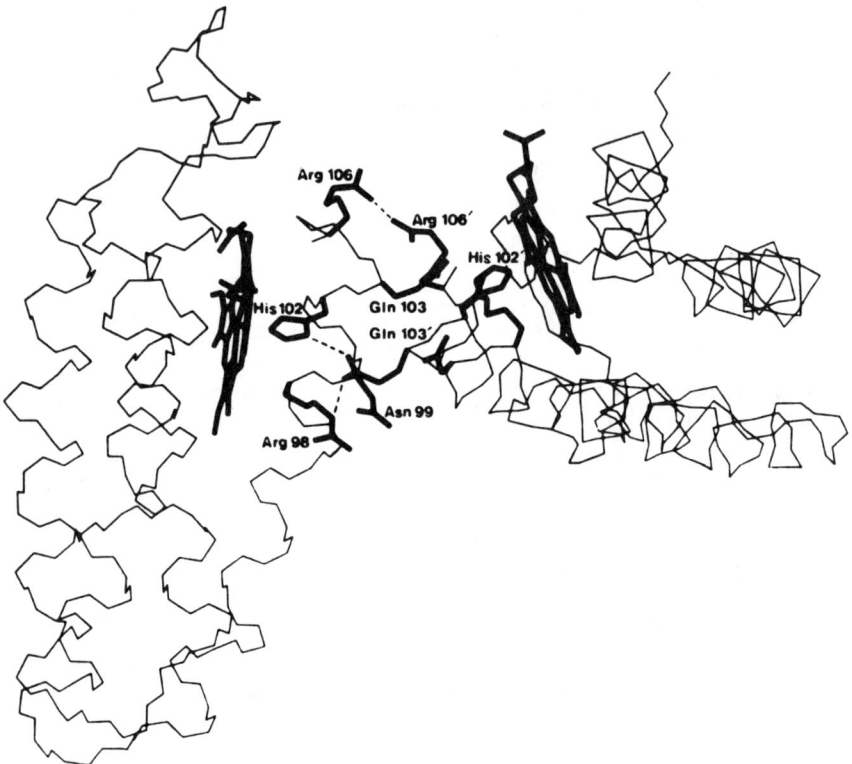

Figure 7. The structure of cytochrome b_{562} (opposite page) showing the groups likely to be responsible for the pH-dependent redox potential (see ref. 26). On this page is shown the binding mode of two molecules in the crystals and the hydrogen-bonding network. (Reproduced from ref. 26.)

At present we can describe the motions of the helices only by analogy with calmodulin and hemoglobin. Here is one major task for the chemist interested in biological electron transfer.

Other Electron-Transfer Centers

The concentration on blue copper centers and heme in electron-transfer proteins has tended to hide the diversity of such centers. Earlier reference was made to Fe_nS_n centers as typical of uncoupled almost-solid-state devices. Other centers are similar, such as Ni–S centers of dihydrogen reactions and cobalamin centers of methane chemistry. (The vitamin B_{12} chemistry in rearrangement enzymes seems to depend on homolytic fission giving free radicals and not on electron transfer; it is based on atom movements.) Yet other centers include Fe–O–Fe (or Mn–O–Mn) for generating tyrosine-free radicals in ribonucleotide reductase; $[MnO]_n$ in the dioxygen-liberating cen-

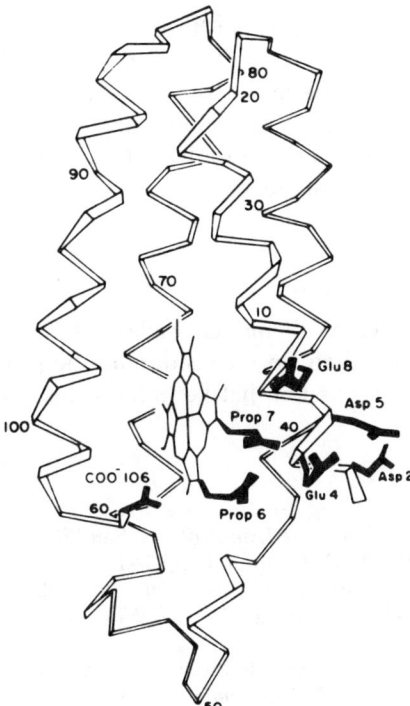

Figure 7.—Continued

ter of photosystem II; type 2 and type 3 copper in copper oxidases; molybdenum in the two-electron reactions of NO_3^-, SO_4^{2-}, and aldehydes; and a considerable number of simple iron enzymes that may well use inner- or outer-sphere electron-transfer mechanisms (e.g., the oxidative enzymes of secondary metabolism).

The most intriguing centers contain two or more metal ions and partake in reactions related to dioxygen reduction or release. The equivalent electronic states are M(II)•O_2•M(II), M(III)•O_2^{2-}•M(III), M(IV)O^{2-}•O^{2-}M(IV), and the various mixed-valent intermediates for the dioxygen molecule. There are also corresponding reduced states such as M(II)•O_2^{2-}•M(II) and M(III)O^{2-}•O^{2-}M(III). However, we must include states of proton reaction such as M(II)(H_2O_2) M(II), M(III)(OH$^-$)•(OH$^-$)M(III), and M(IV)OH$^-$•$^-$HOM(IV). It appears as if these changes must be linked to conformational switches and to movements of protons deep inside proteins, such as cytochrome oxidase.

These series may contain a larger number of metal ions or an uneven number of protons. Choice of metal ion (i.e., copper, manganese, or iron) and choice of protein fold for a given metal ion can provide control over thermodynamic balance between oxygen species. An unsolved problem is the resolution of questions posed by these electron-transfer centers, such as why this or that metal or particular choice of ligands is best for a given function. What is the nature of the protein selected by evolution? I have tried to answer these types of questions for a limited series of electron-transfer proteins, but many other problems in the electron-transfer systems of biology clearly remain untouched.

Acknowledgments

I thank G. R. Moore, G. Williams, G. Pielak, D. Concar, D. Whitford, Y. Gao, and N. Veitch for their very considerable help with both the experimental work and the discussion that lie behind this overview.

References

1. Williams, R. J. P. In *Non-heme Iron Proteins;* San Pietro, A., Ed.; Antioch Press: Yellow Springs, OH, 1965; Chapter 1, pp 7–15.
2. Williams, R. J. P. *Z. Phys. Chem.* **1988**, *269*, 387–402 and references therein.
3. Whitford, D.; Veitch, N.; Concar, D. W.; Williams, R. J. P. *FEBS Lett.* **1988**, *238*, 49–55.
4. Driscoll, P. C.; Hill, H. A. O.; Redfield, C. *Eur. J. Biochem.* **1987**, *170*, 279.
5. King, G. C.; Wright, P. E. *Biochemistry* **1986**, *25*, 2364, and personally communicated work.
6. Tanako, T.; Dickerson, R. E. *J. Mol. Biol.* **1981**, *153*, 95.
7. Colman, P. M.; Freeman, H. C.; Guss, J. M.; Murata, M.; Norris, V. A.; Ramshaw, J. A. M.; Ventatappa, M. P. *Nature* **1978**, *272*, 319–324.
8. Adman, E. T.; Jensen, L. H. *Isr. J. Chem.* **1981**, *21*, 8.
9. Williams, G.; Moore, G. R.; Williams, R. J. P. *Comments Inorg. Chem.* **1985**, *4*, 55.
10. Williams, R. J. P. *Carlsberg Res. Commun.* **1987**, *52*, 1.
11. Vallee, B. L.; Williams, R. J. P. *Proc. Natl. Acad. Sci. U.S.A.* **1968**, *59*, 498.
12. Prang, S. R.; Archarya, K. R.; Goldsmith, E. J.; Stuart, D. I.; Varvill, K.; Fletterick, R. J.; Madsen, N. B.; Johnson, L. N. *Nature* **1988**, *336*, 215–221.
13. Burns, P. S.; Harrod, J. F.; Williams, R. J. P.; Wright, P. E.; *Biochim. Biophys. Acta* **1976**, *428*, 261.
14. Concar, D.; Gao, Y.; Pielak, G.; Whitford, D.; Williams, R. J. P., unpublished work.
15. Wand, A. J.; DiStefano, D. L.; Feng, Y.; Roder, H.; Englander, S. W. *Biochemistry* **1989**, *28*, 186–194 and 195–203.
16. Pielak, G. J.; Concar, D. W.; Moore, G. R.; Williams, R. J. P. *Protein Eng.* **1987**, *1*, 83.
17. Wendoloski, J. J.; Matthew, J. B.; Weber, P. C.; Salemme, F. R. *Science* **1987**, *238*, 794.
18. Moore, G. R.; Veitch, N.; Williams, R. J. P., unpublished results.
19. Arean, C. O.; Moore, G. R.; Williams, G.; Williams, R. J. P. *Eur. J. Biochem.* **1988**, *173*, 607.
20. Burns, P. S.; Williams, R. J. P.; Wright, P. E. *J. Chem. Soc., Chem. Commun.* **1975**, 338–339.
21. Gao, Y.; Williams, R. J. P., unpublished results.

22. Poulos, T. L.; Sheriff, S.; Howard, A. J. *J. Biol. Chem.* **1987,** *272,* 13881.
23. Poulos, T. L.; Finzel, B. C.; Gunzalus, I. C.; Wagner, G. C.; Krant, J. *J. Biol. Chem.* **1985,** *260,* 16122.
24. Wikstrom, M. *Chem. Scr.* **1987,** *27B,* 53.
25. Williams, R. J. P. *FEBS Lett.* **1987,** *226,* 1.
26. Moore, G. R.; Williams, R. J. P.; Peterson, J.; Thomson, A. J.; Mathews, F. S. *Biochim. Biophys. Acta* **1985,** *829,* 83.
27. Williams, R. J. P. In *The Enzymes of Biological Membranes;* Martonosi, A. N., Ed.; Plenum: New York, 1985; Vol. 4, pp 71–110.
28. Williams, R. J. P. *Proc. Roy. Soc. London* **1981,** *B213,* 361–397.
29. Tsukihara, T.; Fukuyama, K.; Nakamura, M.; Katsube, Y.; Tanaka, N.; Kakudo, M.; Wada, K.; Mase, T.; Matsubara, M. *J. Biochem. (Tokyo)* **1981,** *90,* 1763–1773.

RECEIVED for review May 1, 1989. ACCEPTED revised manuscript August 15, 1989.

THEORETICAL ASPECTS OF BIOLOGICAL ELECTRON TRANSFER

2

Formalism for Electron Transfer and Energy Transfer in Bridged Systems

J. R. Reimers and N. S. Hush[1]

Department of Theoretical Chemistry, University of Sydney, Sydney, New South Wales 2006, Australia

There is much current interest in the study of electron transfer or energy transfer between centers linked by a bridge, and a number of basic points need clarification in the interpretation of such processes. For example, it is commonly assumed that communication through the bridge can be described in terms of an effective "two-level" model; under what circumstances is this justified? What are the conditions under which a single-parameter rate constant characterizes the process? With a view to providing some information about these and related questions, a formalism applicable to coherent transfer of energy (including electron transfer and hole transfer in cases where the vibrational motions do not need to be explicitly considered) is developed. This formalism yields a completely general algorithm, which, in particular limits, reduces to a generalized form of Fermi's golden rule and of Rabi's rate equation. In so doing, it unifies a number of existing theories.

ELECTRON (OR HOLE) AND ENERGY TRANSFER between a metal donor center linked to an acceptor by a bridge is increasingly being studied experimentally. Such systems often involve mixed-valent metal states; bioinorganic systems (both naturally occurring (1, 2) and synthetic) are also of much current interest. Design of molecular electronic devices, in which there is also much current interest, frequently involves elements of this general kind. In molecular electronics, one desires to find ways of transferring electrons under controlled conditions between different states of a device. Nature has

[1]Address correspondence to this author.

provided models for this process, and we can learn much from studying processes such as photosynthesis (1, 2), in which light produces an electron in an excited state (the donor state) that reacts to transfer the electron to a geographically distant state (the acceptor state).

Indeed, it is possible to manufacture compounds in which such transfers occur under controlled conditions (3, 4), and such compounds are candidates for molecular electronic devices. Usually, the bridge is composed of organic material and may be either a conductor (e.g., a conjugated bridge) or an insulator (e.g., a saturated hydrocarbon). Some day processes may be developed by which the conductance of the bridge can be modified significantly by an external signal (5, 6). In this chapter we generate a model that describes how the conductance of the bridge varies with respect to some fundamental properties of the bridge. We do not consider the question of how these properties may be modified dynamically. Our model is inspired by the observation that normally insulating bridges can in fact become excellent conductors (see refs. 3 and 4).

The basic formalism is developed in this chapter. Perturbation expansions are performed to describe the kinetics when the bath coupling is weak and when it is strong, and an accurate interpolation scheme is presented for the intermediate region. The relationship of our approach to the effective two-level models of Joachim (39), of Broo and Larsson (45), and of McConnell (24) is shown. As an application, our model is applied to three-level systems containing just one bridge state. The electron transfer is shown to be much faster when the bridge state is resonant with the donor and acceptor states than when it is nonresonant. This system is a model system for electron transfer through bridges. Quantitative features of the modulation of the donor-to-acceptor current are induced by changes made to bridge properties. Finally, a study of the bridge-length dependence of the reaction-rate constant is performed by using McConnell's model Hamiltonian.

Electron-Transfer Theory

Inorganic systems have played a crucial role in the development of electron-transfer theory. This role is exemplified in the work of Taube et al. (7–14) and others (15) on mixed-valence bisruthenium complexes. This chapter presents a formalism for considering electron-transfer problems. This formalism is applied in later papers (16, 17) to study electron transfer in these systems.

More generally, electron- and energy-transfer phenomena have been studied extensively (18–47) and reviewed (46–51). Usually, through-bridge studies (45) implicitly introduce the assumption that the kinetics can be described by using an effective two-level model in which the total Hamiltonian is replaced by an effective donor–acceptor interaction Hamiltonian. We generated a model for the rate constant that is not based on this assumption. Our model provides resolutions for ambiguities in earlier theories and shows how these approaches relate to both each other and fundamental physics.

Two-Level Systems. We started with the energy-transfer model developed by Robinson and Frosch (*18–20*), which describes energy decay in two-level systems and in networks of such systems. Electron-transfer processes form a subset of energy-transfer processes and feature the transfer of an electron (or hole) over some distance, usually large compared to the spatial extent of the donor and acceptor states. In the more general process, the electron could transfer between donor and acceptor states within the same region of space and thus produce negligible dipole moment change.

The language of this chapter is consistent with the more-general phenomenon, and so the terms "donor" and "acceptor" always refer to the electronic states involved and not to chemical species or fragments. The donor and acceptor states are not eigenstates of the complete Hamiltonian, and thus they are not stationary states. Rather, they evolve in time, and it is this time evolution with which we are concerned.

Dependence on the Bath. Robinson and Frosch described a large range of phenomena with a theory containing just two variables: the Hamiltonian matrix element V_{12} connecting the two levels, and the acceptor–solvent coupling λ of the final state to a bath. (These variables were originally (*18*) named β and $\alpha/2$, respectively.) By some means (e.g., electrical, thermal, photochemical, or mechanical), energy is placed in the donor state. As this donor state is not an eigenstate, it evolves in time both coherently and reversibly, thus transferring energy to and from the acceptor state.

The acceptor state, however, is also coupled to a very large number of additional states, which are collectively known as the bath. Physically, the bath may be a surrounding fluid, protein or crystal, and it may even include internal molecular degrees of freedom. Because the bath contains a very large number of states, it is highly improbable that energy will be transferred back from the bath to the acceptor. Excess energy is thus lost, preventing the transfer of energy back from the acceptor to the donor. The final result is that energy is transferred from the donor to the acceptor, and a small amount of excess energy is also transferred to the bath. Our approach extends the model of Robinson and Frosch to include problems in which the donor and acceptor are coupled through a bridge rather than just coupled directly (through space).

Acceptor–Solvent Coupling. An essential feature of this theory is that the electron-transfer rate constant is explicitly dependent upon the acceptor–solvent coupling λ. Thus, properties of the solvent are directly included. Other solvent properties, such as static shifts of state energies and force constants, are included indirectly in the use of solvent-dependent donor–bridge–acceptor potential parameters. If the solvent is a dilute gas, then the coupling λ is probably related to the mean collision frequency ("internal" collisions may need to be included here) and can thus become quite small.

The donor-to-acceptor rate constant cannot exceed $\lambda/\hbar$ ($\hbar$ is the Planck constant). Thus, in a dilute gas the electron-transfer rate constant will be small, independent of the nature of the donor and acceptor. In liquids, λ is usually typical of rotational relaxation times, which, for systems with large dipole moments, can give values of $\lambda/\hbar$ as large as 10^{12} s^{-1}. Solids provide the ability to control λ as, in principle, geometries can be constructed to favor some relaxation mechanisms but disfavor others. Dramatic effects of this nature have been observed in the spectroscopy of mixed crystals. If they could be controlled, then it is possible that a molecular switch could operate by modulating the relaxation time λh^{-1} in order to change the donor-to-acceptor current.

Electron transfer adds an extra degree of complexity to the basic energy-transfer problem in that two different sets of quantum numbers, electronic and vibrational, are involved. Much of the current research on electron-transfer processes reported in the chemical literature is concerned with the special cases in which the vibrational coordinates need to be treated explicitly. Such studies (e.g., refs. 46 and 47 and references therein) involve both vibrational and electronic coordinates; they are often concerned with dependence of the relaxation time λ on the solvent friction. This emphasis is not necessary, however. In this chapter we restrict ourselves to problems describable by just the electronic quantum numbers. A more general theory will be presented elsewhere (52).

General Formalism

Let $|\psi_D\rangle$ and $|\psi_A\rangle$ be wave functions representing the appropriate physical quantity (here, typically an electron or hole function) representing the donor and acceptor states, respectively. Let the n wave functions $|\psi_i\rangle$ for all $i = 1-n$ represent similar quantities representing bridge states. All of these functions are required to be orthogonal. The total Hamiltonian operator $\mathbf{H}$ for the system plus bath is expressed as

$$\mathbf{H} = \mathbf{H}_s - i\lambda|\psi_A\rangle\langle\psi_A| \tag{1}$$

where $\mathbf{H}_s$ is the hermitian Hamiltonian operator for the isolated system modified, if necessary, to account for static effects (such as solvent shifts) of the bath; λ indicates the rate of irreversible decay of the acceptor state into the bath states. If the acceptor is isolated from the bridge and donor, and if the system has probability $P(0)$ of being in the acceptor state at time 0, then it will have probability

$$P(t) = P(0) \exp\left(\frac{-2\lambda t}{\hbar}\right) \tag{2}$$

of being in that state at time t. The eigenvalues ϵ_j and eigenvectors $|\chi_j\rangle$ of the total Hamiltonian are obtained by diagonalizing **H**, which yields

$$|\psi_i\rangle = \sum_j C_{ij}|\chi_j\rangle \tag{3}$$

As **H** is not hermitian, ϵ_j will be complex and the matrix of the eigenvectors χ_j, may be both not unitary and singular. Singularity is a rare phenomenon, and its effects (or, rather, lack of effects) are discussed elsewhere (53). When they are linearly independent, this invertible transformation yields

$$|\chi_j\rangle = \sum_i C_{ji}^{-1}|\psi_i\rangle \tag{4}$$

With the assumption that the system is in the donor state $|\psi_D\rangle$ at time $t = 0$, then the probability $P_i(t)$ of being in state i at time t is

$$P_i(t) = |\langle\psi_i|e^{-i\mathbf{H}t/\hbar}|\psi_D\rangle|^2 = \left|\sum_j C_{Dj}C_{ji}^{-1}\exp\frac{-i\epsilon_j t}{\hbar}\right|^2 \tag{5}$$

and the probability $P(t)$ that the excitation has not leaked to the bath is given by

$$P(t) = \sum_i P_i(t) = \sum_j p_{jj}\exp\frac{2\mathbf{I}(\epsilon_j)t}{\hbar} + 2\mathbf{R}\left(\sum_{j<k} p_{jk}\exp\frac{-i(\epsilon_j - \epsilon_k^*)t}{\hbar}\right) \tag{6}$$

where

$$p_{jk} = p_{kj}^* = C_{Dj}C_{Dk}^*\sum_i C_{ji}^{-1}(C_{ki}^{-1})^* \tag{7}$$

are the intensities of the associated Fourier components of $P(t)$, with real (**R**) and imaginary (**I**) components. The initial normalization of the system is reflected in the condition

$$1 = P(0) = \sum_{j,k} p_{jk} \tag{8}$$

Although $p_{jj} \geq 0$, typically $\mathbf{R}(p_{jk}) \leq 0$ for $j \neq k$, and

$$\frac{dP(0)}{dt} = 0 \qquad (9)$$

as initially no amplitude is present in the decaying mode. In general, a rather complicated time dependence is expected for these reactions as order n exponentially decaying modes contribute, as well as order $n^2/2$ exponentially decaying sinusoidal modes. If none of the (complex) energy levels are degenerate or pseudodegenerate, then the coefficients p_{jk} will all be near zero except for one dominant diagonal term. Simple single-exponential decay of the donor state is thus observed, and it is possible to introduce a single rate constant to describe the kinetics. The decay becomes more complicated if either a bridge level approaches the donor level or the complex energy separation between the donor and acceptor state becomes too small. No uniquely defined rate constant exists when the time-dependence is so complicated. Thus, we define a measure of the overall reaction-rate constant k as

$$k = \tau^{-1} \qquad (10)$$

where τ is the area under the decay curve $P(t)$, obtained by using a weight function W_Φ that limits the quantum yield of the donor-state-to-acceptor-state transfer process to Φ.

$$\tau = \Phi^{-1} \int_0^\infty W_\Phi(P(t))\, dt \qquad (11)$$

It is possible to define W_Φ in many ways; the details of this are discussed elsewhere (53). In this chapter, we mainly restrict ourselves to the simple case in which the full quantum yield $\Phi = 1$ is obtained and $W_1(P(t)) = P(t)$. The integral in eq 11 is then evaluated to give $\tau = \tau_d + \tau_{od}$, where

$$\tau_d = -\frac{\hbar}{2\Phi} \sum_j \frac{p_{jj}}{\mathbf{I}(\epsilon_j)} \qquad (12)$$

and

$$\tau_{od} = \frac{2\hbar}{\Phi} \mathbf{I}\left(\sum_{j<k} \frac{p_{jk}}{\epsilon_j - \epsilon_k^*} \right) \qquad (13)$$

represent contributions from diagonal and off-diagonal terms, respectively, to the probability decay curve $P(t)$. Although τ_d is always >0, τ_{od} is typically <0.

For unit quantum yield, these equations can be solved analytically for two-level systems (i.e., direct through-space coupling between donor and acceptor, not including any bridge). Let the total Hamiltonian be given by

$$\mathbf{H} = \begin{pmatrix} H_{DD} & H_{DA} \\ H_{DA} & H_{AA} - i\lambda \end{pmatrix} \quad (14)$$

where the first basis vector is $|\psi_D\rangle$ and the second basis vector is $|\psi_A\rangle$. The through-space coupling H_{DA} is often referred to in the literature as V_{12}, and we introduce the energy gap $\eta = H_{AA} - H_{DD}$ so that the rate constant is given by (53)

$$\hbar k = \frac{2\lambda H_{DA}^2}{2H_{DA}^2 + \lambda^2 + \eta^2} \quad (15)$$

This equation is completely general in the context of this chapter (i.e., when applied to problems described by just one type of quantum number). Equations similar to this are also known to be important in electron-transfer problems in which both vibrational and electronic coordinates must be considered (54).

As an illustration of the quantities involved, some results for two-level systems are given in Figures 1 and 2. In Figure 1, the probability of being in the acceptor state at time t, $P_A(t)$, and the probability of not decaying to the bath, $P(t)$ from eq 6, are plotted against reduced time for the resonant problem $\eta = 0$. Here, we take $\alpha = |H_{DA}|$. Figure 2 is a plot of eq 15 versus the reduced decay rate λ/α. The rate constant exhibits three regimes: the slow drain regime, in which the rate constant increases in proportion to λ; a maximum regime identified with the maximum rate constant predicted by Rabi; and a regime in which the rate constant slows with increasing bath decay rate. This third regime is identified as a microscopic form of Fermi's golden rule. The time-response functions given in Figure 1 are sampled from each of these three regimes. Rabi's maximum regime is the one that most closely couples the entrance and exit rates through the acceptor state, and in fact only a small amount of amplitude builds up on this state.

A general analytical solution of eq 10 is difficult to obtain, and so we proceed by evoking perturbation theory to determine approximate expressions for $\hbar k$ in the limits of small λ and large λ. Then we combine these results and construct by interpolation an expression appropriate for the entire range of λ.

Small-λ Expansion. When λ is small, we express the total Hamiltonian $\mathbf{H}$ as

$$\mathbf{H} = \mathbf{H}_s + \Delta\mathbf{H}, \quad \Delta\mathbf{H} = -i\lambda|\psi_A\rangle\langle\psi_A| \quad (16)$$

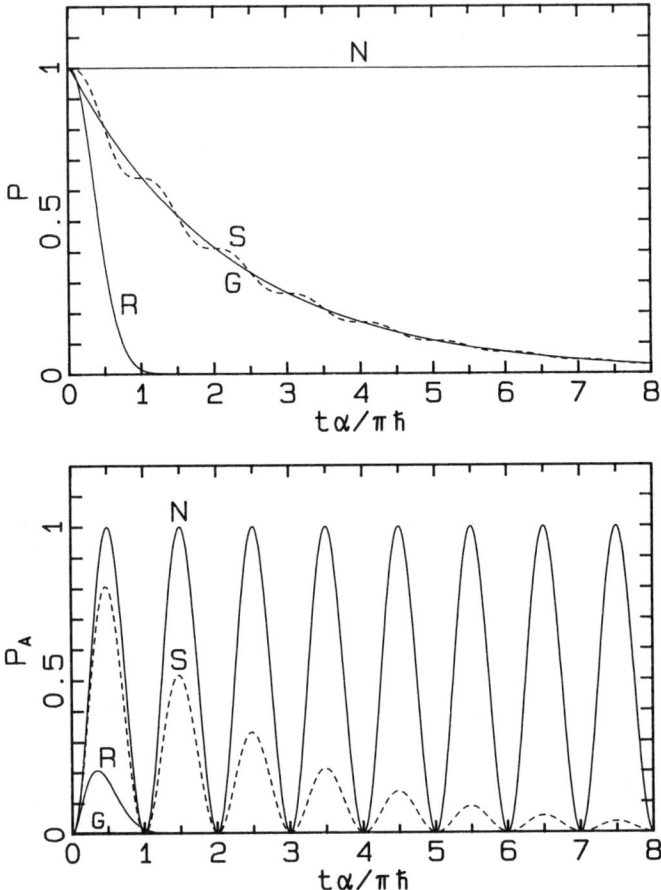

Figure 1. Symmetric resonant two-level system: plots of the probability of being in the acceptor state, P_A, and the probability of not decaying to the bath, P, versus reduced time $t\alpha/(\pi\hbar)$ for various acceptor-to-bath coupling strengths λ. Key: N, no decay, $\lambda = 0$; S, slow drain regime, $\lambda = 0.1 \times 2^{1/2}\alpha$; R, Rabi maximum rate, $\lambda = 2^{1/2}\alpha$; and G, golden rule regime, $\lambda = 10 \times 2^{1/2}\alpha$ (this curve barely exceeds zero).

and determine the unitary eigenvector matrix $\mathbf{U}'$ and the associated diagonal eigenvalue matrix $\mathbf{E}'$ of the system Hamiltonian $\mathbf{H}_s$. Terms containing λ are treated as perturbations and are exposed by transforming $\mathbf{H}$ to produce

$$\mathbf{H}' = (\mathbf{U}')^\dagger \mathbf{H} \mathbf{U}' = \mathbf{E}' - \Delta\mathbf{H}' \qquad (17)$$

where

$$\Delta H_{ij}' = -i\lambda U_{Ai}' U_{Aj}' \qquad (18)$$

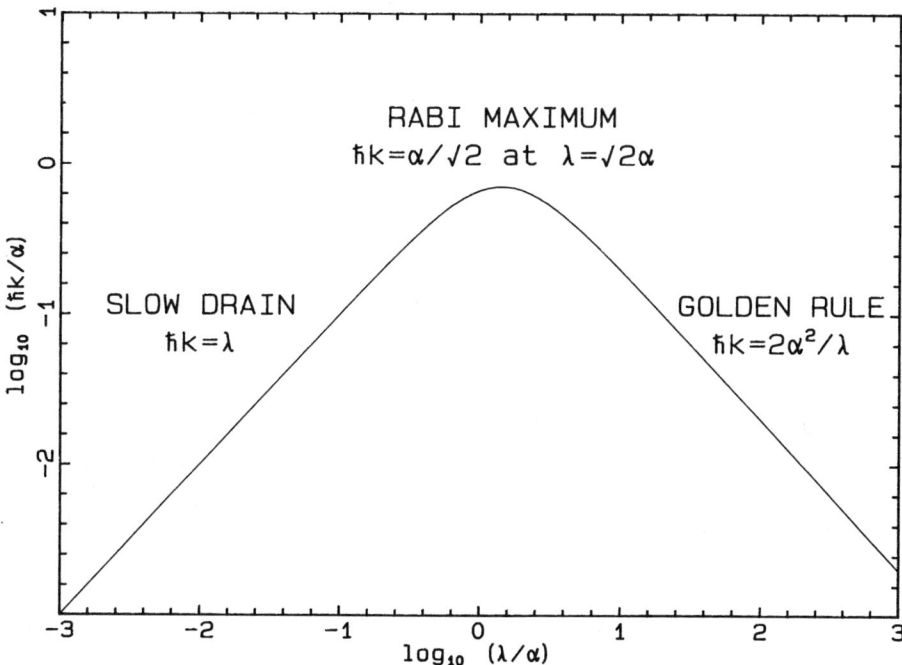

Figure 2. Symmetric resonant two-level system: plot of the reduced rate constant $\hbar k/\alpha$ versus the reduced acceptor-to-bath coupling strength λ/α for full quantum yield ($\Phi = 1$, $\sigma = 0$). (Reproduced with permission from ref. 53. Copyright 1989 Elsevier Science Publishers.)

is the matrix of perturbations. If

$$|\lambda U_{Ai}' U_{Aj}'| \ll |E_i' - E_j'| \forall i,j \tag{19}$$

then the eigenvalues and eigenvectors of $\mathbf{H}'$ are given accurately by a low-order perturbation expansion in λ. From these results (53), the rate constant $\hbar k$ is approximated by

$$\hbar k' = \frac{2\lambda}{n_a' n_r' + 2b \lambda^2} \tag{20}$$

where b is related to the effective HOMO–LUMO gap, n_r' is the effective number of resonant states, and n_a' is an index of the donor–acceptor asymmetry. When the full quantum yield is observed, these quantities are given by

$$b = -4 \sum_{j<k} \frac{U_{Dj}' U_{Dk}' U_{Aj}' U_{Ak}'}{(E_j' - E_k')^2} \quad (21)$$

$$n_r' = n \quad (22)$$

and

$$n_a' = \frac{1}{n_r'} \sum_j \left(\frac{U_{Dj}'}{U_{Aj}'}\right)^2 \quad (23)$$

is unity if the donor and acceptor states are related by symmetry.

Equation 19 is not sufficient to guarantee that $2b\lambda^2 \ll n_a' n_r'$, but for λ sufficiently small, this extra condition also holds, and then eq 20 simplifies to

$$\hbar k_{SD}' = \frac{2\lambda}{n_a' n_r'} \quad (24)$$

The asymmetry parameter n_a' is so called because, when the donor and acceptor are related by symmetry, $U_{Dj}' = \pm U_{Aj}'$ and thus $n_a' = 1$. When no symmetry relation exists, however, for some eigenstates j, U_{Aj}' will become quite small while U_{Dj}' grows large (i.e., the acceptor will contribute much less than the donor to some eigenstates). This difference usually gives rise to quite large values of n_a' and so dramatically reduces the reaction-rate constant (for small asymmetries, it is possible that n_a' falls slightly below unity). Thus, reactions between symmetrically equivalent donor and acceptor usually proceed much faster than corresponding reactions between distinctly asymmetric species.

The term n_r' is an indication of the number of states resonant with the donor state (and includes 1 for the donor state itself). More precisely, it is the maximum number of resolvable diagonal components of the time-decay function $P(t)$ (see eq 6). For high quantum yields less than unity (say $\Phi = 0.99$), only resonant components would be resolvable in $P(t)$, but as $\Phi \to 1$, $n_r' \to n$. This relationship is discussed in detail elsewhere (53).

Large-λ Expansion. When λ is large, a different perturbation expansion is used to approximate the rate constant. The total Hamiltonian **H** is written as

$$\mathbf{H} = \mathbf{H_0} + \mathbf{\Delta H} \quad (25)$$

where $\Delta H_{ij} = \Delta H_{ji} = H_{ij}\delta_{jA}$ for all $i < j$, and $\Delta_{ii} = 0$, i.e.,

$$\mathbf{H}_0 = \begin{pmatrix} H_{DD} & H_{D1} & \cdots & H_{Dn} & 0 \\ H_{1D} & H_{11} & \cdots & H_{1n} & 0 \\ \vdots & \vdots & & \vdots & \vdots \\ H_{nD} & H_{n1} & \cdots & H_{nn} & 0 \\ 0 & 0 & \cdots & 0 & H_{AA} - i\lambda \end{pmatrix} \quad (26)$$

and

$$\Delta\mathbf{H} = \begin{pmatrix} 0 & 0 & \cdots & 0 & H_{DA} \\ 0 & 0 & \cdots & 0 & H_{1A} \\ \vdots & \vdots & & \vdots & \vdots \\ 0 & 0 & \cdots & 0 & H_{nA} \\ H_{AD} & H_{A1} & \cdots & H_{An} & 0 \end{pmatrix} \quad (27)$$

If $\mathbf{E}''$ and $\mathbf{U}''$ are the eigenvalues and eigenvectors, respectively, of $\mathbf{H}_0$, then the perturbation is exposed by transforming the total Hamiltonian $\mathbf{H}$:

$$\mathbf{H}'' = (\mathbf{U}'')^\dagger \mathbf{H} \mathbf{U}'' = \mathbf{E}'' + \Delta\mathbf{H}'' \quad (28)$$

where

$$\Delta\mathbf{H}'' = (\mathbf{U}'')^\dagger \Delta\mathbf{H} \mathbf{U}'' \quad (29)$$

If

$$|E_j'' - H_{AA} + i\lambda| \gg |H_{Aj}| \forall j \neq A \quad (30)$$

then the eigenvalues and eigenvectors of $\mathbf{H}''$ are obtained by using a perturbation expansion in H_{Aj}. Back transformation produces the eigenvalues $\boldsymbol{\epsilon}$

$$\epsilon_A = H_{AA} - i\lambda, \qquad \epsilon_{j \neq A} = E_j'' - \frac{i\lambda}{\eta^2 + \lambda^2}(H_{Aj}'')^2 \quad (31)$$

and eigenvectors

$$C_{ij} = U_{ij}'' + \frac{i\lambda}{\eta^2 + \lambda^2}\left(H_{ij}\delta_{jA} - \sum_{k \neq A} H_{Ak} U_{kj}'' \delta_{iA}\right) \quad (32)$$

of $\mathbf{H}$, where

$$\eta = H_{AA} - H_{DD} \quad (33)$$

$$H_{Aj}'' = \sum_{i \neq A} H_{Ai} U_{ij}'' \quad (34)$$

and δ_{iA} is the delta function. For unit quantum yield, the approximate rate constant calculated by using eqs 10–13 is given by

$$\hbar k_{GR}'' = \frac{2\lambda(\overline{V}_{12}^{(GR)})^2}{\eta^2 + \lambda^2} \tag{35}$$

where

$$\overline{V}_{12}^{(GR)} = \left[\sum_{j \neq A} \left(\frac{U_{Dj}''}{H_{Aj}''}\right)^2\right]^{1/2} \tag{36}$$

When the donor and acceptor states are related by symmetry, $\eta = 0$, and eq 35 simplifies to

$$\hbar k_{GR}'' = \frac{2(\overline{V}_{12}^{(GR)})^2}{\lambda} \tag{37}$$

For problems involving electron transfer, it is often appropriate to consider the acceptor state not as comprising just one level, but rather as comprising a continuum of levels with a Franck–Condon weighted density of states ρ_{FC}. Then, under certain conditions (52) (principally when ρ_{FC} is large), one can integrate over the energy gap η in eq 35, giving

$$\hbar k_{GR}'' = 2\pi(\overline{V}_{12}^{(GR)})^2 \rho_{FC} \tag{38}$$

As described by Robinson and Frosch (18, 19), the λ-dependence disappears because the number of states that contribute to the integral increase in proportion to λ while the individual contributions (eq 37) decrease in proportion to λ. Equation 38 is usually referred to as Fermi's golden rule, and eq 35 is a microscopic form of this rule.

In summary, when λ is larger than *all* of the matrix elements of $\mathbf{H}_s$, the rate constant is given by an expression analogous to Fermi's golden rule. The key feature here is that the rate constant is proportional to the square of some energy-coupling term. However, the decay into the bath *is not* expected to proceed exponentially with time unless the bridge levels are nonresonant with the donor level. The majority of experiments performed involving electron or energy transfer have rather small couplings, much smaller than λ, and thus this limit is most important.

Mid-Range in λ. When we combine the perturbation expressions for both the low-λ and high-λ regimes, we see that the rate constant increases with λ when λ is small, but decreases with λ when λ is large. Thus, the rate constant must attain some maximum value at some optimal λ. Many years ago, Rabi postulated that such a maximum existed, and he advanced some simple formulae for this maximum rate constant in two-level systems. We used two separate approaches to determining the properties of this maximum in bridge systems. First, we extrapolated the low-λ formula, eq 20. Second, we constructed an interpolation function that bridges the low-λ and high-λ results, eqs 24 and 35.

The condition in eq 19 may apply even when λ is of the order of λ_R, the value of the acceptor-to-bath coupling that maximizes the rate constant k. Then eq 20 is a good approximation to the rate constant in the maximum (Rabi) regime. From eq 20, this maximum is

$$\hbar k_R' = (2n_a'n_r'b)^{-1/2} \tag{39}$$

and occurs when

$$\lambda_R' = \left(\frac{n_a'n_r'}{2b}\right)^{1/2} \tag{40}$$

Because λ_R' is reasonably large, one questions the accuracy of eq 19 at this value of λ. When compared to k_R, the exact maximum obtained from eq 15 for two-level systems, eq 39 overestimates the rate constant maximum by a factor of 2½. This discrepancy has minimal effect on through-bridge calculations if it is removed by an empirical renormalization (53), giving:

$$\hbar k_R' = 2^{-1/2} \overline{V}_{12}^{(R)} \tag{41}$$

where

$$\overline{V}_{12}^{(R)} = (2n_a'n_r'b)^{-1/2} \tag{42}$$

The rescaled value of λ_R' is then given by

$$\lambda_R' = n_a'n_r'\hbar k_R' = \left(\frac{n_a'n_r'}{b}\right)^{1/2} \tag{43}$$

It is possible to construct analytical functions of λ that interpolate the rate constant k between the large-λ and small-λ limits described previously. If such an interpolating function provides a good approximation to k for all λ, the problem of taking a perturbation expression beyond its bounds would be avoided, and the function would give a good approximation of the Rabi maximum rate constant. A very useful interpolating function is obtained simply by adding the appropriate low-λ and high-λ parts of the decay area τ. We define such a function as

$$\tau_i = \frac{n_a' n_r'}{2\lambda} + \frac{\lambda}{2(\overline{V}_{12}^{(GR)})^2} \tag{44}$$

where n_r' and n_a' are defined from the small-λ limit by eqs 22 and 23, respectively, and $\overline{V}_{12}^{(GR)}$ is defined from the golden rule limit by eq 36. This function leads to a definition of an approximate rate constant k_i,

$$\hbar k_i = \frac{2\lambda (\overline{V}_{12}^{(GR)})^2}{(\overline{V}_{12}^{(GR)})^2 n_a' n_r' + \lambda^2} \tag{45}$$

which does in fact have the correct limits as $\lambda \to 0$ and as $\lambda \to \infty$. The maximum in this function (at full quantum yield) is

$$\hbar k_{iR} = (n_a' n_r')^{-1/2} \overline{V}_{12}^{(GR)} \tag{46}$$

and occurs at $\lambda = \lambda_{iR}$, where

$$\lambda_{iR} = (n_a' n_r')^{1/2} \overline{V}_{12}^{(GR)} \tag{47}$$

so that

$$\lambda_{iR} = n_a' n_r' \hbar k_{iR} \tag{48}$$

and

$$\hbar k_{iR} \lambda_{iR} = (\overline{V}_{12}^{(GR)})^2 \tag{49}$$

When eq 46 is applied to two-level systems, eq 15 is obtained (even for arbitrary η) so that this formula in fact gives exact results. Empirically, it is observed that eq 46 is always exact to within numerical precision when full quantum yield is obtained. Thus, the entire λ dependence of the rate constant may be expressed explicitly. All of the other quantities in eq 46 can be obtained from a knowledge of just the isolated system Hamiltonian, providing a considerable philosophical and computational advance. This general result is interesting, but no formal proof of this equivalence is yet known.

From a practical standpoint, the method of extrapolation is simpler to implement than the interpolation method because eq 46 contains parameters derived from two different perturbation expansions using two different reference Hamiltonians, eqs 16 and 26. The advantage of using k_i is, however, that it is more accurate. For almost all problems of practical importance, k_R' is sufficiently accurate, and thus this extra accuracy is immaterial. If accuracy forces one to use k_i, then, in certain important limits, the eigenstates of one of the two Hamiltonians can be expressed in terms of the eigenstates of the other, thus greatly simplifying the theory.

Relation to the Work of Joachim

The Rabi maximum regime is very useful in that it reveals the maximum rate constant, irrespective of the nature of the environment. Its analogous regime for problems in which both electronic and vibrational coordinates must be considered is treated successfully by Beats's generalized method (29). In a well-designed system, the "on" state of some molecular switch will have a transfer rate close to this maximum rate, and so this regime is important to the design of molecular electronic devices.

Joachim (39) developed a theory for this regime in bridged systems that has provided much insight into the causes of distance-independent transfer processes. His theory is inspired, but ad hoc, in that no physical basis exists for the implicit postulates. As a result, nonphysical effects such as discontinuities in the maximum rate constant as a function of the Hamiltonian are introduced. This chapter demonstrates that, under certain conditions, our general formula for the Rabi rate constant reduces to a form very similar to the equations of Joachim, but containing none of the nonphysical effects.

We worked from the extrapolated formula, eq 39, and assumed that no bridge states are resonant with either the donor or acceptor states. (Joachim's theory is intended to treat resonant situations.) Now only one off-diagonal p-matrix element (*see* eq 7) remains significant, and this contributes

$$\frac{1}{(E_D' - E_A')^2} \tag{50}$$

to b (*see* eq 21), where D and A are the indices of the eigenstates whose character is dominated by the donor and acceptor basis states, respectively. The effective coupling element for the Rabi rate, $\overline{V}_{12}^{(R)}$ (*see* eq 42), simplifies in this nonresonant limit to

$$\overline{V}_{12}^{(NR)} = \frac{|E_D' - E_A'|}{(2n_a' n_r')^{1/2}} \tag{51}$$

If we consider only problems in which the donor and acceptor are related by symmetry, then $n_a' = 1$. Also, in these circumstances, it is shown elsewhere (53) that when the quantum yield is high but <1 (say 0.99), then the number of resonant modes $n_r' \to 2$. Equation 39 thus simplifies to

$$\hbar k_{SNR}' = 2^{-1/2} \overline{V}_{12}^{(SNR)} \tag{52}$$

where

$$\overline{V}_{12}^{(SNR)} = \frac{1}{2} |E_D' - E_A'|$$

Joachim (39) proposed two expressions for the Rabi rate constant and called them the "H_{eff}" method and the "Exact" method. The exact method involved the solution of the time-dependent Schrödinger equation for the system Hamiltonian H_s, and the H_{eff} method is derived as an approximation to the exact method. Rather, we demonstrate that his H_{eff} method is closer to physical reality and that his exact method usually provides only a rough numerical approximation to it. Joachim applied his methods to systems described by matrices of the form suggested by McConnell (24)

$$H_s = \begin{pmatrix} 0 & \beta & 0 & : & 0 & 0 & 0 \\ \beta & e_0 & \alpha & : & 0 & 0 & 0 \\ 0 & \alpha & e_0 & : & 0 & 0 & 0 \\ : & : & : & & : & : & : \\ 0 & 0 & 0 & : & e_0 & \alpha & 0 \\ 0 & 0 & 0 & : & \alpha & e_0 & \beta \\ 0 & 0 & 0 & : & 0 & \beta & 0 \end{pmatrix} \tag{53}$$

where $H_{DA} = 0$, $H_{DD} = H_{AA}$, $H_{BB} = H_{AA} + e_0$, $H_{D1} = H_{An} = \beta$, and $H_{i,i+1} = \alpha$ for all i and $i+1$ on the bridge.

The exact method is based on a simple extension of Rabi's original concept. Rabi's idea, appropriate for two-level systems, is that the maximum possible rate is the inverse of the time. In the no-reaction case, where amplitude just pendulates from donor to acceptor and back again, it takes the original amplitude in the donor to first reach the acceptor. This process is illustrated in Figure 3 where, for a two-level system, the probability of being in the acceptor state P_A is plotted against reduced time $t\overline{V}_{12}/(\pi\hbar)$.

It is easy to show that the pendulation frequency is $\omega = 2\overline{V}_{12}/\hbar$, so the time of the first donor-to-acceptor transition is π/ω, and the rate constant is $\hbar k_R = 2\pi^{-1} V_{12}$. Joachim's extension (39) of this idea to bridged systems is to define the rate constant from the reciprocal of the time at which the first donor-to-acceptor transition is a maximum; this maximum is determined numerically by solving the time-dependent Schrödinger equation. An example is given in Figure 3, where, for the situation in which $n = 8$, $e_0 = 1$, $\alpha = 0.8$, and $\beta = 0.08$, the time-response function is plotted and Joachim's

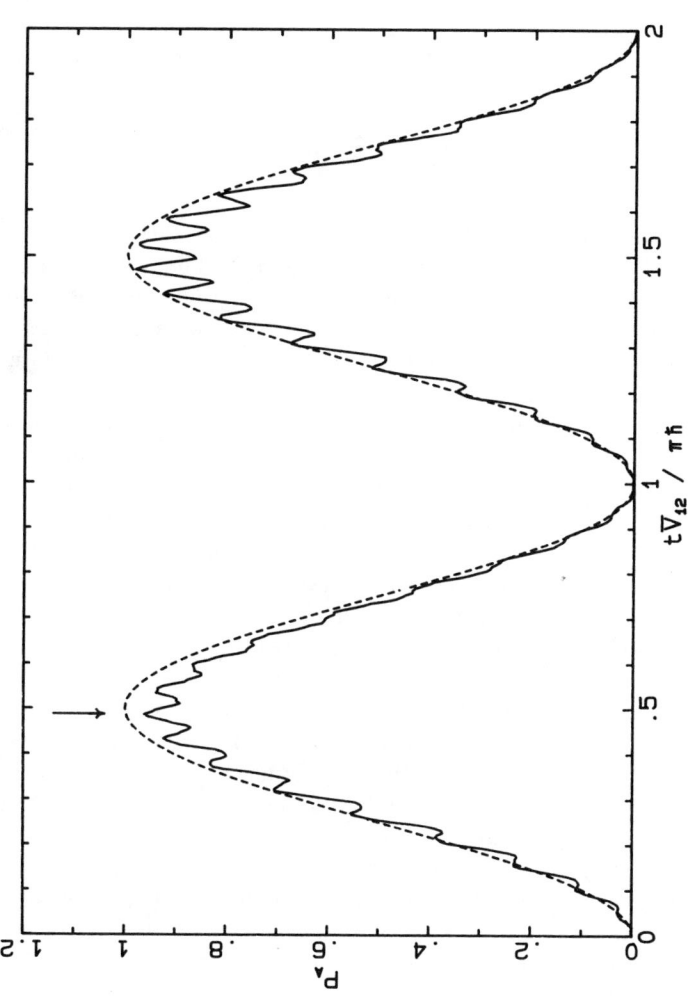

Figure 3. Plot of the acceptor state probability P_A versus reduced time $t\bar{V}_{12}/\pi\hbar$. Key: ——, for a 10-level system; and - - -, for a two-level system. The arrow indicates the time chosen by Joachim's algorithm.

maximum is indicated by an arrow. For Rabi's two-level system, the time-response function is sinusoidal, but for Joachim's bridged systems, the time-response function contains additional contributions from the many high-frequency Fourier components. Thus, the actual time at which the maximum occurs is very sensitive to the phasing of these high-frequency components. The center of the band of the first recurrence is a good indication of the rate constant; the actual maximum itself is this quantity plus a noise term.

Joachim's H_{eff} method (39) is designed to determine the center of the band of the first transition. It is thus a much better measure of the reaction rate than is his exact method, and it gives the rate constant as

$$\hbar k_J = \frac{2}{\pi} \overline{V}_{12}^{(SNR)} \qquad (54)$$

This equation is very similar to our nonresonance equation, eq 52. The difference between the prefactors is only minor. That such a difference should exist is expected, as the prefactor in Rabi's original equation was not obtained from fundamental theory, but was just an estimate. More important, eq 52 is valid only in the nonresonance limit, whereas Joachim intends that eq 54 be applied universally.

An ambiguity arises in Joachim's equation at resonance, because it is no longer clear which two eigenstates j,k should be chosen to represent the donor and acceptor basis states. He resolved this ambiguity by selecting the D and A eigenstates as the states with the particular j and k that maximize the product $|U_{Dj}'U_{Dk}'U_{Aj}'U_{Ak}'|$. In the light of the derivation presented here, this step is tantamount to including only the largest off-diagonal contribution to τ (see the definition of b, eq 21). His definition works well off-resonance, but fails at resonance, as then the product will be equally as large for at least two different sets of j and k.

As a resonance is crossed, the identity of the "D" and "A" eigenstates changes so that eq 54 is, in general, discontinuous. Joachim himself may not have been aware of this problem. In his original paper (39) he used only a coarse abscissa grid for his plots and drew lines through the discontinuities. Figure 4 from ref 39 is redrawn in our Figure 4 on a much finer grid to highlight the discontinuities; for this figure, $n = 8$ and $\beta/\alpha = 0.1$. The open circles in Figure 4 are the (continuous) results k_R obtained by numerical solution of eq 10 (employing a quantum yield of about 0.99; see ref. 53). The analytical approximation eq 46 provides results that are very similar to k_R itself. To highlight the effects, this figure is redrawn in Figure 5, along with a plot of the real part of the eigenvalues of our complex matrix H, eq 1. When $e_0 = 0$, the eight bridge eigenstates are symmetrically distributed about the two donor levels. Increasing e_0 breaks this symmetry and, in turn, forces each of the four low-lying bridge eigenstates to become resonant with the donor and acceptor levels. These resonances give rise to strong interactions, faster rate constants, and discontinuities in k_J.

For the problem chosen by Joachim, the effect of the discontinuity is quite small, but its effect is highly system-dependent. It is possible that k_J itself remains continuous and only its first derivative is discontinuous (53), but in other circumstances, it is possible to see a 10-fold change in the value of k_J as the discontinuity is crossed.

Although Joachim's H_{eff} expression is clearly quite useful away from resonance, it was designed specifically to handle resonant problems. Joachim used this expression to study bridge-length dependences. We show later that, although Joachim's approach usually produces good results, it also produces undesirable unphysical effects (unlikely to correspond to reality). As it is no more difficult to evaluate eq 46 than it is to implement Joachim's algorithm, its use is preferred to the more approximate equation. Indeed, solution of the fundamental equation, eq 10, is possible for quite large bridge sizes. Joachim's H_{eff} expression is an effective two-level model. It assumes that the rate constant for a bridged system can be expressed in the same functional form as the rate constant for a two-level system. This, however, is only the case when no bridge levels are nonresonant. In general, effective two-level models are inappropriate for bridged systems.

Relation to the Work of Broo and Larsson

Larsson (43–45) introduced an effective two-level model to estimate the effective donor-to-acceptor coupling for situations in which the bridge is nonresonant. Our formalism gives results very similar to those of the recent work of Broo and Larsson (45) in the nonresonance limit.

Golden Rule. First, we investigated the golden rule regime and simplified the general expressions for the rate constant eq 10 in the limit that both the bridge levels are not resonant with either the donor or acceptor levels, and that λ is large. Here, "large" takes on a different meaning from that of eq 30, as the introduction of the nonresonance condition allows the perturbation expansion to be performed for much smaller values of λ than was previously possible. First, the total Hamiltonian **H** is written as

$$\mathbf{H} = \mathbf{H}_0 + \mathbf{\Delta H} \tag{55}$$

where $\Delta H_{ij} = \Delta H_{ji} = H_{ij}\delta_{jA} + \Delta H_{ij}\delta_{jD}$ for all $i < j$, and $\Delta H_{ii} = 0$; that is,

$$\mathbf{H}_0 = \begin{pmatrix} H_{DD} & 0 & \cdots & 0 & 0 \\ 0 & H_{11} & \cdots & H_{1n} & 0 \\ \vdots & \vdots & & \vdots & \vdots \\ 0 & H_{n1} & \cdots & H_{nn} & 0 \\ 0 & 0 & \cdots & 0 & H_{AA} - i\lambda \end{pmatrix} \tag{56}$$

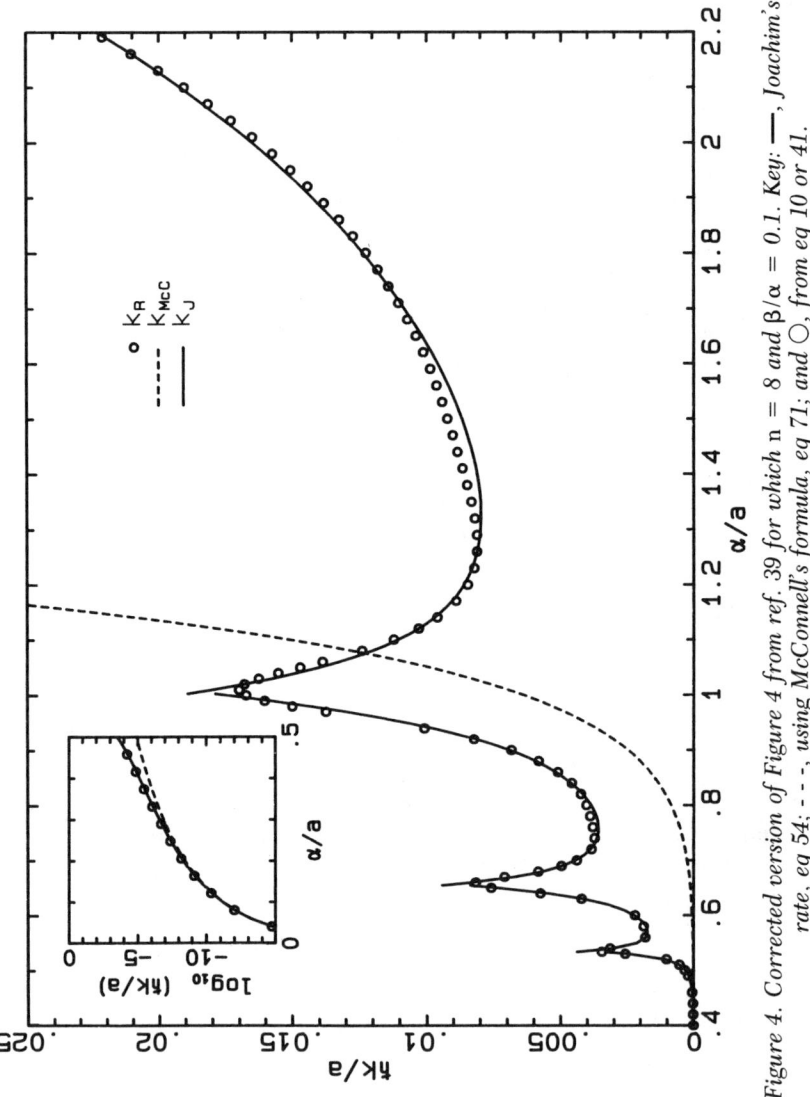

Figure 4. Corrected version of Figure 4 from ref. 39 for which n = 8 and β/α = 0.1. Key: ——, Joachim's H_{eff} rate, eq 54; - - -, using McConnell's formula, eq 71; and ○, from eq 10 or 41.

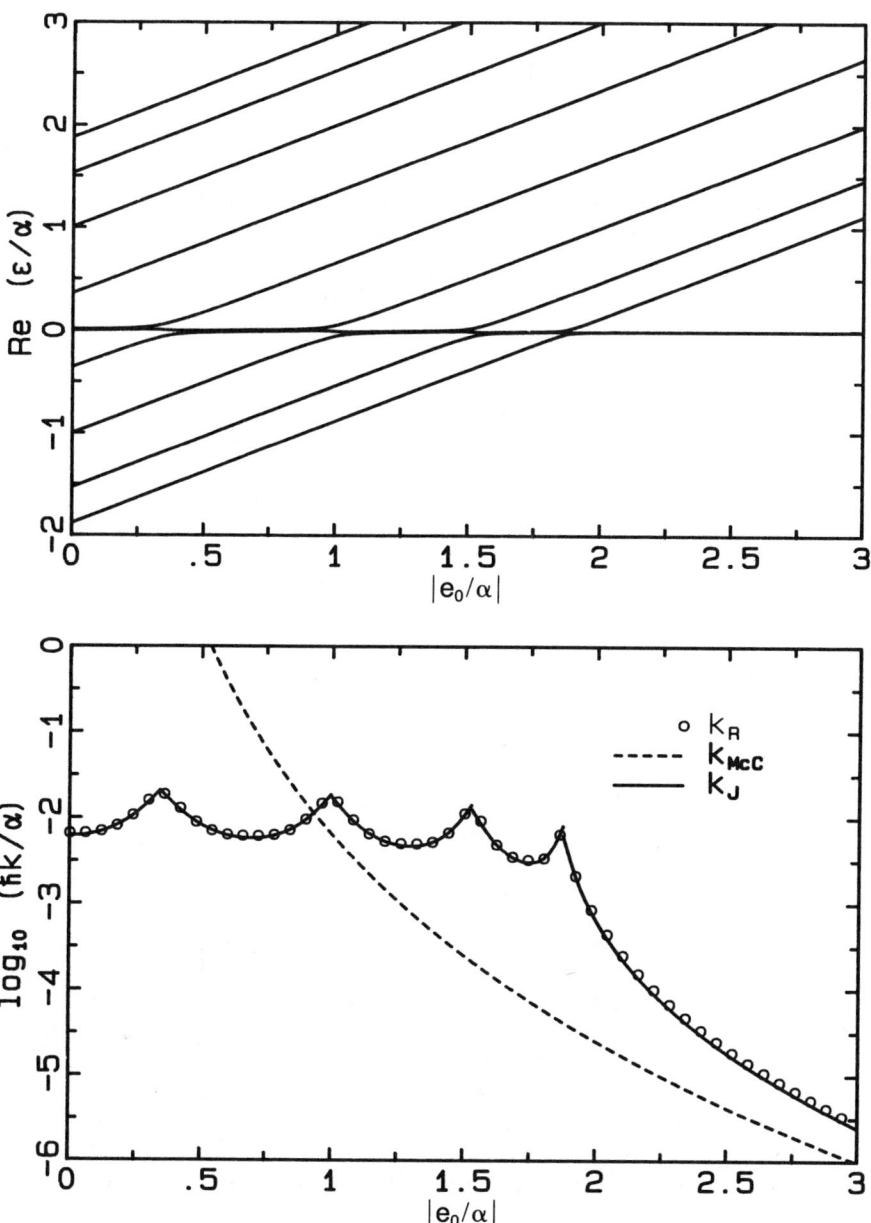

Figure 5. Bottom, Figure 4 rescaled. Top, reduced real component of the eigenvalues of **H** plotted against the reduced bridge energy gap, e_0/α.

and

$$\Delta \mathbf{H} = \begin{pmatrix} 0 & H_{D1} & \cdots & H_{Dn} & H_{DA} \\ H_{1D} & 0 & \cdots & 0 & H_{1A} \\ \vdots & \vdots & & \vdots & \vdots \\ H_{nD} & 0 & \cdots & 0 & H_{nA} \\ H_{AD} & H_{A1} & \cdots & H_{An} & 0 \end{pmatrix} \quad (57)$$

If $\mathbf{E}'''$ and $\mathbf{U}'''$ are the eigenvalues and vectors, respectively, of $\mathbf{H_0}$, then the perturbation is exposed by transforming the total Hamiltonian $\mathbf{H}$:

$$\mathbf{H}''' = (\mathbf{U}''')^\dagger \mathbf{H} \mathbf{U}''' + \Delta \mathbf{H}''' \quad (58)$$

where

$$\Delta \mathbf{H}''' = (\mathbf{U}''')^\dagger \Delta \mathbf{H} \mathbf{U}''' \quad (59)$$

If the nonresonance condition

$$|E_j''' - H_{AA}| \gg |H_{Aj}| \quad (60a)$$

and

$$|E_j''' - H_{DD}| \gg |H_{Dj}| \ \forall \ j \neq A, D \quad (60b)$$

where

$$H_{Aj}''' = \sum_i H_{Ai} U_{ij}''' \quad \text{and} \quad H_{Dj}''' = \sum_i H_{Di} U_{ij}''' \quad (61)$$

holds and the weak condition on λ

$$|\eta + i\lambda| \gg \left| H_{DA} + \sum_{j \neq A, D} \frac{H_{Dj}''' H_{Aj}'''}{H_{DD} - E_j'''} \right| \quad (62)$$

also holds, then the eigenvalues and eigenvectors of $\mathbf{H}'''$ are obtained by using a perturbation expansion of up to fourth order in H_{Dj}''', H_{Aj}''', and H_{DA}. Back transformation produces the eigenvalues ϵ and eigenvectors of $\mathbf{H}$. Fourth-order perturbation terms are included, as the direct coupling H_{DA} is often very small in bridged systems. This perturbation makes the fourth-order through-bridge coupling

$$\sum_{j \neq A, D} \frac{H_{Dj}''' H_{Aj}'''}{H_{DD} - E_j'''} \quad (63)$$

of paramount importance.

The large-λ condition, eq 62, ensures that higher terms in the perturbation expansion are small. It is a much weaker condition than that used previously in eq 30, as the direct-coupling and through-bridge terms are often many orders of magnitude smaller than a typical matrix element H_{ij}. Thus, eq 62 may be valid for much smaller λ than eq 30 is valid for.

Evaluation of the rate constant from eq 10–13 is straightforward and is described elsewhere (53). Basically, a single eigenvalue is dominated by the donor state so that the rate-constant decay is dominated by just one of its diagonal Fourier components. Thus, the decay is single exponential, and it obeys the rate law for two-level systems in the golden rule regime, eq 15.

$$\hbar k_{GR}''' = \frac{2\lambda(\overline{V}_{12}^{(NGR)})^2}{\overline{\eta}^2 + \lambda^2} \tag{64}$$

The effective two-state coupling is defined in the nonresonant golden rule limit as

$$\overline{V}_{12}^{(NGR)} = H_{DA} + \sum_{j \neq A,D} \frac{H_{Dj}''' H_{Aj}'''}{H_{DD} - E_j'''} \tag{65}$$

and the effective energy gap is

$$\overline{\eta} = H_{AA} - H_{DD} + \sum_j \left(\frac{(H_{Aj}''')^2}{H_{AA} - E_j'''} - \frac{(H_{Dj}''')^2}{H_{DD} - E_j'''} \right) \tag{66}$$

For the important case in which the donor and acceptor are related by symmetry, $\overline{\eta} = 0$; eq 64 simplifies to the familiar golden rule form

$$\hbar k_{GR}''' = \frac{2(\overline{V}_{12}^{(NGR)})^2}{\lambda} \tag{67}$$

Broo and Larsson (43–45) derived a quantity similar to $\overline{V}_{12}^{(NGR)}$, which we call $\overline{V}_{12}^{(E)}$, by constructing a two-level Hamiltonian $\underline{H}(E)$. This Hamiltonian's eigenvalues equal the approximate eigenvalues of the donor state and the acceptor state of the system Hamiltonian $\mathbf{H}_s$, which was obtained by using low-order perturbation theory. Their approach is very similar to ours, except that we applied perturbation theory to evaluate the rate constant. In contrast, they first introduced an effective two-state model and then applied the perturbation theory. This approach gives

$$\underline{H}(E) = \begin{pmatrix} H_{DD} + \sum_{j \neq A,D} \frac{(H_{Dj}''')^2}{E - E_j'''} & H_{DA} + \sum_{j \neq A,D} \frac{H_{Dj}''' H_{Aj}'''}{E - E_j'''} \\ H_{DA} + \sum_{j \neq A,D} \frac{H_{Dj}''' H_{Aj}'''}{E - E_j'''} & H_{AA} + \sum_{j \neq A,D} \frac{(H_{Aj}''')^2}{E - E_j'''} \end{pmatrix} \tag{68}$$

after one cycle of Löwdin partitioning (55, 56).

Unfortunately, the parameter E in this equation is ill-defined, and this construction yields a unique result only for the very important case when $H_{DD} = H_{AA}$. In general, Broo and Larsson (45) expressed this parameter in an ad hoc fashion as the average $E = (H_{DD} + H_{AA})/2$. Other attempts (57) to resolve this ambiguity generate models with other adverse properties such as nonhermitian system-only Hamiltonians $\mathbf{H}_s$. Typically, E is much larger than the energy difference $H_{AA} - H_{DD}$, and so the precise definition of E is not of great practical significance. Our approach provides a unique resolution of this ambiguity, but eq 64 also contains an adverse physical property. The microscopic forward and reverse rate constants calculated by using eq 64 are different because of the unique role given to H_{DD} in eq 65. Broo and Larsson's (45) resolution of the ambiguity in E does, correctly, show microscopic reversibility. In our results, the asymmetry results from the very different physical roles ascribed to the donor and acceptor states. If necessary, an improved rate constant could be defined as the reciprocal of half of the time required to complete microscopically both the forward and reverse reactions, $k = 2k_F k_R/(k_F + k_R)$.

Rabi Maximum. Next, we investigated the Rabi maximum regime. In Larsson's approach (43–45), one effective coupling element is generated and used in a two-level kinetics theory to determine the reaction-rate constant. We proceeded by expanding our general formula for the Rabi rate constant, eq 52, in the limit of a nonresonant bridge. For problems in which the donor and acceptor states are related by symmetry, our approach and the approach of Broo and Larsson (45) give the same results. When η is large, application of fourth-order perturbation theory yields

$$E_D' = H_{DD} - \eta^{-1}\left(H_{DA} + \sum_{j \neq A,D} \frac{H_{Dj}'' H_{Aj}''}{H_{DD} - E_j''}\right)^2 \tag{69}$$

and

$$E_A' = H_{AA} - \eta^{-1}\left(H_{DA} + \sum_{j \neq A,D} \frac{H_{Dj}'' H_{Aj}''}{H_{AA} - E_j''}\right)^2 \tag{70}$$

where E_j'', H_{Dj}'', and H_{Aj}'' are given in eqs 58 and 61. These equations preserve microscopic reversibility, and they cannot be written in the form given by an effective two-level theory. However, as E is much larger than its uncertainty, very similar results are expected from the two approaches.

In practical computational terms, no great difference exists between the method of Broo and Larsson (43–45) and the methods presented here. Both sets of expressions appear expensive to evaluate for large bridges, as they require the knowledge of all of the eigenvalues and eigenvectors of the system

Hamiltonian. However, the equations can be rewritten (53) in a form that can be solved iteratively without requiring any eigenvalues to be found or even the entire matrix to be simultaneously held in core memory. This approach makes practical calculations involving many thousands of bridge states.

The concepts underlying Larsson's approach have been around for a long time and have proved very successful in interpreting nonresonant electron-transfer processes. Broo and Larsson (45) cemented these ideas into a strong simple unit. From the foregoing discussion, it is reasonable to conclude that our approach reduces to Broo and Larsson's in the nonresonant limit, as it should.

Relation to the Work of McConnell

Many years ago, McConnell (24) presented a very useful formula for interpreting spin transfer processes in situations where the bridge levels are nonresonant with the donor and acceptor levels. In a well-designed system, the "off" state of some molecular electronics device will most likely display kinetics in this regime (5, 6); thus, it is important to understand the properties of this regime.

Today, chemical physics interest is often directed toward problems in which resonance occurs, and for these problems McConnell's approach is inapplicable. Authors often compare the results of their advanced theories for resonant rate constants to the results of McConnell's theory without discussion. From reading, one is left with the feeling that McConnell's formula should be discarded and only the new formula retained. In ref. 39, a factor of 2 has adventitiously entered into McConnell's formula, and so Joachim failed to notice that his formula does in fact give the correct nonresonance results. In this chapter, we reinterpret McConnell's work in the light of modern discussion and show how his simple analytical results can be included at the appropriate limit of more-general theories.

The usefulness of McConnell's formula stems from the practical difficulties in evaluating equations like eqs 51, 65, and 69 for large bridges. An efficient iterative method for evaluating these equations is described elsewhere (53). McConnell's formula provides an analytical solution for the rate constant of systems described by $n + 2$ dimensional Hamiltonian matrices of the form of eq 53. This model Hamiltonian is excellent for studying through-bridge kinetics. McConnell (24) showed that, for such systems,

$$\frac{\overline{V}_{12}}{\alpha} = -2 \left(\frac{\beta}{\alpha}\right)^2 \left(\frac{-\alpha}{e_0}\right)^n \tag{71}$$

This formula is extremely useful, as it directly relates matrix elements to rate constants. Clearly, it is appropriate only in the nonresonant bridge limit, and methods such as those of Joachim (39) and of Broo and Larsson (45) reduce to McConnell's formula in this limit.

Model of Resonant Bridge Coupling

Equation 46 is observed numerically to provide exact results for the rate constant when full quantum yield is obtained. It is much simpler than the general expression, and it may be applied to obtain analytical expressions for the rate constant of complicated systems. In designing a molecular electronics device to have particular desired properties, one has a great advantage if an analytical expression is available for the current (rate constant) as a function of the molecular parameters. We adopted a three-level system as the basic model for electron transfer through a resonant bridge. The method applied here may be applied to generate analytical models for more complicated systems, if desired. We neglected the effects of through-space coupling, and wrote the three-level Hamiltonian as

$$\mathbf{H} = \begin{pmatrix} H_{DD} & H_{DB} & 0 \\ H_{DB} & H_{BB} & H_{BA} \\ 0 & H_{BA} & H_{AA} - i\lambda \end{pmatrix} \quad (72)$$

For simplicity, we eliminated the zero of energy by introducing the variables $e_0 = H_{BB} - H_{DD}$ and $\eta = H_{AA} - H_{DD}$. This step leaves five independent variables.

To evaluate eq 46, we first obtained $\overline{V}_{12}^{(GR)}$ by using eq 36. This calculation gives

$$\overline{V}_{12}^{(GR)} = \frac{H_{DB}H_{BA}}{(2H_{DB}^2 + e_0^2)^{1/2}} \quad (73)$$

The effective coupling matrix element decreases as the bridge energy is raised or lowered, and the coupling strength is independent of the acceptor energy. This independence is a general property of eq 36. Next, we evaluated $n_a' n_r'$ by using eq 23. This step gives

$$n_a' n_r' = \frac{2H_{DB}^4 + H_{BA}^4 + \eta^2(2H_{DB}^2 + e_0^2) + 2\eta e_0(H_{DB}^2 - H_{BA}^2)}{H_{DB}^2 H_{BA}^2} \quad (74)$$

Finally, the rate constant for all through-bridge-coupled three-level systems at full quantum yield is given by eq 46 as

$$\hbar k = \frac{2\lambda H_{DB}^2 H_{BA}^2}{2H_{DB}^4 + H_{BA}^4 + 2\eta e_0(H_{DB}^2 - H_{BA}^2) + (2H_{DB}^2 + e_0^2)(\eta^2 + \lambda^2)} \quad (75)$$

For symmetric systems where $\alpha = H_{DB} = H_{BA}$ and $\eta = 0$, the rate constant becomes

$$\hbar k = \frac{2\lambda\alpha^4}{3\alpha^4 + (2\alpha^2 + e_0^2)\lambda^2} \tag{76}$$

The rate constant is a maximum when $e_0 = 0$ (i.e., when there is a complete resonance between the bridge level and the donor and acceptor levels). At this maximum, the value of the rate constant is

$$\hbar k = \frac{2\lambda\alpha^2}{3\alpha^2 + 2\lambda^2} \tag{77}$$

This result is very similar to the rate law for symmetric two-level systems, if we associate the coupling H_{DA} in eq 15 with H_{DB} and H_{BA} in eq 72. Thus, adding the resonant bridging state decreases the Rabi rate constant by only a small amount, from $2^{-1/2} H_{DA}$ to $2^{-1/6} H_{DB}$. However, it may have added a significant spatial distance between the regions in which the donor and acceptor states are localized, and thus it is resonant coupling that gives rise to small distance dependences of electron-transfer rate constants.

It is possible to find the equation of the line upon which the rate constant is simultaneously a maximum with respect to η, e_0, λ, and H_{DB}. This line is near

$$\eta = e_0 = 0, \quad \left|\frac{H_{DB}}{H_{BA}}\right| = 2^{-1/4}, \quad \frac{\lambda}{|H_{BA}|} = 2^{1/4} \tag{78}$$

and gives the rate constant as approximately

$$\frac{\hbar}{k_{max}} = 2^{-5/4} |H_{BA}| \tag{79}$$

Thus, the fastest rates occur when the donor and acceptor states are isoenergetic with the bridge state, and the donor-to-bridge coupling is slightly less than the bridge-to-acceptor coupling. This result is not physically correct as, because of the privileged role of H_{BA}, a different microscopic rate constant is evaluated for the forward reaction and for the reverse reaction. It appears only when the coupling is asymmetric, and the difference between the calculated rate constants is usually only small. Calculations of bridge modulation of the electron-transfer rates should not be significantly affected.

Bridge-Length Dependence

The model Hamiltonian eq 53 introduced by McConnell (24) is very useful for studying the bridge-length dependence of electron-transfer reaction rates. Joachim (39) used this Hamiltonian to good advantage and demonstrated that the available parameter space allowed regions in which the rate constant oscillated as the bridge length increased. Because of the importance of this result and the small problems found with the method, we repeated Joachim's calculations by using the Rabi rate k_R obtained by numerically maximizing eq 10 with respect to bath interaction λ, using a high quantum yield of about $\Phi = 0.99$. The results from our analytical expression eq 49 usually differ insignificantly from these numerical results.

Figure 6 corresponds to Figure 3 in ref. 39. In our version, Joachim's rate constant k_J and k_R are plotted on a log scale as a function of the bridge length n for the situations in which the Hamiltonian parameters $\beta/\alpha = 10$, 1, or 0.1 while $e_0/\alpha = 0$, 0.5, 2, or 10. In general, there is excellent agreement between the two sets of rate constants k_R and k_J. Minor differences occur for the situation in which $\beta/\alpha = 1$, $e_0/\alpha = 0.5$ where k_J/α oscillates while k_R/α decreases monotonically; and for the situation in which $\beta/\alpha = 10$, $e_0/\alpha = 0.5$ where k_J/α initially increases before decreasing, while k_R/α decreases monotonically.

Joachim (39) appears to have plotted his figure too coarsely to observe this nonmonotonic behavior in his rate constant. The behavior arises because the nature of the two eigenstates selected by Joachim's algorithm to represent the donor and acceptor states changes as n increases. The only major difference occurs for the situation in which $\beta/\alpha = 10$, $e_0/\alpha = 0$ where k_J/α increases monotonically toward a limiting value while k_R/α decreases monotonically. Joachim did not investigate this region of the function space and so failed to observe this effect. The effect arises because Joachim's algorithm locks onto one pair of eigenstates, which happen to be the lowest and highest eigenstates. Their energy separation is $\sim |2\beta|$, independent of bridge length. A large number of eigenstates contribute about equally to the rate constant, k_R. As n increases, this magnitude dilutes the effect of the largest contribution and so reduces the rate constant.

Figure 7 corresponds to Figure 5 in ref. 39. Unfortunately, Joachim did not give sufficient information for his original figure to be reproduced, so the correspondence is only indirect. We evaluated the rate constant at $n = 1$ to 5 and extracted the bridge-length dependence parameter γ by fitting the data to

$$k_R(n) = k_R(0) \exp(-\gamma n) \qquad (80)$$

The results are plotted in Figure 7 as contour lines for which $\gamma = 0.1, 0.2, 0.5,$ and 1.0 as functions of the reduced Hamiltonian parameters $|\beta/\alpha|$ and $|e_0/\alpha|$. Unfortunately, extracting the decay constants is not always a well-defined procedure. Figure 6 shows that the rate constant may oscillate as a function of n and, even if it decays monotonically, the decay is not always exponential. We thus plot on Figure 7 the regions (labeled "A") in which

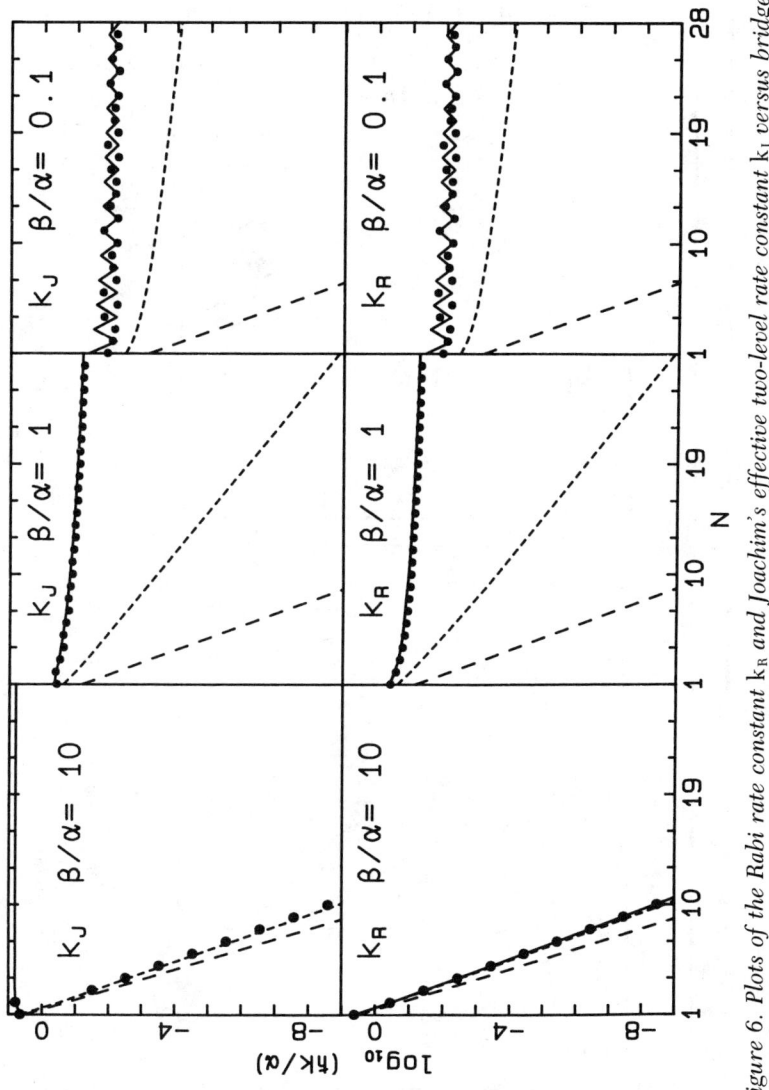

Figure 6. Plots of the Rabi rate constant k_R and Joachim's effective two-level rate constant k_J versus bridge length N. Key: —, at $e_0/\alpha = 0$; ●, at $e_0/\alpha = 0.5$; - - -, at $e_0/\alpha = 2$; and – –, at $e_0/\alpha = 10$.

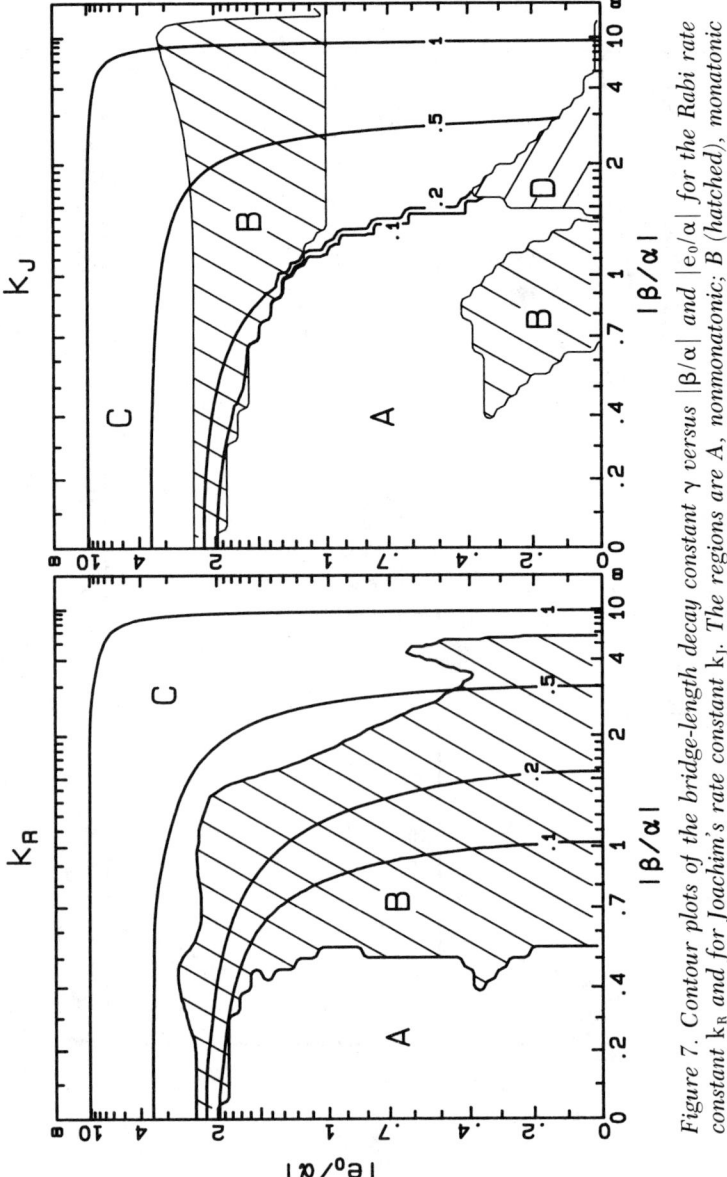

Figure 7. Contour plots of the bridge-length decay constant γ versus $|\beta/\alpha|$ and $|e_0/\alpha|$ for the Rabi rate constant k_R and for Joachim's rate constant k_J. The regions are A, nonmonatonic; B (hatched), monatonic decreasing, nonexponential; C, exponential decay; and D (hatched), monatonic increasing, nonexponential.

nonmonotonic decay is observed; the regions (hatched and labeled "B") in which monotonic but nonexponential decay is observed; the regions (labeled "C") in which exponential decay is observed; and the regions (hatched and labeled "D") in which a monotonic increase in decay rate is observed.

As a criterion for exponential decay, we selected as exponential decays those in which the error in the slope is less than 5% to within the 90% t-test confidence level. The scales in Figure 7 are linear in the arc tangent of the variable. This approach contracts the range $0 \rightarrow \infty$ to a finite length and emphasizes the central region in which the variables are both near 1.

The plot for k_R shows that the oscillatory region A is that for which

$$\left|\frac{\beta}{\alpha}\right| < 0.5, \quad \left|\frac{e_0}{\alpha}\right| < 2 \qquad (81)$$

Here, the rate constant decays only very slowly with bridge length, and it is conceivable to build a highly conducting bridge by exploiting this region. Extending beyond it is the nonexponential decay region B, which contains the decay contours with $\gamma = 0.1$ and 0.2. Clearly, these exponents cannot be interpreted literally, but they do give a qualitative feel to the decay over the region $n = 1$–5. The nonexponential region extends up to about

$$\left|\frac{\beta}{\alpha}\right| + 2\left|\frac{e_0}{\alpha}\right| < 4.5 \qquad (82)$$

The exponentially decaying region C extends to infinity beyond region B, and includes both McConnell's region with large $|e_0/\alpha|$ and the region of large $|\beta/\alpha|$.

Qualitatively, these results suggest that if a value of $\gamma < 0.5$ is ever deduced, one should check for nonexponential behavior. Further, if a value of $\gamma < 0.2$ is deduced, one should check for oscillatory behavior. No monotonic increasing regions exist for k_R.

The plot for k_J looks rather similar to the plot for k_R in the nonresonance region with large $|e_0/\alpha|$. Analysis is hampered when $|e_0/\alpha| < 1$ because the discontinuities in Joachim's expression cause nonmonotonic behavior to be observed in areas that are not intrinsically oscillatory (see Figure 6 with $\beta/\alpha = 10$, $e_0/\alpha = 0.5$ as an example). Further, the monotonically increasing region D is unphysical and all four contours of γ end discontinuously in this region. Therefore, although Joachim's theory is insightful and generally quite accurate, it is not always applicable. The rate constant k_R and its accurate analytical approximates are thus preferred to k_J.

The effect of a bridge state is small when the bridge state is resonant with the donor and acceptor states. This principle can be applied to explain the results for k_R seen in Figure 7. First, we transformed the Hamiltonian matrix into a form in which the bridge states do not interact with themselves. We wrote

$$\mathbf{H} = \begin{pmatrix} 0 & \beta & 0 & \cdots & 0 & 0 & 0 \\ \beta & e_0 & \alpha & \cdots & 0 & 0 & 0 \\ 0 & \alpha & e_0 & \cdots & 0 & 0 & 0 \\ \vdots & \vdots & \vdots & & \vdots & \vdots & \vdots \\ 0 & 0 & 0 & \cdots & e_0 & \alpha & 0 \\ 0 & 0 & 0 & \cdots & \alpha & e_0 & \beta \\ 0 & 0 & 0 & \cdots & 0 & \beta & i\lambda \end{pmatrix} \quad (83)$$

diagonalized the diagonal blocks, and used these eigenvectors to transform **H** and produce

$$\mathbf{H'} = \begin{pmatrix} \epsilon_+ & 0 & H_{1,3}' & \cdots & H_{1,n}' & 0 & 0 \\ 0 & \epsilon_- & H_{2,3}' & \cdots & H_{2,n}' & 0 & 0 \\ H_{1,3}' & H_{2,3}' & & & & H_{n+1,3}' & H_{n+2,3}' \\ \vdots & \vdots & & \mathbf{E}_H & & \vdots & \vdots \\ H_{1,n}' & H_{2,n}' & & & & H_{n+1,n}' & H_{n+2,n}' \\ 0 & 0 & H_{n+1,3}' & \cdots & H_{n+1,n}' & \epsilon_- & 0 \\ 0 & 0 & H_{n+2,3}' & \cdots & H_{n+2,n}' & 0 & \epsilon_+ \end{pmatrix} \quad (84)$$

where

$$\epsilon_\pm = \frac{\alpha \pm (\alpha^2 + 4\beta^2)^{1/2}}{2} \quad (85)$$

are the eigenvalues of the 2×2 blocks, and $\mathbf{E}_H$ is the $n \times n$ diagonal matrix of the eigenvalues of the Hückel Hamiltonian. This Hamiltonian appears in many band theories, and its eigenvalues cover quite evenly the range $e_0 - 2\alpha$ to $e_0 + 2\alpha$. Complex couplings such as $H_{1,j}$ contain the couplings of the donor and acceptor levels to the bridge eigenstates, and we chose not to express them explicitly.

Qualitatively, resonances of the donor and acceptor states with the bridge states are avoided if $\epsilon_\pm$ fall outside the band given by $\mathbf{E}_H$. When this condition is not met, then oscillations in $k_R(n)$ are possible. Increasing the bridge length n can tune different bridge levels onto the donor and acceptor levels in much the same fashion as e_0 tuned resonances in Figure 5. If, however, the coupling matrix elements $H_{1,j}'$ and $H_{2,j}'$ associated with the tuned resonance are much smaller than other coupling matrix elements, then the effect of the tuning may be insufficient to induce oscillations in k_R, but it will produce nonexponential decay. Thus, we see that the distinct regions A (oscillatory), B (nonexponential), and C (exponential) should exist in Figure 7.

The donor (or acceptor) state may be represented by just one of the two eigenstates $\epsilon_\pm$ or it may be evenly partitioned into both of the eigenstates. If

$$e_{0^2} - \beta^2 \gg 0 \tag{86}$$

then it can be represented by just one eigenstate. Thus, in order to avoid resonances, only one of the eigenvalues must lie outside of the bridge band. This situation occurs when

$$\beta^2 + 2|\alpha e_0| - 4\alpha^2 \gg 0 \tag{87}$$

When both eigenstates contain significant donor (or acceptor) character, then the second eigenvalue also must not fall within the bridge band, and this condition requires that

$$\beta^2 - 2|\alpha e_0| - 4\alpha^2 \gg 0 \tag{88}$$

The loci of the three curves for which these functions actually equal 0 are plotted in Figure 8. Complicated bridge-length dependences are expected in the hatched region of this figure, as well as in the nearby surrounding areas. Also plotted in Figure 8 are the regional boundaries from Figure 7. The oscillatory–monotonic boundary lies completely within the hatched region, but the monotonic–exponential boundary mostly lies slightly outside of it.

That the hatched region underestimates the extent of the nonexponential effects is expected, as this region is obtained by replacing the $\gg$ operators in eq 86–88 with $>$ operators. In one small area, the observed monotonic–exponential boundary actually lies inside the unstable hatched area. To some extent this location is brought about because the effects of conditions 86 and 88 constructively interfere in the region where both conditions almost hold, and so the effect of the resonances is weakened.

A more significant effect, however, is that the observed boundaries are evaluated by using only a finite range of n from 1 to 5. In this region, the decay curves typically display a small amount of curvature, which becomes more obvious as larger ranges of n are considered. The same problem arises in experimental bridge-length dependence studies, and any observed result is limited by the accuracy and extent of the available data.

Conclusions

The formalism that was introduced allows the rate of energy-transfer (in particular, electron-transfer) processes to be interpreted in terms of the coupling present both within the system of interest and between the system

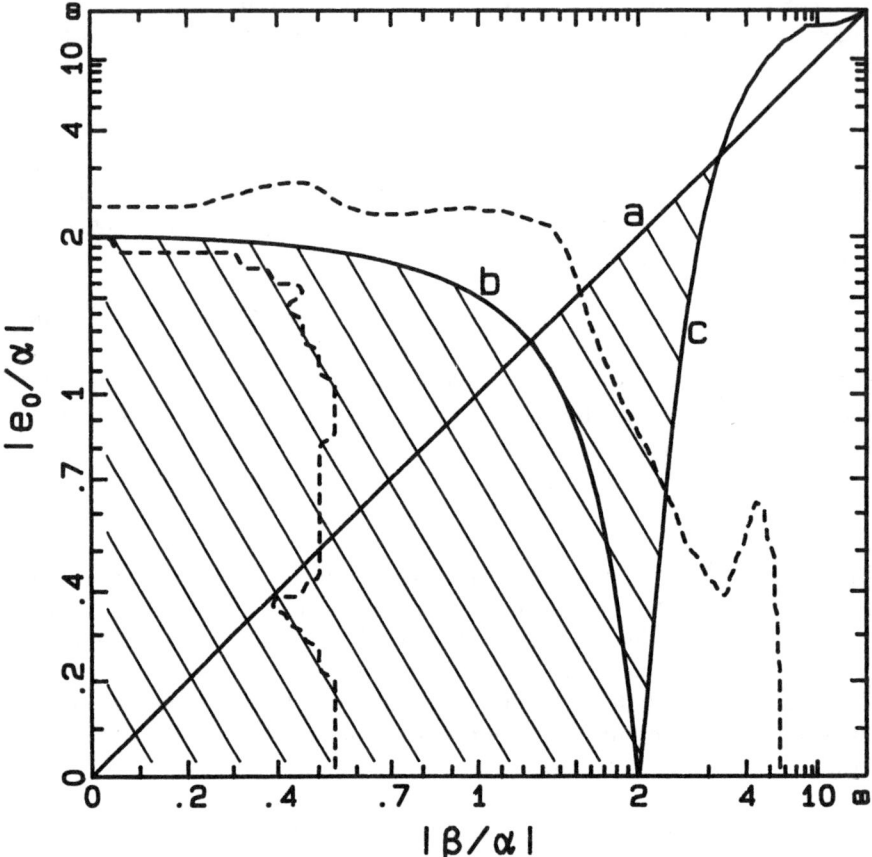

Figure 8. Plot of the region in which nonexponential bridge-length dependence is expected: Curve a is $e_0^2 - \beta^2 = 0$; curve b is $\beta^2 + 2|\alpha e_0| - 4\alpha^2 = 0$; and curve c is $\beta^2 - 2|\alpha e_0| - 4\alpha^2 = 0$. The dotted lines are the nonmonatonic-to-monatonic boundary and the monotonic-to-exponential boundary for k_R, as shown in Figure 7.

and its surrounding bath. This interpretation is particularly relevant to problems in molecular electronics as the model describes the modulation of the donor-to-acceptor current by properties of the connecting bridge. Interactions between the molecular system and the surrounding bath are shown to be very important in determining the current, and the current is shown to increase with bath coupling when the coupling is low, to reach some maximum (the Rabi regime), and to decrease with coupling when the coupling is high (the microscopic golden rule regime).

Considerable enhancement of the electron-transfer rate is shown to occur when the bridge levels are degenerate with the donor and acceptor levels. For three-level systems, the reaction rate is shown to decrease in proportion to e_0^{-2}, where e_0 is the donor-to-bridge energy gap. The donor-to-acceptor current is most sensitive to this bridge parameter. For systems containing n bridge levels that are described by McConnell's Hamiltonian, to a good approximation the decay of the reaction rate constant with bridge length is shown to be exponential if

$$\beta^2 + 2|\alpha e_0| - 4\alpha^2 \gg 0 \qquad (89)$$

and monotonic but nonexponential if

$$\beta^2 + 2|\alpha e_0| - 4\alpha^2 \approx 0 \qquad (90)$$

and nonmonotonic (possibly oscillatory) if

$$\beta^2 + 2|\alpha e_0| - 4\alpha^2 \ll 0 \qquad (91)$$

These regions correspond somewhat to the regions in which the bridge-length dependence decay constant γ is $\gg 0.2$, ~ 0.2, and $\ll 0.2$, respectively.

Exponential time-decay of the donor to acceptor states is shown only to occur when the donor state is nonresonant with both the bridge states and the acceptor state. This situation occurs when either the real or imaginary component of the energy gap is large. (A large imaginary component actually implies a large uncertainty in the real component of the energy gap, and averaging over the implied distribution results in a large average effective real component). Fast reaction rates rely upon the presence of resonances, and so they are expected to proceed nonexponentially with time. The number of Fourier components of the time decay increases with the square of the number of states resonant with the donor state.

Effective two-state models are shown to be appropriate only in the nonresonant bridge limit. Usually, these models are applied when the bridge is resonant as well. In such a case, these models are shown to be quite reliable except for ambiguities that arise, which are resolved by ad hoc assumptions, and lead to nonphysical effects. A solid foundation is built for methods such as that of Broo and Larsson (45), the insightful studies of Joachim are verified and extended, and the significance of McConnell's formula to current research is stressed.

Acknowledgments

We are indebted to the Australian Research Council for the provision of a Senior Research Fellowship to J. R. Reimers.

References

1. Martin, J. L.; Breton, J.; Hoff, A. J.; Migus, A.; Antonetti, A. *Proc. Natl. Acad. Sci. U.S.A.* **1986**, *83*, 957.
2. Breton, J.; Martin, J. L.; Hoff, A. J.; Migus, A.; Antonetti, A. *Proc. Natl. Acad. Sci. U.S.A.* **1986**, *83*, 5121.
3. Penfield, K. W.; Miller, J. R.; Paddon-Row, M. N.; Cotsaris, E.; Oliver, A. M.; Hush, N. S. *J. Am. Chem. Soc.* **1987**, *109*, 5061.
4. Oevering, H.; Paddon-Row, M. N.; Heppener, M.; Oliver, A. M.; Cotsaris, E.; Verhoeven, J. W.; Hush, N. S. *J. Am. Chem. Soc.* **1987**, *109*, 3258.
5. Aviram, A. *J. Am. Chem. Soc.* **1988**, *110*, 5687.
6. Hush, N. S.; Wong, A. T.; Bacskay, G. B.; Reimers, J. R. *J. Am. Chem. Soc.*, submitted.
7. Ford, P.; Rudd, D. F. P.; Gaunder, R.; Taube, H. *J. Am. Chem. Soc.* **1968**, *90*, 1187.
8. Tom, G. M.; Creutz, C.; Taube, H. *J. Am. Chem. Soc.* **1974**, *96*, 7827.
9. Sen, J.; Taube, H. *Acta. Chem. Scand.* **1977**, *A33*, 125.
10. Sutton, J. E.; Sutton, P. M.; Taube, H. *Inorg. Chem.* **1979**, *18*, 1017.
11. Sutton, C. A.; Taube, H. *Inorg. Chem.* **1981**, *20*, 3125.
12. Stein, C. A.; Taube, H. *J. Am. Chem. Soc.* **1981**, *103*, 693.
13. Richardson, D. E.; Taube, H. *J. Am. Chem. Soc.* **1983**, *105*, 40.
14. Stein, C. A.; Taube, H. *J. Am. Chem. Soc.* **1987**, *100*, 1635.
15. Woitellier, S.; Launay, J. P.; Spangler, C. W. *Inorg. Chem.* **1989**, *28*, 758.
16. Reimers, J. R.; Hush, N. S. *Inorg. Chem.*, submitted.
17. Reimers, J. R.; Hush, N. S. *Inorg. Chem.*, submitted.
18. Robinson, G. W.; Frosch, R. P. *J. Chem. Phys.* **1962**, *37*, 1962.
19. Robinson, G. W.; Frosch, R. P. *J. Chem. Phys.* **1963**, *38*, 1187.
20. Robinson, G. W. In *Excited States, Vol. 1*; Lim, E. C., Ed.; Academic: New York, 1974.
21. Condon, E. U. *Phys. Rev.* **1928**, *32*, 858.
22. Marcus, R. A. *Discuss. Faraday Soc.* **1960**, *29*, 21.
23. Hush, N. S. *Trans. Faraday Soc.* **1961**, *57*, 577.
24. McConnell, H. M. *J. Chem. Phys.* **1961**, *35*, 508.
25. Marcus, R. A. *J. Chem. Phys.* **1965**, *43*, 679.
26. Levich, V. G. In *Physical Chemistry—An Advanced Treatise, IX*; Eyring, H., Ed.; Academic: New York, 1970; p 985.
27. Hopfield, J. J. *Proc. Nat. Acad. Sci. U.S.A.* **1974**, *71*, 3640.
28. Jortner, J. *J. Chem. Phys.* **1976**, *64*, 4860.
29. Cribb, P. H.; Nordholm, S.; Hush, N. S. *Chem. Phys.* **1978**, *29*, 31.
30. Cribb, P. H.; Nordholm, S.; Hush, N. S. *Chem. Phys.* **1979**, *44*, 315.
31. Wertheimer, R.; Silbey, R. *Chem. Phys. Lett.* **1980**, *75*, 243.
32. Friesner, R.; Wertheimer, R. *Proc. Natl. Acad. Sci. U.S.A.* **1982**, *79*, 2138.
33. Friedman, H. L.; Newton, M. D. *J. Chem. Phys.* **1983**, *78*, 4086.
34. Hale, P. D.; Ratner, M. A. *Int. J. Quant. Chem. Symp.* **1984**, *18*, 195.
35. Beratan, D. N.; Onuchic, J. N.; Hopfield, J. J. *J. Chem. Phys.* **1985**, *83*, 5325.
36. Parris, P. E.; Silbey, R. *J. Chem. Phys.* **1985**, *83*, 5619.
37. Newton, M. D. *J. Phys. Chem.* **1986**, *90*, 3734.
38. Parris, P. E.; Silbey, R. *J. Chem. Phys.* **1987**, *86*, 6381.
39. Joachim, C. *Chem. Phys.* **1987**, *116*, 339.
40. Bertran, P. *Chem. Phys. Lett.* **1987**, *140*, 57.
41. Marcus, R. A. *Chem. Phys. Lett.* **1987**, *133*, 471.
42. Baňackŷ, P.; Zajac, A. *Chem. Phys.* **1988**, *123*, 267.
43. Larsson, S. *J. Am. Chem. Soc.* **1981**, *103*, 4034.

44. Larsson, S. *J. Chem. Soc. Faraday Trans. II* **1983**, *79*, 1375.
45. Broo, A.; Larsson, S., to be published.
46. Onuchic, J. N.; Wolynes, P. G. *J. Phys. Chem.* **1988**, *92*, 6495.
47. Mikkelsen, K. V.; Ratner, M. A. *J. Phys. Chem.* **1989**, *93*, 1759.
48. Mikkelsen, K. V.; Ratner, M. A. *Chem. Rev.* **1987**, *87*, 113.
49. Marcus, R. A. In *Light-Induced Charge Separation in Biology and Chemistry*; Gerischer, H.; Katz, J. J., Eds.; Verlag Chemie: Berlin, 1979.
50. DeVault, D. *Quantum-Mechanical Tunnelling in Biological Systems*; Cambridge University Press: Cambridge, England, 1984.
51. Newton, N. D.; Sutin, N. *Ann. Rev. Phys. Chem.* **1984**, *35*, 437.
52. Reimers, J. R.; Hush, N. S. *Chem. Phys.*, in preparation.
53. Reimers, J. R.; Hush, N. S. *Chem. Phys.* **1989**, *134*, 323.
54. Onuchic, J. N.; Beratan, D. N.; Hopfield, J. J. *J. Phys. Chem.* **1986**, *90*, 3707.
55. Löwdin, P. *J. Math. Phys.* **1962**, *3*, 969.
56. Löwdin, P. *J. Molec. Spectrosc.* **1963**, *10*, 12.
57. Brandow, B. H. *Int. J. Quant. Chem.* **1979**, *XV*, 207.

RECEIVED for review May 1, 1989. ACCEPTED revised manuscript October 9, 1989.

3

Some Aspects of Electron Transfer in Biological Systems

Norman Sutin and Bruce S. Brunschwig

Department of Chemistry, Brookhaven National Laboratory, Upton, NY 11973

Electron transfer occurs over relatively long distances in a variety of systems. In interpreting the measured electron-transfer rates, it is usually assumed that the rate constants depend exponentially on the distance separating the two redox sites and that this distance dependence arises from the decrease in the electronic coupling of the redox sites with increasing separation. Although the electronic coupling is an important factor determining the distance dependence of the rate, theoretical considerations show that the nuclear factors are also important. Another factor, of particular importance in biological systems, is the accessibility of different protein conformations. Such conformational changes afford a mechanism for controlling electron-transfer rates and can lead to directional electron transfer under suitable conditions. The effects of conformational changes can be particularly marked, and even unexpected, in the inverted free-energy region.

FACTORS DETERMINING ELECTRON-TRANSFER RATES in biological systems are currently of considerable interest (1–5). Recent studies have focused on two aspects–the distance dependence of the rates and the role of protein conformational changes. Both are discussed in this chapter.

A number of experimental approaches are being used to obtain electron-transfer rates as a function of the distance separating the redox sites. One approach involves the measurement of intramolecular electron-transfer rates in synthetic systems in which the two redox sites are separated by a rigid bridging group or spacer of varying length (6–11). Another approach involves the modification of naturally occurring systems. Typically, one of the metal centers in a multisite metalloprotein is replaced by a different metal center (e.g., one of the iron atoms in hemoglobin is replaced by a zinc atom), which is then photoexcited to initiate electron transfer (*12, 13*). Alternatively, a redox center or cofactor may be covalently or electrostatically attached to a specific region of a single-site metalloprotein; the latter may be the native protein (e.g., cytochrome c or myoglobin) or a derivative (e.g., zinc-substituted cytochrome c) (*14–23*).

In interpreting the experimental data, it is frequently assumed that the distance dependence of the observed rate arises solely from the decrease in the electronic coupling of the redox sites with increasing separation. Although the decrease in electronic coupling is an important consideration, other distance-dependent factors also need to be considered (*24, 25*). One aim of this chapter is to draw attention to the importance of (nuclear) reorganization energy in determining distance dependence.

Another aspect considered in this chapter is the role of protein conformational changes. Such conformational changes may precede or follow the actual electron transfer and can lead to directional electron transfer in suitable systems. Directional electron transfer has recently been invoked to rationalize an "anomalous" rate of intramolecular electron transfer between the iron and ruthenium centers in a modified cytochrome c (*17*). The oxidation of the Fe(II) heme by $Ru^{III}(NH_3)_4(isn)$ (isn is isonicotinamide) attached to the histidine-33 residue of cytochrome c is much slower (*17*) than the reduction of the Fe(III) heme by a histidine-bound $Ru^{II}(NH_3)_5$ moiety (*16*), despite the very similar driving forces for the two reactions.

Isied and coworkers (*17*) interpreted this apparent dependence of the rate on direction by proposing that the iron(II) protein undergoes a conformational change prior to its oxidation to iron(III). The conditions under which directional electron transfer may occur are discussed, as are some remarkable consequences of conformational changes on electron-transfer rates in the so-called inverted region.

Theoretical Model

For the present purpose, we adopt a semiclassical model in which the electron transfer occurs at the intersection of two harmonic (free-energy) curves, one for the reactant(s) and the other for the product(s) (*1, 26–30*). We further restrict the discussion to intramolecular electron-transfer processes. The electron-transfer rate constants for such systems are given by (*1, 26–30*)

$$k = \kappa_{el}\nu_n\Gamma_n e^{-\Delta G^*/RT} \quad (1)$$

$$\Delta G^* = \frac{(\lambda + \Delta G^0)^2}{4\lambda} \quad (2)$$

$$\lambda = \lambda_{in} + \lambda_{out} \quad (3)$$

where κ_{el} is the electronic transmission coefficient or adiabaticity factor, ν_n is the frequency of the nuclear motion that takes the system through the intersection region (destroying the transition-state configuration), Γ_n is a correction for nuclear tunneling, ΔG^* is the contribution to the free energy of activation from the nuclear reorganization, ΔG^0 is the free-energy change for the electron transfer, and λ is the reorganization parameter. The reorganization parameter is the energy difference between the reactants' and products' free-energy surfaces at the reactants' equilibrium nuclear configuration for the case where $\Delta G^0 = 0$ (see Figure 1).

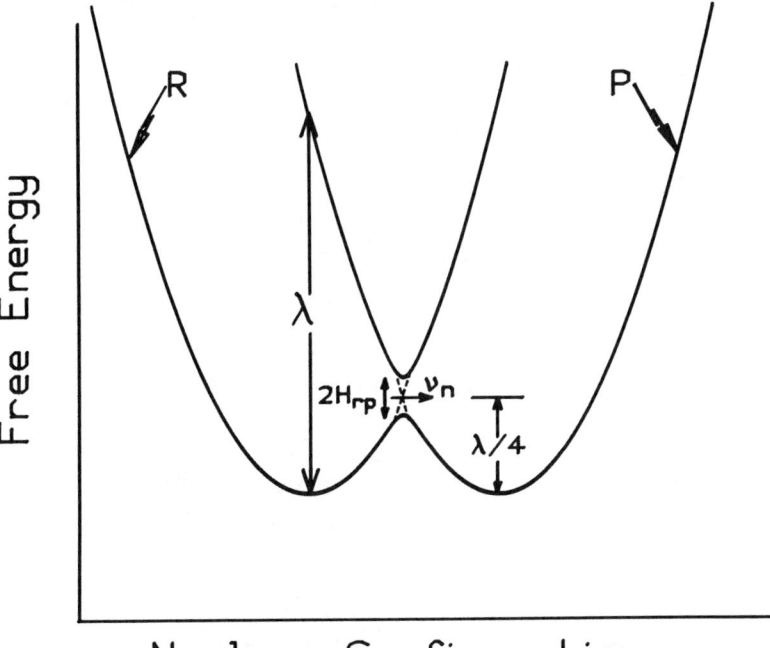

Figure 1. *Plot of the free energy of the reactants (left parabola) and products (right parabola) as a function of nuclear configuration (reaction coordinate) for an electron-exchange reaction. The splitting at the intersection of the curves is equal to* $2H_{rp}$, *where* H_{rp} *is the electronic coupling matrix element.* λ *is the vertical difference between the free energies of the reactants and products at the equilibrium nuclear configuration of the reactants. Thermal electron transfer occurs at the nuclear configuration appropriate to the intersection of the reactant and product curves.*

The reorganization parameter contains a contribution from each mode that undergoes a displacement as a consequence of the electron transfer. Two classes of contributions to λ are generally distinguished: one, λ_{in}, is associated with generally fast changes in the intramolecular (inner-shell) bond distances and angles; the other, λ_{out}, is associated with slow changes in the polarization of the surrounding medium (outer-shell). λ_{out} depends upon solvent polarity, the shape of the molecule, and the separation of the redox sites. In the case of proteins or other macromolecules, conformational changes that are coupled to the electron transfer may be included in either λ_{in} or λ_{out}, depending on their time scale.

The free energies of the reactants and products are functions of their nuclear configurations. Sections through the parabolic basins obtained by plotting the free energies of the reactants and products versus the configurations of their inner- and outer-coordination shells are shown in Figure 2. The straight line joining the reactant (R) and product (P) minima is the reaction coordinate; this line is the abscissa for the plot in Figure 1. The dashed line is the path of steepest descent from the activated complex to the R and P minima. This pathway is relevant to the description of the detailed dynamics of the reaction.

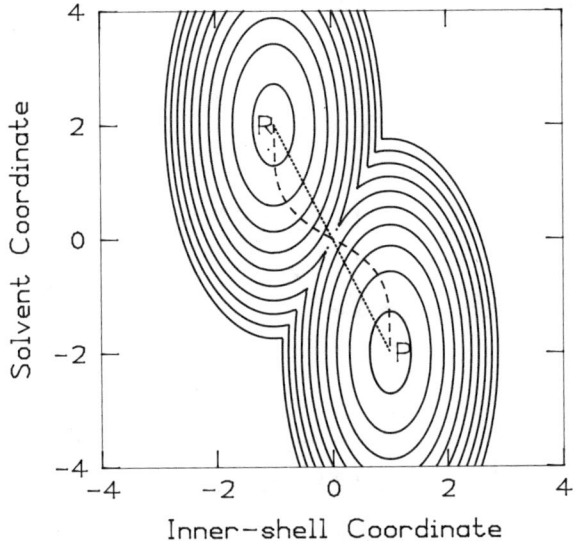

Figure 2. Sections through the parabolic basins obtained by plotting the free energies of the reactants and products as a function of the intramolecular (inner-shell) and solvent coordinates. The reactants' minimum R is in the upper left corner; the products' minimum P is in the lower right corner. The solid lines are equally spaced energy contours, the dashed curve is the classical reaction path, and the dotted line is the reaction coordinate. The surface is drawn with the assumption that both ΔG^0 and H_{rp} are equal to zero.

Two free-energy regimes can be distinguished, depending on the relative magnitudes of ΔG^0 and λ (eq 2) (1, 24, 26–30). In the "normal" regime ($|\Delta G^0| < \lambda$), ΔG^* decreases and the rate constant increases as the driving force for the electron transfer becomes more favorable. When $|\Delta G^0| = \lambda$, $\Delta G^* = 0$, and the reaction is barrierless. If the driving force is increased even further ($|\Delta G^0| > \lambda$), ΔG^* will increase and the rate constant will decrease with increasing driving force. This set of circumstances is the "inverted" regime. The dependence of the logarithm of the rate constant on the driving force for the electron transfer (at constant λ and κ_{el}) is shown in Figure 3.

The difference between the normal and inverted regimes can be further illustrated by considering the variation of the rate constant with λ at constant ΔG^0. Again the rate is maximal when $\lambda = |\Delta G^0|$. When $\lambda > |\Delta G^0|$, increasing λ increases ΔG^* and decreases the rate. By contrast, when $|\Delta G^0| > \lambda$, increasing λ decreases ΔG^* and increases the rate. Thus, the rate responds oppositely to changes in the intrinsic barrier λ in the normal and inverted regions (1, 24, 26–30). Because diminished reactant separation, solvent polarity, and structural differences decrease the intrinsic barrier, such changes promote rapid electron transfer in the normal region but lead to decreased rates in the inverted region. This situation is illustrated in Figure 4.

The Electronic Factor

Within the Landau–Zener treatment of avoided crossings, the electronic transmission coefficient in eq 1 is given by

$$\kappa_{el} = \frac{2\left[1 - \exp\left(\frac{-\nu_{el}}{2\nu_n}\right)\right]}{2 - \exp\left(\frac{-\nu_{el}}{2\nu_n}\right)} \tag{4a}$$

$$\nu_{el} = \frac{H_{rp}^2}{\hbar}\left(\frac{\pi}{\lambda RT}\right)^{1/2} \tag{4b}$$

where ν_{el} is the electron-hopping frequency in the intersection region, H_{rp} is the electronic coupling matrix element, and $\hbar = h/2\pi$ (h is the Planck constant). Equation 4 shows that $\kappa_{el} = 1$ (i.e., the reaction is adiabatic), when $\nu_{el} \gg 2\nu_n$ and that $\kappa_{el} = \nu_{el}/\nu_n \ll 1$ (i.e., the reaction is nonadiabatic) when $\nu_{el} \ll 2\nu_n$. Within this framework, the rate constant for a nonadiabatic reaction is given by

$$k_{na} = \frac{H_{rp}^2}{\hbar}\left(\frac{\pi}{\lambda RT}\right)^{1/2} e^{-\Delta G^*/RT} \tag{5}$$

where the nuclear tunneling factor in eq 1 has been neglected. For a nonadiabatic reaction, the product $\kappa_{el}\nu_n$ is independent of ν_n; that is, the electron-transfer rate is determined by the electron-hopping frequency ν_{el} and not by a nuclear-vibration frequency.

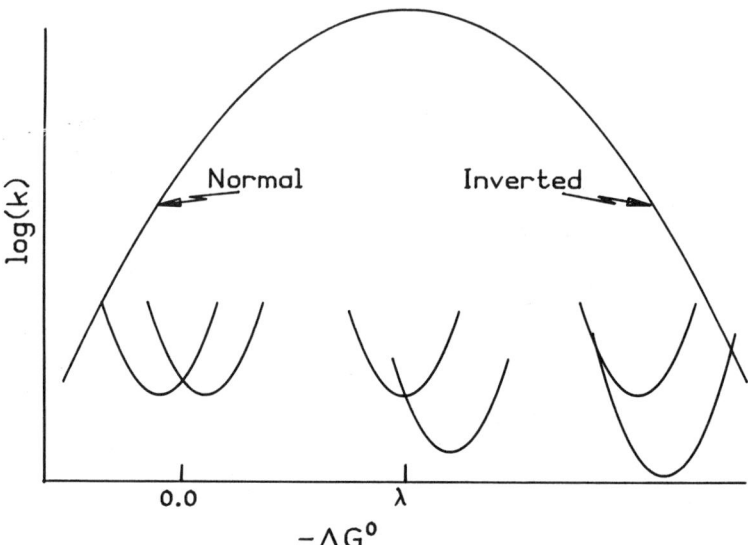

Figure 3. Variation of the logarithm of the rate constant for an electron-transfer reaction with driving force at constant λ: $|\Delta G^0| < \lambda$ defines the normal region, and $|\Delta G^0| > \lambda$ defines the inverted region.

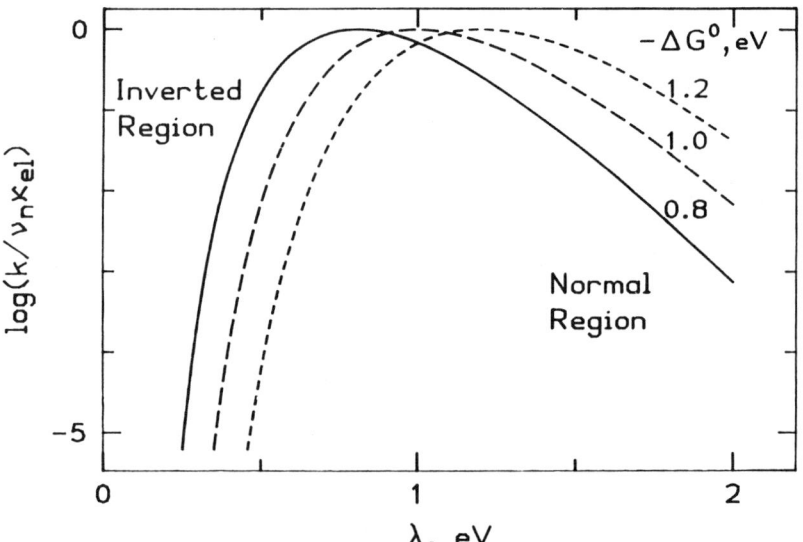

Figure 4. Variation of the logarithm of the rate constant for an electron-transfer reaction with the reorganization parameter. The three curves are for driving forces of 0.8, 1.0, and 1.2 eV. Note the opposite effect of changes in λ on rate constants in the normal and inverted regions and the relatively large effect of λ changes on reaction rates in the inverted region.

Although there is general agreement that H_{rp} will decrease with increasing separation of the redox sites, the exact dependence of H_{rp} on separation distance (r) is still an open question. For many systems, H_{rp} appears to decrease exponentially with increasing separation distance (eq 6)

$$H_{rp} = H_{rp}^0 \exp\left(\frac{-\beta(r - r_0)}{2}\right) \tag{6}$$

where H_{rp}^0 is the value of H_{rp} at $r = r_0$, r_0 is the close-contact separation of the redox sites, and β measures the rate of decrease of the electronic coupling with separation. Because of the decreased coupling at large separations, most reactions will be nonadiabatic at large separations of the redox sites. However, there is a caveat. Equation 4 shows that the relevant parameter determining the degree of adiabaticity is not H_{rp} but the ratio ν_{el}/ν_n. As a consequence, a reaction may remain adiabatic even up to large r if ν_n is sufficiently small. For example, certain long-range electron transfers in metalloproteins might be adiabatic if coupled to a sufficiently slow protein-conformational change.

This description (and eqs 1–3, apart from the nuclear tunneling correction) is based on transition-state theory, in which an equilibrium distribution between the initial and transition-state configurations is assumed. This equilibrium assumption may break down when a particular nuclear motion leading to (or from) the transition-state configuration is very slow. It then becomes necessary to use a steady-state or a more elaborate treatment (31–33).

Electronic Coupling Element. For concreteness, we restrict the discussion of the electronic coupling element to the case in which the redox sites are metal centers. In suitable mixed-valence systems, H_{rp} (at the equilibrium nuclear configuration of the reactants) can be estimated from the intensities of the metal-to-metal charge-transfer transitions (34). A variety of theoretical approaches have also been used to estimate H_{rp} and its distance dependence.

Electron transfer proceeding by direct overlap of the d orbitals of the two metal centers decreases very rapidly with increasing separation of the sites. Consequently, rapid electron transfer over large distances requires mediation by the intervening medium or ligands. In fact, mediation by the ligands is an important factor in close-contact (or even interpenetrating) encounter complexes formed between "simple" aquo or ammine metal complexes (35, 36).

Mediation by ligands or bridging groups is conveniently considered in terms of electron and hole superexchange formalisms. The former mechanism involves imparting to the ground states of the reactants and products some metal-to-ligand charge-transfer (MLCT) excited-state character (deriving from electron promotion to the empty π^* or σ^* antibonding orbitals of the ligands or bridging groups). By contrast, the hole superexchange mechanism involves mixing of ligand-to-metal charge-transfer (LMCT) excited states (formed by electron withdrawal from the filled π or σ bonding orbitals of the ligands or bridging groups). These two conduction mechanisms are depicted in Scheme I. Depending upon the nature of the metal centers and of the ligands, either the electron- or the hole-conduction mechanism may predominate.

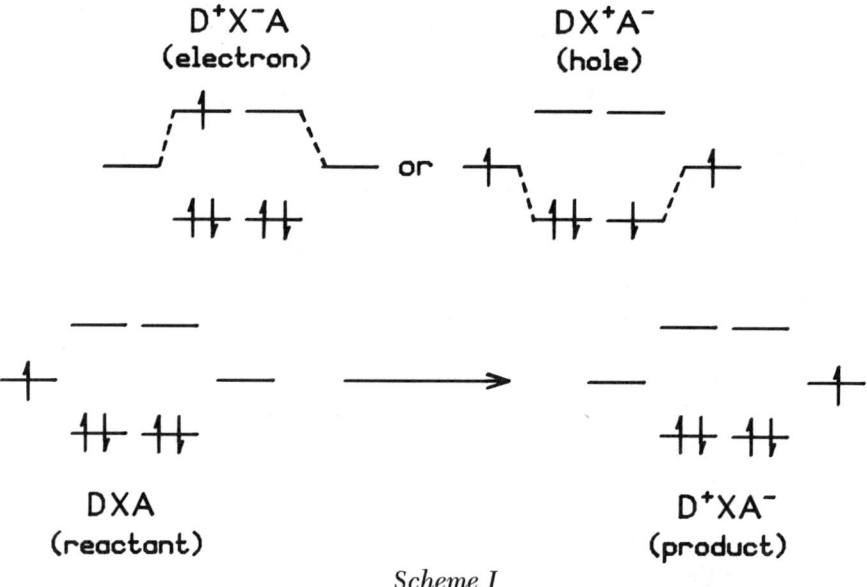

Scheme I

The electronic coupling in outer-sphere electron-exchange reactions of metal complexes can be treated in terms of molecular orbitals formed from linear combinations of metal- and ligand-based orbitals of the appropriate symmetry and energy. By using perturbation theory and neglecting metal–ligand overlap, the molecular orbitals for the metal–ligand bonds can be written as

$$\Psi_{ML} = \Psi_M + \gamma \Psi_L \tag{7}$$

$$\gamma = \frac{\Psi_M H \Psi_L}{E_M - E_L} \tag{8}$$

where Ψ_M and Ψ_L are the (unperturbed) metal and ligand wave functions, and E_M and E_L are their energies at the nuclear configurations corresponding to the intersection region (transition state), respectively. The coefficient γ is the metal–ligand mixing or covalency parameter. Newton (35, 36) showed that, provided the ligands are π or σ donors and the electron transfer proceeds by overlap of the orbitals of the ligands of the two reactants, the electron-coupling matrix element for a (symmetrical) exchange reaction can be estimated from

$$H_{rp} \sim \gamma'^2 H_{LL} \quad (9)$$

$$H_{LL} = \Psi_{LD} H \Psi_{LA} \quad (10)$$

where H_{LL} is the one-electron transfer integral for the ligand orbitals of the two reactants in contact in the transition state, and γ' is the covalency parameter for the oxidized complex (λ' in Newton's notation (35, 36)).

The electronic coupling elements for the $Fe(H_2O)_6^{2+/3+}$ ($t_{2g}(\pi)$ transfer), $Ru(NH_3)_6^{2+/3+}$ ($t_{2g}(\pi)$ transfer), and low-spin $Co(NH_3)_6^{2+/3+}$ ($e_g(\sigma)$ transfer) electron exchanges (eq 11) have recently been calculated

$$ML_6^{2+} + ML_6^{3+} \rightleftarrows ML_6^{3+} + ML_6^{2+} \quad (11)$$

by using this approach and ab initio wave functions for the redox pairs modeled as supermolecule clusters (35, 36). Because the ligands are electron donors, the hole variety of superexchange mechanism dominates. The electronic coupling in these systems is treated in terms of the ligand-to-metal charge-transfer (LMCT) interaction. For the apex-to-apex approach of the two reactants, H_{LL} is calculated to be about 5000 cm^{-1} with the normalized H_{rp} ranging from 12 cm^{-1} for the $Fe(H_2O)_6^{2+/3+}$ exchange to 940 cm^{-1} for the low-spin $Co(NH_3)_6^{2+/3+}$ exchange. Moreover, for both H_2O and NH_3 ligands, $\sigma > \pi$ mediation (35, 36).

For the aquo and ammine complexes, hole conduction is favored over electron conduction because of the relatively poor reducibility of the saturated H_2O and NH_3 ligands. With unsaturated ligands such as pyridine (pyr), which has relatively low-lying π* orbitals, metal-to-ligand charge transfer will be more important than with saturated ligands. Thus, in the $Ru(NH_3)_5pyr^{2+/3+}$ exchange (eq 12), both metal-to-ligand (Ru(II) to pyr π*) and ligand-to-metal (pyr π to Ru(III)) charge-transfer interactions may be significant.

$$Ru(NH_3)_5pyr^{2+} + Ru(NH_3)_5pyr^{3+} \rightleftarrows$$

$$Ru(NH_3)_5pyr^{3+} + Ru(NH_3)_5pyr^{2+} \quad (12)$$

If the electronic coupling of two pentaammine pyridine complexes is predominantly through the edge-to-edge C3···C3 interaction of the two pyridines, then the following expression for H_{rp} can be derived by using simple molecular orbital (MO) theory (37–41)

$$H_{rp} = H_{D,N} \left(\frac{a_N^D}{(E_{d\pi} - E_{pyr})_D} \right) a_3^D a_3^A H_{C3,C3} \left(\frac{a_N^A}{(E_{d\pi} - E_{pyr})_A} \right) H_{A,N} \quad (13a)$$

where $H_{D,N}$ and $H_{A,N}$ are the coupling elements between the dπ orbitals of the donor and acceptor metal sites and the p orbitals of their coordinated nitrogen atoms, respectively; $H_{C3,C3}$ is the coupling element between the C-3 atoms of the two pyridine rings in the edge-to-edge approach of the reactants; a_3 is the coefficient of the p_z orbital of C-3 in the LCAO (linear combination of atomic orbitals) expansion of the pyridine molecular orbital; a_N is the coefficient of the pyridine nitrogen in the LCAO expansion; and $E_{d\pi}$ and E_{pyr} are the energies of the metal dπ and pyridine molecular orbitals, respectively. (In this treatment, the matrix element and coefficients for the donor site are averages for the oxidized and reduced donor, and similarly for the acceptor site.) For exchange reactions such as eq 12, eq 13a reduces to eq 13b, which is equivalent to eq 9.

$$H_{rp} \sim \left(H_{M,N} \frac{a_N a_3}{E_{d\pi} - E_{pyr}} \right)^2 H_{C3,C3} \qquad (13b)$$

Equation 13 shows that, for efficient conduction, the orbital coefficients should be large both for the atom coordinated to the metal center and for the peripheral atom providing the coupling to the other reactant.

Electron Transfer in Metalloproteins. Long-range electron transfer in metalloproteins can be treated by an extension of the same approach. Cytochrome c will be considered as an example. In cytochrome c the iron atom is coordinated to the four nitrogens of the porphyrin ring and is bonded to two axial ligands (histidine-18 and methionine-80). The porphyrin ring is covalently bonded to the protein via two thioether bridges. In a derivatized cytochrome c, a ruthenium pentaammine is attached to histidine-33 of the protein (*14*, *15*). The derivatized metalloprotein can then be modeled as two metal (iron and ruthenium) complexes attached to a bridging group (the polypeptide chain) via their ligands (the porphyrin or histidine-18, methionine-80, and histidine-33). Such complex–bridge–complex systems, in which the bridge is attached to the periphery of the coordinated ligands, can also be considered in terms of extended Hückel theory. Equation 14a has been proposed by Larsson (*37–41*) for the case in which the two redox sites are connected by n sequential bridging units.

$$H_{rp} = \frac{\prod_{i=0}^{n} (H_{i,i+1})}{\prod_{i=1}^{n} (\Delta E_i)} \qquad (14a)$$

In eq 14a the bridging units are either coupled to one another ($H_{i,i+1}$) or, in the case of the terminal units, to the redox sites ($H_{0,1}$ and $H_{n,n+1}$). ΔE_i is the vertical difference between the energies of the electron donor (or acceptor) and bridging unit i. The superexchange mixing again occurs at the nuclear configuration appropriate to the intersection region. If all the $H_{i,i+1}$ and ΔE_i are similar, then the decrease of H_{rp}^2 with increasing separation distance is exponential, with β given by

$$\beta = \frac{2}{l} \ln \left(\frac{\Delta E}{H} \right) \tag{14b}$$

where l is the length of the bridging unit. In the present context, $l = (r - r_0)/n$, where n is the number of bridging units (30, 37–41).

This discussion shows that β is likely to vary from system to system and to depend both on the nature of the intervening medium and on the nature of the redox sites. The donor electron density at the (peripheral) ligand atom, to which the bridging group is attached, is an important parameter. If this density is sufficiently large so that H_{rp}^0 (eq 6) ~ 50 cm^{-1}, then the reaction will be adiabatic at $r = r_0$ (the exact condition is $\nu_{el} = \nu_n$, which gives $\kappa_{el} = 0.6$). The magnitude of β can then be estimated from

$$k = \nu_n e^{-\beta (r-r_0)} e^{-\Delta G^*/RT} \tag{15}$$

provided that the appropriate nuclear vibration frequency is used and allowance is made for the distance dependence ΔG^*. Reported values of β range from ~0.7 to 1.5 Å^{-1}, with typical values around 1.1–1.2 Å^{-1} (1). For apex-to-apex contact of the aquo and ammine complexes considered earlier, H_{rp} varies by over two orders of magnitude, 10–1000 cm^{-1} (r ~ 7.0–7.3 Å), depending on the nature of the metal center. For ruthenium-attached cytochrome c (21), H_{rp} ~ 0.1 cm^{-1} at r ~ 16–18 Å (n = 14 amino acid residues) and H_{rp} ~ 0.7 cm^{-1} for a comparable separation of osmium(II) and ruthenium(III) centers in a polyproline-bridged (n = 3) binuclear complex (25). Here, as before, r is the direct, straight-line separation distance.

An important consideration in the metalloprotein systems is whether the electronic coupling is through-bond (i.e., along the tortuous path provided by the polypeptide chain) or through-space. The latter situation involves coupling across polypeptide chains, perhaps via the amino acid side chains, van der Waals contacts, or hydrogen-bonding networks (2). The similar magnitudes of H_{rp} for the protein (n = 14) and polyproline (n = 3) systems suggests that the electron transfer in the derivatized cytochrome c proceeds predominantly through space. This aspect, as well as orientation and orbital symmetry effects, is currently under active investigation.

Reorganization Energies

Distance Dependence. As discussed earlier, electron-transfer rates depend on nuclear reorganization energy, as well as on electronic coupling.

Both factors vary with the distance separating the redox sites. The distance dependence of the electronic coupling was considered in the preceding discussion. Here we focus on the distance dependence of the reorganization energy (eq 2).

The ΔG^0 in eq 2 is the driving force for the electron transfer at separation distance r. Thus, depending on the charges on the redox sites and also on the neighboring charge distribution, the driving force for the electron transfer will change as the distance separating the redox sites is varied (1). Although such changes in ΔG^0 can be important, particularly in low-dielectric-constant media (6, 7), in polar media the ΔG^0 changes are likely to be overshadowed by those in λ_{out}.

The value of λ_{out} depends on the shape of the molecule and, for a given molecular shape, on the distance separating the redox sites. Expressions for λ_{out} have been derived for systems that can be represented by two spheres (for bimolecular reactions) or by an ellipsoidal or spherical cavity (for intramolecular electron transfer) in a dielectric continuum (26, 27, 42, 43). These expressions have been tested experimentally (24). The λ_{out} values calculated from the ellipsoidal model are in fairly good agreement with the values calculated from the energies of the metal-to-metal charge-transfer transitions in mixed-valence complexes (E_{op}, eq 16) (42, 43). In particular, the distance dependence of the calculated λ_{out} parallels that of E_{op}.

$$(NH_3)_5Ru^{II}(bridge)Ru^{III}(NH_3)_5 \xrightarrow{h\nu = E_{op}} [(NH_3)_5Ru^{III}(bridge)Ru^{II}(NH_3)_5]^* \quad (16)$$

λ_{out} can also be obtained from the temperature dependence of thermal electron-transfer rates in systems with small (or known) λ_{in} values (24, 25). Even if λ_{in} is not known, the distance dependence of the activation energy should reflect that of λ_{out} in suitably chosen systems. This approach has been exploited by Isied and coworkers (25) in a study of the thermal intramolecular electron transfer in a mixed-metal system (eq 17).

$$(NH_3)_5Os^{II}\text{--iso(proline)}_n\text{--}Ru^{III}(NH_3)_5 \xrightarrow{k}$$
$$(NH_3)_5Os^{III}\text{--iso(proline)}_n\text{--}Ru^{II}(NH_3)_5 \quad (17)$$

Values of the activation energy and entropy as a function of the distance separating the two metal centers were determined from the temperature dependence of the rate as a function of the number of proline units in the polypeptide chain. Somewhat unexpectedly, the distance dependence of the activation enthalpy ($\Delta H^{\ddagger}$), which reflects the distance dependence of λ_{out}, is larger (slope of $\Delta H^{\ddagger}/RT$ versus $r = 0.91$ Å^{-1}) than that of the activation entropy ($\Delta S^{\ddagger}$), which reflects the distance dependence of ln κ_{el} (slope of $-\Delta S^{\ddagger}/R$ versus $r = 0.68$ Å^{-1}). For these systems, the distance dependence of the nuclear factor is larger than that of the electronic factor! Thus, the assumption that the decrease in rate with distance arises entirely through the electronic factor can be in serious error.

The single-sphere model was used to model λ_{out} for ruthenium-derivatized cytochrome c; λ_{out} is calculated to be ~ 0.7 eV for an iron–ruthenium separation of 18 Å (24). In systems of this type, $r(M–M)$ can be varied by attaching the $Ru(NH_3)_5$ moiety to different residues on the protein surface. Although it is generally assumed that λ_{out} variations can be neglected for such systems, calculations (based on a dielectric continuum model with $D_{in'} = 1.8$) show that λ_{out} is quite sensitive to the position of the ruthenium on the protein surface. Thus, even in derivatized metalloproteins, the distance dependence of λ_{out} must be taken into account when interpreting measured intramolecular electron-transfer rates.

Consecutive Conformation Changes and Electron Transfer.

Metalloproteins may have thermally accessible conformers with electron-transfer properties that differ from those of the stable protein. When λ is large or H_{rp} is small, the direct electron transfer will be slow and more complicated mechanisms may operate with unstable conformers. Here we discuss two such mechanisms (eqs 18 and 19). In the first, an unstable reactant conformer R* is formed prior to the electron transfer. In the second, the electron transfer yields an unstable product conformer P*. (Equations 18 and 19 have also been considered by Hoffman and Ratner (44); see ref. 45 for a detailed discussion.)

$$R \underset{k_{r^*r}}{\overset{k_{rr^*}}{\rightleftharpoons}} R^* \underset{k_{pr^*}}{\overset{k_{r^*p}}{\rightleftharpoons}} P \qquad (18)$$

$$R \underset{k_{p^*r}}{\overset{k_{rp^*}}{\rightleftharpoons}} P^* \underset{k_{pp^*}}{\overset{k_{p^*p}}{\rightleftharpoons}} P \qquad (19)$$

In terms of these equations $K_{rr^*} = k_{rr^*}/k_{r^*r}$ and $K_{pp^*} = k_{pp^*}/k_{p^*p}$ are the equilibrium constants for the conformational changes, K_{rp} is the equilibrium constant for the overall reaction, k_{r^*p} and k_{rp^*} are the rate constants for the forward electron transfers, and k_{pr^*} and k_{p^*r} are those for the reverse electron transfers. The spectra of the stable and unstable conformers are assumed to be similar, so their interconversion cannot be observed directly. Because $K_{rr^*} \ll 1$, only very small amounts of R* are ever present. As a consequence, the assumption regarding the similarity of the spectra can be relaxed for the R* mechanism.

Equations 18 and 19 are of the same general form, and their kinetics are treated elsewhere (45). Two cases can be distinguished, depending on whether the rates are controlled by the electron transfer (case 1) or by the conformational change (case 2). In the latter case, the overall reaction is said to be "gated". The rate constants for these two cases are summarized in Table I.

Table I. Rate Constants and Free-Energy Barriers (ΔG^*) for Direct and Two-Step Electron-Transfer Reactions

Mechanism	Case 1,[a] Electron-Transfer Controlled	Case 2,[b] Gated
$R \to P$	k_{rp} $$\frac{(\Delta G_{rp}^0 + \lambda_{rp})^2}{4\lambda_{rp}}$$	—
$R \rightleftharpoons R^* \rightleftharpoons P$	$K_{rr^*}k_{r^*p}$ $$\Delta G_{rr^*}^0 + \frac{(\Delta G_{rp}^0 - \Delta G_{rr^*}^0 + \lambda_{r^*p})^2}{4\lambda_{r^*p}}$$	k_{rr} $$\frac{(\Delta G_{rr^*}^0 + \lambda_{rr^*})^2}{4\lambda_{rr^*}}$$
$R \rightleftharpoons P^* \rightleftharpoons P$	k_{rp^*} $$\frac{(\Delta G_{rp}^0 + \Delta_{pp^*}^0 + \lambda_{rp^*})^2}{4\lambda_{rp^*}}$$	$K_{rp^*}k_{p^*p}$ $$\Delta G_{rp^*}^0 + \frac{(\Delta G_{pp^*}^0 + \lambda_{pp^*})^2}{4\lambda_{pp^*}}$$

[a] For Case 1, the R* mechanism requires that $k_{r^*r} \gg k_{rr^*}, k_{r^*p}$, and k_{pr^*}. The P* mechanism requires that k_{p^*p} or $k_{rp^*} \gg k_{p^*r}$.
[b] For Case 2, the R* mechanism requires that $k_{r^*p} \gg k_{rr^*}, k_{r^*r}$, and k_{pr^*}. The P* mechanism requires that $k_{p^*r} \gg k_{rp^*}, k_{pp^*}$, and k_{p^*p}.

We will assume that the electron-transfer reaction can be represented in one-dimensional nuclear-configuration space, as shown in Figure 5. Despite this restrictive assumption, the simplified model has all the important features of the more general case (45). We further assume that the energy surfaces are harmonic, with identical force constants. The change in the equilibrium nuclear configurations for the electron-transfer reaction involving the intermediate can be either smaller or larger than that for the direct $R \to P$ reaction (i.e., for the R* mechanism, either $|q_n^{r^*} - q_n^p| < |q_n^r - q_n^p|$ or $|q_n^{r^*} - q_n^p| > |q_n^r - q_n^p|$, where q_n^i is the equilibrium nuclear configuration of species i). In the former case, the equilibrium configurations of R* and P* lie between those of R and P, and the reorganization parameters for the $R^* \to P$ and $R \to P^*$ reactions will be smaller than that for $R \to P$ (i.e., λ_{rp^*} or $\lambda_{r^*p} < \lambda_{rp}$, where λ_{r^*p} and λ_{rp^*} are the reorganization parameters for $R^* \to P$ and $R \to P^*$, respectively). This type of unstable conformer is referred to as a low-λ intermediate (Figure 5, top). When the change in the equilibrium nuclear configurations for the reaction involving the interme-

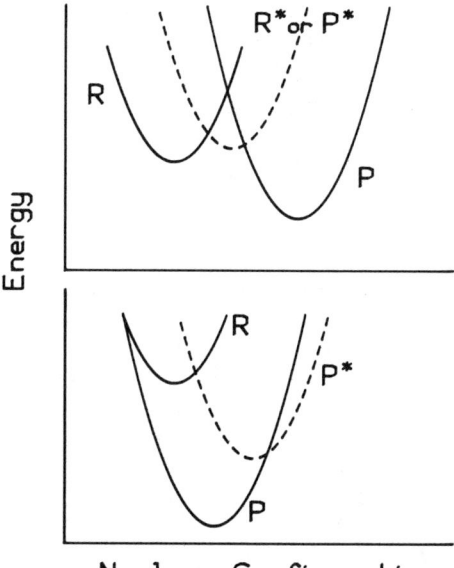

Figure 5. Plot of the energy of the reactants (left parabola), an intermediate (dashed parabola), and the products (lowest parabola) as a function of nuclear configuration. The parabolas are assumed to have identical force constants. The top figure represents a reaction in the normal region with a low-λ R* or P* intermediate. The bottom figure represents a reaction in the inverted region with a high-λ P* intermediate.

diate (R → P* or R* → P) is larger than that for the direct R → P reaction, the unstable conformer is called a high-λ intermediate. In this case, the intermediate lies either to the left of R or to the right of P (Figure 5, bottom).

The free-energy changes for the individual steps are related by eqs 20 and 21

$$\Delta G_{rp}^0 = \Delta G_{rr^*}^0 + \Delta G_{r^*p}^0 \qquad (20)$$

$$\Delta G_{rp}^0 = \Delta G_{rp^*}^0 - \Delta G_{pp^*}^0 \qquad (21)$$

where $\Delta G_{rr^*}^0$ and $\Delta G_{pp^*}^0$ are the free-energy changes for R → R* and P → P*, respectively; and $\Delta G_{r^*p}^0$ and $\Delta G_{rp^*}^0$ are the free-energy changes for the electron-transfer reactions R* → P and R → P*, respectively. The expressions for the reorganization (free-energy) barriers for the electron transfers and the conformational changes can be obtained from eq 2 with the appropriate reorganization parameters and driving forces. The results are included in Table I. The barriers for the case 1 R* and case 2 P* mechanisms include the contribution from the preequilibrium step.

Comparison of the expressions for the barriers in case 1 (electron-transfer control) allows one to predict whether the R* mechanism, the P* mechanism, or the direct R → P reaction will be favored under a given set of conditions. The equations can be solved to give expressions for the energies of R* and P*, $\Delta G_{rr*}^{\pm}$ and $\Delta G_{pp*}^{\pm}$, respectively, at which the two-step and the direct mechanisms have the same barriers. The relevant expressions for reaction via R* and P* are, respectively,

$$\frac{\Delta G_{rr*}^{\pm}}{\lambda_{rp}} = (1 \pm \sqrt{\rho}) \left(\frac{\Delta G_{rp}^{0}}{\lambda_{rp}}\right) - (\rho \pm \sqrt{\rho}) \tag{22}$$

$$\frac{\Delta G_{pp*}^{\pm}}{\lambda_{rp}} = -(1 \pm \sqrt{\rho}) \left(\frac{\Delta G_{rp}^{0}}{\lambda_{rp}}\right) - (\rho \pm \sqrt{\rho}) \tag{23}$$

$$\rho = \frac{\lambda_{i*}}{\lambda_{rp}} \tag{24}$$

where λ_{i*} is λ_{r*p} or λ_{rp*} where appropriate. A two-step mechanism is favored when the energy of the intermediate, ΔG_{rr*}^{0} or ΔG_{pp*}^{0}, falls in the range between the two solutions for $\Delta G_{rr*}^{\pm}$ or $\Delta G_{pp*}^{\pm}$

$$0 \leq \Delta G_i^+ < \Delta G_i^0 < \Delta G_i^- \tag{25}$$

where i stands for either $rr*$ or $pp*$ and the appropriate roots of eq 22 or 23 are used.

Equations 22 and 23 behave differently for low-λ ($\rho < 1$) and high-λ ($\rho > 1$) intermediates. In Figure 6 values of $\Delta G_{rr*}^{\pm}/\lambda_{rp}$ (long dashed lines) and $\Delta G_{pp*}^{\pm}/\lambda_{rp}$ (short dashed lines) for the low-λ intermediates ($\rho = 0.50$) are plotted versus $\Delta G_{rp}^{0}/\lambda_{rp}$. The total shaded area is the region where the P* mechanism is favored; the R* mechanism is favored in the heavily shaded area. For all other regions the direct path has the lowest energy barrier. The low-λ R* pathway is favored only at low driving force and for $\Delta G_{rr*}^{0} < \lambda_{rp}/4$, although the low-$\lambda$ P* pathway is favored over a wider range of $\Delta G_{rp}^{0}/\lambda_{rp}$ and ΔG_{pp*}^{0} values. As ρ approaches unity, the intercept approaches zero and the region where the R* mechanism is favored disappears. As ρ decreases ($\lambda_{rp} \gg \lambda_{i*}$), the region where the two-step mechanisms are favored (the shaded area) becomes larger. For reactions with only a small driving force, the energy of R* or P* must be less than $\sqrt{\lambda_{rp}\lambda_{i*}} - \lambda_{rp}$ for the two-step pathway to be energetically favorable. The two-step reactions utilizing high-λ intermediates are never favorable in the normal free-energy region. However, they can become favorable in the inverted region.

Inverted Region. In the inverted region ($|\Delta G_{rp}^{0}| > \lambda_{rp}$), the low-$\lambda$ R* pathway is always unfavorable because both the increased driving force

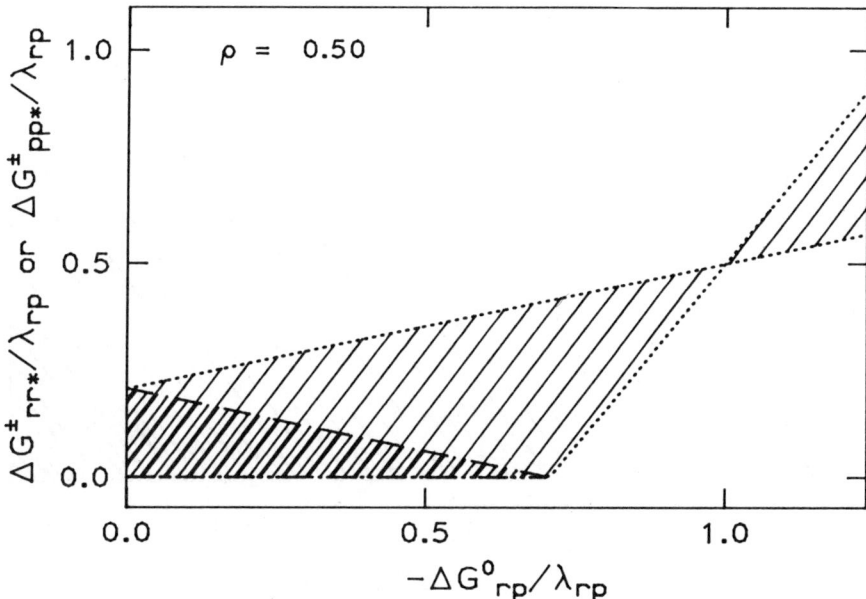

Figure 6. A plot of $\Delta G_{rr^*}^{\pm}/\lambda_{rp}$ (or $\Delta G_{pp^*}^{\pm}/\lambda_{rp}$) versus $\Delta G_{rp}^{0}/\lambda_{rp}$ where $\Delta G_{rr^*}^{\pm}$ (or $\Delta G_{pp^*}^{\pm}$) is the free energy of R* (or P*) when the barriers for the two-step and the direct pathways are equal. The figure is calculated with $\rho = 0.50$ (low-λ intermediates). The heavily shaded area is the region in which the R* mechanism is favorable, and the total shaded area is the region in which the P* mechanism is favorable. The direct route has the lowest energy barrier in the unshaded region.

and the smaller reorganization parameter make the R* → P reaction more inverted than the direct reaction. For the low-λ P* pathway, the driving force and the reorganization parameter for the electron transfer are both smaller than for the direct reaction. The lower driving force reduces the barrier by making the reaction less inverted (or even normal), but the lower reorganization parameter increases the barrier by making the reaction more inverted. Equations 22–24 can again be used to predict when the P* mechanism is favored over the direct pathway. In general, the low-λ P* mechanism is favored only at large values of $\Delta G_{pp^*}^{0}$ [$> (\lambda_{rp} - \lambda_{rp^*})$], as seen in Figure 7. However, these large values of $\Delta G_{pp^*}^{0}$ can result in very large barriers for the P* → P reaction.

For the high-λ intermediates, the increase in the reorganization parameter can make these pathways favorable in the inverted region. For the high-λ P* pathway, both the increase in λ and the decrease in driving force (relative to the direct reaction) make the electron transfer less inverted or even normal. Figure 7 shows the regions of $\Delta G_{rp}^{0}/\lambda_{rp}$ and $\Delta G_{i}^{0}/\lambda_{rp}$ for a given ρ where the two-step pathways are favored.

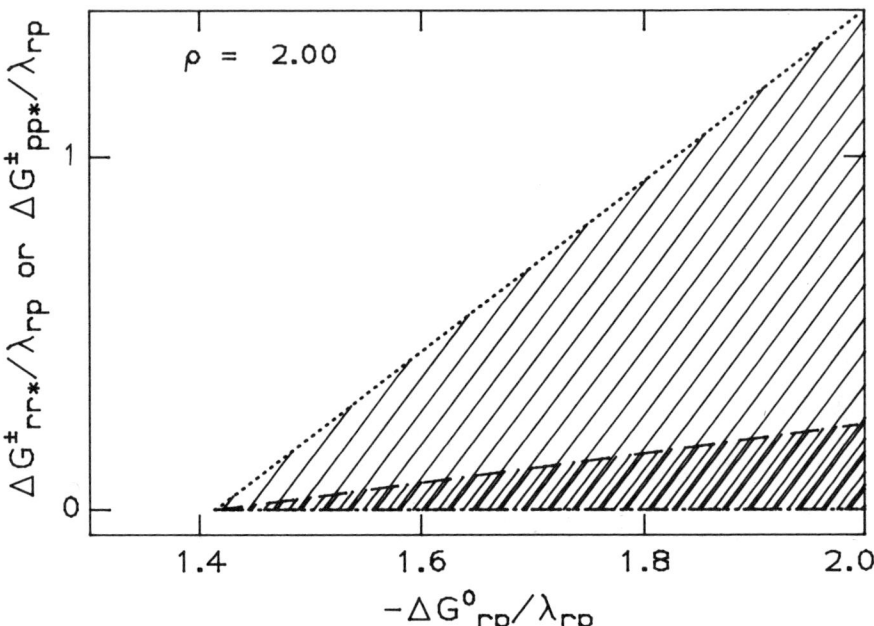

Figure 7. Same plot as Figure 6, except that $\rho = 2.0$ (high-λ intermediates).

In metalloprotein systems there will often be many conformational isomers of the reactants or products. Most of these conformers will have unfavorable electron-transfer properties (i.e., they will have large reorganization parameters or be relatively unstable) and will not be electron-transfer-active in the normal free-energy region. However, these unfavorable properties become an advantage in the inverted region. The conformers can become active intermediates in the electron transfer, particularly if they are formed relatively rapidly. As a consequence, metalloprotein systems can use such unfavorable conformers to achieve rapid electron transfer when the direct reaction is slow by virtue of being in the inverted region.

Three-Step Mechanisms. In the preceding discussion it was assumed that only a single intermediate, either R* or P*, was formed in the overall reaction. More complicated mechanisms can also exist (46). For example, eq 26 shows a three-step mechanism in which both an R* and a P* intermediate are involved.

$$R \rightleftarrows R^* \rightleftarrows P^* \rightleftarrows P \tag{26}$$

We define the three-step mechanism in eq 26 as a high-λ pathway when $\lambda_{r^*p^*} > \lambda_{rp}$, where $\lambda_{r^*p^*}$ is the reorganization parameter for the R* → P* electron transfer, and as a low-λ pathway when $\lambda_{r^*p^*} < \lambda_{rp}$. In conventional bimolecular electron transfers, R* and P* are often referred to as the precursor and successor complexes, respectively. A similar notation may be useful for intramolecular electron transfers. The case in which the high-λ intermediate corresponds to a large separation of the redox sites (i.e., a high λ_{out}) in bimolecular electron transfers has been considered previously (47).

When the R* → P* step is rate determining and $\Delta G_{rr*}^0 = \Delta G_{pp*}^0$, the rate of the three-step intramolecular mechanism can readily be compared with the rate for the corresponding direct reaction (as well as with the two-step mechanisms). Proceeding as in the two-step case, eq 27 is derived for ΔG_{rr*}, the energy of R* (or P*) when the direct and three-step intramolecular mechanisms have equal barriers.

$$\frac{\Delta G_{rr*}}{\lambda_{rp}} = \left(\frac{\rho - 1}{4\rho}\right)\left(\frac{\Delta G_{rp}^0}{\lambda_{rp}}\right)^2 + \frac{1 - \rho}{4} \tag{27a}$$

$$\rho = \frac{\lambda_{r*p*}}{\lambda_{rp}} \tag{27b}$$

Equation 27a is a parabola centered on the $\Delta G_{rr*}/\lambda_{rp}$ axis with an intercept of $(1 - \rho)/4$. The shaded area in Figure 8 shows the region where the three-

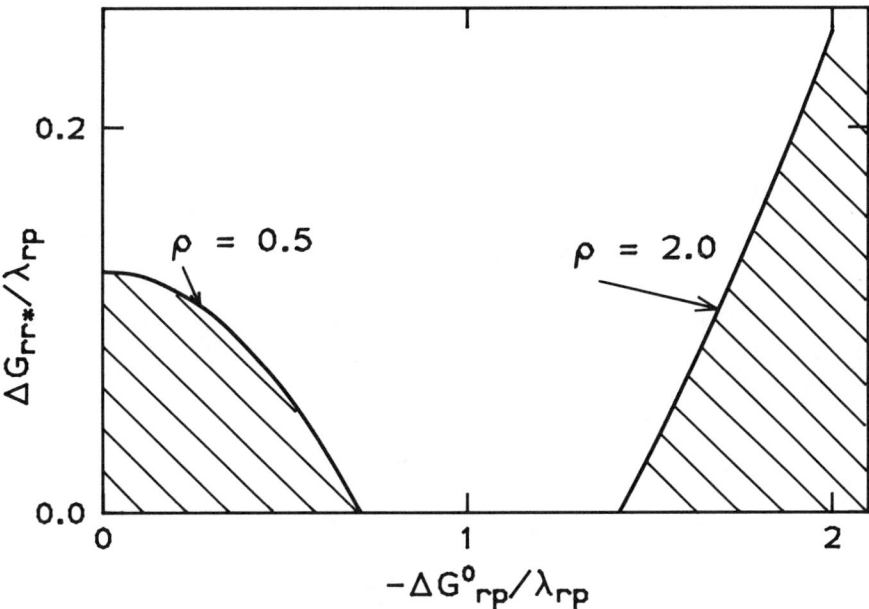

Figure 8. A plot of $\Delta G_{rr*}^{\pm}/\lambda_{rp}$ versus $\Delta G_{rp}^0/\lambda_{rp}$, where $\Delta G_{rr*}^{\pm}$ is the free energy of R* when the barriers for the three-step and the direct pathways are equal. The shaded area to the left of the figure ($\rho = 0.5$) is the region in which the low-λ three-step mechanism is favorable; the shaded area to the right of the figure ($\rho = 2.0$) is the region in which the high-λ three-step mechanism is favorable. The direct route has the lowest energy barrier in the unshaded regions. The figure is drawn with the assumption that $\Delta G_{rr*}^0 = \Delta G_{pp*}^0$.

step mechanism is favorable for both high- and low-λ pathways. For low-λ pathways (ρ < 1), the three-step mechanism is favored only in the normal free-energy region and at low values of $|\Delta G_{rr*}^0|/\lambda_{rp}$ (the parabola is inverted). For the high-λ pathways (ρ > 1), the three-step mechanism is favored only in the inverted region (the parabola is upright).

Low-λ Intermediates and Directional Electron Transfer

Directional electron transfer has been invoked for ruthenium-modified cytochrome c (17), and may in fact exist in a variety of conformationally labile metalloprotein systems. In terms of the model discussed here, the cytochrome c results may be interpreted by postulating that Fe(II) cytochrome c has a low-λ conformer. The reactions of the pentaammine–ruthenium derivative of cytochrome c may then be represented by eq 28. The electron transfer in the forward direction proceeds by a two-step mechanism in which a rapid (but unobserved) conformational change, eq 28b, succeeds the electron-transfer step, eq 28a. The concerted electron transfer is slower than the two-step mechanism in this case.

$$(cyt)Fe^{III}-Ru^{II}(NH_3)_5 \xrightarrow{k_{for}*} *[(cyt)Fe^{II}-Ru^{III}(NH_3)_5] \quad (28a)$$

$$*[(cyt)Fe^{II}-Ru^{III}(NH_3)_5] \xrightarrow{1/K_i*} (cyt)Fe^{II}-Ru^{III}(NH_3)_5 \quad (28b)$$

The reverse electron-transfer reaction for the tetraammine isonicotinamide–ruthenium derivative of cytochrome c may proceed either by the concerted reaction, eq 29,

$$(cyt)Fe^{II}-Ru^{III}(NH_3)_4(isn) \xrightarrow{k_{rev}'} (cyt)Fe^{III}-Ru^{II}(NH_3)_4(isn) \quad (29)$$

or by a two-step mechanism in which a rapid conformational change on the iron(II), eq 30a, precedes the electron transfer, eq 30b. The prime denotes a reaction of the isonicotinamide (isn) derivative.

$$(cyt)\check{F}e^{II}-Ru^{III}(NH_3)_4(isn) \xrightarrow{k_{i*}'} *[(cyt)Fe^{II}-Ru^{III}(NH_3)_4(isn)] \quad (30a)$$

$$*[(cyt)Fe^{II}-Ru^{III}(NH_3)_4(isn)] \xrightarrow{k_{rev}*'} (cyt)Fe^{III}-Ru^{II}(NH_3)_4(isn) \quad (30b)$$

Consider the following parameters: $\Delta G_{for}^0 \sim \Delta G_{rev}^{0'} \sim -3.5$ kcal mol^{-1}, $\Delta G_{i*}^0 = 6$ kcal mol^{-1} ($K_{i*} = K_{i*}' = 4 \times 10^{-5}$), $\lambda_{rp} = 36$ kcal mol^{-1} for the direct reactions, and $\lambda_{i*} = 5$ kcal mol^{-1} for the electron-transfer reaction in eq 28a or 30b. Calculations using these values yield barriers for eqs 28a, 29, and 30 of 2.8, 7.3, and 7.0 kcal mol^{-1}, respectively. Provided that all the electron-transfer steps have similar (electronic) prefactors, the observed rate constants for the forward and reverse reactions will differ by about three orders of magnitude, despite the similar driving forces for the overall reactions (Figure 9). The observed rate constants depend strongly on the direction of the electron transfer. The barrier for the electron transfer, eq 30b, is about 1 kcal mol^{-1}, and the rate constant for this step is more than four orders of magnitude faster than that for the direct reaction, eq 29; however, the preequilibrium step in eq 30a is unfavorable by 6 kcal mol^{-1}. This barrier results in a net rate for eq 30 that is only slightly faster than that for the direct reaction. If it is assumed that $k_{for} = k_{rev}' \sim 10^{-2}$ s^{-1}, then $k_{for}* \sim 20$ s^{-1}, $k_{rev}*' \sim 400$ s^{-1}, and $k_{for}(obs) \sim 20$ s^{-1}, $k_{rev}'(obs) \sim 10^{-2}$ s^{-1}. The latter rate constants are similar to those seen for the cytochrome c derivatives.

Finally, in this discussion, the electronic factor has been assumed to be the same for the different electron-transfer pathways. Additional subtleties are introduced when this factor is allowed to vary. Recently Isied and co-workers. (48) extended their studies of forward and reverse electron-transfer rates in derivatized cytochrome c. Although their new results are still pre-

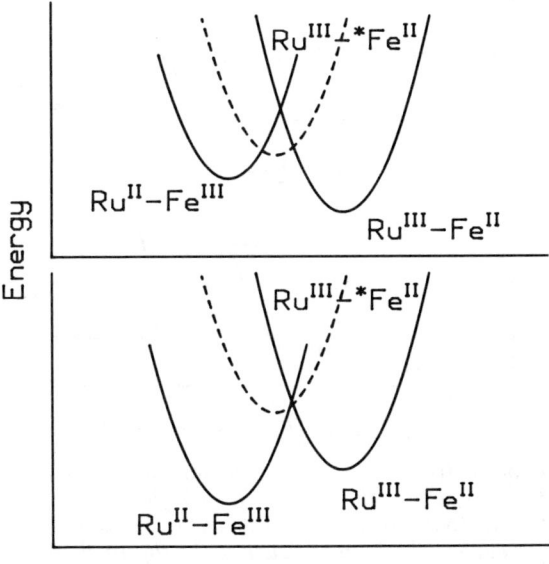

Nuclear Configuration

Figure 9. Plot of the free energy of the reactants, intermediate, and products for electron transfer within the ruthenium-modified cytochrome c. The top figure is for the Ru^{II}(NH$_3$)$_5$-cyt-c-Fe^{III} system; the reaction proceeds from the left to the right curve. The figure shows the lowering of the reaction barrier by the involvement of the P* intermediate (the conformationally modified cyt-c-Fe^{II}). The bottom figure is for the Ru^{III}(NH$_3$)$_4$isn-cyt-c-Fe^{II} system; the reaction proceeds from the right to the left curve. The bottom figure shows the negligible effect of the R* intermediate on the barrier for electron transfer within the isonicotinamide derivative.

liminary, the rates for all the systems studied cannot be reproduced by a scheme that considers only λ changes. However, the rate constants can be approximated by a scheme in which a low-λ conformer of the Fe(II) cytochrome c has an H_{r^*p} (or H_{rp^*}) that differs from the H_{rp} of the stable conformer. The following parameters yield electron-transfer rate constants close to the observed values: $(\kappa_{el}\nu_n)_{rp} \sim 10^3$, $(\kappa_{el}\nu_n)_{r^*p} = (\kappa_{el}\nu_n)_{rp^*} \sim 10^7$, $\lambda_{rp} \sim 1.5$ eV, $\lambda_{r^*p} = \lambda_{rp^*} \sim 0.5$ eV, and $\Delta G_{rr^*}^0 = \Delta G_{pp^*}^0 \sim 0.5$ eV. Figure 10 illustrates the free-energy surfaces for these parameters. The intersection of the Fe(III) and *Fe(II) cytochrome c surfaces lies above that of the Fe(III) and Fe(II) surfaces. However, despite the unfavorable nuclear barrier for reaction via the intermediate, because $(\kappa_{el}\nu_n)_{rp} \ll (\kappa_{el}\nu_n)_{rp^*}$ the reaction still proceeds through the unstable conformer of the Fe(II) cytochrome c.

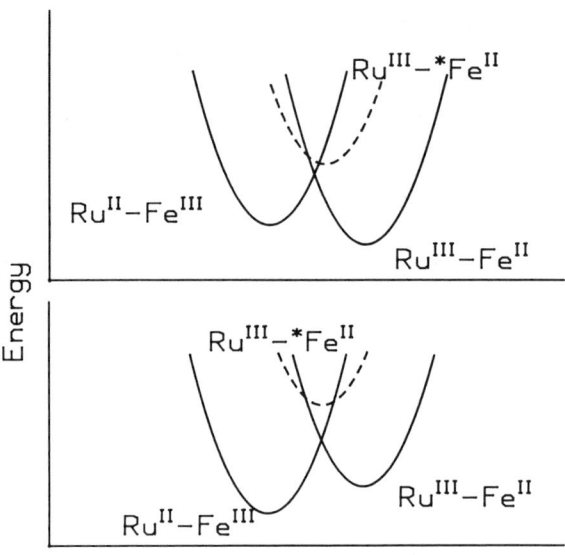

Nuclear Configuration

Figure 10. Plot of the free energy of the reactants, intermediate, and products for electron transfer within the ruthenium-modified cytochrome c. The top figure is for the $Ru^{II}(NH_3)_5$-cyt-c-Fe^{III} system; the reaction proceeds from the left to the right curve. The bottom figure is for the $Ru^{III}(NH_3)_4isn$-cyt-c-Fe^{II} system. Here the reaction proceeds from the right to the left curve. The figure is drawn with $\lambda_{rp} = 1.5$ eV, $\lambda_{r^*p} = \lambda_{rp^*} = 0.5$ eV, and $\Delta G_{rr^*}^0 = \Delta G_{pp^*}^0 = 0.5$ eV.

Acknowledgments

We acknowledge very helpful discussions with Carol Creutz and Marshall Newton. This research was carried out at Brookhaven National Laboratory under Contract DE–AC02–76CH00016 with the U.S. Department of Energy and supported by its Division of Chemical Sciences, Office of Basic Energy Sciences.

References

1. Marcus, R. A.; Sutin, N. *Biochim. Biophys. Acta* **1985**, *811*, 265.
2. Beratan, D. N.; Onuchic, J. N.; Hopfield, J. J. *J. Chem. Phys.* **1987**, *86*, 4488.
3. Closs, G. L.; Miller, J. R. *Science* **1988**, *240*, 440.
4. McLendon, G. *Acc. Chem. Res.* **1988**, *21*, 160.
5. Larsson, S. *Chem. Scripta* **1988**, *28A*, 15.
6. Oevering, H.; Paddon-Row, M. N.; Heppener, M.; Oliver, A. M.; Cotsaris, E.; Verhoeven, J. W.; Hush, N. S. *J. Am. Chem. Soc.* **1987**, *109*, 3258.
7. Paddon-Row, M. N.; Oliver, A. M.; Warman, J. M.; Smit, K. J.; de Haas, M. P.; Oevering, H.; Verhoewen, J. W. *J. Phys. Chem.* **1988**, *92*, 6958.
8. Isied, S. S.; Vassilian, A.; Magnuson, R. H.; Schwarz, H. A. *J. Am. Chem. Soc.* **1986**, *107*, 7432.
9. Isied, S. S. *Prog. Inorg. Chem.* **1984**, *32*, 443.
10. Wasielewski, M. R.; Niemczyk, M. P.; Svec, W. A.; Pewitt, E. B. *J. Am. Chem. Soc.* **1985**, *107*, 1080.
11. Wasielewski, M. R.; Niemczyk, M. P.; Svec, W. A.; Pewitt, E. B. *J. Am. Chem. Soc.* **1985**, *107*, 5562.
12. Ho, P. S.; Sutoris, C.; Liang, N.; Margoliash, E.; Hoffman, B. M. *J. Am. Chem. Soc.* **1985**, *107*, 1070.
13. Petterson-Kennedy, S. E.; McGourty, J. L.; Kalweit, J. A.; Hoffman, B. M. *J. Am. Chem. Soc.* **1986**, *108*, 1739.
14. Winkler, J.; Nocera, D. G.; Yocom, K. M.; Bordignon, E.; Gray H. B. *J. Am. Chem. Soc.* **1982**, *104*, 5782.
15. Nocera, D. G.; Winkler, J. R.; Yocom, K. M.; Bordignon, E.; Gray, H. B. *J. Am. Chem. Soc.* **1984**, *106*, 5145.
16. Isied, S. S.; Kuehn, C.; Worosila, G. *J. Am. Chem. Soc.* **1984**, *106*, 1722.
17. Bechtold, R.; Kuehn, C.; Lepre, C.; Isied, S. S. *Nature* **1986**, *322*, 286.
18. Conklin, K. T.; McLendon, G. *Inorg. Chem.* **1986**, *25*, 4084.
19. Conklin, K. T.; McLendon, G. *J. Am. Chem. Soc.* **1988**, *110*, 3345.
20. Brunschwig, B. S.; DeLaive, P. J.; English, A. M.; Goldberg, M.; Gray, H. B.; Mayo, S. L.; Sutin, N. *Inorg. Chem.* **1985**, *24*, 3473.
21. Elias, H.; Chou, M. H.; Winkler, J. R. *J. Am. Chem. Soc.* **1988**, *110*, 429.
22. Axup, A. W.; Albin, M.; Mayo, S. L.; Crutchley, R. J.; Gray, H. B. *J. Am. Chem. Soc.* **1988**, *110*, 435.
23. Karas, J. L.; Lieber, C. M.; Gray, H. B. *J. Am. Chem. Soc.* **1988**, *110*, 559.
24. Sutin, N.; Brunschwig, B. S.; Creutz, C.; Winkler, J. R.; *Pure Appl. Chem.* **1988**, *60*, 1817.
25. Isied, S. S.; Vassilian, A.; Wishart, J. F.; Creutz, C.; Schwarz, H. A.; Sutin, N. *J. Am. Chem. Soc.* **1988**, *110*, 635.
26. Marcus, R. A. *Annu. Rev. Phys. Chem.* **1964**, *15*, 155.
27. Marcus, R. A. *J. Chem. Phys.* **1965**, *43*, 679.
28. Sutin, N. *Acc. Chem. Res.* **1982**, *15*, 275.
29. Sutin, N. *Prog. Inorg. Chem.* **1983**, *30*, 441.
30. Newton, M. D.; Sutin, N. *Annu. Rev. Phys. Chem.* **1984**, *35*, 437.
31. Zusman, L. D. *Chem. Phys.* **1980**, *49*, 259.
32. Hynes, J. T. *J. Phys. Chem.* **1986**, *90*, 3701.
33. Nadler, W.; Marcus, R. A. *J. Chem. Phys.* **1987**, *86*, 3906.
34. Creutz, C. *Prog. Inorg. Chem.* **1983**, *30*, 1.

35. Newton, M. D. *J. Phys. Chem.* **1988**, *92*, 3049.
36. Newton, M. D. *J. Phys. Chem.* **1966**, *90*, 3734.
37. Larsson, S. *J. Am. Chem. Soc.* **1981**, *103*, 4034.
38. Larsson, S. *Chem. Phys. Lett.* **1982**, *90*, 136.
39. Larsson, S. *J. Phys. Chem.* **1984**, *88*, 1321.
40. Larsson, S. *J. Chem. Soc. Faraday Trans. 2* **1983**, *79*, 1375.
41. Onuchic, J. N.; Beratan, D. N. *J. Am. Chem. Soc.* **1987**, *109*, 6771.
42. Brunschwig, B. S.; Ehrenson S.; Sutin, N. *J. Phys. Chem.* **1986**, *90*, 3657.
43. Brunschwig, B. S.; Ehrenson, S.; Sutin, N. *J. Phys. Chem.* **1987**, *91*, 4714.
44. Hoffman, B. M.; Ratner, M. C. *J. Am. Chem. Soc.* **1987**, *109*, 6237.
45. Brunschwig, B. S.; Sutin, N. *J. Am. Chem. Soc.* **1989**, *111*, 7454.
46. Sutin, N. In *Bioinorganic Chemistry II*; Raymond, K. N., Ed.; Advances in Chemistry 162; American Chemical Society: Washington, DC, 1977; pp 156–172.
47. Brunschwig, B. S.; Ehrenson, S.; Sutin, N. *J. Am. Chem. Soc.* **1984**, *106*, 6858.
48. Wishart, J. F.; Cho, M.; Isied, S. S. In *Abstracts of the Thirteenth DOE Solar Photochemistry Research Conference*; Argonne National Laboratory: Argonne, IL, 1989; p 188.

RECEIVED for review May 1, 1989. ACCEPTED revised manuscript July 24, 1989.

EXPERIMENTAL APPROACHES TO BIOLOGICAL ELECTRON TRANSFER:

PEPTIDES AND PROTEINS

4

Directional Electron Transfer in Ruthenium-Modified Cytochrome c

Stephan S. Isied

Department of Chemistry, Rutgers, The State University of New Jersey, New Brunswick, NJ 08903

> *Studies of intramolecular oxidation and reduction in cytochrome c complexes covalently modified at the His-33 residue with a variety of ruthenium amine and ruthenium polypyridine complexes are presented. The redox potential of the ruthenium complexes vary over a potential range above and below the redox potential of the native cytochrome c. These studies show that the reduction of cytochrome c with these ruthenium complexes proceeds with a rate-limiting electron-transfer step that changes with the driving force of the reaction, as expected. Oxidation of cytochrome c proceeds with rates significantly lower than those expected on the basis of the driving force of the reaction. A mechanism to interpret the directional electron-transfer behavior of these ruthenium cytochrome c complexes on the basis of conformational changes of the reduced cytochrome c is described.*

RAPID ELECTRON TRANSFER CAN BE OBSERVED OVER LONG DISTANCES ($\sim$10–20 Å) (*1–11*), as shown by studies on electron transfer with organic and inorganic donor–acceptor complexes. The factors that control the rate of electron transfer in these systems are driving force, reorganization energy, and distance and orientation.

Polypeptide Donor–Acceptor Complexes

In an attempt to gain further insight into these factors, we have synthesized and studied a series of bridged polyproline complexes (*12*) of the type $[(NH_3)_5Os\text{-}L\text{-}Ru(NH_3)_5]^{5+}$, where L (the ligand) is $isn(Pro)_n$, isn is the isonicotinyl group, and n = 0, 1, 2, 3, or 4 (Table I).

Table I. Rates and Distances of Intramolecular Electron Transfer in [(NH$_3$)$_5$Os-L-Ru(NH$_3$)$_5$]$^{5+}$

Complex	Rate Constant, s^{-1}	Os–Ru Distance, Å
	$\geq 5 \times 10^9$	9.0
	3.1×10^6	12.1–12.3
	3.7×10^4	14.4–15.1
	3.2×10^2	17.8–18.3
	~ 50	20.9–21.5

NOTE: L is isn(Pro)$_n$, isn is isonicotinyl group, and $n = 0, 1, 2, 3,$ or 4.

The C- and N-terminal residues in Table I are derivatized with [(NH$_3$)$_5$Ru^{II}-] and [(NH$_3$)$_5$Os^{II}isn-], respectively. The redox potentials of the Os(II–III) couple and the Ru(II–III) couple are such that electron transfer in this series of complexes occurs from Os(II) → Ru(III) with a driving force of ~150 mV (13). In this series of complexes, the driving force (the difference in redox potential between the Ru(II–III) and the Os(II–III) couple) and the inner-sphere reorganization energy are kept constant, while the distance between the Os and Ru centers increases by 3.2 Å per proline. Under the conditions used to carry out these experiments (~0.1 M CF$_3$COOH), the proline oligomers are predominantly in the all-*trans* configuration and therefore act as a rigid spacer separating the metal ions (12).

The rate of intramolecular electron transfer in this series of molecules decreases by more than 8 orders of magnitude as proline residues are introduced. Analysis of the temperature dependence of these rate constants showed that the decrease in rate with distance in these molecules is attributed to both the increase in outer-sphere reorganization with distance and the decrease in electronic coupling between the donor and acceptor as proline residues are introduced (12, 13).

Intramolecular electron transfer for the tetraproline complex occurs with a rate constant ~ 50 s^{-1} at a metal-to-metal distance of ~ 21 Å (12, 14). Rapid electron transfer at these long distances is observed at very low driving forces. Hence, increasing the driving force or decreasing the outer-sphere reorganization energy is expected to yield rapid intramolecular electron-transfer rates at even longer distances. Extrapolation of these results indicates that intramolecular electron transfer will be observable at metal–metal separations of 30–40 Å in the millisecond time scale. We are currently pursuing this goal by synthesizing molecules that have 6–10 prolines separating the donor and acceptor metal ions.

Protein Donor–Acceptor Complexes

One of the techniques that has led to a new understanding of the mechanism of electron transfer in electron-transfer proteins is the use of the protein as a donor–acceptor complex by covalently attaching to these proteins a well-defined transition metal complex that binds to a specific amino acid site. Although this technique had been used to modify ribonuclease (15, 16), the synthetic breakthrough in modifying cytochrome c occurred when the reaction of $[(NH_3)_5Ru(OH_2)]^{2+}$ with cyt c was carried out at high protein concentrations with high metal-to-protein molar ratios (17). Exhaustive characterization of this modified protein showed that the ruthenium is covalently bound to the His 33 side chain of cyt c and that the protein did not undergo any measurable perturbation as a result of the modification (18–20).

When the modified protein is prepared in the Ru(III)cyt c(III) oxidation state and then reduced with a variety of radicals generated by pulse radiolysis techniques (21, 22), intramolecular electron transfer from the ruthenium site to the heme site occurs with a rate constant $k = 53$ s^{-1} (reduction potential, $E°$, for cyt c = 0.26 V and $E°$ for $[(NH_3)_5Ru^{II-III}(His)]$ = 0.10 Vs. NHE) (23, 24). The temperature dependence, concentration dependence, and pH dependence of this electron-transfer reaction were investigated. The results of this investigation showed that the rate of electron transfer is independent of concentration and moderately sensitive to temperature (enthalpy change $\Delta H^{\ddagger} \sim 3.5$ kcal M^{-1} and entropy change $\Delta S^{\ddagger} \sim -39$ eu). The electron-transfer reaction is independent of pH between pH 5 and 9, and then increases below pH 5 as the native conformation of the cyt c changes (23).

The observation of intramolecular electron transfer between the ruthenium site and the heme site occurring at distances of 12–15 Å (Figure 1) is extremely significant, because it represents the first observation of an intramolecular electron-transfer reaction within a ruthenium-modified electron-transfer protein. The magnitude of the rate constant (53 s^{-1}) is similar to the rate constants for other dynamical processes that are known to occur within the native cyt c protein. This finding led us to question whether the unimolecular rate observed is rate limiting in electron transfer (as in equa-

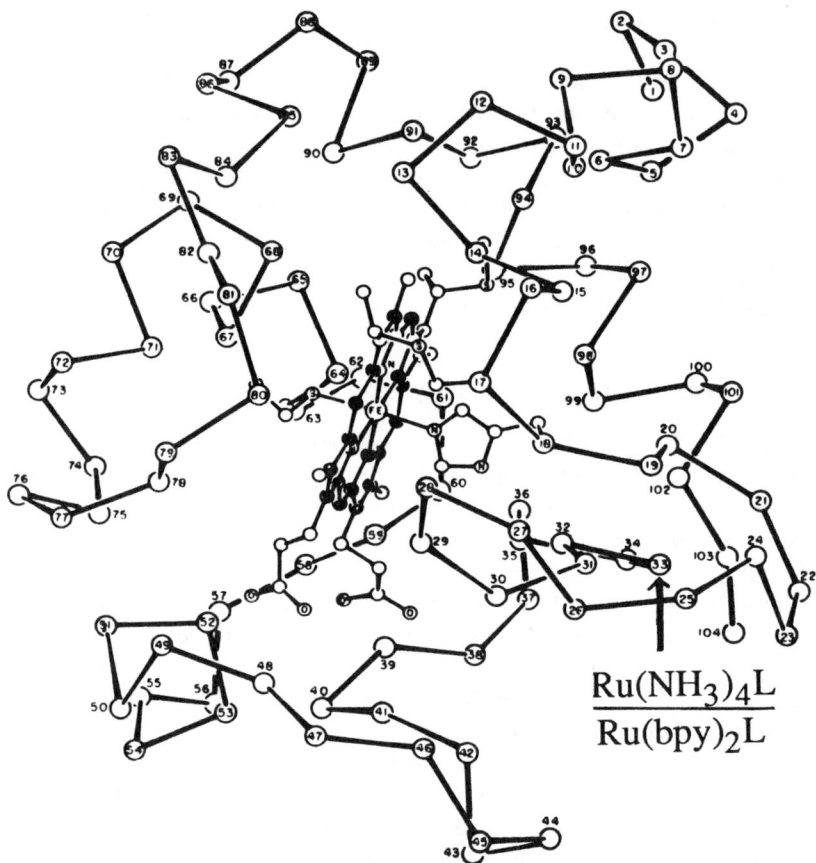

Figure 1. Ruthenium-modified cytochrome c, showing the relative position of the heme and the ruthenium sites.

tion 1) or in a protein-associated conformational change (as in equations 2a and 2b).

$$Ru^{II}cyt\ c^{III} \xrightarrow{k_{ET}} Ru^{III}cyt\ c^{II} \tag{1}$$

where k_{ET} is the rate constant for intramolecular electron transfer.

$$Ru^{II}cyt\ c^{III} \xrightarrow{k_{CC}} Ru^{II*}cyt\ c^{III} \tag{2a}$$

$$Ru^{II*}cyt\ c^{III} \xrightarrow{k_{CC}} Ru^{III}cyt\ c^{II}\ (fast) \tag{2b}$$

where k_{CC} is the rate constant for a protein conformational change.

To answer this question, we designed a series of related ruthenium molecules that are more oxidizing than cyt c and would therefore allow us to reverse the direction of electron transfer in the modified cyt c. Thus, we could change the heme of cyt c from an electron acceptor to an electron donor. The rationale behind these experiments is rather simple. If the unimolecular rate observed is rate limiting in electron transfer, then oxidation as well as reduction should be observed within these ruthenium-modified proteins.

The remaining part of this chapter describes the results of these experiments and proposes a kinetic scheme for the electron-transfer reactions of ruthenium-modified cytochrome c.

Results

The ruthenium-modified proteins listed in Table II were prepared and purified by procedures similar to those published earlier (19, 23, 25). The complexes were characterized by difference visible spectra, circular dichroism, tryptic digestion, Ru–Fe analysis, cyclic voltammetry, and differential pulse polarography. Results of these characterization studies clearly showed that all the ruthenium complexes were bound to the His 33 site and that there was no measurable difference in the conformation of the modified and the native proteins. Table II lists the reduction potential of the ruthenium ammine site in the ruthenium-modified proteins. For the bipyridine series, the reduction potentials have been obtained only for the corresponding ruthenium complexes.

Figure 1 shows the structure of cyt c and the relative positions of the heme to the ruthenium-modified sites. Table II summarizes the rates of

Table II. Rates of Intramolecular Electron Transfer and Reduction Potential of Ruthenium–Cytochrome c Complexes

Ruthenium-Modified cyt c	$E°$, V	Rate Constant, Intramolecular ET, k, s^{-1}	Direction of ET
Native HH cytochrome c	0.26	—	—
c-[(NH$_3$)$_4$Ru(OH)]-(II/III)	−0.01	5×10^2	Ru → heme
[(NH$_3$)$_5$Ru]-(II/III)	0.13	55	Ru → heme
c-[(NH$_3$)$_4$Ru(py)]-(II/III)	0.36	2.0	Ru → heme
t-[(NH$_3$)$_4$Ru(py)]-(II/III)	0.37	1.5	Ru → heme
c-[(NH$_3$)$_4$Ru(isn)]-(II/III)	0.44	$<10^{-2}$	Heme → Ru
t-[(NH$_3$)$_4$Ru(isn)]-(II/III)	0.44	$<10^{-2}$	Heme → Ru
c-[(NH$_3$)$_4$Ru(Mepz)]-(II/I)	−0.02	6×10^2	Ru → heme
c-[(NH$_3$)$_4$Ru(Mepz)]-(II/III)	0.72	~1.5	Heme → Ru
[Ru(bpy)$_2$(py)]-(II/I)	−1.3	2.8×10^5	Ru → heme
[Ru(bpy)$_2$(im)]-(II/I)	−1.3	2.0×10^5	Ru → heme
[Ru(bpy)$_2$(py)]-(II/III)	1.1	40	Heme → Ru
[Ru(bpy)$_2$(im)]-(II/III)	1.0	55	Heme → Ru

NOTE: $E°$ was taken versus the normal hydrogen electrode. ET is electron transfer.

intramolecular electron transfer for the reduction and the oxidation of cyt c by these ruthenium reagents. Chart I shows the ligands in the ruthenium reagents.

Discussion

Table II shows that the rate of reduction of cyt c can be changed by more than 5 orders of magnitude, depending on the redox potential and the reorganization energy of the ruthenium-modified protein. Two types of complexes coordinated to His 33 of cytochrome c can be identified. In the the first type, electron transfer takes place from (or to) a ruthenium t_{2g}-type orbital. This condition is true for all the oxidation reactions of cytochrome c by the different ruthenium complexes.

For the intramolecular reduction of cytochrome c, the reducing agent can be either a ruthenium t_{2g} metal-centered electron as in the [$(NH_3)_4Ru^{II}$-L] (L is *cis* or *trans* OH^-, NH_3, py, isn) complexes or a ligand-centered π^* orbital as in the *cis*-[$(NH_3)_4Ru^{II}$(Mepz)] and the [$(bpy)_2Ru^{II}$-L] (L is py or im) complexes. The role of the ruthenium(II) attached to the cytochrome c is either as a reducing agent or as a covalent linker to hold the Mepz and bpy ligand in the proximity of the His 33 site. Thus, the distance for electron transfer from these organic radicals to the cytochrome closely resembles the distances for reduction from the ruthenium orbitals. The reorganization energy for electron transfer for the ligand-centered reductions could be different from the reorganization energy from the ruthenium-centered orbitals. This difference should be taken into account when comparing rates of reduction of cytochrome c from ruthenium-centered orbitals vs. ligand-centered orbitals.

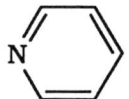 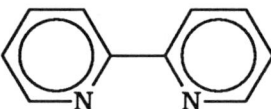 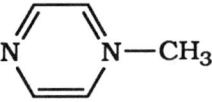

py: pyridine   bpy: 2,2'-bipyridine   Mepz: methylpyrazinium

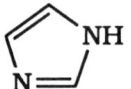

Isn: isonicotinamide   Im: imidazole

Chart I. Ligands in ruthenium reagents.

The oxidized ruthenium(III) form of the $[(NH_3)_4Ru\text{-}Mepz]^{3+}$ and the $[(bpy)_2Ru\text{-}L]^{3+}$ are intramolecularly reduced by cytochrome c to the corresponding ruthenium(II) species (a metal-centered orbital). In these cases, reorganizational energies of the ruthenium center are associated with the rate of electron transfer.

The novel property of the second type, the $[(NH_3)_4Ru^{II}\text{-}Mepz]$ and the $[(bpy)_2Ru^{II}\text{-}L]$ cytochrome c species, is that oxidation to the ruthenium(III) center can be observed, as well as reduction from the ligand-centered radicals. Thus both the oxidation and reduction of cytochrome c can be observed from the same inorganic modifier, even though it is metal-centered in one direction and ligand-centered in the other. Correction for the reorganizational energies between the metal center and the ligand center allows a direct comparison of oxidation and reduction at these similar distances.

Binuclear ruthenium complexes of the bridging pyrazine and bipyridine type can be used to observe metal-centered oxidation and reduction. Complexes of the type $[(NH_3)_5Os\text{-}LL\text{-}Ru(NH_3)_4\text{-}L\text{-}(OH_2)]^{4/5+}$ (where LL is pyrazine and L is 4,4'bpy) have multiple oxidation–reduction properties that will allow the oxidation and reduction of cytochrome c to metal-centered orbitals. In these cases, correction for reorganizational energies because of the origin of the orbital for the electron-transfer reaction will not be necessary.

"One-Direction" Electron Transfer

Table II shows that the rates of oxidation and reduction of cytochrome c by these covalently modified ruthenium complexes do not proceed in a simple reversible elementary step. The fact that the rate of intramolecular oxidation of cis- and trans-$[(NH_3)_4Ru^{II}\text{-}L]$ (L is isn) by cytochrome c is much slower than the reduction of cytochrome c with $[(NH_3)_5Ru^{II}\text{-}]$ indicates that more complex chemistry is associated with electron transfer in the protein. Association of this complexity with the protein rather than with the ruthenium label is inferred from the variety of ruthenium complexes that exhibit the same behavior. We interpreted this in 1986 as "directional electron transfer", where protein conformational states play a role in the intramolecular electron-transfer reaction (24).

Subsequent to our work, a paper on "gated electron-transfer reactions" by Hoffman et al. (26) was published to interpret a related observation in their work on photoinduced electron transfer in protein electron-transfer complexes. Sutin et al. (27, in this volume) worked out the general theoretical formalisms for the many possible types of "directional electron transfer". This theoretical work opens up more avenues for designing new types of donor–acceptor complexes that exhibit directional electron transfer.

The chemistry of cytochrome c can be used to aid in the interpretation of this directional electron transfer in ruthenium-modified cytochrome c.

The reduction of cytochrome c with ruthenium reagents covalently attached to His 33 is dependent on the driving force of the reaction. Table II shows that this rate can be varied over 5 orders of magnitude by changes in the ruthenium complexes. On the other hand, the intramolecular oxidation of cytochrome c is 4–5 orders of magnitude slower than its reduction after correction for driving force and reorganizational energy. A mechanism to interpret these results is shown in Scheme I. Thus cytochrome c^{III} is reduced to an activated intermediate, cyt c^{*II}, which undergoes a conformational change to the stable form of cyt c^{II}. Therefore, in the reduction of cyt c we measure the rate of formation of this activated intermediate, k_1. For the oxidation of cytochrome c, the pre-equilibrium to form the same activated intermediate is required first. This requirement depresses the observed rate of intramolecular oxidation to k_{-1}/K_{eq}, and therefore we observe a significantly decreased intramolecular rate of oxidation for the cytochrome c. This mechanism is one of many that can be used to interpret the observed results. The attractiveness of this mechanism is its simplicity. Other mechanisms involving more intermediates and associated rate constants can also be used to interpret the results.

Another attractive feature of this mechanism is that it is very similar to the mechanisms proposed for electron transfer at solid electrodes. In the language of electrochemistry, the mechanism (Scheme I) is referred to as the EC mechanism (i.e., a chemical reaction [an equilibrium or conformational change, etc.] following the electron-transfer step in the reduction process).

Finally, it is of interest to define the molecular event that leads to this conformational change. Further work from NMR and time-dependent res-

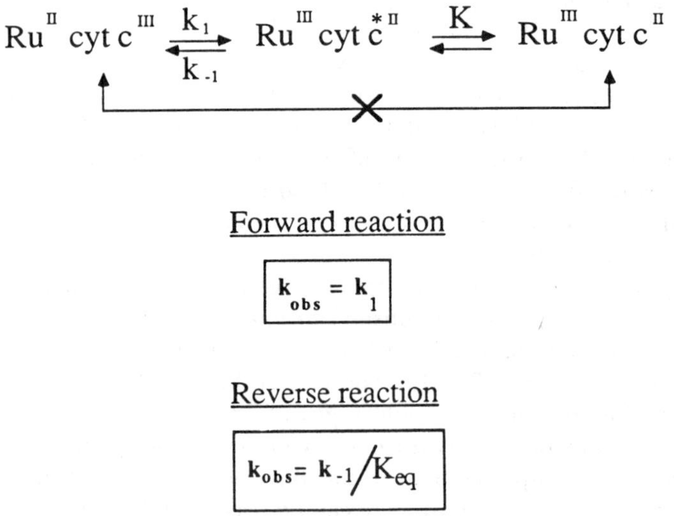

$$Ru^{II}\ cyt\ c^{III} \underset{k_{-1}}{\overset{k_1}{\rightleftharpoons}} Ru^{III}\ cyt\ c^{*II} \overset{K}{\rightleftharpoons} Ru^{III}\ cyt\ c^{II}$$

Forward reaction

$$k_{obs} = k_1$$

Reverse reaction

$$k_{obs} = k_{-1}/K_{eq}$$

Scheme I. *Directional electron-transfer mechanism.*

onance Raman spectroscopy might shed light on these molecular events. The other important question that should be addressed is whether "one-direction electron transfer" can be observed from different sites of the protein surface to the heme and vice versa. Chemical modification of cytochrome c and site-directed mutagenesis experiments are required to generate specific binding sites in different regions of the cytochrome c so that experiments similar to the ones outlined in this chapter can be carried out. Comparison of the intramolecular oxidation and reduction properties of the heme site from different regions of the protein should provide answers to these questions.

Acknowledgments

I am indebted to many collaborators who contributed significantly to this research: Harold Schwarz and James Wishart (Brookhaven National Laboratories) for their collaboration on the pulse radiolysis work, and A. Vassilian, Mary Gardineer, Rolf Bechtold, C. Kuehn, and M. Cho, from the Rutgers group. I would also like to acknowledge helpful discussions with H. Taube, N. Sutin, C. Creutz, and B. Brunschwig.

References

1. Beitz, J. V.; Miller, J. R. In *Tunneling in Biological Systems*, Academic: New York, 1979; pp 269–280.
2. Closs, G. L., Miller, J. R. *Science* **1988**, *240*, 440–247.
3. Isied, S. S. *Progr. Inorg. Chem.* **1984**, *32*, 443–517.
4. Marcus, R. A.; Sutin, N. *Biochim. Biophys. Acta* **1985**, *811*, 265.
5. Mayo, S. L.; Ellis, W. R., Jr.; Crutchley, R. J.; Gray, H. B. *Science* **1988**, *233*, 948.
6. Gray, H. B. *Chem Soc. Rev.* **1986**, *15*, 17.
7. Nocera, D. G.; Winkler, J. R.; Yocum, K. M.; Bordignon, E.; Gray, H. B. *J. Am. Chem. Soc.* **1984**, *106*, 5145.
8. Larsson, S. *Disc. Faraday Soc.* **1984**, *106*, 1584.
9. McLendon, G. *Acc. Chem. Res.* **1988**, *21*, 160.
10. Verhoeven, J. W.; Paddon-Row, M. N.; Hush, N. S.; Oevering, H.; Heppener, M. *Pure Appl. Chem.* **1986**, *58*, 1285.
11. Wasielewski, M. R.; Niemczyk, M. P.; Svec, W. A.; Pewitt, E. B. *J. Am. Chem. Soc.* **1980**, *107*, 1080; ibid., 5562.
12. Isied, S. S.; Vassilian, A.; Magnuson, R.; Schwarz, H. *J. Am. Chem. Soc.* **1985**, *107*, 7432–7438.
13. Isied, S. S.; Vassilian, A.; Wishart, J.; Creutz, C.; Schwarz, H.; Sutin, N. *J. Am. Chem. Soc.* **1988**, *110*, 635.
14. Vassilian, A.; Wishart, J.; van Hemelryk, B.; Schwarz, H.; Isied, S. S., manuscript in preparation.
15. Recchia, J.; Matthews, C. R.; Rhee, M. J.; Horrocks, W. D., Jr., *Biochim. Biophys. Acta* **1982**, *702*, 105.
16. Matthews, C. R.; Erikson, P. M.; Froebe, C. L. *Biochim. Biophys. Acta* **1980**, *624*, 499.

17. Gulka, R., M.S. Thesis, Rutgers University, 1979.
18. Isied, S. S.; Worosila, G.; Atherton, S. J. *J. Am. Chem. Soc.* **1982**, *104*, 7659–7661.
19. Isied, S. S.; Kuehn, C.; Worosila, G. *J. Am. Chem. Soc.* **1984**, *106*, 5145.
20. Yocum, K. M.; Shelton, J. B.; Schroeder, W. A.; Worosila, G.; Isied, S. S.; Bordignon, E.; Gray, H. B. *Proc. Natl. Acad. Sci. U.S.A.* **1982**, *79*, 7052.
21. *Radiation Chemistry;* Farhataziz; Rogers, M., Eds.; VHC Publishers: New York, 1987.
22. Schwarz, H. A.; Creutz, C. *Inorg. Chem.* **1983**, *22*, 707–713.
23. Bechtold, R.; Gardineer, M. B.; Kazmi, A.; van Hemelryck, B.; Isied, S. S. *J. Phys. Chem.* **1986**, *90*, 3800.
24. Bechtold, R.; Kuehn, C.; Lepre, C.; Isied, S. S. *Nature (London)* **1986**, *322*, 286.
25. Isied, S. S.; Taube, H. *Inorg. Chem.* **1976**, *15*, 3070.
26. Hoffman, B. M.; Ratner, M. A. *J. Am. Chem. Soc.* **1987**, *109*, 6237.
27. Sutin, N.; Brunschwig, B. S. In *Electron Transfer in Biology and the Solid State: Inorganic Compounds with Unusual Properties;* Johnson, M. K.; King, R. B.; Kurtz, D. M.; Kutal, C.; Norton, M. L.; Scott, R. A., Eds.; Advances in Chemistry 226; American Chemical Society: Washington, DC, 1990; Chapter 3.

RECEIVED for review May 1, 1989. ACCEPTED revised manuscript August 1, 1989.

Photoinduced Electron Transfer Across Peptide Spacers

Leonardo A. Cabana and Kirk S. Schanze[1]

Department of Chemistry, University of Florida, Gainesville, FL 32611

*A series of molecules was prepared in which a metal complex center, [(bpy)Re(I)(CO)$_3$(pyr)]$^+$ (bpy is 2,2'-bipyridine, pyr is 4-aminopyridine), is linked to a 4-(N,N-dimethylamino)benzoate unit (DMAB) by a covalent bridge consisting of 0, 1, and 2 L-proline units (compounds **0**, **1**, and **2**, respectively). In these molecules, electron transfer from DMAB to Re is initiated by photoexcitation of the Re → bpy metal-to-ligand charge-transfer (MLCT) excited state.*

$$(bpy)Re^I\cdots DMAB \xrightarrow{h\nu}$$

$$(bpy^-)Re^{II}\cdots DMAB \xrightarrow{k_{ET}} (bpy^-)Re^I\cdots DMAB^+$$
$$\text{MLCT}$$

*Electron-transfer kinetics were studied in two solvents by monitoring the MLCT emission. Rate constants for intramolecular electron transfer (k_{ET}) for **0**, **1**, and **2** in CH$_3$OH at 20 °C are 9.8 × 10^7, 5.3 × 10^6, and 5.6 × 10^5 s^{-1}, respectively. The temperature dependence of k_{ET} also was determined in CH$_3$OH. The results are consistent with a nonadiabatic, long-range electron-transfer mechanism. Emission-decay kinetics of **2** in CH$_3$CN suggest the presence of two conformational isomers with dramatically different electron-transfer rates. The emission data is supported by the observation of two conformational isomers in the ^{13}C NMR spectrum of **2**. The strong dependence of rate on conformation for **2** is consistent with a through-space mechanism for electron transfer.*

THE MECHANISM OF ELECTRON-TRANSFER (ET) REACTIONS has been studied by using photochemistry for many years. Early kinetic studies focused

[1]Address correspondence to this author.

on bimolecular quenching of photoexcited chromophores by electron donors or acceptors (quenchers). Through these studies a significant amount of information has been obtained concerning the relationship between the driving force (ΔG_{ET}) and the rate (k_{ET}) for endothermic and weakly exothermic ET reactions (1–8).

However, a disadvantage of bimolecular systems is that the rate of the intrinsic ET step is not directly obtained from the observed quenching rate because the chromophore and quencher must diffuse together to form an encounter complex prior to ET (8). This limitation is severe for strongly exothermic reactions because diffusion is the rate-determining step for quenching in this case (1, 4). In addition, no information is available concerning the structure of the encounter complex during ET for bimolecular reactions.

To overcome the disadvantages of bimolecular systems, an increasing number of studies have focused on photoinduced intramolecular ET in covalently linked chromophore–quencher (C–Q) compounds (9–30). In C–Q systems, an electron donor or acceptor is covalently attached to a chromophore to allow measurement of k_{ET} without the complication of diffusion effects. In these systems, it is possible in principle to measure k_{ET} directly, even in cases where the rate is exceedingly rapid. This measurement has become possible through advances in fast time-resolved methods for the study of the kinetics of photoinduced processes (31, 32).

Spacer Structures

In many early C–Q systems, the chromophore and quencher were attached via flexible methylene chains (9–11). Some information was obtained with these systems regarding the effect of relative orientation of the donor and acceptor on k_{ET} (9). However, because of the flexibility of the methylene spacers, little information was obtained concerning the effect on ET of separation distance or the molecular structure of the spacer. More recently, attention has turned to C–Q systems, in which the chromophore and quencher are attached by rigid spacers in an effort to provide detailed information about structural factors that control ET reaction rates. In these studies chromophore and quencher sites are held together by rigid organic spacers, peptides, and protein matrices (14, 15, 19–22, 27, 28, 33–37).

The well-defined spacer structure makes it possible to determine k_{ET} under conditions in which the separation distance between and the relative orientation of the donor and acceptor are known. When the rigid spacer systems are used, it becomes possible to address several questions of interest:

1. What is the effect of distance on the rate for ET?
2. How does the molecular and electronic structure of the spacer affect the rate for long-distance ET?

3. How does k_{ET} depend on the relative orientation of the donor and acceptor?

In addition to addressing these questions, studies of photoinduced intramolecular ET across rigid spacers have provided experimental evidence for the Marcus inverted region (13) and have led to the development of synthetic systems that mimic the primary events in photosynthesis (14, 15, 17, 18, 38).

Structural and kinetic studies of proteins involved in biological redox reactions (e.g., the photosynthetic reaction center protein complex, cytochrome c, blue copper proteins) (39–45) have led to the realization that the protein-bound redox sites are frequently separated by 5–20 Å when ET occurs. In many cases the transferring electron "tunnels" across an intervening protein matrix (45, 46). This realization has led to an increasing number of experimental and theoretical investigations focused on understanding the role of the intervening protein in mediating long-range ET. One experimental approach to this problem has been to study ET in structurally well-defined native and modified proteins (33–37, 40–42, 45–47). However, a disadvantage of this approach is that it is difficult to systematically vary structural elements in the protein-based systems.

An alternative approach is to study intramolecular ET in systems in which a donor and acceptor are separated by synthetic peptide spacers. This approach was pioneered by Isied, who studied thermally activated ET between metal centers in several systems that use a series of rigid oligo-L-proline peptide spacers (48–50). With this method, the separation distance between the redox sites can be systematically varied to allow investigation of the distance dependence of k_{ET} across a well-defined peptide spacer. Oligoproline peptides are well-suited for use as spacers because in protic solvents the peptide chain adopts a helical form that exists predominantly in a single conformation and undergoes isomerization at a comparatively slow rate. This fact has been established in NMR spectroscopic experiments (51–54) and through studies of end-to-end Forster excited-state energy transfer across oligoprolines ranging from 6 to 10 amino acid units (55, 56).

Photoinduced Intramolecular ET

We initiated a series of investigations in which we apply the technique of photoinduced intramolecular ET to study the kinetics of ET across a series of rigid oligoproline spacers. These studies were designed to obtain quantitative information concerning the distance dependence of k_{ET} across peptides for the photochemically activated process. One goal is to provide data that will allow a direct comparison of the distance dependence for a photoinduced ET with the thermally activated ET processes studied by Isied. It is hoped that such comparisons will lend insight concerning the mechanism for electron tunneling across spacers that have structural characteristics in common with redox proteins.

In the peptide-linked electron-donor–acceptor systems described in this chapter (*see* structures), the photoexcited Re(I) metal complex acts as an electron acceptor and the dimethylaminobenzoate (DMAB) moiety acts as an electron donor. The Re(I) complex was chosen primarily because it has a well-characterized metal-to-ligand charge-transfer (MLCT) excited state that is luminescent, relatively long-lived, and a strong oxidant (*24, 25, 57, 58*). The DMAB donor was chosen for the relative ease with which it can be incorporated into the peptide synthesis and because the amido linkage that binds the donor to the peptide chain restricts dynamic motion.

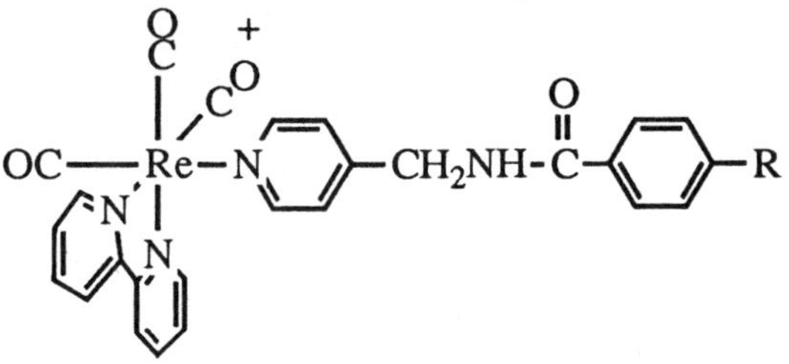

0-m : R = H

0 :   R = NMe_2

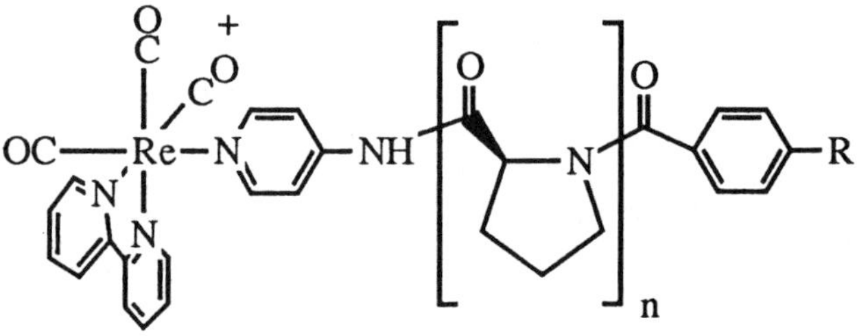

1-m :   n = 1, R = H   **1 :**   n = 1, R = NMe_2

2-m :   n = 2, R = H   **2 :**   n = 2, R = NMe_2

In compounds **0–2**, photoexcitation of the Re(I) complex into the MLCT excited state initiates intramolecular ET from the DMAB donor to the metal center.

$$(bpy)Re^{I} \cdots DMAB \xrightarrow{h\nu}$$
$$\underset{MLCT}{(bpy^-)Re^{II} \cdots DMAB} \xrightarrow{k_{ET}} (bpy^-)Re^{I} \cdots DMAB^+ \quad (1)$$

where bpy is 2,2'-bipyridine. The rate constant for ET in **0–2** has been examined in two solvents by both steady-state and time-resolved fluorescence spectroscopy. The results indicate that the rate of ET is strongly dependent upon the number of peptide spacers. In addition, temperature-dependence experiments indicate that ET in **0–2** is nonadiabatic and thus suggest that long-range ET occurs in each case.

Experimental Procedures

The peptide ligands and metal complexes were prepared by standard methods (25, 27). Each complex was purified by repeated column chromatography. The structures of the complexes were confirmed by ^{1}H and ^{13}C NMR spectroscopy and by elemental analysis.

Cyclic voltammetry was carried out on a voltammograph (Bioanalytical Systems CV-2). Experiments were carried out in a two-compartment cell in which Pt disk working electrodes and Pt wire auxiliary electrodes were separated from the reference electrode (saturated calomel, SCE) by a medium-porosity glass frit. Tetraethylammonium perchlorate (TEAP) (Kodak, recrystallized) was used at a concentration of 0.1 M as supporting electrolyte.

Solvents used in emission experiments were Kodak spectroquality. In all experiments, sample concentrations were $\sim 10^{-5}$ M, with optical densities at 400 nm ~ 0.10. Stern–Volmer experiments demonstrated that this concentration was sufficiently low to preclude bimolecular quenching. Temperature control during the luminescence experiments was maintained to within ± 1 °C by using a recirculating bath (Hakke D3). Steady-state luminescence spectra were obtained on a spectrophotometer (Spex Industries F-112A). Quantum-yield measurements were made relative to an emission actinometer consisting of Zn(II)-[5,10,15,20-tetraphenylporphyrin] in air-saturated benzene (emission quantum yield, Φ_{em} = 0.030) (59). The metal-complex samples and the actinometer had matched optical densities at the excitation wavelength (nm 400).

Emission-decay measurements were carried out by using time-correlated single-photon counting (60). Excitation light was filtered with a near-UV bandpass filter (Schott, UG-11, maximum transmittance at 350 nm), and emission light was filtered with a 600-nm interference filter. In all cases, data acquisition was carried out until 10,000 counts were obtained in the maximum channel. Analysis of emission-decay data was carried out by the nonlinear least squares method; the analysis provided for deconvolution of the excitation lamp response (60). In every case, the reduced chi-squared value for the fit of the experimental data was acceptable ($\chi^2 \leq 1.3$).

Results and Discussion

Thermodynamics and Kinetics of Intramolecular ET. The thermodynamic driving force for photoinduced DMAB to Re intramolecular ET (the second step in eq 1) can be approximated by using eq 2 (20).

$$\Delta G_{ET} = E_{1/2}(\text{DMAB/DMAB}^+) - E_{1/2}(\text{bpy/bpy}^-) - E_{MLCT} - \frac{14.45}{\epsilon R_{DA}} \quad (2)$$

In this equation, $E_{1/2}(\text{DMAB/DMAB}^+)$ and $E_{1/2}\text{bpy/bpy}^-)$ are the half-wave oxidation and reduction potentials for the DMAB donor and the 2,2'-bipyridine ligand, respectively; E_{MLCT} is the excited-state energy; and the last term represents the coulombic interaction between the electron and electron hole in the charge-separated state.

In order to estimate ΔG_{ET} for each of the Re–DMAB complexes, the relevant electrochemical data were obtained in CH_3CN solution by using cyclic voltammetry (see Table I). The MLCT excited-state energy for the complexes was obtained by fitting the emission spectra with a spectral sim-

Table I. Electrochemical Potentials, Emission Energies, and ΔG_{ET}

Complex	E_p (DMAB/DMAB$^+$), V	$E_{1/2}(bpy/bpy^-)$, V	E_{MLCT}, eV	ΔG_{ET}, eV
0	+0.98	−1.18	2.34	−0.18
1	+0.92	−1.20	2.31	−0.19
2	+0.93	−1.20	2.31	−0.18

NOTE: $E_{1/2}$ and E_p values were obtained from cyclic voltammetry in CH_3CN solution with the SCE reference. E_{MLCT} is emission energy calculated from spectral fit (ref. 59) of emission from complex in CH_3OH. ΔG_{ET} was calculated from eq 2 and data in Table I. Error in ΔG_{ET} is estimated to be ±0.1 eV.

ulation program that allows approximation of the 0–0 band energy from the broad structureless spectra that are typical of Re(I) MLCT states (57, 61). The estimated E_{MLCT} energies are also listed in Table I. By using these data and eq 2, ΔG_{ET} values were calculated for **0–2** (see Table I). The data show that intramolecular ET is weakly exothermic and is nearly constant across the series.

Scheme I presents an excited-state scheme that is used to model the kinetic behavior of the Re–DMAB complexes. Briefly, the model assumes that the photon is absorbed by the metal-complex chromophore and rapidly produces the $d\pi$ Re $\rightarrow$ π^* bpy MLCT excited state (24, 25, 57, 58). The MLCT state subsequently decays, by either radiative or nonradiative decay, back to the ground state ($k_d^°$) or by electron transfer (k_{ET}) to produce a charge-transfer state. Because ET competes with decay of the MLCT state

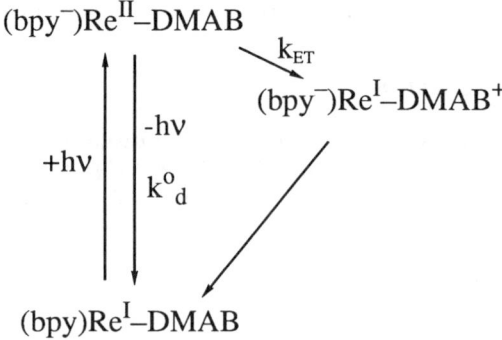

Scheme I. Energy-level diagram for **0–2**, including kinetic rate constants.

by normal paths, the kinetics can be determined from the MLCT luminescence by using either steady-state or time-resolved techniques.

Equations 3 and 4 have been used to calculate ET rate constants (11, 16, 19–21, 26, 27),

$$k_{ET}(\Phi) = \left[\frac{\Phi°}{\Phi} - 1\right]\frac{1}{\tau°} \quad (3)$$

$$k_{ET}(\tau) = \frac{1}{\tau} - \frac{1}{\tau°} \quad (4)$$

where Φ and τ are the emission yield and lifetime for the donor-substituted complexes (0–2), and $\Phi°$ and $\tau°$ are the same parameters for the corresponding unsubstituted model complexes (0-m, 1-m, and 2-m). These equations are derived from the kinetic scheme (Scheme I) under the assumption that $k_d°$ is the same for the donor-substituted and unsubstituted complexes. The assumption of a low electronic interaction between the Re center and the DMAB moiety is implicit.

This model provides an accurate representation of the excited-state manifold of the Re–DMAB complexes, as shown by comparison of the electronic absorption spectra of the donor-substituted complexes with the corresponding unsubstituted complexes. The spectra of 1-m, 1, and N,N-diethyl-4-(N,N-dimethyl)benzamide (DE-DMAB) are shown in Figure 1. Each of the metal complexes displays a low-intensity absorption band ($\epsilon \sim 4000$ M^{-1} cm^{-1}) near 360 nm. This band is assigned to the dπ Re $\rightarrow \pi^*$ bpy MLCT absorption (24, 25, 57, 58). The only significant difference between the spectra of 1 and 1-m is the enhanced absorption noted in the 200–330-nm region for 1 compared to 1-m. This region is where the DMAB π,π^* absorption occurs, as indicated by the spectrum of DE–DMAB. The fact that the absorption spectrum of 1 is nearly the sum of the component parts (1-m and DE–DMAB) indicates that the electronic interaction between the DMAB donor and the Re center is very weak.

Conformations of the Peptide Spacers and Structures for 0-2.

The peptide bond in proline oligomers can exist in two conformations, as shown in structure I. This conformational isomerization has been studied in various proline-containing peptides by using ^{13}C NMR spectroscopy (51–54). This technique is useful because the chemical shift for each of the four proline ring carbons is different for the *cis* and *trans* conformations. Carbon NMR spectroscopic experiments were carried out on 1 and 2 in two solvents to examine the *cis*–*trans* conformational equilibrium for the peptide spacers. Figure 2a shows the ^{13}C NMR spectra of 1 and 2 in the 20–70-ppm region in CD$_3$CN. As expected, in the spectrum of 1 the proline ring carbons appear as four resonances. However, in the spectrum of 2 the proline ring carbons

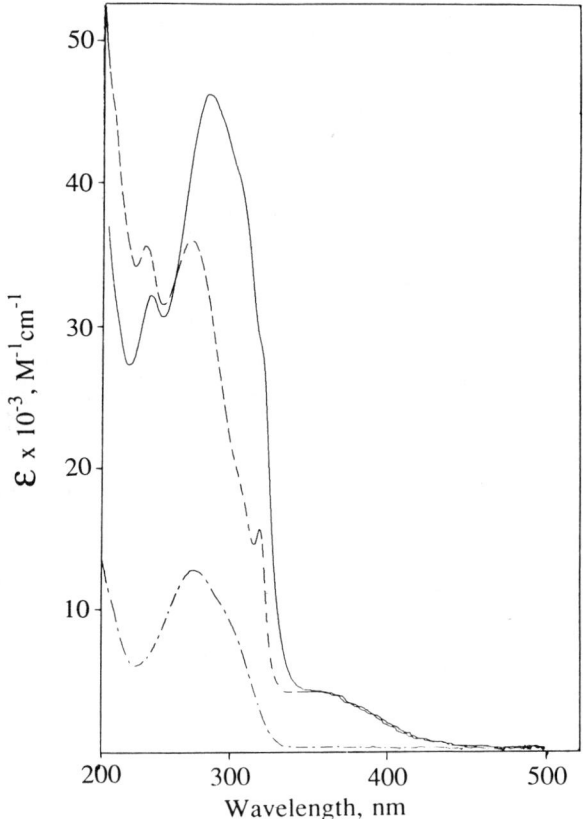

Figure 1. UV–visible absorption spectra in CH_3OH solution. Key: —, **1**; ---, **1-m**; – • – • –, DE–DMAB.

I

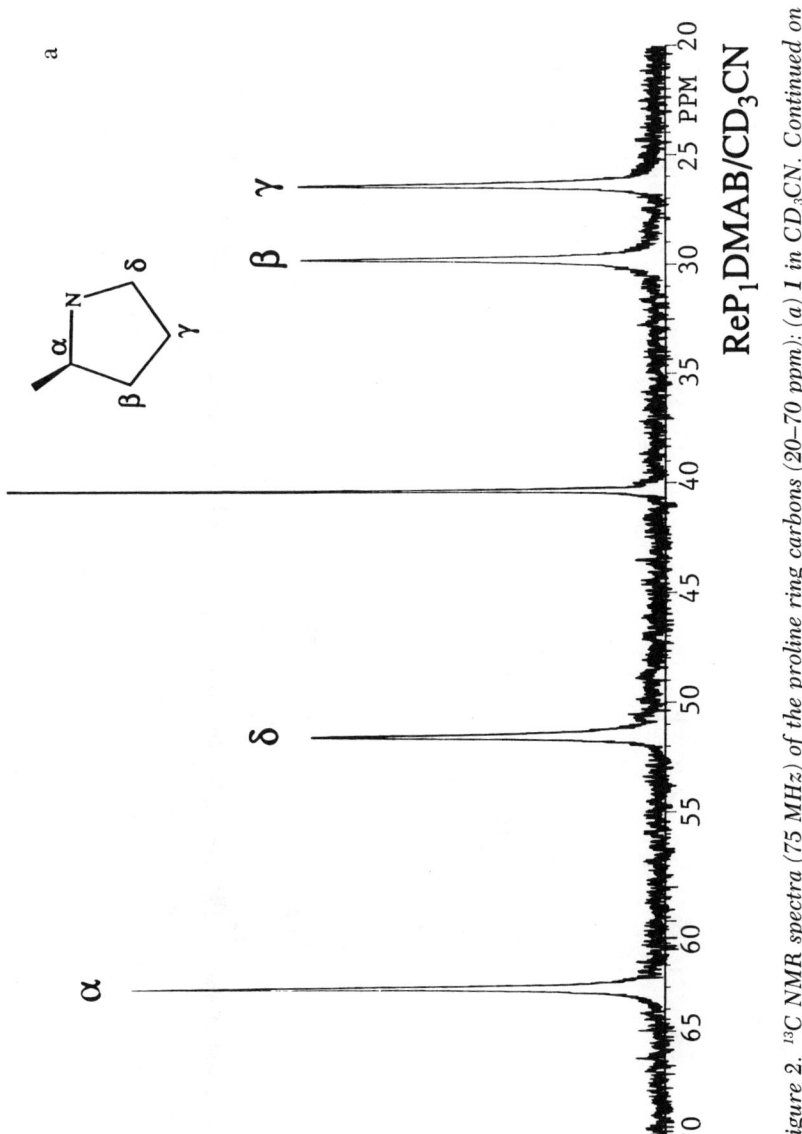

Figure 2. ^{13}C NMR spectra (75 MHz) of the proline ring carbons (20–70 ppm): (a) 1 in CD_3CN. Continued on next page.

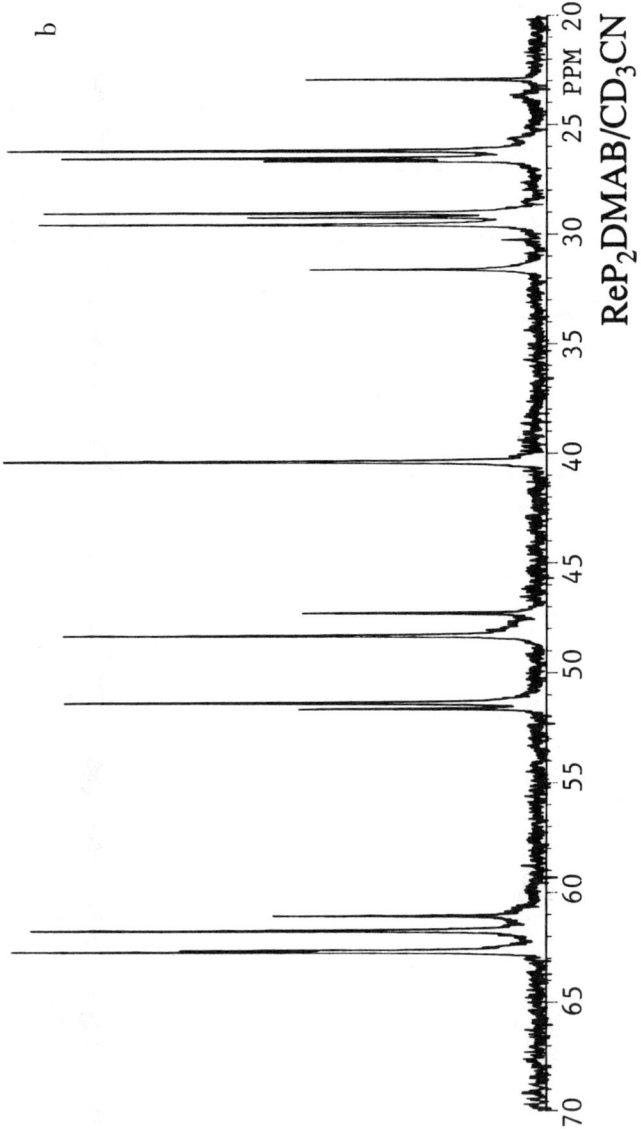

Figure 2. (b) **2** in CD_3CN.

5. CABANA & SCHANZE *Peptide Spacers* 111

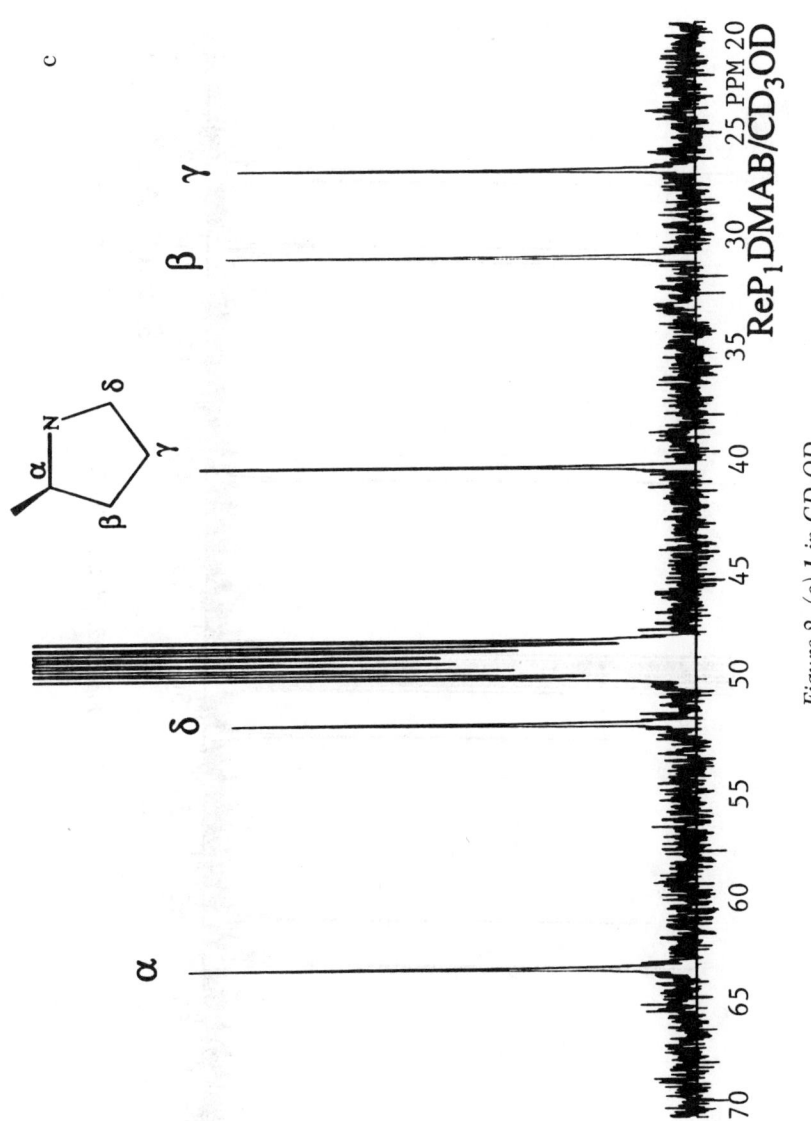

Figure 2. (c) 1 in CD$_3$OD.

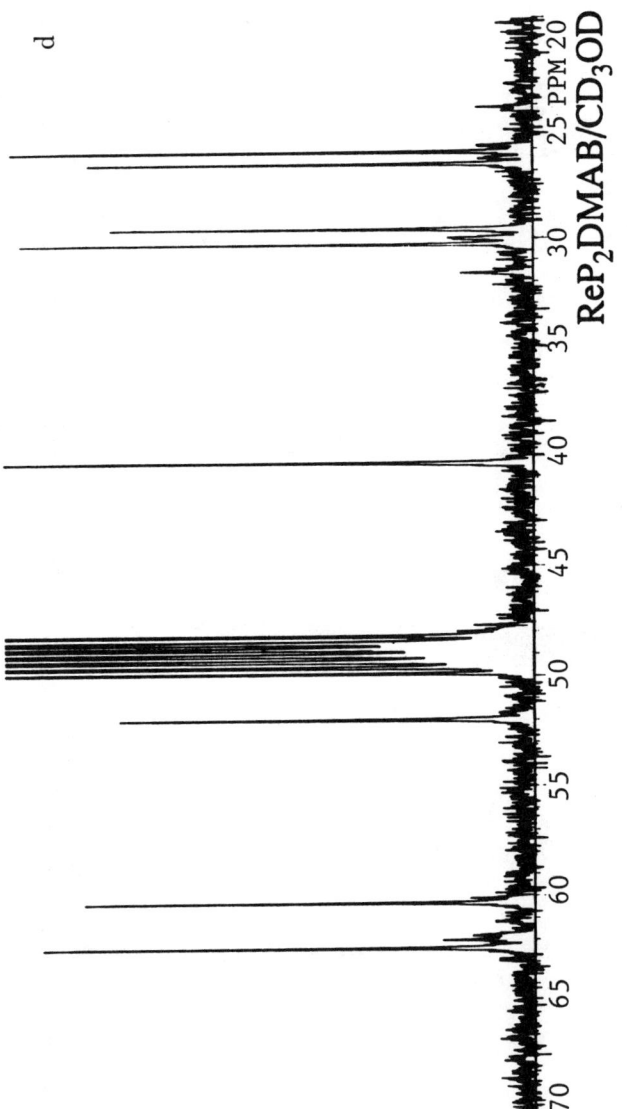

Figure 2. (d) 2 in CD_3OD.

appear as 16 resonances, rather than the expected 8. This splitting of the proline ring carbon resonances is attributed to the existence of both *cis* and *trans* conformations at the proline–proline peptide bond for **2** in CD$_3$CN. The less-intense resonances that are shifted from the stronger resonances are assigned to molecules in which the proline–proline bond is in the *cis* conformation (*51–54*). Figure 2b shows the same region of the carbon NMR spectrum for **1** and **2** in CD$_3$OD solution. Again in the spectrum of **1**, the four ring carbons appear as singlets. Interestingly, the carbon spectrum of **2** in CD$_3$OD is much simpler than in CD$_3$CN; resonances caused by molecules in the *cis* conformation are virtually absent. This result indicates that, in CD$_3$OD solution, **2** exists predominantly in the conformation with the proline–proline bond in the *trans* conformation.

The structures of **0**, **1**, and **2** were approximated by a PC-based molecular modeling program employing the following technique (*62*). The coordinates for the [(bpy)Re(CO)$_3$(4-aminopyridine)]$^+$ complex were input with the coordinates from the X-ray structure of [(bpy)Re(CO)$_3$(4,4'-bipyridine)]$^+$ (*63*). The coordinates for the proline spacers were input with the coordinates from the X-ray structure of *trans*-poly-L-proline (*64*). The structure of DMAB was obtained with the structure-generating function of the modeling program. The DMAB-to-proline, proline-to-4-aminopyridine, and DMAB-to-4-aminomethylpyridine amide bonds were all assumed to be in the *trans* conformation. The structures thus generated are shown in Chart I. From these structures, an estimate can be made for the Re-to-DMAB distance in the three complexes by taking the distance from the Re atom to C-1 in the DMAB aromatic ring: **0**, 9.2 Å; **1**, 11.4 Å; and **2**, 14.9 Å.

Intramolecular ET in Methanol Solution. The quantum yield (Φ) and lifetime (τ) of the Re → bpy MLCT luminescence of **0-m**, **1-m**, **2-m**, and **0–2** were measured at 20 °C in methanol. The data listed in Table II represent the average of three independent measurements of each parameter, and the errors are the experimental standard deviations. Comparison of the data shows that in each case Φ and τ are quenched in the donor-substituted complexes compared to the unsubstituted complexes. Furthermore, the extent of quenching is strongly dependent upon the number of proline spacers between the Re complex and the DMAB donor. The origin of this quenching is assumed to be solely competition of DMAB-to-Re ET with decay of the MLCT state by normal decay paths.

Quenching. Several lines of evidence support the premise that MLCT quenching results from ET.

1. A flash photolysis experiment was conducted on a solution containing **0-m** (10^{-4} M) and DE–DMAB (0.01 M). The solution was excited with an excimer-pumped dye laser into the

Table II. Emission Yields and Lifetimes in CH$_3$OH Solution

Complex	Φ_{em}	τ, ns
0-m	—	178 ± 1
1-m	0.034 ± 0.001	121 ± 1
2-m	0.032 ± 0.001	120.4 ± 0.6
0	—	9.62 ± 0.04
1	0.022 ± 0.001	73.7 ± 0.7
2	0.030 ± 0.001	115.2 ± 0.6

NOTE: Argon outgassed solutions were at 20 °C. Each value is the average of three independent measurements; errors are standard deviation in measurements.

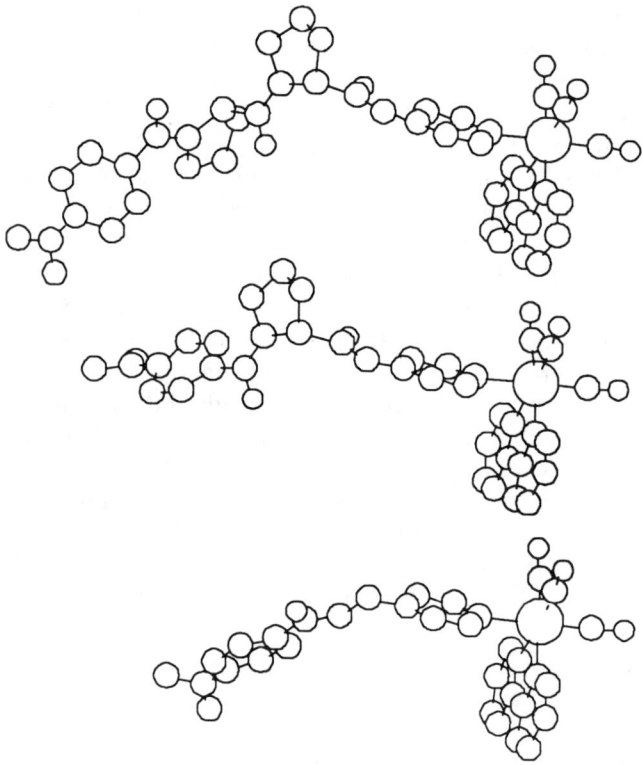

Chart I. Possible solution conformations for *0*, *1*, and *2*, determined as described in the text.

Re MLCT absorption band of **0-m**. A long-lived transient absorption (lifetime >30 μs) that was produced is assigned to the charge-transfer products formed via bimolecular ET from DE–DMAB to the photoexcited Re complex.

2. The singlet and triplet energies of ethyl *p*-(*N*,*N*-dimethyl)-aminobenzoate are 82 and 70 kcal, respectively (*65*, *66*). (These energies are based on vertical excitation to the ^1L_b and ^3L_b excited states.) The singlet and triplet energies of the DMAB moiety are expected to be very similar. Therefore, energy transfer from the Re-based MLCT state (55 kcal) to DMAB is endothermic and will be comparatively slow.

3. Studies of the kinetics for bimolecular quenching of [(bpy)-Re(CO)$_3$L]$^+$ (L is CH$_3$CN) by a series of organic amines (including DE–DMAB) and methoxybenzenes with varying oxidation potentials are entirely consistent with electron-transfer quenching in each case. The data are similar to those observed in studies of bimolecular quenching of Ru(bpy)$_3$$^{2+}$ by reductive ET from a series of organic amines (*4*), except that the Re complex is also quenched efficiently by the weak methoxybenzene reductants because of the more positive excited-state reduction potential (*Re$^+$/Re^0).

Taken together, these experiments provide compelling evidence that the predominant mechanism for quenching of the Re-based MLCT state by the DMAB moiety is ET.

ET Rate Constants. By using the luminescence data, ET rate constants were calculated for **0** with eq 4 and for **1** and **2** by using both eqs 3 and 4 (*see* Table III). In determining ET rates, the model used for complex **0** was **0-m** and the models used for complexes **1** and **2** were **1-m** and **2-m**, respectively. There is relatively good agreement in the rate constants determined by using the quantum yield and lifetime data for **1** and **2**. This

Table III. Electron-Transfer Rate Constants and Activation Parameters in CH$_3$OH Solution

Complex	$k_{ET}(\Phi)$, s^{-1} [a]	$k_{ET}(\tau)$, s^{-1} [a]	$\Delta H^{\ddagger}(\tau)$, kcal/mol[b]	$\Delta S^{\ddagger}(\tau)$, kcal/mol[b]
0	—	(9.81 ± 0.04) × 10^7	3.29 ± 0.07	−10.7 ± 0.2
1	(4 ± 1) × 10^6	(5.3 ± 0.2) × 10^6	4.04 ± 0.05	−14.0 ± 0.2
2	(6 ± 3) × 10^5	(3.7 ± 0.6) × 10^5	4.1 ± 0.5	−19 ± 2

[a] $k_{ET}(\Phi)$ and $k_{ET}(\tau)$ were calculated at 20 °C with eqs 3 and 4, respectively. Each rate constant is the average of three independent determinations; errors were determined by propagating the errors in the fluorescence measurements.
[b] Activation parameters were determined from an Eyring plot of rate constants (derived from lifetime data) at six temperatures ranging from −7 to 41 °C. Rate constants at each temperature were measured in two independent runs, and averages were used to construct the Eyring plots.

agreement indicates that the assumptions used in deriving the kinetic scheme are valid (compare $k_{ET}(\Phi)$ and $k_{ET}(\tau)$ in Table III). However, because of the lower precision in the quantum-yield data, the $k_{ET}(\tau)$ values are used in the following analysis. Comparison of the data for **0–2** demonstrates that the ET rate, which decreases by a factor of 10–20 with each added residue, is strongly dependent on the number of proline spacers.

In the semiclassical treatment of ET, the rate constant for ET is given by

$$k_{ET} = \nu_n \kappa_{el} \kappa_n \qquad (5)$$

where ν_n is a nuclear-vibration frequency, κ_{el} is the electronic transmission coefficient, and κ_n is the nuclear transmission factor (67). This equation suggests that the electron-transfer rate constant can be separated into two terms: one that describes the electronic interaction between the donor and acceptor (κ_{el}) and one that contains the dependence of the rate on nuclear reorganization and free-energy effects (κ_n, the Franck–Condon terms). An expression that is commonly used to describe the distance dependence of nonadiabatic ET is

$$k(r) = k_0 \exp\{-\beta(r - r_0)\} \qquad (6)$$

where k_0 is the rate when the donor (D) and acceptor (A) are at the van der Waals contact distance, r_0 is the sum of the radii of D and A, and r is the center-to-center separation distance between D and A (47, 68). Comparison of these equations shows that $k_0 \sim \nu_n \kappa_n$ and suggests that the exponential term in eq 6 describes the distance dependence of κ_{el}. In other words, eq 6 implies that the decay of rate with distance arises only from a decrease in the electronic coupling between D and A with increased separation.

Figure 3a shows a plot of the 20 °C rate data according to the treatment suggested by eq 6 for **0–2**, where r is taken as the distance from the Re center to C-1 of the DMAB ring. A least squares analysis of the data yields a value of $\beta = 0.96$ Å^{-1} for the slope of the rate plot. It is of interest to compare this distance dependence with that observed for other systems that have been examined to date. A survey of experimental studies concerning the distance dependence of ET in proteins and in synthetic systems indicates that β ranges from 0.8 to 2.0 Å^{-1} (20, 21, 26, 35, 47–50, 69–71). Despite the wide variability, most of the data tend to cluster in the range 0.8–1.2 Å^{-1}. Some specific examples include:

1. ET from an aromatic hydrocarbon donor to organic acceptors separated by rigid organic glasses ($\beta = 1.2$ Å^{-1}) (71).

2. ET from an aromatic hydrocarbon donor to an aromatic hydrocarbon acceptor across a series of saturated cyclic organic spacers ($\beta = 1.0$ Å^{-1}) (70).

3. ET from a photoexcited dimethoxynaphthalene donor to a cyanoethylene acceptor across a series of saturated bicyclic organic spacers (β = 0.88 Å^{-1}) (*21*).

Distance Dependence. These examples show that the distance dependence for DMAB to Re ET across the oligoproline spacers is very similar to that observed for ET across saturated organic spacers. Barring complications involved with the interpretation of β values, this result suggests that the ability of the peptide to mediate electronic coupling between a donor and an acceptor is not very different from that of a saturated hydrocarbon. This finding is consistent with the predictions of theoretical models that have been used to describe the electronic structure of peptides. The models suggest that ET (electronic coupling) along the peptide backbone via unfilled

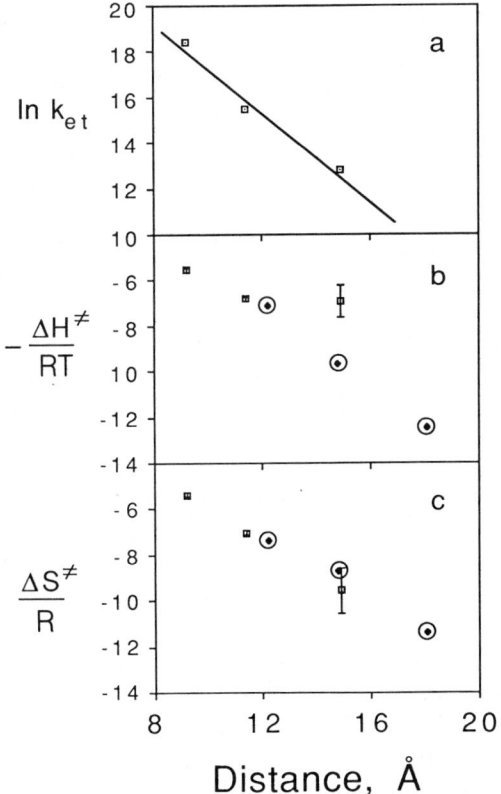

Figure 3. (a) Plot of ln k_{ET} vs. distance for **0, 1,** and **2**; (b) plot of $-\Delta H^{\ddagger}/RT$ vs. distance; (c) plot of $\Delta S^{\ddagger}/R$ vs. distance. Squares with error bars represent data for **0, 1** and **2**; circled polygons represent data for Os(II) → Ru(III) ET (data from ref. 48). Distance values for **0–2** are estimated as described in the text.

(π^*) levels is improbable because of the comparatively high energy of the orbitals involved (72). Thus, in the peptide-bridged systems, ET must occur by paths that involve (electronic coupling via) σ orbitals similar to those of saturated-hydrocarbon spacers.

The data presented herein, along with data from work of Isied, allow comparison of the distance dependence of the rate for ET in three different oligoproline-bridged systems. Isied studied thermally activated ET in two-metal dimer systems (49, 50)

$$(NH_3)_5M^{2+}-iso-(L-pro)_n-M^{3+}(NH_3)_5 \rightarrow$$

$$(NH_3)_5M^{3+}-iso-(L-pro)_n-M^{2+}(NH_3)_5 \quad (7)$$

where iso is isonicotinic acid and pro is proline. For Os(II) → Co(III) ET ($\Delta G < -0.1$ V), $\beta \sim 2.0$ Å^{-1}; for Os(II) → Ru(III) ($\Delta G \sim -0.15$ V), $\beta \sim 1.5$ Å^{-1} (49, 50). Remarkably, comparison of these data and the data on the Re–DMAB system show that for ET across the same oligoproline spacer, β values range from 1.0 to 2.0 Å^{-1}. This range would be reflected by a rate difference of approximately 10^4 at a separation distance of 10 Å. These are not subtle differences!

The large disparity may arise from one or more of several factors that may differ in the systems. The decay of the electronic coupling term (κ_{el}) with distance may be dependent on the molecular and electronic structure of the donor and acceptor. Several theoretical studies using superexchange theory support this hypothesis (73, 74). The treatment of distance-dependence data obtained by eq 5 implies that the only term that varies with distance is the electronic coupling (κ_{el}). However, the outer-sphere reorganization energy is expected to increase with distance (50, 68, 75); this increase would lead to a decrease of the Franck–Condon term (κ_n), in addition to the expected decrease in the electronic coupling term (κ_{el}). The size of the increase in the outer-sphere reorganization energy with distance may depend on at least two factors that are different in the oligoproline systems studied to date: the size of the donor and acceptor ions and the polarity of the solvent (50).

Temperature Dependence. One way to test the possibility that both κ_{el} and κ_n vary with distance would be to analyze the rate data in a manner that allows separation of the distance dependence of the two terms. For an ET reaction in which the entropy change (ΔS_{ET}) is zero, the activation parameters obtained from temperature-dependence studies allow separation of nuclear and electronic factors as follows (50, 76):

$$\ln \kappa_n \propto \frac{-\Delta H^\ddagger}{RT} \qquad \ln \kappa_{el} \propto \frac{\Delta S^\ddagger}{R} \quad (8)$$

where $\Delta H^{\ddagger}$ is activation enthalpy, $\Delta S^{\ddagger}$ is activation entropy, R is the gas constant, and T is temperature.

To provide the data necessary for analysis as suggested by eq 8, the temperature dependence of k_{ET} in methanol has been determined from the MLCT luminescence lifetime of complexes **0-m**, **1-m**, **2-m**, and **0–2** at six temperatures ranging from −7 to 40 °C. ET rate constants at each temperature were determined by using eq 4. The entire temperature-dependence experiment was repeated two times for each complex, and the set of average rate constants thus obtained were used to construct Eyring plots. $\Delta H^{\ddagger}$ and $\Delta S^{\ddagger}$ values were determined from the Eyring plots, and the values are listed in Table III.

Activation Entropy and Enthalpy. The data in Table III show that for **0–2** the activation entropy is less than zero, and it becomes increasingly negative as peptide spacers are added. Negative $\Delta S^{\ddagger}$ has been observed for ET reactions in a number of model systems and for intramolecular ET in modified protein systems. In these cases, negative $\Delta S^{\ddagger}$ values have been attributed to a nonadiabatic ET mechanism arising from separated donor and acceptor sites with weak electronic coupling between them (*40, 41, 49, 50, 77*). The observation of negative $\Delta S^{\ddagger}$ for ET in **0–2** provides evidence that ET is nonadiabatic. This observation supports the premise that the proline peptide spacers are rigid and hold the Re and DMAB centers apart during the ET process. Intramolecular ET is forced to occur by a through-bond (or possibly through-space) mechanism across the peptide spacer. On the other hand, no clear trend emerges from the activation enthalpy data; this term is nearly constant within experimental error across the series. A systematic increase in $\Delta H^{\ddagger}$ with added spacers is expected because of the dependence of the outer-sphere reorganization energy on distance (*67, 68*). The reason for the lack of a systematic variation of $\Delta H^{\ddagger}$ with distance in the Re–DMAB system is unclear at present.

It is of interest to carry out an analysis of the activation parameter data as suggested by eq 8 to separate the distance dependence of the nuclear and electronic terms. Experimental data are not available concerning ΔS_{ET} for the excited-state ET reaction given in eq 1. However, on the basis of the following reasoning, we suggest that the entropy change is small or nearly zero for the Re–DMAB system. The largest contribution to the overall entropy change for the DMAB → Re ET reaction is expected to be associated with solvent reorganization around the donor and acceptor that is driven by charge transfer. Because the photoinduced ET process involves the overall charge rearrangement Re^{+}–DMAB → Re–DMAB^{+}, changes in entropy associated with solvent reorganization around the Re center concurrent with its reduction are expected to be closely balanced by the entropy associated with solvent reorganization around DMAB concurrent with its oxidation. In addition, the overall entropy changes associated with $M^+ \to M^0$ in bipyridine complexes and for $D \to D^+$ in organic donors have been shown to be comparatively small (78, 79).

In Figure 3b and 3c, $-\Delta H^\ddagger/RT$ and $\Delta S^\ddagger/R$ are plotted as a function of distance for **0–2**. The activation enthalpy plot shows no systematic variation, although the activation entropy plot shows a pronounced negative slope. The appearance of the plots suggests that the distance dependence of k_{ET} arises mainly from variation of κ_{el} (50). An interesting correlation arises upon comparison of the activation parameter data for photoinduced ET in **0–2** and for thermal Os(II) → Ru(III) ET in $(NH_3)_5Os^{II}$-iso-(L-pro)$_n$-Ru$^{III}(NH_3)_5$ (n is 1, 2, and 3). The data points for $\Delta H^\ddagger$ and $\Delta S^\ddagger$ in the Os–Ru system are included in the correlations shown in Figures 3b and 3c (50).

Comparison of the plots shows that the slope of the $\Delta H^\ddagger$ plot for the Os–Ru system is considerably greater than the apparent slope for the Re–DMAB system. A possible explanation for the difference in slope is that the distance dependence of the outer-sphere reorganization energy is greater for the Os–Ru system than for **0–2**. This distance dependence could result from the fact that the experiments were carried out in H_2O for the Os–Ru system and in CH_3OH for **0–2**. By contrast, the $\Delta S^\ddagger/R$ points for the Os–Ru system and for **0–2** follow the same correlation. This remarkable correspondence suggests that the distance dependence of the electronic transmission term is identical for the two systems. Such correspondence is not surprising in light of the fact that the molecular structure of the spacer in the two systems is identical. The overall distance dependence of the ET rate is quite different for the two systems. The temperature-dependence data suggest that this difference arises mainly from a difference in the distance dependence of the nuclear terms in the two systems.

Studies in CH_3CN Solution. The luminescence decays for **0-m**, **1-m**, and **0–2** were monitored in CH_3CN solution. Lifetimes obtained from analysis of the experimental decays are listed in Table IV. Decays for **0-m**, **1-m**, **0**, and **1** were fitted adequately with a single exponential decay function. Rate constants for intramolecular ET in each complex were calculated with eq 4 by using **0-m** and **1-m** as the models for **0** and **1**, respectively. In contrast to the situation for **0** and **1**, the luminescence decay for **2** was biphasic. In accord with this condition, a biexponential equation was used to fit the data.

$$I(t) = \alpha_1 \exp\frac{-t}{\tau_1} + \alpha_2 \exp\frac{-t}{\tau_2} \quad (9)$$

In this expression, $I(t)$ is the instantaneous emission intensity, t is time, and τ_i and α_i are the lifetime and amplitude, respectively, of the ith decay component. Table IV contains the parameters obtained from the fit of the decay for **2** by using this equation. The decay displays two clearly resolved

Table IV. Emission Lifetimes and Electron-Transfer Rate Constants in CH_3CN Solution

Complex	τ, ns (%)[a]	k_{ET}, s^{-1}[a]
0-m	188	—
0	10.6	8.9×10^7
1-m	143	—
1	57.5	1.0×10^7[b]
2	16.3 (51%)	5.4×10^7
	98.9 (49%)	3.1×10^6[b]

[a] Argon outgassed solutions were at 20 °C. The estimated errors in lifetimes and rate constants are ±1% and ±5%, respectively.
[b] These rate constants were calculated by using eq. 4 and the corresponding lifetime component. The model complex used for the calculation was 1-m.

components with roughly equal amplitudes. By using the lifetimes of the two components and the lifetime of the model complex (1-m), two ET rate constants were calculated with eq 4.

An interesting correlation is noted between the luminescence data for 2 and the carbon NMR spectroscopic data presented in Figure 2b. For 2, two sets of resonances resolved in the CD_3CN NMR spectrum were assigned to species in which the proline–proline bond was in the *trans* and *cis* conformations. Molecular models of 2 indicate that the Re–DMAB distance is strongly dependent on the conformation of the proline–proline peptide bond. In the *cis* conformer (*cis*-2), the closest Re-to-DMAB separation distance is less than 10 Å; in the *trans* conformer (*trans*-2), this distance is estimated to be greater than 14 Å. Perhaps the two lifetime components observed in the luminescence decay correspond to these two conformers. The short and long lifetime components could be due to *cis*-2 and *trans*-2, respectively.

This interpretation of the biphasic decay for 2 can be used in interpreting the ET rate data obtained for the Re–DMAB complexes in CH_3CN. The ET rate decreases systematically along the series 0, 1, and *trans*-2. However, the ET rate for *cis*-2 is faster than for 1. The strong dependence of ET rate on conformation for 2 suggests that ET may occur via a through-space mechanism, at least in the case of *cis*-2. Further insight concerning this hypothesis may come from temperature-dependence experiments on 2 under conditions in which the decay for each conformation can be resolved.

Conclusions

The Re–DMAB system provides interesting information concerning the distance dependence of photoinduced ET. The temperature-dependent rate data in CH_3OH solution suggests that ET across the peptide spacers is strongly nonadiabatic. This situation is consistent with the hypothesis that the oligoproline spacers are rigid and that the Re center and the DMAB moiety are not in close contact during ET. The activation entropy data for 0–2 correlate remarkably well with data on thermal ET in $(NH_3)_5Os^{II}$-iso-$(L\text{-pro})_n$-$Ru^{III}(NH_3)_5$. This preliminary evidence indicates that the distance dependence of the electronic transmission factor for thermal and photochemical ET is similar in these two proline-bridged systems. Studies of 2 with two slowly interconverting conformers present indicate that the ET rate is strongly dependent on conformation and suggest the possibility that ET occurs via a through-space mechanism.

Work in progress seeks to vary the driving force for ET by placing substituents on the 2,2'-bipyridine ligand at the Re center. Efforts are also underway to better define the structure of the oligoproline spacers in 1 and 2 by using molecular mechanics calculations and X-ray crystallography.

Acknowledgments

We thank Russ Schmehl for conducting the transient absorption experiments at Tulane University. Acknowledgment is made to the Donors of The Petroleum Research Fund, administered by the American Chemical Society, for support of this research.

References

1. Rehm, D.; Weller, A. *Israel J. Chem.* **1970**, *8*, 259.
2. Balzani, V.; Moggi, L.; Manfrin, M. F.; Bolletta, F.; Lawrence, G. S. *Coord. Chem. Rev.* **1975**, *15*, 321.
3. Whitten, D. G. *Acc. Chem. Res.* **1980**, *13*, 83.
4. Bock, C. R.; Connor, J. A.; Guitierrez, A. R.; Meyer, T. J.; Whitten, D. G.; Sullivan, B. P.; Nagle, J. K. *J. Am. Chem. Soc.* **1979**, *101*, 4815.
5. Meyer, T. J. *Acc. Chem. Res.* **1978**, *11*, 94.
6. Meyer, T. J. *Prog. Inorg. Chem.* **1983**, *30*, 389.
7. Sutin, N. *Acc. Chem. Res.* **1982**, *15*, 275.
8. Newton, M. D.; Sutin, N. *Ann. Rev. Phys. Chem.* **1984**, *35*, 437.
9. Mataga, N. In *The Exciplex*; Gordon, M.; Ware, W. R., Eds.; Academic: New York, 1975.
10. Siemiarczuk, A.; McIntosh, A. R.; Ho, T.-F.; Stillman, M. J.; Roach, K. J.; Weedon, A. C.; Bolton, J. R.; Connolly, J. C. *J. Am. Chem. Soc.* **1983**, *105*, 7224.
11. Schmidt, J. A.; McIntosh, A. R.; Weedon, A. C.; Bolton, J. R.; Connolly, J. S.; Hurley, J. K.; Wasielewski, M. R. *J. Am. Chem. Soc.* **1988**, *110*, 1773.
12. Wasielewski, M. R.; Niemczyk, M. P. *J. Am. Chem. Soc.* **1984**, *106*, 5043.
13. Wasielewski, M. R.; Niemczyk, M. P.; Svec, W. A.; Pewitt, E. B. *J. Am. Chem. Soc.* **1985**, *107*, 1080.
14. Wasielewski, M. R.; Johnson, D. G.; Svec, W. A.; Kersey, K. M.; Minsek, D. W. *J. Am. Chem. Soc.* **1988**, *110*, 7219.
15. Wasielewski, M. R.; Niemczyk, M. P.; Svec, W. A.; Pewitt, E. B. *J. Am. Chem. Soc.* **1985**, *107*, 5562.
16. Gust, D.; Moore, T. A.; Liddell, P. A.; Nemeth, G. A.; Makings, L. R.; Moore,

A. L.; Barrett, D.; Pessiki, P. J.; Bensasson, R. V.; Rougee, M.; Chachaty, C.; DeSchryver, F. C.; Van der Auweraer, M.; Holzwarth, A. R.; Connolly, J. S. *J. Am. Chem. Soc.* **1987**, *109*, 846.
17. Gust, D.; Moore, T. A.; Moore, A. L.; Barrett, D.; Harding, L. O.; Makings, L. R.; Liddell, P. A.; DeSchryver, F. C.; Van der Auweraer, M.; Bensasson, R. V.; Rougee, M. *J. Am. Chem. Soc.* **1988**, *110*, 321.
18. Gust, D.; Moore, T. A.; Moore, A. L.; Makings, L. R.; Seely, G. R.; Ma, X.; Trier, T. T.; Gao, F. *J. Am. Chem. Soc.* **1988**, *110*, 7567.
19. Leland, B. A.; Joran, A. D.; Felker, P. M.; Hopfield, J. J.; Zewail, A. H.; Dervan, P. B. *J. Phys. Chem.* **1985**, *89*, 5571.
20. Oevering, H.; Paddon-Row, M. N.; Heppener, M.; Oliver, A. M.; Cotsaris, E.; Verhoeven, J. W.; Hush, N. S. *J. Am. Chem. Soc.* **1987**, *109*, 3258.
21. Paddon-Row, M. N.; Oliver, A. M.; Warman, J. M.; Smit, K. J.; de Haas, M. P.; Oevering, H.; Verhoeven, J. W. *J. Phys. Chem.* **1988**, *92*, 6958.
22. Hush, N. S. In *Supramolecular Photochemistry;* Balzani, V., Ed.; D. Reidel: Boston, 1987; p 53.
23. Cooley, L. F.; Headford, C. E. L.; Elliott, C. M.; Kelley, D. F. *J. Am. Chem. Soc.* **1988**, *110*, 6673.
24. Westmoreland, T. D.; Schanze, K. S.; Neveaux, P. E., Jr.; Danielson, E.; Sullivan, B. P.; Chen, P.; Meyer, T. J. *Inorg. Chem.* **1985**, *24*, 2597.
25. Chen, P.; Westmoreland, T. D.; Danielson, E.; Schanze, K. S.; Anthon, D.; Neveaux, P. E., Jr.; Meyer, T. J. *Inorg. Chem.* **1987**, *26*, 1116.
26. Finckh, P.; Heitele, H.; Volk, M.; Michel-Beyerle, M. E. *J. Phys. Chem.* **1988**, *92*, 6584.
27. Schanze, K. S.; Sauer, K. *J. Am. Chem. Soc.* **1988**, *110*, 1180.
28. Farraggi, M.; DeFillippis, M. R.; Klapper, M. H. *J. Am. Chem. Soc.* **1989**, *111*, 5141.
29. McLendon, G. L. *Acc. Chem. Res.* **1988**, *21*, 160.
30. Heiler, D.; McLendon, G. L.; Rogalskyj, P. *J. Am. Chem. Soc.* **1987**, *109*, 604.
31. *Chemical Applications of Ultrafast Spectroscopy;* Fleming, G. R., Ed.; Oxford University Press: New York, 1986.
32. Hilinski, E. F.; Rentzepis, P. M. *Anal. Chem.* **1983**, *55*, 1121A.
33. McGourty, J. L.; Blough, N. V.; Hoffman, B. M. *J. Am. Chem. Soc.* **1983**, *105*, 4470.
34. Zemel, H.; Blough, N. V.; Margoliash, E.; Hoffman, B. M. *Coord. Chem. Rev.* **1985**, *64*, 125.
35. Mayo, S. L.; Ellis, W. R.; Crutchley, R. J.; Gray, H. B. *Science (Washington, D.C.)* **1986**, *233*, 948.
36. Axup, A. W.; Albin, M.; Mayo, S. L.; Crutchley, R. J.; Gray, H. B. *J. Am. Chem. Soc.* **1988**, *110*, 435.
37. Elias, H.; Chou, M. H.; Winkler, J. R. *J. Am. Chem. Soc.* **1988**, *110*, 429.
38. Chen, P.; Danielson, E.; Meyer, T. J. *J. Phys. Chem.* **1988**, *92*, 3708.
39. *Antennas and Reaction Centers of Photosynthetic Bacteria;* Michel-Beyerle, M. E., Ed.; Springer-Verlag: New York, 1985.
40. Isied, S. S.; Kuehn, C.; Worosila, G. *J. Am. Chem. Soc.* **1984**, *106*, 1722.
41. Kostic, N. M.; Margalit, R.; Che, C.-M.; Gray, H. B. *J. Am. Chem. Soc.* **1983**, *105*, 7765.
42. McLendon, G.; Miller, J. R. *J. Am. Chem. Soc.* **1985**, *107*, 7811.
43. Cummins, D.; Gray, H. B. *J. Am. Chem. Soc.* **1977**, *99*, 5158.
44. Holwerda, R. A.; Knaff, D. B.; Gray, H. B.; Clemmer, J. D.; Crowley, R.; Smith, J. M.; Mauk, A. G. *J. Am. Chem. Soc.* **1980**, *102*, 1142.
45. *Tunneling in Biological Systems;* Chance, B.; DeVault, D.; Frauenfelder, H.; Marcus, R. A.; Schrieffer, J. R.; Sutin, N., Eds.; Academic: New York, 1979.

46. Peterson-Kennedy, S. E.; McGourty, J. L.; Hoffman, B. M. *J. Am. Chem. Soc.* **1984**, *106*, 5010.
47. Marcus, R. A.; Sutin, N. *Biochim. Biophys. Acta* **1985**, *811*, 265.
48. Isied, S. S.; Vassilian, A. *J. Am. Chem. Soc.* **1984**, *106*, 1732.
49. Isied, S. S.; Vassilian, A.; Magnuson, R. H.; Schwartz, H. A. *J. Am. Chem. Soc.* **1985**, *107*, 7432.
50. Isied, S. S.; Vassilian, A.; Wishart, J. F.; Creutz, C.; Schwartz, H. A.; Sutin, N. *J. Am. Chem. Soc.* **1988**, *110*, 635.
51. Chao, Y.-Y.; Bersohn, R. *Biopolymers* **1978**, *17*, 2761.
52. Grathwohl, C.; Wuthrich, K. *Biopolymers* **1976**, *15*, 2025.
53. Grathwohl, C.; Wuthrich, K. *Biopolymers* **1976**, *15*, 2043.
54. Deber, C. M.; Bovey, F. A.; Carver, J. P.; Blout, E. R. *J. Am. Chem. Soc.* **1970**, *92*, 6191.
55. Stryer, L.; Haughland, R. P. *Proc. Natl. Acad. Sci. U.S.A.* **1967**, *58*, 719.
56. Gabor, G. *Biopolymers* **1968**, *6*, 809.
57. Caspar, J. V., Ph.D. Dissertation, University of North Carolina at Chapel Hill, 1982.
58. Reitz, G. A.; Demas, J. N.; DeGraff, B. A.; Stephens, E. M. *J. Am. Chem. Soc.* **1988**, *110*, 5051.
59. Gouterman, M. In *The Porphyrins;* Dolphin, D., Ed.; Academic: New York, 1978; vol 3, p 1.
60. O'Connor, D. V.; Phillips, D. *Time-Correlated Single Photon Counting;* Academic: New York, 1984.
61. Caspar, J. V.; Westmoreland, T. D.; Allen, G. H.; Bradley, D. G.; Meyer, T. J.; Woodruff, W. H. *J. Am. Chem. Soc.* **1984**, *106*, 3492.
62. *Alchemy II;* Tripos Associates: St. Louis, MO.
63. Chen, P.; Curry, M.; Meyer, T. J. *Inorg. Chem.* **1989**, *28*, 2271.
64. Burge, R. E.; Harrison, P. M.; McGavin, S. *Acta Crystallogr.* **1962**, *15*, 914.
65. Rettig, W.; Lippert, E. *J. Mol. Struc.* **1980**, *61*, 17.
66. Rettig, W.; Wermuth, G. *J. Photochem.* **1985**, *28*, 351.
67. Sutin, N. *Prog. Inorg. Chem.* **1983**, *30*, 441.
68. Hush, N. S. *Coord. Chem. Rev.* **1985**, *64*, 135.
69. Guarr, T.; McLendon, G. *Coord. Chem. Rev.* **1985**, *68*, 1.
70. Closs, G. L.; Calcaterra, L. T.; Green, N. J.; Penfield, K. W.; Miller, J. R. *J. Phys. Chem.* **1986**, *90*, 3673.
71. Miller, J. R.; Beitz, J. V.; Huddleston, R. K. *J. Am. Chem. Soc.* **1984**, *106*, 5051.
72. Pethig, R. *Dielectric and Electronic Properties of Biological Materials;* Wiley: New York, 1979.
73. Beratan, D.; Hopfield, J. J. *J. Am. Chem. Soc.* **1984**, *106*, 1584.
74. Beratan, D. *J. Am. Chem. Soc.* **1986**, *108*, 4321.
75. Lee, G. H.; Della Ciana, L.; Haim, A. *J. Am. Chem. Soc.* **1989**, *111*, 2536.
76. Sutin, N. In *Supramolecular Photochemistry;* Balzani, V., Ed.; D. Reidel: Boston, 1987; p 73.
77. Taube, H. In ref. 45; Chance, B. et al., Eds.; p. 173.
78. Hupp, J. T.; Weaver, M. J. *Inorg. Chem.* **1984**, *23*, 3639.
79. Jaworski, J. S. *J. Electroanal. Chem.* **1989**, *23*, 3639.

RECEIVED for review May 1, 1989. ACCEPTED revised manuscript September 11, 1989.

6

Energetics and Dynamics of Gated Reactions

Control of Observed Rates by Conformational Interconversion

Brian M. Hoffman, Mark A. Ratner, and Sten A. Wallin

Department of Chemistry, Northwestern University, Evanston, IL 60208

> *Several unusual aspects of reactions in larger molecules, especially biological systems in the condensed phase, are discussed. In particular, we are concerned with processes involving "soft" (overdamped, low-frequency) degrees of freedom, such as might be associated with a conformational interconversion in a macromolecule. The role of subsidiary stable minima on the reaction surface is considered; subsidiary stable minima can provide an indirect transfer route whose activation barriers can be substantially smaller than the direct, concerted reaction path from precursor to successor. When these considerations are applied to a cyclic electron-transfer scheme, "gating" behavior, in which conformational changes control the overall process, emerges naturally. This chapter also considers quite a different dynamic effect, where anisotropy in the diffusion of the system on the potential energy surface substantially changes the dynamic behavior and can lead to transient decays that cannot be fit to a single exponential law.*

CONVENTIONAL TRANSITION-STATE THEORY MUST BE EXTENDED when a reacting system has certain key features associated with its potential-energy surface. These theoretical considerations were initiated (1, 2) to explore what consequences might arise when an electron-transfer system exhibits multiple stable conformational states that can interconvert at rates competitive with electron transfer (3–5). As noted earlier (1), this situation is general and by no means limited to electron-transfer reactions (6–8).

The following section describes a cyclic kinetic scheme appropriate for conventional intramolecular electron transfer (ET), where reactant and product each exhibit a single stable conformation and conformational dynamics are not important. After that the role of conformational interconversion is discussed, and the kinetic scheme that applies to a system in which electron transfer is coupled to the conversion between two stable conformations is developed (1). Then these equations are solved, and their most interesting limiting forms are analyzed.

Finally, the alternate situation (2) is described, in which conformational dynamics qualitatively change the observed kinetics of a reaction. The reacting system that is considered undergoes anisotropic diffusion on a two-dimensional energy surface, one dimension of which can be considered a conformational coordinate. This case corresponds to a generalization of Kramers' description of chemical reactions in the strongly coupled regime and leads to nonexponential relaxations, even for surfaces with one stable conformation for each of the reactant and product.

Conventional Intramolecular Electron Transfer

A general kinetic scheme for simple intramolecular electron-transfer reactions is given in Figure 1. A system consisting of an intramolecular (linked) donor–acceptor [d,a] pair in its ground state, A, is excited to a reactive state, A*. Most often this is done by flash photolytic excitation of d. Electron

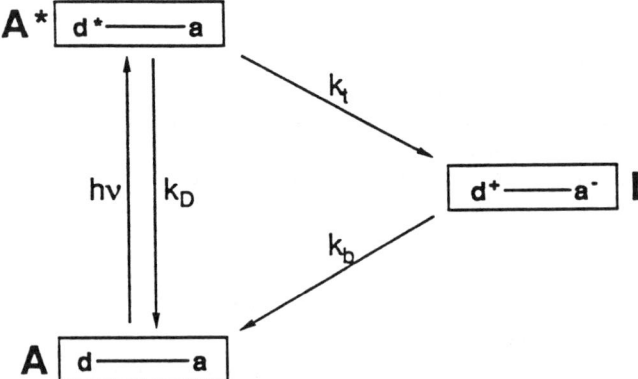

Figure 1. Simple kinetic scheme for electron transfer within a linked donor–acceptor pair. State A represents the ground-state [d,a] pair. A* is the state formed at t = 0 by excitation to the donor excited state (d*,a). State A* can decay back to A (rate constant, k_D) or can react to form the charge-transfer intermediate, I (rate constant, k_t), composed of the (d$^+$,a$^-$) pair. In turn, reverse charge transfer re-forms A from I (rate constant, k_b).

transfer, **d*** → **a** (or **d** → **a***), then produces the charge-separated intermediate, I = [**d**$^+$,**a**$^-$], at rate constant, k_t. This intermediate can return to the ground state by the reverse electron-transfer process, **a**$^-$ → **d**$^+$, with rate constant, k_b. Within this scheme, the time course of the species A* and I are as follows (9):

$$A^*(t) = A^*(0)e^{-k_p t} \tag{1a}$$

$$I(t) = A^*(0) \left(\frac{k_t}{k_b - k_p}\right)(e^{-k_p t} - e^{-k_b t}) \tag{1b}$$

where A*(0) is the initial concentration of [**d***,**a**]. According to eq 1a, A* decays exponentially with the rate constant,

$$k_p = k_D + k_t \tag{1c}$$

where k_D is the decay rate of A* in the absence of ET. Thus, k_t can be determined from k_p by independent measurement of k_D.

Equation 1b represents the creation of I from A* at a rate constant k_t and its reaction to form A at rate constant k_b. This equation corresponds to an exponential rise and fall of I, with its concentration maximum at time τ: I(τ) = A*(0)($k_t/k_>$) exp(–$k_<$τ); τ = (ln ($k_>/k_<$))/($k_> - k_<$), where $k_>$ and $k_<$ refer to the larger and smaller of k_b and k_t. If $k_b < k_p$, then k_b can be determined from the slow decay of intermediate I; if $k_b > k_p$, then k_b is obtained by measuring the rapid appearance of I.

The first portion of this scheme also applies to reactions where A* represents a state in which a rapid bimolecular event has reduced **d** or oxidized **a**. Such preparation can be done by pulse radiolysis or by flash photolysis of a sacrificial ET reagent (10); a subsequent intramolecular charge-transfer step produces I as the end product. These two cases differ from the flash photolytic excitation of **d** (or **a**) in that $k_D = 0$.

Introduction of Conformational Variation

An electron-transfer event is controlled by the Franck–Condon principle, and thus is highly sensitive to the accompanying nuclear rearrangements. In the standard theory of Marcus and Hush (11, 12), the coupled nuclear modes of motion are divided into inner- and outer-sphere modes. Inner-sphere modes include bond-length changes in the redox centers themselves, and contribute a reorganization energy, λ_i; the surrounding medium is treated as a continuous dielectric, with reorganization energy λ_o, and the total reorganization energy is $\lambda = \lambda_o + \lambda_i$. A potential-energy diagram depicting this situation is presented in Figure 2. In most cases, and particularly for small inorganic ions reacting near room temperature, inner- and outer-sphere reorganizations are treated on equal footing.

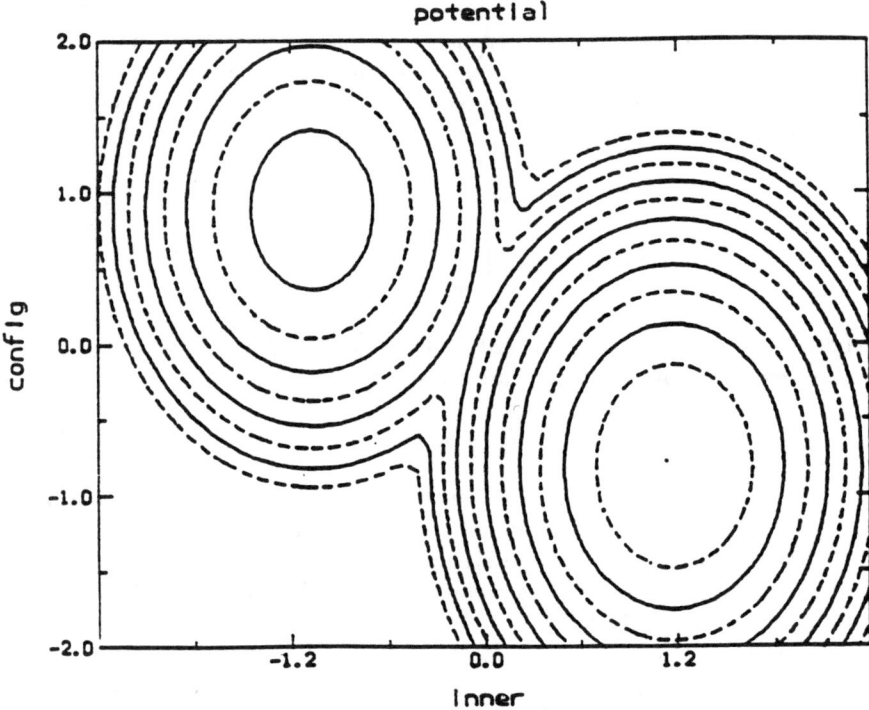

Figure 2. Schematic potential-energy surface for ordinary electron transfer without conformational state change. The abscissa is the inner-sphere vibrational coordinate, the ordinate is either the outer-sphere coordinate or an internal low-frequency motion. The reaction proceeds over a saddle point located on a line joining precursor minimum (upper left) to successor minimum (lower right). (Reproduced from ref. 1. Copyright 1987 American Chemical Society.)

This formalism assumes that, in an ET process such as the forward-transfer (A* → I) and back-transfer (I → A) reactions, there is but a single stable conformational form for each of the precursor (P) and successor (S) electron-transfer states (Figure 2). However, there must be many cases in which an "outer-sphere" coordinate is associated with a conformational mode that has two (or more) minima corresponding to alternative stable conformations (Figure 3) and the substates of the P and S species display a dynamic conformational equilibrium that can modulate the ET rates.

Moreover, major protein conformational changes can occur at rates that are competitive with observed rates of ET (13, 14). Such "gating" may occur in the complex between zinc cytochrome c peroxidase (ZnCcP) and cytochrome c (3, 9, 15–17) or for cytochrome c itself (18). More generally, it has been proposed that a variety of other chemical reactions are conformationally

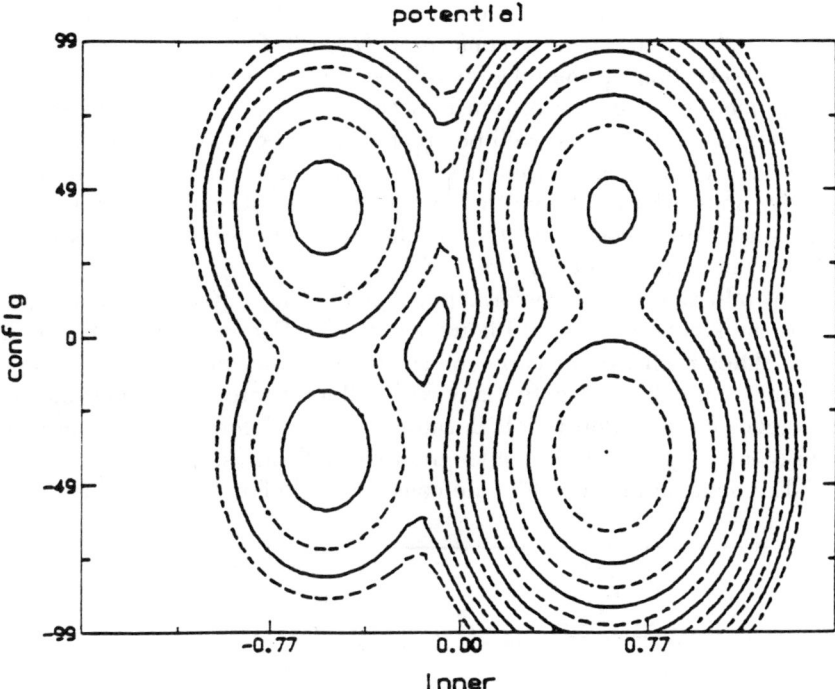

Figure 3. Schematic potential-energy surface for electron transfer with conformational equilibrium. The ordinate is a conformational coordinate, and the surface describes processes such as the A* → I transfer described in the text and in Figure 4. Transfer from the conformationally favored precursor state (upper left) to the stable successor conformer (lower right) occurs by a two-step process (either across and then down or down and then across), rather than by the combined upper left to lower right path, which has a higher barrier. (Reproduced from ref. 1. Copyright 1987 American Chemical Society.)

gated (19–21). A two-state case, which represents the simplest model for any and all such reactions, was presented previously (1). This chapter gives a more general solution to the kinetic equations that arise.

Consider a donor–acceptor pair, [d,a], as part of a system that exhibits two conformations, (B) boat and (C) chair. For concreteness, we may imagine d and a as attached 1,4 on a cyclohexane ring; the same formal situation will be realized in far more interesting ways with protein complexes. In this case, each of the three system states shown in Figure 1, (A, A*, and I) is composed of two conformational substates (A = A_B + A_C; A* = A_B* + A_C*; I = I_B + I_C). In the illustration, the distance and the through-bond relationships that govern electron transfer between d and a will differ in the B and C conformations of the ring.

For nonadiabatic electron transfer, this variation would cause the matrix element that enters as a prefactor into the (nonadiabatic limit) rate expressions to differ in the B and C conformers (11, 12). Generally, the energetics of charge creation or annihilation will be coupled to conformation. In the illustration, this coupling obviously occurs, for the coulombic stabilization of a charge-separated state will be greater in the boat conformer, where d^+ and a^- are close, than in the chair. Moreover, the process of forming the reactive state A* can change the conformational energetics. By linkage relationships, this means that the equilibrium between conformers can be different in system states A, A*, and I.

Introduction of the conformational equilibria implicit in Figure 3 expands the kinetic scheme of Figure 1 to that of Figure 4. Here, the A, A*, and I states each exhibit two conformations of the linked donor–acceptor system (B and C), each potentially with different interconversion rates and energetics (equilibrium ratio). Because of these differences, the activation energy for ET is not the same for the two conformers.

This scheme includes ET rate constants only for the $d^* \to a$ and $a^- \to$

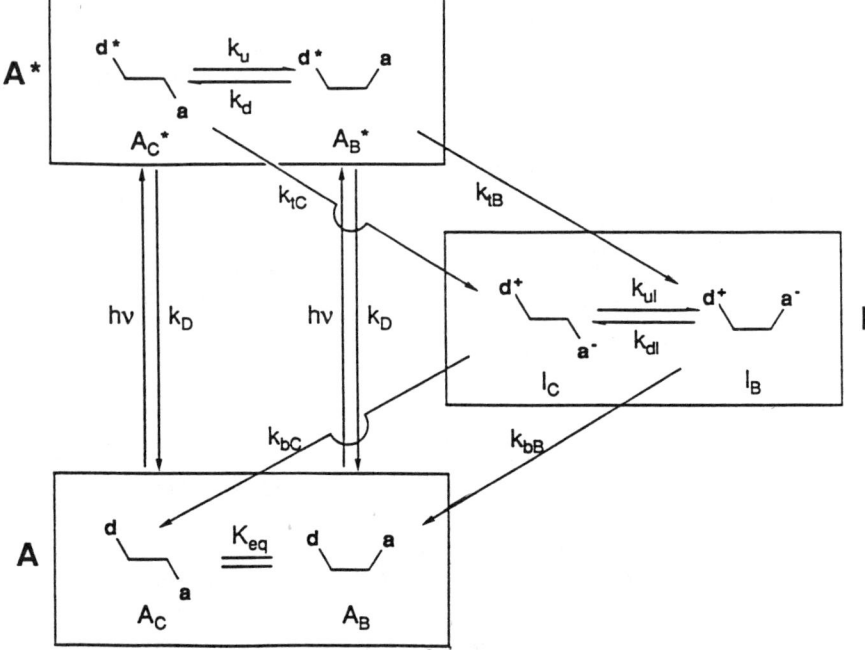

Figure 4. Kinetic scheme for electron transfer for a donor–acceptor couple within a system where the states A, A*, and I each undergo conversion between two stable conformers C and B. The symbols and rate constants are as shown in Figure 1, with the following additions. The conformational substates and associated electron-transfer rate constants are labeled by conformer; the conformational conversions are defined by the equilibrium constant in A (K_{eq}) and by rate constants in A* (k_u, k_d) and in I (k_{uI}, k_{dI}).

d^+ electron-transfer processes, in which the system conformation is conserved (the "horizontal" reactions in Figure 3), and conformational and ET steps only occur sequentially. Intuitively, it might be expected that a kinetic scheme for ET in the presence of an energetic coupling between charge state and conformation must include ET that is synchronous with a conformational change in the medium coordinate, namely the "diagonal" processes in Figure 3, $P_B \rightarrow S_C$, and $P_C \rightarrow S_B$. This in turn would open the issue of "friction" along the coupled diagonal path, particularly in protein systems where large molecular masses are involved and at low temperatures.

However, in the cases of interest here it is not necessary to include the synchronous process (e.g., $A_C^* \rightarrow I_B$) for reasons that can be discerned upon examining Figure 3. On a potential-energy surface in which the precursor, A* or I, and also the successor, I or A, exhibit two stable minima, the transition states on the diagonal reaction paths (e.g., P_C to S_B) lie at a higher energy than all other transition states, including those for the isoconformational ET reactions ($P_C \rightarrow S_C$ and $P_B \rightarrow S_B$) and those for the conformational conversions ($P_C \rightarrow P_B$ and $S_C \rightarrow S_B$) (Figure 3). There can be no compensation for this increased activation energy by the preexponential term when ET is adiabatic, nor is effective compensation likely in the nonadiabatic, long-range ET processes of interest here. Thus, in this case where the conversion is between stable conformers, the rates for the sequential conformational and ET processes are greater than those for the synchronous reactions, which can be omitted.

In some cases the conformational step will be slower than the ET step, and these deserve to be called gated reactions. This point perhaps represents the most interesting feature of our analysis and will be shown later to have profound consequences for efforts to ascertain the degree of conformational control in electron-transfer reactions.

Our original publication (1) speculated that the synchronous reaction always can be omitted. Brunschwig and Sutin (22) confirm that this condition is true for the case of concern in our original paper and here, where the minima are stable. They further show that it is not true in the limiting case where the second minimum is barely stable; this limitation is wholly unimportant in the present context.

Electron-Transfer Kinetics with Conformational Dynamics

We will discuss the kinetic scheme shown in Figure 4 in terms of the photoinitiated ($d^* \rightarrow a$) electron-transfer and thermal ($d^+ \leq$- a^-) processes in a donor–acceptor pair coupled within a molecule or complex that exhibits two stable conformations. Reactions initiated by a second-order reduction of d are described by the first half of this scheme. The population of A* in the C conformation is defined as (A_C^*), in the B conformation as (A_B^*), and the total population of A* as $(A^*) \equiv (A_B^*) + (A_C^*)$. The population of the electron-transfer intermediate, I, and its conformational substates are defined analogously. In the ground state, A, the conformational equilibrium is described by the equilibrium constant, $(A_B/A_C)_{eq} \equiv K_{eq}$. However, coupling between electronic state and conformation might alter the conformational equilibrium in A* or I: $(A_B^*/A_C^*) \equiv K_{eq}^* = k_u/k_d$; $(I_B/I_C)_{eq} \equiv K_{eq}^I = k_{ul}/k_{dl}$.

An important feature of most experiments is that the detection methods employed do not distinguish between conformers. Thus, we require equations for the total populations, A* and I, following the formation of A* at $t = 0$ by flash excitation or second-order reduction (oxidation) of A. Generally, the $d^* \to a$ transfer process is experimentally characterized by following the loss of A*, the $d^+ \leq\text{-} a^-$ process by following the time course of I. We will examine the two processes in turn.

Reaction of Initial State: A* → I: Exact Solution. The kinetic schemes in Figures 3 and 4 correspond to the following kinetic equations for the A* populations:

$$\frac{d(A_C^*)}{dt} = -(k_{tC} + k_u + k_D)(A_C^*) + k_d(A_B^*) \tag{2a}$$

$$\frac{d(A_B^*)}{dt} = k_u(A_C^*) - (k_{tB} + k_d + k_D)(A_B^*) \tag{2b}$$

For the general case, the initial conditions for the A* state are determined by the equilibrium distribution of the A_C and A_B ground states: $[A_C^*(t = 0)] = (1 - F)A^*(0)$; $[A_B^*(t = 0)] = FA^*(0)$. Here $A^*(0) \equiv [A_C^*(t = 0)] + [A_B^*(t = 0)]$ and $F = k_u/(k_u + k_d) = K_{eq}/(1 + K_{eq})$. Solving eqs 2a and 2b for the general case via a Laplace–Carson transformation (23) yields

$$\frac{A^*(t)}{A^*(0)} = c_1 e^{-ft} + c_2 e^{-gt} \tag{3}$$

$$\frac{A_C^*(t)}{A^*(0)} = (1 - F)(c_{1C} e^{-ft} + c_{2C} e^{-gt}) \tag{4a}$$

$$\frac{A_B^*(t)}{A^*(0)} = F(c_{1B} e^{-ft} + c_{2B} e^{-gt}) \tag{4b}$$

with the definitions

$$(f, g) = 1/2[2k_D + k_d + k_u + k_{tB} + k_{tC} \mp \sqrt{(k_u + k_d)^2 + (k_{tB} - k_{tC})(k_{tB} - k_{tC} + 2k_d - 2k_u)}] \tag{5}$$

$$c_1 = \frac{k_x - f}{g - f} \tag{6a}$$

$$c_2 = \frac{g - k_x}{g - f} \tag{6b}$$

$$c_{1C} = \frac{k_{xC} - f}{g - f} \tag{7a}$$

$$c_{2C} = \frac{g - k_{xC}}{g - f} \tag{7b}$$

$$c_{1B} = \frac{k_{xB} - f}{g - f} \tag{7c}$$

$$c_{2B} = \frac{g - k_{xB}}{g - f} \tag{7d}$$

where

$$k_x = k_D + k_u + k_d + (1 - F)k_{tB} + Fk_{tC} \tag{8}$$

$$k_{xC} = k_D + k_{tB} + \frac{k_d}{1 - F} \tag{9a}$$

$$k_{xB} = k_D + k_{tC} + \frac{k_u}{F} \tag{9b}$$

In contrast to the exponential dependence of $A^*(t)$ for the simple ET scheme (Figure 2, eq 1a), the general form of the overall $A^*(t)$ population, eq 3, is biexponential, with the two rate constants being the sum (g) and difference (f) of two positive terms: $0 \leq f \leq g$. The long-time behavior of (A^*) is governed by the rate constant, f; the short-time behavior by the larger rate constant, g. In the following discussion, it will frequently be useful to consider situations where a single exponential term controls the behavior of $(A^*(t))$. The effective rate, k_{obs}, can be written as the sum of the decay-rate constant, k_D, and an effective electron-transfer term, k_{ET}.

$$k_{obs} = k_D + k_{ET} \tag{10}$$

There are several illuminating limiting cases of the general solution. A most interesting situation occurs when $K_{eq} \ll 1$ and thus the only state initially prepared is A_C^*. If A_C^* does not exhibit rapid ET, namely $k_{tC} \sim 0$, then ET only occurs subsequent to the $A_C^* \rightarrow A_B^*$ conversion. This is "gated" ET in the traditional sense that until the "gate" has been opened (that is, until A_B^* is formed from A_C^*), there is no conversion to product I. When the state, A_C^*, initially populated cannot undergo ET, there must be an initial lag period, with $k_{ET} = 0$ (eq 10); the early-time ($t = 0$) rate constant for decay of A*

$$k_{obs} = k_D \qquad (11a)$$

represents only the $A_C^* \rightarrow A_C$ return to ground state. Subsequently, k_{obs} increases with time as A_B^* is formed and undergoes the $A_B^* \rightarrow I_B$ reaction.

In this gating situation, the long-time behavior of eq 3 exhibits two special cases of physical interest. When the conformational equilibrium is fast compared to the ET rate, corresponding to gated ET with a fast gate, at long time, the $A_C^* \leftrightarrow A_B^*$ preequilibrium is established, giving the steady-state result,

$$k_{obs} \approx k_D + [k_{tB}(F) + k_{tC}(1 - F)] \qquad (11b)$$

Thus, the apparent electron-transfer rate k_{ET} is a population-weighted average. The opposite extreme arises when conformational conversion $A_C^* \rightarrow A_B^*$ is slow compared to the subsequent ET; this is the traditional gated ET with a slow gate. At long time, the $A_C^* \xrightarrow{k_u} A_B^*$ conformational step is rate-limiting. Each complex attaining the high-energy conformation of the precursor crosses with essentially unit efficiency to the product state I; the overall ET rate constant is just $k_{ET} = k_u$, and

$$k_{obs} = k_D + k_u \qquad (11c)$$

Reaction of ET Intermediate: I → A (Scheme II): Exact Solution. The ET intermediate state denoted I in Figure 4 is created from A* and simultaneously is being converted to A by ET. The two conformers of A* can undergo parallel reactions to the corresponding I conformers. In this section the term successor complex refers to A, and precursor complex is I.

The kinetic schemes in Figure 3 and Figure 4 yield the following kinetic equations for the conformers of I:

$$\frac{d(I_C)}{dt} = -(k_{ul} + k_{bC})(I_C) + k_{dl}(I_B) + k_{tC}(A_C^*) \qquad (12a)$$

$$\frac{d(I_B)}{dt} = k_{ul}(I_C) - (k_{dl} + k_{bB})(I_B) + k_{tB}(A_B^*) \qquad (12b)$$

Once again, an exact solution may be obtained via the Laplace–Carson transformation. The solution to the general case for the time course of $I(t)$ is a rather unwieldy four-exponential function that does not lend itself to a discussion of its salient points. (For the sake of completeness, this solution is given in the Appendix.) We will instead consider here an illuminating subcase of the general case, where I is formed only through I_B and where $k_u \gg k_d \gg k_D$ and $k_{tC} = 0$. Thus, the $A^* \to I$ process can be described by a single process, $A_B^* \to I_B$, and by a single exponential decay with a rate $k_{obs} = k_D + k_{tB}$. In effect, we ignore gating in A^* by considering only state A_B^* in order to see most clearly the effect of $I_B \leftrightarrow I_C$ on the $I \to A$ process. The kinetic equations for the conformers of I obtained for this case are

$$\frac{d(I_C)}{dt} = -(k_{ul} + k_{bC})(I_C) + k_{dl}(I_B) \tag{13a}$$

$$\frac{d(I_B)}{dt} = k_{ul}(I_C) - (k_{dl} + k_{bB})(I_B) + k_{tB}(A_B^*) \tag{13b}$$

By solving the coupled equation set (eqs 13a and 13b) for the boundary conditions $[I_C(t=0)] = [I_B(t=0)] = 0$ and by using

$$A_B^*(t) = A^*(0)e^{-k_{obs}t} \tag{14}$$

we find

$$\frac{I(t)}{A^*(0)} = \frac{k_{tB}}{n-m}\left[\frac{k_z - m}{k_{obs} - m}(e^{-mt} - e^{-k_{obs}t}) + \frac{n - k_z}{n - k_{obs}}(e^{-k_{obs}t} - e^{-nt})\right] \tag{15}$$

where

$$(m,n) = \frac{1}{2}[k_{bB} + k_{bC} + k_{dl} + k_{ul} \mp \sqrt{(k_{dl} + k_{ul})^2 + (k_{bB} - k_{bC})(k_{bB} - k_{bC} + 2k_{dl} - 2k_{ul})}] \tag{16}$$

and

$$k_z = k_{bC} + k_{ul} + k_{dl} \tag{17}$$

The time course of (I) is determined by a three-exponential function involving the $A^* \to I$ rate constant, k_{obs}, and the $I \to A$ rate constants, m and n. The rate constants m and n are composites only of the fundamental rate constants associated with state I; the fundamental rate constants describing interconversion and decay of the A^* substates do not appear in m and n.

To illustrate this exact description of the $I \to A$ process, we examined the situation in which gating within I has the maximal effect. I is populated only through the $A_B^* \to I_B$ channel, but the I_B conformer is unreactive, with $k_{bB} = 0$. This situation represents idealized conformational control of

ET because the $d^+ \leq\text{-} a^-$ ET process can occur only after an $I_B \rightarrow I_C$ conversion through the sequential process,

$$I_B \xrightarrow{k_{dI}} I_C \xrightarrow{k_{bC}} A$$

where the $I_B \rightarrow I_C$ step is the "gate". There are two interesting subcases. When the gating conformational change, $I_B \rightarrow I_C$, is slow compared to the rate of ET within the reactive B conformer ($k_{dI} \ll k_{bC}$),

$$\frac{I(t)}{A^*(0)} = \frac{k_{tB}}{k_{dI} - k_{obs}} (e^{-k_{obs}t} - e^{-k_{dI}t}) \tag{18}$$

When the gate is rapid compared to the subsequent ET step ($k_{bC} \ll k_{dI}$), then

$$\frac{I(t)}{A^*(0)} = \frac{k_{tB}}{k_{eff} - k_{obs}} (e^{-k_{obs}t} - e^{-k_{eff}t}) \tag{19}$$

where

$$k_{eff} = \frac{k_{bC}}{k_{bC} + k_{ul}} \tag{20}$$

Both of these subcases yield a biexponential functional form identical to that of the ungated kinetic scheme, eq 1b. One of the rate constants is k_{obs}, associated with creation of I; the other describes the ET reaction, I $\rightarrow$ A. When the gate is slow, the apparent rate constant of the ET reaction is not that for the ET step, but rather is k_{dI}, a conformational parameter. Conversely, when the conformational change is fast, eq 19 says that the $d^+ \leq a^-$ ET reaction within the I_C conformer is the rate-determining step for the I $\rightarrow$ A process.

Another interesting limit can be illustrated by again considering the situation when I is populated only through the $A_B^* \rightarrow I_B$ channel, but now the I_B conformer is reactive and $k_{bB} > k_{dI} \sim k_{obs} > k_{ul}, k_{bC}$. In this limit, if eq 15 can be written in the form

$$\frac{I(t)}{A^*(0)} = c_1 e^{-k_{obs}t} + c_2 e^{-mt} + c_3 e^{-nt} \tag{21}$$

where

$$-c_3 \approx c_2 > c_1 \tag{22}$$

This result implies that the intermediate appears with rate constant $n \sim k_{bB} + k_{dI}$ and decays with rate constant $m \sim k_{ul} + k_{bC}$.

Multimode Reaction Dynamics

In condensed-phase chemical reactions, including but not restricted to electron transfer, the relationship between the Born–Oppenheimer potential energy surfaces (or free-energy surfaces) and the rate processes occurring on those surfaces is nearly always discussed in terms of the activation energies (or activation free energies) involved in surmounting barriers along a reaction path leading from a reactant to a product configuration (11, 12, 24). At high temperatures the rate constant will then be given in a generalized transition-state theory form

$$k = k_o \kappa e^{-\Delta G^\ddagger/RT} \qquad (23)$$

where k is the rate constant, k_o is an attempt frequency for crossing the barrier, κ is a transmission coefficient describing recrossing of the barrier, and $\Delta G^\ddagger$ is the activation free energy. The dominant temperature dependence arises from the exponential term, and one generally expects a single-exponential time-course for the reactant concentration. The standard ET rate theory, in the adiabatic limit, is exactly of this form, with κk_o replaced by a universal frequency near 10^{13} s^{-1}, and $\Delta G^\ddagger = (\lambda + \Delta)^2/4\lambda$. For nonadiabatic ET, the prefactor κk_o becomes

$$\kappa k_o = \frac{2\pi V_{DA}^2}{\hbar(4\pi\lambda kT)^{1/2}} \qquad (24)$$

with V_{DA} as the electron tunneling matrix element. Formulas 23 and 24 would be appropriate for the individual electron-transfer rate constants (k_{tB}, k_{tC}, k_{bB}, and k_{bC}) that appear in the earlier discussion in which the simple transition-state theory was extended to include multiple stable minima.

We have outlined the modifications that must be made in the standard nonadiabatic rate expressions of eqs 23 and 24 when multiple stable minima occur on the potential surface. For other situations of real importance, dynamics on the potential-energy surface, rather than the surface itself, require extension of these formulas. The first of these dynamic extensions to receive extensive recent experimental and theoretical work arises from dynamic solvent relaxation effects (25–41); it is important for both large- and small-molecule dynamics, especially at short times. Solvent relaxation is important in several situations, most obviously when it is required to trap a successor complex that might otherwise convert back to precursor. A simplified picture (33) of these dynamic corrections replaces the standard form for nonadiabatic transfer

$$k_{ET} = k_{NA} \qquad (25)$$

by the corrected version

$$k_{ET} = \frac{k_{NA}}{1 + K} \tag{26}$$

where we have

$$k_{NA} = \frac{2\pi}{\hbar} \frac{V_{DA}^2}{(4\pi\lambda k_B T)^{1/2}} e^{-E_A/RT} \tag{27}$$

$$K = \frac{4\pi V_{ab}^2}{\lambda \hbar} \tau_l \tag{28}$$

$$\tau_l = \frac{\epsilon_\infty}{\epsilon_o} \tau_D \tag{29}$$

with τ_l, τ_D, ϵ_∞, and ϵ_o, respectively, the longitudinal and Debye relaxation times and the optical and static dielectric constants. For slow solvent relaxation,

$$k_{ET} \rightarrow \left[\frac{k_{NA}\lambda}{16\pi k_B T}\right]^{1/2} \exp\frac{-E_A}{RT} \cdot \frac{1}{\tau_l} \quad \begin{array}{l} K \ll 1 \\ K \gg 1 \end{array} \tag{30}$$

Thus, in this limit, the rate is simply the product of the activation energy to attain the barrier top and the rate of solvent relaxation required to stabilize the product configuration.

We treated a second set of circumstances in which the simple activated complex picture requires dynamical extension. In condensed phase, excited translational or vibrational states undergo frequent energy-relaxing collisions with the solvent. Under these conditions, the dynamic behavior of the system point moving on the potential surface will generally be diffusive, with mean free path short compared to characteristic distances along the potential surface. Kramers long ago studied the implications of this diffusive transport for chemical rate processes (*42*). In the past 15 years a large number of studies, both theoretical and experimental, have used generalized Kramers models to analyze rates for processes as different as isomerization, ionic conduction, and enzyme–substrate binding (*43*, *48*).

The original Kramers formulation, and the majority of its extensions, considered pictures in which motion occurs along a single reaction coordinate, and thus the diffusive dynamics of barrier crossing occurs in one dimension. Situations can frequently occur, however, in which the one-dimensional picture is inadequate. Agmon and Hopfield (*49*, *50*) noted that for reactions in proteins, it may be necessary to treat internal motion of the protein as defining the "medium" coordinate (Figure 2). When the protein is highly conformationally mobile, there are no new effects.

However, as motion in this coordinate is slowed, for example by a decrease in temperature, reorganization of the protein becomes difficult and finally impossible; the system no longer can reach the classical transition state (Figure 2), and the reaction rate decreases. Examples might include CO rebinding (21, 51), the case of ET with stable intermediate states, or isomerization about a double bond, in which extension of the bond vibration coordinate substantially changes the barrier height. Under such conditions, the Kramers picture should be extended to two coordinates, and just such an analysis of the isomerization problem has been provided by Agmon and Kosloff (52).

In many systems, however, particularly in proteins with slow conformational motions characterized by very large effective masses and small force constants, the characteristic times for the coordinates may be very different. Additionally, the different degrees of freedom may have totally different effective frictions, corresponding to differing diffusion constants along the differing motion directions. As a result, the evolution of the system on the potential surface can be governed not only by the potential surface itself (that is, by the forces arising from the gradient of the potential), but also from nonuniform frictional forces for differing coordinates.

A simple example might help to clarify this discussion. Figure 5A presents a schematic potential surface for CO rebinding to myoglobin. The abscissa corresponds to translational motion of the CO away from the Fe site, and the ordinate measures protein conformational motion. The conformational ("outer-sphere") motions will occur very slowly; the small-species ("inner-sphere") motions along the abscissa are far faster. Agmon and Hopfield (49, 50) therefore assumed that the effective reactive motion corresponds to straight-line motions along the x coordinate, starting from some distribution of probability (just a Boltzmann-type distribution) along the y axis. Physically, this condition will ideally correspond to a distribution of barrier heights, corresponding to crossing the ridge line R at various cuts along the y axis. This distribution, in turn, can mean that dynamics about the simple saddle point S of Figure 5A can become irrelevant for the overall rate because barrier crossing will occur all along the ridge line, not only at the saddle point itself. Then the overall rate will be given by an average of the rate for each straight-line trajectory parallel to x—that is, at each conformation, a rate of CO rebinding can be defined, and the overall observed rate will be an average over all accessible conformations.

The appealing and physically reasonable picture of Agmon and Hopfield represented an important step in interpreting reactions occurring in situations such as a protein where the characteristic frictions and motion time scales are substantially different along various coordinates. However, the straight-line trajectories parallel to the x axis, corresponding to fast-coordi-

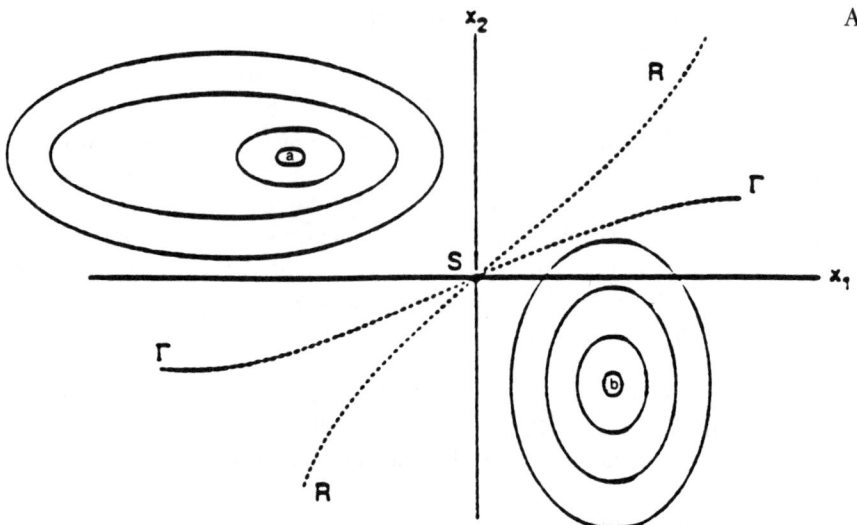

Figure 5. (A) *A schematic sketch of the potential surface for the ligand protein rebinding reaction.* R *denotes the ridge,* Γ *the separatrix, and* S *the saddle point. The stable reactant and product configurations are denoted by* a *and* b, *respectively. The shapes and orientations of* R *and* Γ *are schematic and other possibilities exist as discussed in the text.*

nate motion at fixed position along the slow coordinate, clearly represents a great over-simplification that cannot be generally valid. The issues raised, however, including the differences between reactions determined by ridge-line crossing as opposed to saddle-point crossing, the effects of very different frictions along the various motions, and possible nonexponential rate behavior arising from short-time transients on the product side of the ridge line but not yet stable at product-type geometries, merit extended consideration (53).

Recently, this problem was considered by using a Smoluchowski equation approach (2). The equations of motion for the system are then

$$\dot{x} = \frac{-1}{m\gamma_1} \frac{\partial V}{\partial x} + \sqrt{2k_B T/m\gamma_1}\ \dot{w}_1 \qquad (31a)$$

$$\dot{y} = \frac{-1}{m\gamma_2} \frac{\partial V}{\partial y} + \sqrt{2k_B T/m\gamma_2}\ \dot{w}_2 \qquad (31b)$$

with γ_1 and γ_2 the friction coefficients for x and y motions, respectively; and k_B, T, m, w_1, and w_2 are, respectively, Boltzmann constant, temperature, mass, and two independent white noise (random modulation) terms. This equation is expected to hold when the motion is strongly damped, so that

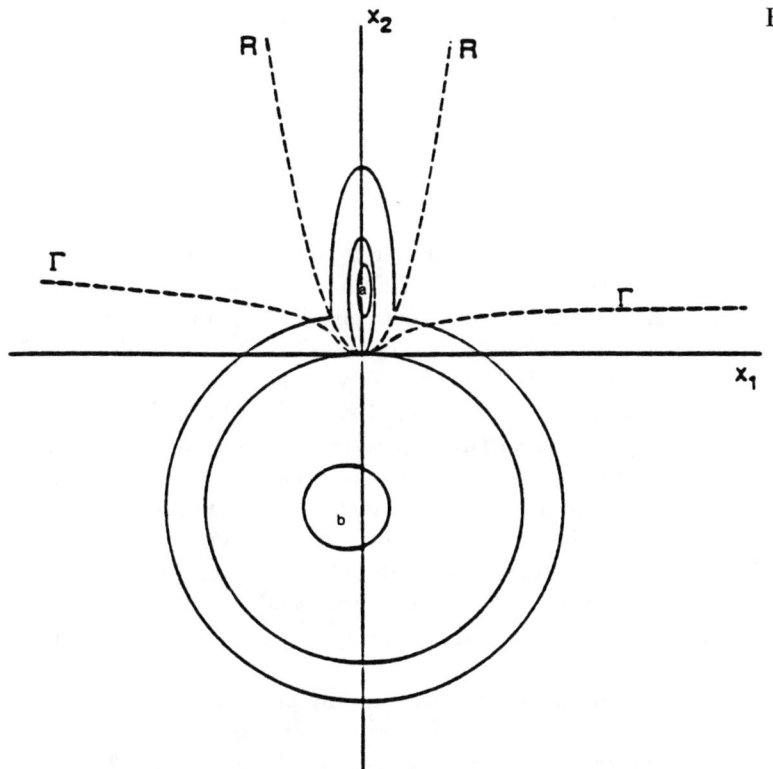

Figure 5. (B) A plot of the potential corresponding to the lower eigenvalue of the 2 × 2 vibronic coupling problem. The ratio of the diffusion coefficients is $\delta = 0.1$. The ridge and separatrix shown in the figure have been calculated for this ratio. Unlike the situation in the schematic picture (A), R and Γ touch (at the saddle point), but do not cross each other. (Reproduced with permission from ref. 2. Copyright 1989 American Institute of Physics.)

γ_1 and γ_2 exceed all important frequencies; this should be true in protein ET reactions.

Anisotropy in the motion along the potential surface can have two causes. The first cause is the potential $V(x,y)$ itself, whose structure can cause the dominant evolution to skirt the saddle point, following instead a route through stable intermediates, as already discussed. The second cause is anisotropic diffusion or friction that arises because the effective damping, or relaxation, is stronger along one direction (say y) than along the other. The latter case introduces anisotropic behavior on the rhs of eq 31, and results in faster motion along one coordinate than along the other, just as in the Agmon–Hopfield picture. The Agmon–Hopfield model, however, corresponds to the limit where there is no diffusive flow along the y coordinate (that is, to $\gamma_2/\gamma_1 \to \infty$), which is not physically justified. Solution of eq 31, then, extends the Agmon–Hopfield picture to yield the correct result for general anisotropic diffusion D_{xx}/D_{yy}, where D_{xx}, the diffusion coefficient along x, is given by

$$D_{xx} = \frac{k_B T}{m\gamma_1} \qquad (32)$$

The results of mathematical analysis of the Smoluchowski equations, 31a and 31b, are best understood in terms of Figures 5A and 5B, which show cases of small and large anisotropy, respectively. Figure 5A is a schematic representation of the CO-rebinding problem. Figure 5B is a plot of a 2 × 2 vibronic coupling (Jahn–Teller) problem with large anisotropy (D_{xx}/D_{yy} = 0.10). In both cases, the curve labeled R denotes the ridge line—that is, the line on the potential surface separating reactant and product regions. The curve labeled Γ is called the separatrix, and it is the set of points that separates dynamic motions corresponding to reactant and product. The ridge line is determined solely by the potential surface, $V(x,y)$, and is the curve that separates the potential wells of the chemically and physically distinct reactant and product (for instance, precursor and successor states in an ET process). The separatrix Γ is dynamically determined as the "line of no return".

Trajectories that pass over Γ irreversibly proceed to form product, but a trajectory that has not reached Γ may recross the ridge line (indeed, will recross it in long-time, anisotropic situations). If $D_{xx} = D_{yy}$, the separatrix and ridge become identical and a normal, one-dimensional reaction-path treatment is applicable. However, they can become very different if the diffusion coefficient is highly anisotropic, as the figures indicate, and in this case quite unusual kinetics are predicted.

When the ridge line and separatrix diverge, if the rate constant is measured by appearance of the system point near the bottom of the product well, then at very long times the reaction dynamics is exponential in time, with the activation energy fixed by the height of the saddle barrier, E_s. But short-time transients can appear, corresponding to the system crossing back and forth over the ridge line into the product region before irreversibly passing the separatrix Γ. An actual measurement of the reaction progress corresponds (as it nearly always will, given our definition of precursor and successor in terms of geometry) to an observable phenomenon that changes when the system crosses the ridge line (e.g., CO bound or unbound; charges separated or not). Thus, the kinetics observed will, in general, be multiexponential, because of the multiple crossings and recrossings of the ridge.

The Smoluchowski model for anisotropic diffusive reactions on a two-dimensional potential surface (eqs 31a and 31b) leads, as Klosek-Dygas et al. show (2), to a proper mathematical description of the physical notion expressed by Agmon and Hopfield (and by Grote and Hynes (45) and van der Zwan and Hynes (*46*)). The actual time dependence of the precursor concentration thus contains not only energetic information about the barrier at the saddle point (from the temperature dependence of its long-time asymptotic value), but also information on the anisotropy of the diffusion coefficient (from its shorter-time dynamics). Frauenfelder and coworkers have studied extensively the nonexponential transients in the CO rebinding situation (*21, 51*), but we are aware of no such in-depth experimental study of transient nonexponential behavior in ET reactions.

Discussion

This chapter examines two generalizations of the standard activated complex rate theory that are needed to discuss dynamic processes occurring on potential surfaces on which multiple stable minima occur, or on which the system evolves by anisotropic diffusion.

We analyzed the simplest kinetic scheme for a system that displays more than one stable conformational state and undergoes intramolecular electron transfer. In particular, we focused on the situation in which the conformational conversion rates and equilibrium can change upon excitation or charge transfer.

Despite this energetic coupling, the synchronous processes (e.g., $P_C \rightarrow S_B$) can be omitted from consideration because the activation free energy along the synchronous pathway is greater than that for electron transfer within a single configuration. The fact that ET and conformational reactions are sequential, and not concerted, is a major factor in efforts to disentangle conformational and electron-transfer influences. This factor is of key importance because standard detection methods monitor only the ET event, and not conformational changes within one electronic state. In many, if not most, instances the measured time course of a single gated ET reaction is likely to be indistinguishable from a reaction without gating.

Fortunately, the partial decoupling of ET and conformational processes afforded by the absence of synchronous events in principle and in practice allows for the identification of an observed decay rate constant. If one constructs a series of homologous systems in which the ET energetics (or electronic coupling) is modified without change in the conformational equilibrium, thus leaving the conformational rates unchanged, then the observed rate constants will be unchanged if the reaction is controlled by a conformational rate, but will vary if this is not so.

With the aims of specificity and clarity, we discussed conformational effects in terms of intramolecular electron-transfer phenomena, focusing on a donor–acceptor complex activated by flash photolysis or a second-order redox event. However, the general set of rate equations (eqs 2a and 9) and the kinetic schemes of Figure 4 are applicable to other types of rate process. For example, either proton-transfer or isomerization reactions can be controlled by gating—that is, by conversion between two conformational geometries that have different reactivities. The presence of additional stable minima on a two-dimensional potential-energy surface that involves a conformational coordinate (the vertical coordinate of Figure 4) distinguishes this class of reactions.

We also briefly recounted the behavior to be expected when the motion from precursor to successor involves no secondary stable minima, but rather anisotropic diffusion, arising from substantially different friction, or damping, along conformational and inner-sphere vibrational modes. Under such circumstances, the system trajectories will stray substantially from the steepest-descents pathway from reactant over the saddle point to product geometries. In this case the initial population of reactant will exhibit a multiexponential decay because the ridge line, which separates reactant and product states on the potential surface, no longer is coincident with the separatrix, which separates reactant and product regions of reactive trajectories. Study of the transient decay behavior can provide important insights into the differing frictions along the two modes, as well as the shape of the potential surface.

Acknowledgments

We acknowledge stimulating discussions of ET processes with A. Nitzan, J. Hupp, and H. B. Gray. This work has been supported by the National Institutes of Health (HL 13531, BMH) and the National Science Foundation (DBM86–06575, BMH and CHE8805585, MAR).

Appendix. Exact Solution to the General Case of I → A

The solution to the coupled equation set (12) was found by using the Laplace–Carson transformation. The boundary conditions are $[I_C(t = 0)] = [I_B(t = 0)] = 0$; $A_c^*(t)$ and $A_B^*(t)$ are given in eq 4, and the total population in state I at time t is

$$\frac{I(t)}{A^*(0)} = \left[\frac{af^2 + bf + c}{(g - f)(m - f)(n - f)} e^{-ft} + \frac{ag^2 + bg + c}{(f - g)(m - g)(n - g)} e^{-gt} + \frac{am^2 + bm + c}{(f - m)(g - m)(n - m)} e^{-mt} + \frac{an^2 + bn + c}{(f - n)(g - n)(m - n)} e^{-nt} \right] \quad (A.1)$$

where

$$a = k_{tC} + (k_{tB} - k_{tC}) F \quad (A.2)$$

$$-b = k_{tB}[k_u + F(k_{uI} + k_{dI} + k_{bC} + k_{tC} + k_D)] + k_{tC}[k_d + (1 - F)(k_{uI} + k_{dI} + k_{bB} + k_{tB} + k_D)] \quad (A.3)$$

$$c = k_{tB}(k_{uI} + k_{dI} + k_{bC})[k_u + F(k_{tC} + k_D)] +$$
$$k_{tC}(k_{uI} + k_{dI} + k_{bB})[k_d + (1 - F)(k_{tB} + k_D)] \quad (A.4)$$

f and g are as defined in eq 5, and m and n are as defined in eq 16. As in the subcase examined earlier, m and n are composites only of the fundamental rates of state I.

References

1. Hoffman, B. M.; Ratner, M. A. *J. Am. Chem. Soc.* **1987**, *109*, 6237–6243. Erratum: *J. Am. Chem. Soc.* **1988**, *110*, 8267.
2. Klosek-Dygas, M. M.; Hoffman, B. M.; Matkowsky, B. J.; Nitzan, A.; Ratner, M. A.; Schuss, Z. *J. Chem. Phys.* **1989**, *90*, 1141–1148.
3. Liang, N.; Pielak, G. J.; Mauk, A. G.; Smith, M.; Hoffman, B. M. *Proc. Natl. Acad. Sci. U.S.A.* **1987**, *84*, 1249–1252.
4. Lieber, C. M.; Karas, J. L.; Gray, H. B. *J. Am. Chem. Soc.* **1987**, *109*, 3778–3779.
5. McLendon, G.; Pardue, K.; Bak, P. *J. Am. Chem. Soc.* **1987**, *109*, 7540–7541.
6. Bernardo, M. M.; Robandt, P. V.; Schroeder, R. R.; Rorabacher, D. B. *J. Am. Chem. Soc.* **1989**, *111*, 1224–1231.
7. Bond, A. M.; Oldham, K. B. *J. Phys. Chem.* **1983**, *87*, 2472–2502.
8. O'Connell, K. M.; Evans, D. H. *J. Am. Chem. Soc.* **1983**, *105*, 1473–1481.
9. Liang, N.; Kang, C. H.; Ho, P. S.; Margoliash, E.; Hoffman, B. M. *J. Am. Chem. Soc.* **1986**, *108*, 4665–4666.
10. Mayo, S.; Ellis, W. R.; Crutchley, R. J.; Gray, H. B. *Science (Washington, D.C.)* **1986**, *233*, 948–952.
11. Marcus, R. A.; Sutin, N. *Biochim. Biophys. Acta* **1985**, *811*, 265–322.
12. Newton, M. D.; Sutin, N. *Annu. Rev. Phys. Chem.* **1984**, *35*, 437–480.
13. Williams, G.; Moore, G. R.; Williams, R. J. P. *Comments Inorg. Chem.* **1985**, *4*, 55–98.
14. Marden, M. C.; Hazard, E. S.; Gibson, Q. H. *Biochemistry* **1986**, *25*, 7591–7596.
15. McGourty, J. L.; Blough, N. V.; Hoffman, B. M. *J. Am. Chem. Soc.* **1983**, *105*, 4470–4472.
16. Peterson-Kennedy, S. E.; McGourty, J. L.; Kalweit, J. A.; Hoffman, B. M. *J. Am. Chem. Soc.* **1986**, *108*, 1739–1746.
17. Ho, P. S.; Sutoris, C.; Liang, N.; Margoliash, E.; Hoffman, B. M. *J. Am. Chem. Soc.* **1985**, *107*, 1070–1071.
18. Bechtold, R.; Kuehn, C.; Lepre, C.; Isied, S. *Nature (London)* **1986**, *322*, 286–288.
19. Fleming, G. R. *Chemical Applications of Ultrafast Spectroscopy*; Oxford: New York, 1986; pp 179ff.
20. Northrup, S. H.; McCammon, J. A. *J. Am. Chem. Soc.* **1984**, *106*, 930–934.
21. Debrunner, P. G.; Frauenfelder, H. *Annu. Rev. Phys. Chem.* **1982**, *33*, 283–299.
22. Brunschwig, B.; Sutin, N. *J. Am. Chem. Soc.* **1989**, *111*, 7454–7465.
23. Rodiguin, B. M.; Rodiguina, E. N. *Consecutive Chemical Reactions: Mathematical Analysis and Development*; Van Nostrand: Princeton, NJ, 1964.
24. Mikkelsen, K. V.; Ratner, M. A. *Chem. Rev.* **1987**, *87*, 113.
25. Zusman, L. D. *Chem. Phys.* **1980**, *49*, 295.
26. Alexandrov, I. V.; Gabrielyan, R. G. *Mol. Phys.* **1963**, *37*, 1963.
27. Friedman, H. L.; Newton, M. *Faraday Discuss. Chem. Soc.* **1982**, *74*, 73.

28. Calef, D. F.; Wolynes, P. G. *J. Phys. Chem.* **1983,** *87,* 3387.
29. Hynes, J. B. *J. Phys. Chem.* **1986,** *90,* 3701.
30. Sumi, H.; Marcus, R. A. *J. Chem. Phys.* **1987,** *84,* 4272.
31. Nadler, W.; Marcus, R. A. *J. Chem. Phys.* **1987,** *86,* 3906.
32. Sparpaglione, M.; Mukamel, S. *J. Chem. Phys.* **1988,** *88,* 3263.
33. Rips, I.; Jortner, J. *J. Chem. Phys.* **1987,** *87,* 2090.
34. Dakhnovskii, Yu. I.; Ovchinikov, A. A. *Mol. Phys.* **1986,** *58,* 237.
35. Zusman, L. D. *Chem. Phys.* **1988,** *51,* 119.
36. Murillo, M.; Cukier, R. I. *J. Chem. Phys.* **1988,** *89,* 6736.
37. McManis, G. E.; Weaver, M. J. *J. Chem. Phys.* **1989,** *90,* 912.
38. Gennett, T.; Milner, D. F.; Weaver, M. J. *J. Phys. Chem.* **1985,** *89,* 2787.
39. Simon, J. D.; Su, S. G. *J. Chem. Phys.* **1987,** *87,* 7016.
40. Su, S. G.; Simon, J. D. *J. Chem. Phys.* **1988,** *89,* 908.
41. Kahlow, M. A.; Jarzeba, W.; Kang, T. J.; Barbara, P. F. *J. Chem. Phys.* **1989,** *90,* 151.
42. Kramers, H. A. *Physica (Utrecht)* **1940,** *7,* 284.
43. Hynes, J. T. *Annu. Rev. Phys. Chem.* **1985,** *36,* 573–597.
44. Skinner, J. L.; Wolynes, P. G. *J. Phys. Chem.* **1980,** *72,* 4913–4919.
45. Grote, R. F.; Hynes, J. T. *J. Chem. Phys.* **1980,** *73,* 2715–2732.
46. van der Zwan, G.; Hynes, J. T. *J. Chem. Phys.* **1983,** *78,* 4174–4185.
47. Calef, D. F.; Wolynes, P. G. *J. Chem. Phys.* **1983,** *78,* 470–482.
48. Nitzan, A. *Adv. Chem. Phys.* **1988,** *47,* 489.
49. Agmon, N.; Hopfield, J. J. *J. Chem. Phys.* **1983,** *78,* 6947–6959.
50. Agmon, N.; Hopfield, J. J. *J. Chem. Phys.* **1983,** *79,* 2042–2053.
51. Frauenfelder, H.; Young, R. D. *Comments Mol. Cell. Biophys.* **1986,** *3,* 347.
52. Agmon, N.; Kosloff, R. *J. Phys. Chem.* **1987,** *91,* 1988.
53. Onuchic, J. N. *J. Chem. Phys.* **1987,** *86,* 3925–3943.

RECEIVED for review June 6, 1989. ACCEPTED revised manuscript October 13, 1989.

Electronic Coupling and Protein Dynamics in Biological Electron-Transfer Reactions

J. S. Bashkin and G. McLendon[1]

Department of Chemistry, University of Rochester, Rochester, NY 14627

Long-distance electron transfer between proteins has received extensive experimental and theoretical treatment over the past several years. Large discrepancies exist between the reported values for the electronic damping factor, β, for protein–protein complexes and ruthenium-modified proteins. In this chapter, we employ triplet–triplet energy transfer between cytochrome c (cyt c) and cytochrome c peroxidase (ccp) to provide a more direct measure of β for this system than was previously obtainable. We also examine protein dynamics on picosecond–nanosecond time scales by using the time-resolved fluorescence of an arylaminonaphthalene dye complexed to apomyoglobin. The time-dependent shift of the fluorescence is analyzed to evaluate the potential influence of protein dynamics on long-distance electron-transfer rates, and hence on the values of β extracted from rate data.

LONG-DISTANCE ELECTRON TRANSFER BETWEEN PROTEINS has received extensive experimental and theoretical treatment over the past several years (1–7). The rate constant for such a reaction can be written as:

$$k_{et} = \nu \cdot \text{FCWD} \qquad (1)$$

where ν is $(\pi/\hbar^2\lambda_s k_B T)^{1/2}|V(R_0)|^2 \exp(-\beta R)$; FCWD is the Franck–Condon weighted density of states, equal to $(\Delta G - \lambda)^2/4\lambda$; R is the closest do-

---
[1]Address correspondence to this author.

nor–acceptor approach distance; and $V(R_0)$ is the electronic coupling matrix element at contact distance R_0. (Also, *see* list of symbols.) The term ν is a frequency factor, containing both an electronic coupling–distance dependence term, $\exp(-\beta R)$, and a nuclear frequency term $(\nu_N = (\pi/\hbar^2 \lambda_s k_B T)^{1/2})$. It has been pointed out (8–19) that, where the rate of solvent reorientation is slow relative to electron transfer,

$$k_{et} \leq \nu_N = \tau_L^{-1} \qquad (2)$$

where τ_L is the longitudinal relaxation time of the solvent. The Franck–Condon weighted density of states (FCWD) contains the dependence of rate on driving force (ΔG) and most of the dependence on reorganization energies ($\lambda = \lambda_{\text{inner sphere}} + \lambda_{\text{solvent}}$). Most experimental efforts have been directed toward understanding the Franck–Condon (FC) factors and observing the "inverted region" predicted by Marcus (20) (decreasing rates with increasing

Symbols

FCWD	Franck–Condon weighted density of states
$\hbar$	Planck constant
k_B	Boltzmann constant
k_{et}	rate constant for an electron-transfer reaction
R	closest donor–acceptor approach distance
R_0	contact distance
T	absolute temperature
T_1	lowest-energy triplet state
T_2	excited triplet state
V_{et}	electron-transfer coupling matrix element
V_{ex}	energy-transfer coupling matrix element
$V(R_0)$	electronic coupling matrix element
β	electronic damping factor
β_+	coupling term for hole transfer
β_-	coupling term for electron transfer
ΔG	driving force
ΔG_i^*	inner-sphere (intramolecular) barrier
ΔG_s^*	outer-sphere (solvent) barrier
ΔG_T^*	total barrier
λ	reorganization energy
λ_s	solvent reorganization energy
ν	frequency factor
ν_i	characteristic intramolecular (reactant) frequencies
ν_N	nuclear frequency term
ν_s	frequency of solvent reorganization
τ_L	longitudinal relaxation time of solvent

driving force). By measuring how electron-transfer rates vary with thermodynamic driving force (ΔG), the reorganization energy (λ) is calculated, assuming a constant $\nu \sim 10^{13}$ s^{-1}. The electronic damping factor, β, may then be extracted if the donor–acceptor distance is known.

Few experimental efforts have been reported that measure directly the parameters contained within ν for proteins (β and ν_N), and the assumption of a value of 10^{13} s^{-1} is open to question. Hoffman and co-workers (21) performed a temperature-dependence study of electron-transfer rates between Zn$_2$Fe$_2$ hemoglobin subunits and extracted from these data values for $V(R_0)$, β, and λ, but no separation of λ into λ_i and λ_s was possible. Consequently, much less is known about the frequency factor term than the FC terms. We now present two experiments designed specifically to address this issue.

Two experimental approaches for studying electron transfer between proteins have been developed. McLendon and co-workers (22–27), Hoffman and co-workers (28–32), and others (33) have studied protein–protein couples in which the donor and acceptor (heme) proteins form stable, noncovalent 1:1 complexes. Metal substitutions in the heme moiety are employed to vary ΔG and to permit photoinitiation of the electron-transfer process. With this approach, the observed rate at optimal ΔG may be used to estimate β, by assuming $\nu_N \sim 10^{13}$ s^{-1}, FCWD ~ 1, and that R is known. Such estimates vary over a significant range, from $\beta = 1.5$ Å^{-1} for the cyt c : cyt b$_5$ couple (27) to $\beta = 1.2$ Å^{-1} for the cyt c : ccp (25) and Zn$_2$Fe$_2$ hemoglobin couples (cyt c is cytochrome c; ccp is cytochrome c peroxidase) (34–36). This approach suffers from some uncertainty in donor–acceptor (D–A) distances, which are relatively well defined for the Zn$_2$Fe$_2$ hemoglobin (Hb) hybrids (34–36), but less well defined for the other systems. The existence of multiple protein binding sites and the dynamics of a protein–protein complex may give rise to a distribution of D–A separations. These aspects are currently under study in our laboratory. The assumptions of $\nu_N \sim 10^{13}$ s^{-1} and FCWD ~ 1 are also questionable.

Gray and co-workers (5, 37–39) and others (40–45) have pursued a different approach, in which a RuL$_5$ (L is NH$_3$, pyridine) fragment is covalently bound to a specific histidine residue on a protein surface, with a consequently well-defined donor–acceptor separation. Potential difficulties may arise with this methodology because FC factors, which can depend on D–A distance, are not completely known for all derivatives. The number of available His residues is limited, thus the number of D–A separations is also limited. A series of L$_5$Ru–Mb derivatives has yielded $\beta = 0.85 \pm 0.1$ Å^{-1} (5), which is significantly smaller than the estimates obtained from metal substitutions. Such discrepancies have dramatic consequences for predicted reaction rates. For example, with $R = 16$ Å, a change in β from 0.8 to 1.2 s^{-1} results in a 10^3 difference in the calculated rate constant for electron transfer. An independent technique for determination of β is required to understand this disparity.

Triplet–Triplet Energy Transfer

The first project to be discussed here was suggested by the work of Closs et al. (46, 47) comparing electron transfer and triplet–triplet energy-transfer rates. Dexter (48) showed that, although the energy-transfer reaction $^3D/^1A \rightarrow {}^1D/^3A$ is rigorously dipole forbidden, the process can still occur by an exchange mechanism that is formally a simultaneous double electron transfer. By exact analogy with electron-transfer reactions, the rate of the triplet–triplet energy transfer is given by (46):

$$k_{ex} = \nu \cdot \exp[-(\beta_+ + \beta_-)R] \cdot \text{FCWD} \qquad (3)$$

where β_+ is the coupling term for hole transfer and β_- for electron transfer. When $\beta_+ \sim \beta_-$, then $|V_{ex}| \simeq 2|V_{et}|$. Unlike electron transfer, however, no charge buildup occurs during triplet–triplet transfer because electrons are simultaneously exchanged between the donor and acceptor. Consequently, solvent reorganization energy (λ_s) and solvent dynamics (τ_L), which may dominate the FC factors and ν_N in electron transfer, can be neglected in energy transfer. Therefore, less ambiguous information on β might be obtained.

The equivalence of electron transfer and triplet–triplet energy transfer implied by eq 3 was beautifully confirmed experimentally by Closs et al. (46, 47). By using a common series of spacer molecules, and therefore equivalent electronic coupling terms, they found that $\ln(k_{et}) \simeq 2 \ln(k_{ex})$, as suggested by eq 2. We have now applied this concept to estimate the electronic damping parameter β via triplet energy transfer for the electron-transfer protein couple, cyt c : ccp.

Zinc cytochrome c (Zn cyt c), which acts as a triplet donor, and H_2 porphyrin cytochrome c peroxidase (H_2ccp), which acts as a triplet acceptor, were used. When it is free in solution, Zn cyt c has a triplet lifetime of ~14 ms (25). A similar value of 13.5 ± 0.5 ms was determined in this work, monitoring delayed fluorescence at 570 nm. In like manner, H_2ccp has a triplet lifetime of ~1.4 ms (25). Moreover, cyt c and ccp form a strong noncovalent complex ($K = 10^8$ M^{-1}). On binding to H_2ccp, the lifetime of the Zn cyt c triplet state decreases to 10 ms (Figure 1), although that of the H_2ccp increases to a similar value. This behavior is consistent with direct energy transfer from Zn cyt c to H_2ccp with $k_{ex} = 20$ s^{-1}.

In light of Zemel and Hoffman's study (35) of allowed triplet $T_1 \rightarrow T_2$ processes in Zn_4 Hb, we varied the excitation power by 10-fold with no change in the observed rate, thereby eliminating a (two-photon) $T_1 \rightarrow T_2$ process involving the acceptor. Taking the previously estimated donor–acceptor (edge–edge) distance for the cyt c : ccp couple as $R = 16$ Å, and assuming (maximal) values of $\nu' = 10^{13}$ s^{-1} and FCWD = 1 gives $\beta = 0.85 \pm 0.05$ Å^{-1}. Any smaller values of ν' or FCWD would further lower β. This value agrees surprisingly well with the estimates of Gray and co-

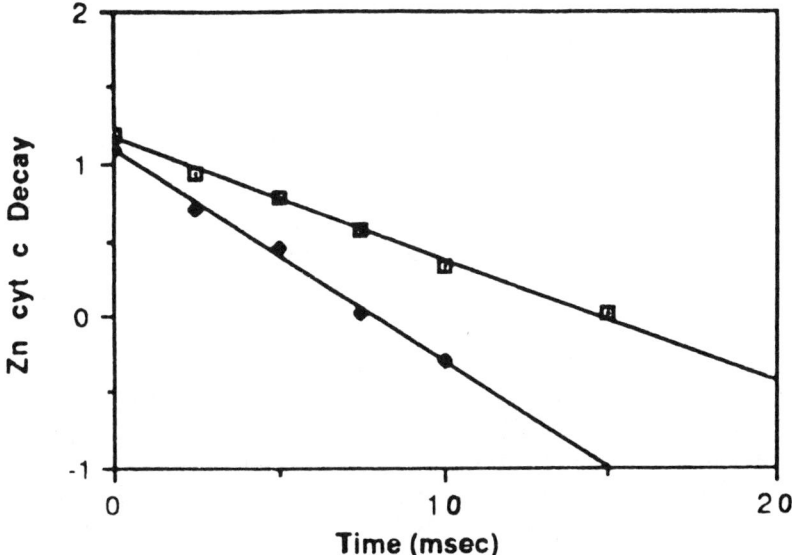

Figure 1. Decay data for ^{3}Zn cyt c delayed emission ($\lambda = 575$ nm). Top curve: Zn cyt c alone (pH 7.0, 10 mM Pi, T = 298 K, conc. = 10^{-5} M); k = 80 s^{-1}. Lower curve: Zn cyt c + H_2ccp (1 equivalent), all other conditions as for top curve; k = 140 s^{-1}.

workers (5, 38) for myoglobin, but is significantly smaller than the values previously obtained by us and others. It remains to be seen whether this result is a general one, and further studies are in progress.

This result leads to a second curious dilemma. If β is indeed low (β ~ 0.9 Å^{-1}), then a low nuclear frequency factor is required to explain the slow electron-transfer rates. Gray and co-workers (5, 38) derive a value from their data of $\nu_N \sim 10^{10}$ s^{-1}. We do not know why ν_N should be so low. One possible explanation may lie in detailed consideration of solvent dynamics, as outlined in the following section.

Solvent Dynamics

The dependence of electron-transfer rates in condensed media on solvent relaxation and nonequilibrium solvation has been the subject of numerous studies (8–19, 46, 47, 49–66). Weaver and Gennett (10) showed rates of electrochemical exchange between organometallic redox couples to be correlated to τ_L, the longitudinal solvent relaxation time. Kosower (9, 11), Fleming (50–53), and others (54, 55) investigated the photophysics of organic dyes and the influence of solvent dynamics on dye fluorescence. Using coumarin and arylaminonaphthalene derivatives, these researchers all observed a correlation between τ_L and the formation and decay rates for charge-transfer excited states.

The interpretations, based on the theories of Hynes, Wolynes, and coworkers and others (12–17) suggest that for an adiabatic electron transfer, the nuclear frequency factor is dominated by solvent dynamics, $v_N \sim 1/\tau_L$, and the expression from transition-state theory ($v_N \sim k_B T/\hbar$) is not appropriate (12–18). The frequency nuclear factor v_N derived from transition-state theory is usually written (10, 18):

$$v_N = \left(\frac{v_s^2 \Delta G_s^* + v_i^2 \Delta G_i^*}{\Delta G_s^* + \Delta G_i^*} \right)^{1/2} \qquad (4)$$

where v_s is frequency of solvent reorganization, v_i is characteristic intramolecular (reactant) frequencies, ΔG_s^* is outer-sphere (solvent) barrier, ΔG_i^* is inner-sphere (intramolecular) barrier, and $\Delta G_T^* = \Delta G_s^* + \Delta G_i^*$.

Weaver and Gennett (10) showed that eq 4 underestimates the contribution of damped solvent motion when λ_i is small. Rather, he found that the following relation, derived for exchange reactions (14, 15), gave the best agreement with his data:

$$v_N \approx (\tau_L)^{-1} \cdot \left(\frac{\Delta G_T^*}{4\pi k_B T} \right)^{1/2} \qquad (5)$$

In contrast, Efrima and Bixon (16) and McGuire and McLendon and coworkers (18, 19) noted that, in the nonadiabatic limit, the frequency factor will be controlled not by solvent relaxation, but by electronic coupling, $V(R)$, and thus be independent of τ_L:

$$\text{adiabatic limit} \qquad k_{et}^{ad} \propto \frac{1}{\tau_L} \qquad (6)$$

$$\text{nonadiabatic limit} \qquad k_{et}^{d} \propto [V(R)]^2$$

In the simplest treatment (18, 19), the two limiting rates can be treated independently, and the observed rate will then be

$$k_{obs} = \frac{k^{ad} \cdot k^d}{k^{ad} + k^d} \qquad (7)$$

Hence, k_{obs} may actually vary with $(1/\tau_L)n$, with n varying from 0 to 1, depending on the relative values of τ_L and $V(R)$. McGuire's (18, 19) experimental results on nonadiabatic electron transfer in glycerol (over a 10^7 change in τ_L) gave $n = 0.6$.

Protein Dynamics

None of these studies, however, focused on the implications of the dynamics of the condensed phase for electron transfer between proteins. There has been no direct measure of τ_L for a protein, and the question arises whether the dynamics of protein response to charge separation may modulate electron-transfer rates in biological systems.

The following project follows much of the same methodology employed by Brand and co-workers (67–71) several years ago. Brand used time-resolved emission spectroscopy (TRES) to study the interaction of fluorescent probes to apomyoglobin. These probes bind directly to the vacant heme pocket of apomyoglobin, as shown by competition experiments with heme ($k_B \sim 10^5$) (67–71) Brand observed spectral shifts on a nanosecond time scale by using N-(-p-tolyl)-2-aminonaphthalene-6-sulfonate (2,6-TNS).

We have proceeded on the assumption that the fluorescence of such dyes is modulated by protein dynamics in a manner analogous to solvent relaxation. We selected analino-2-aminonaphthalene-6-dimethylsulfonamide (2,6-ANSDMA) as the fluorescent probe of choice. This particular dye has convenient spectral properties and binds to apomyoglobin similarly to 2,6-TNS. It removes any potential contributions of hydrogen bonding effects by the sulfonate group to the protein dynamics. The photophysics of 2,6-ANSDMA in various organic solvents has been extensively studied by Kosower (9, 11, 56–66). For example, Huppert and Kosower (56–66) showed that τ_L^{-1} (solvent) $\sim k_{em}$, the risetime for long wavelength emission.

Representative fluorescence-decay curves and theoretical fits for the dye–protein complex are shown in Figure 2. The decays could all be fit well to double exponential decay functions. The lifetimes, dependent on emission wavelength, fell in the ranges 1–3 and 5–8 ns. The TRES were reconstructed from the fitted decay curves and the static fluorescence spectrum. Typical time-resolved spectra are shown in Figure 3. The results clearly show a shift of the 2,6-ANSDMA fluorescence to lower energy with time, along with a broadening in spectral shape. This change is highly nonexponential and occurs on a nanosecond time scale. The effect is best illustrated in Figure 4, in which the half-height points on the high-energy side of the peaks are plotted against time.

Overall, the spectral shift is approximately 1100 cm^{-1} over 20 ns. These results are consistent with those obtained by Brand (67–71). Comparison of our data with those of Kosower (56–66) for 2,6-ANSDMA in organic solvents shows myoglobin to have relaxation properties similar to those of 1-decanol. Following the arguments of Van der Zwan and Hynes (13), the fastest components of τ_L will provide the protein relaxation necessary for electron transfer to occur. Therefore, if we take $\tau_L = 50$ ps (approximating the fluorescence risetime at long wavelength, where $\Delta G \sim 0$), and $\lambda = 0.8$ eV, then with $\Delta G_T^* = (\Delta G - \lambda)^2/4\lambda \sim 0.2$ eV, eq 5 yields a lower limit of $\nu_N = 1.5 \times$

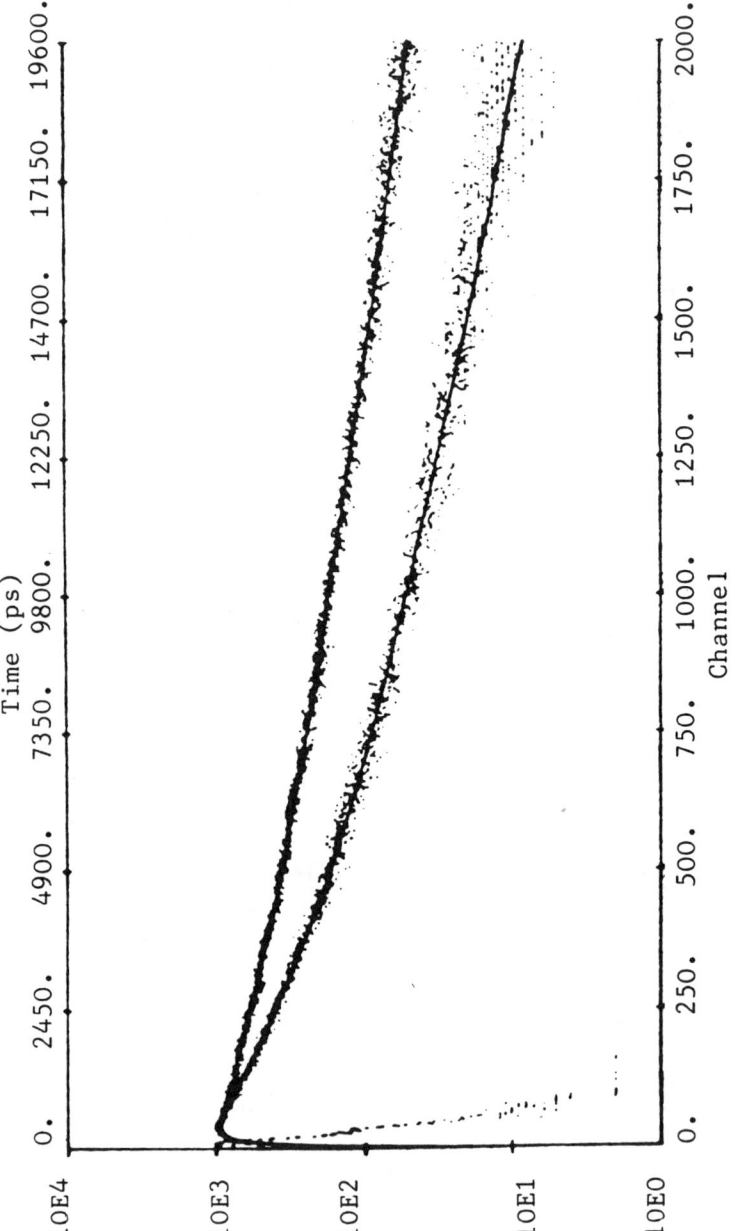

Figure 2. Typical fluorescence decay curves and theoretical fits for the 2,6-ANSDMA–apomyoglobin complex, measured at 470 (upper) and 420 (lower) nm. Parameters used in the fits: at 470 nm: $\tau_1 = 2.09 \pm 0.06$, $amp = 0.0323$; $\tau_2 = 7.79 \pm 0.4$, $amp = 0.038$; $\chi^2 = 1.030$; Durban–Watson = 1.882; and at 400 nm: $\tau_2 = 1.57 \pm 0.02$ ps, $amp = 0.0536$; $\tau_2 = 5.14 \pm 0.2$ ps, $amp = 0.0205$; $\chi^2 = 1.104$; Durban–Watson = 1.796.

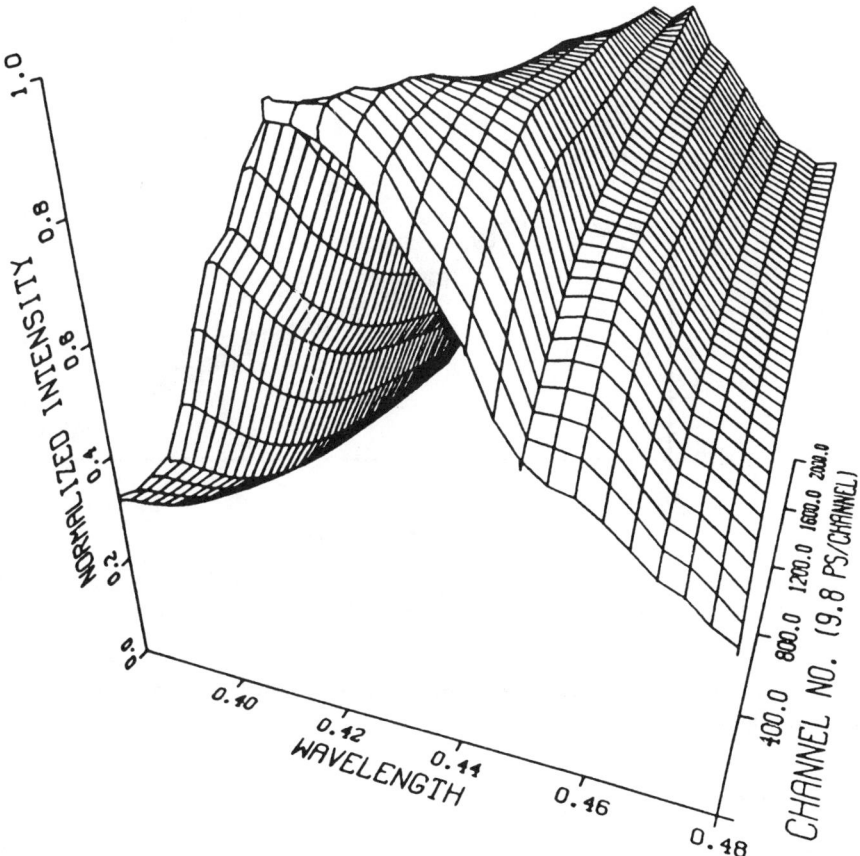

Figure 3. Three-dimensional plot of time-resolved fluorescence emission spectra (TRES) of the 2,6-ANSDMA–apomyoglobin complex. The peak shift is ~415–440 nm.

10^{10} s^{-1}. This value of ν_N is 2–3 orders of magnitude below the value generally accepted and used for the evaluation of various dynamic parameters in protein electron-transfer reactions.

A consequence of this analysis is that the same nuclear prefactor will not be appropriate for all protein systems. As noted by McGuire (18, 19), the dependence of reaction rates on τ_L will depend on the relative magnitudes of $V(R)$ and τ_L. Comparison of electron-transfer rates from different protein couples, or even different metal–heme derivatives of the same proteins, must be made with caution. Not all variation in rates may be attributable to changes in ΔG for a particular series, but as $V(R)$ changes relative to τ_L, protein dynamics will exert varying degrees of influence on electron-transfer rates. This may explain some of the discrepancies in β and λ values

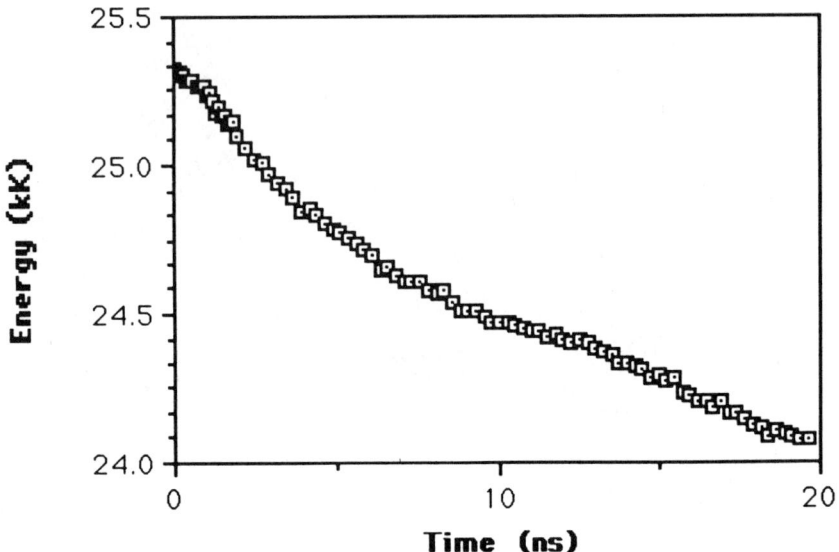

Figure 4. Shift in fluorescence of 2,6-ANSDMA–apomyoglobin complex. Data are half-height energies measured on the high-energy side of the TRES.

determined for different protein couples and through different experimental methodologies.

Experimental Section

Energy Transfer. Zn cyt c (72) was prepared by a variation of the method of Vanderkooi. A 2.0- × 2.5-cm column (Bio-Rex 70, Biorad) was used in place of a 150- × 2.5-cm column (G–50 Sephadex) to remove polymeric forms of the protein. H_2 porphyrin cytochrome c peroxidase (H_2ccp) was prepared by reconstitution of the apoprotein (73) with protoporphyrin IX. The Zn cyt c triplet lifetime was determined by monitoring delayed fluorescence at 570 nm. The H_2ccp triplet lifetime was determined similarly, monitoring at 620 nm. Samples were prepared at 10 mM ionic strength (pH 7 phosphate), conditions that ensure formation of a strong noncovalent complex ($K = 10^8$ M^{-1}). The ^{3}Zn excited states were prepared by excitation with a Q-switched Nd:YAG laser (Quanta-Ray model DCR–2, 7-ns pulse width, second harmonic, 532 nm) or a N_2-pumped dye laser. Laser firing and data collection were controlled with a computer (IBM–XT) and software written in ASYST by D. Heiler. Further details of the apparatus have been published elsewhere (26, 74).

Protein Dynamics. Apomyoglobin was prepared by a variation of the method of Teale (75). The apoprotein was then dialyzed twice against distilled water for 0.5 h each, followed by two dialyses for 1 h each against pH 6.4, 20 mM BIS–TRIS buffer. The protein was further dialyzed against doubly distilled deionized water and then passed down a 2-cm column (Sephadex G–25, 0.01 mM Kphos, pH 6.8) immediately prior to use. 2,6-ANSDMA was prepared and purified by published procedures (66). Samples (20–40 μM protein, 1:4 dye:protein ratio) were degassed by gently stirring over 2 h in a continuously purged (N_2) glove box. Data collection took 4–5 h and was performed at 24 °C.

The fluorescence decay curves were obtained using standard single-photon counting (SPC) techniques and fit with a least-squares fitting routine. Excitation pulses (295 nm) were produced from a mode-locked Nd:YAG pumped dye laser (Coherent model 700). The cavity-dumped dye laser, fitted with a single-plate birefringent filter and a saturable absorber, produced 1-ps pulses as measured by an autocorrelator (Femtochrome model FR103). Subsequent second harmonic generation was produced with β-barium borate. The detector was a broad-band 6-μm proximity focus microchannel plate (Hamamatsu R2809U-11). A 0.25 M emission monochromator (Spex) was used with a 600-line/nm grating. Slits on either side of the monochromator were opened to 2 mm or less. The total instrument response was approximately 50 ps.

Sample integrity was monitored by comparing decay curves measured periodically at one wavelength. The time-resolved emission spectra (TRES) of the protein–dye complex were reconstructed (as described by O'Conner and Phillips (76), software adapted by Mark Prichard) from the static fluorescence spectrum of the protein–dye complex and the fluorescence-decay curves. Temporal resolution of the TRES was 9.8 ps, spectra were followed out to 20 ns past excitation, and wavelength resolution varied between 2 and 5 nm. The static fluorescence spectrum was recorded on fluorescence spectrometers (Spex DM1B or Perkin Elmer MPF-44A model).

Acknowledgments

We gratefully acknowledge Mark Prichard for assistance with the SPC system and programming necessary for the TRES.

References

1. Marcus, R.; Sutin, N. *Biochim. Biophys. Acta* **1985**, *811*, 265.
2. Marcus, R. *Nouv. J. Chim.* **1987**, *11*, 79.
3. McLendon, G. *Acc. Chem. Res.* **1988**, *21*, 160.
4. Guarr, T.; McLendon, G. *Coord. Chem. Rev.* **1985**, *68*, 1.
5. Mayo, S. L.; Ellis, W. R., Jr.; Crutchley, R. J.; Gray, H. B. *Science (Washington, D.C.)* **1986**, *233*, 948.
6. Newton, M. D.; Sutin, N. *Ann. Rev. Phys. Chem.* **1984**, *35*, 437.
7. Barbara, P. F.; Jarzeba, W. *Acc. Chem. Res.* **1988**, *21*, 195.
8. Miller, J. R.; Beitz, J. V.; Huddleston, R. K. *J. Am. Chem. Soc.* **1984**, *106*, 5057.
9. Kosower, E. M. *Acc. Chem. Res.* **1982**, *15*, 259.
10. Weaver, M. J.; Gennett, T. *Chem. Phys. Lett.* **1985**, *113*, 213.
11. Kosower, E. M. *J. Am. Chem. Soc.* **1985**, *107*, 1114.
12. Calef, D. F.; Wolynes, P. G. *J. Chem. Phys.* **1983**, *78*, 470.
13. Van der Zwan, G.; Hynes, J. T. *J. Chem. Phys.* **1982**, *76*, 2993.
14. Zusman, L. D. *Chem. Phys.* **1980**, *49*, 295.
15. Calef, D. F.; Wolynes, P. G. *J. Phys. Chem.* **1983**, *87*, 3387.
16. Efrima, S.; Bixon, M. *J. Chem. Phys.* **1979**, *70*, 3531.
17. Van Duyne, R. P.; Fischer, S. F. *Chem. Phys.* **1974**, *5*, 183.
18. McGuire, M. E., Ph.D. Thesis, University of Rochester, Rochester, NY, 1985.
19. McGuire, M.; McLendon, G. *J. Am. Chem. Soc.* **1986**, *90*, 2549.
20. Marcus, R. *J. Chem. Phys.* **1956**, *24*, 966.

21. Peterson-Kennedy, S. E.; McGourty, J. L.; Hoffman, B. M. *J. Am. Chem. Soc.* **1984**, *106*, 5010.
22. Hazzard, J. T.; McLendon, G.; Cusanovich, M. A.; Tollin, G. *Biochem. Biophys. Res. Commun.* **1988**, *151*, 429.
23. Taylor-Conklin, K.; McLendon, G. *J. Am. Chem. Soc.* **1988**, *110*, 3354.
24. McLendon, G.; Cheung, E.; Taylor, K.; Kornblatt, J. A.; English, A. M.; Miller, J. R. *Proc. Natl. Acad. Sci. U.S.A.* **1986**, *83*, 1330.
25. Taylor-Conklin, K.; McLendon, G. *Inorg. Chem.* **1986**, *25*, 4804.
26. McLendon, G. L.; Winkler, J. R.; Nocera, D. G.; Mauk, M. R.; Mauk, A. G.; Gray, H. B. *J. Am. Chem. Soc.* **1985**, *107*, 739.
27. McLendon, G.; Miller, J. R. *J. Am. Chem. Soc.* **1985**, *107*, 7811.
28. Liang, N.; Mauk, A. G.; Pielak, G. J.; Johnson, J. A.; Smith, M.; Hoffman, B. M. *Science (Washington, D.C.)* **1988**, *240*, 311.
29. Liang, N.; Pielak, G. J.; Mauk, A. G.; Smith, M.; Hoffman, B. M. *Proc. Natl. Acad. Sci. U.S.A.* **1987**, *84*, 1249.
30. Hoffman, B. M.; Ratner, M. A. *J. Am. Chem. Soc.* **1987**, *109*, 6237.
31. Liang, N.; Kang, C. H.; Ho, P. S.; Margoliash, E. Hoffman, B. M. *J. Am. Chem. Soc.* **1986**, *108*, 4665.
32. Ho, P. S.; Sutoris, C.; Liang, N.; Margoliash, E.; Hoffman, B. M. *J. Am. Chem. Soc.* **1985**, *107*, 1070.
33. Dolla, A.; Bruschi, M. *Biochim. Biophys. Acta* **1988**, *932*, 26.
34. McGourty, J.; Blough, N.; Hoffman, B. M. *J. Am. Chem. Soc.* **1983**, *105*, 4470.
35. Zemel, H.; Hoffman, B. M. *J. Am. Chem. Soc.* **1981**, *103*, 1192.
36. Faraggi, M.; Klapper, M. H. *J. Am. Chem. Soc.* **1988**, *110*, 5753.
37. Karas, J. L.; Leiber, C. M.; Gray, H. B. *J. Am. Chem. Soc.* **1988**, *110*, 599.
38. Axup, A. W.; Albin, M.; Mayo, S. L.; Crutchley, R. J.; Gray, H. B. *J. Am. Chem. Soc.* **1988**, *110*, 435.
39. Crutchley, R. J.; Ellis, W. R., Jr.; Gray, H. B. *J. Am. Chem. Soc.* **1985**, *107*, 5002.
40. Bechtold, R.; Kuehn, C.; Lepre, C.; Isied, S. S. *Nature (London)* **1986**, *322*, 286.
41. Elias, H.; Chou, M. H.; Winkler, J. R. *J. Am. Chem. Soc.* **1988**, *110*, 429.
42. Osvath, P.; Salmon, G. A.; Sykes, A. G. *J. Am. Chem. Soc.* **1988**, *110*, 7114.
43. Jackman, M. P.; McGinnis, T.; Powls, R.; Salmon, G. A.; Sykes, A. G. *J. Am. Chem. Soc.* **1988**, *110*, 5880.
44. Jackman, M. P.; Lim, M.-C.; Salmon, G. A.; Sykes, A. G. *J. Chem. Soc., Chem. Commun.* **1988**, *179*.
45. Jackman, M. P.; Lim, M.-C.; Salmon, G. A.; Sykes, A. G. *J. Chem. Soc., Chem. Commun.* **1988**, *65*.
46. Closs, G. L.; Piotrowiak, P.; MacInnis, J. M.; Fleming, G. R. *J. Am. Chem. Soc.* **1988**, *110*, 2652.
47. Closs, G. L.; Calcaterra, L. T.; Green, N. J.; Penfield, K. W.; Miller, J. R. *J. Phys. Chem.* **1986**, *90*, 3673.
48. Dexter, D. L. *J. Chem. Phys.* **1953**, *21*, 836.
49. Weaver, M. J.; Gennett, T. *Chem. Phys. Lett.* **1985**, *113*, 213.
50. Castner, E. W., Jr.; Bagchi, B.; Maroncelli, M.; Webb, S. P.; Ruggiero, A. J.; Fleming, G. R. *Ber. Bunsenges. Phys. Chem.* **1988**, *92*, 363.
51. Maroncelli, M.; Fleming, G. R. *J. Chem. Phys.* **1987**, *86*, 836.
52. Fleming, G. R. *Ann. Rev. Phys. Chem.* **1986**, *37*, 81.
53. Robinson; G. W., Robbins, R. J.; Fleming, G. R.; Morris, J. M.; Knight, A. E. W.; Morrison, R. J. S. *J. Am. Chem. Soc.* **1978**, *100*, 7145.
54. Lessing, H. E.; Reichert, M. *Chem. Phys. Lett.* **1977**, *46*, 111.
55. Demenchenko, A. P. *Biophys. Chem.* **1982**, *15*, 101(c).

56. Huppert, D.; Kanety, H.; Kosower, E. M. *Chem. Phys. Lett.* **1981**, *84*, 48.
57. Kosower, E. M.; Kanety, H.; Dodiuk, H.; Striker, G.; Jovin, T.; Boni, H.; Huppert, D. *J. Phys. Chem.* **1983**, *87*, 2479.
58. Kosower, E. M.; Tanizawa, K. *Chem. Phys. Lett.* **1972**, *16*, 419.
59. Dodiuk, H.; Kosower, E. M. *J. Am. Chem. Soc.* **1977**, *99*, 859.
60. Dodiuk, H.; Kosower, E. M. *J. Am. Chem. Soc.* **1974**, *96*, 6195.
61. Kosower, E. M.; Dodiuk, H. *J. Am. Chem. Soc.* **1978**, *100*, 4173.
62. Kosower, E. M.; Dodiuk, H.; Kanety, H. *J. Am. Chem. Soc.* **1978**, *100*, 4179.
63. Kosower, E. M.; Dodiuk, H.; Tanizawa, K.; Ottolenghi, M.; Orbach, N. *J. Am. Chem. Soc.* **1975**, *97*, 2167.
64. Kosower, E. M.; Dodiuk, H. *J. Phys. Chem.* **1978**, *82*, 2013.
65. Kosower, E. M.; Huppert, D. *Ann. Rev. Phys. Chem.* **1986**, *37*, 127.
66. Dudiuk, H.; Kosower, E. M. *J. Phys. Chem.* **1977**, *81*, 50.
67. Gafni, A.; DeToma, R. P.; Manrow, R. E.; Brand, L. *Biophys. J.* **1977**, *17*, 155.
68. Brand, L.; Gohlke, J. R. *Ann. Rev. Biochem.* **1972**, *41*, 43.
69. Selsikar, C. J.; Brand, L. *J. Am. Chem. Soc.* **1971**, *93*, 5414.
70. Selsikar, C. J.; Brand, L. *J. Am. Chem. Soc.* **1971**, *93*, 5405.
71. Brand, L.; Gohlke, J. R. *J. Biophys. Chem.* **1971**, *246*, 2317.
72. Vanderkooi, J. M.; Erecinska, M. *Eur. J. Biochem.* **1975**, *60*, 199.
73. Yonetani, T. *J. Biol. Chem.* **1967**, *242*, 5008.
74. Magner, E., Ph.D. Thesis, University of Rochester, Rochester, NY, 1988.
75. Teale, F. W. J. *Biochim. Biophys. Acta* **1959**, *35*, 543.
76. O'Connor, D. V.; Phillips, D. *Time-Correlated Single Photon Counting*; Academic: New York, 1984; pp 211–251.

RECEIVED for review May 1, 1989. ACCEPTED revised manuscript August 9, 1989.

8

Electrostatic, Steric, and Reorganizational Control of Electron Self-Exchange in Cytochromes

Dabney White Dixon and Xiaole Hong

Department of Chemistry, Georgia State University, Atlanta, GA 30303

To probe the factors that control biological electron transfer, we analyzed the electron self-exchange reactions of cytochromes c, c_{551}, and b_5. Numerical analysis of the crystal structures gave heme exposures and dipole moments of the proteins. Molecular modeling of the electron-transfer complexes allowed calculation of the heme–heme distance. These values, in conjunction with the experimental rate constants as a function of temperature and ionic strength, give the reorganizational energies, λ, which are 0.7, 0.5, and 1.2 eV for cytochromes c, c_{551}, and b_5, respectively. The numerical values of the reorganizational energies are sensitive to the assumptions made about the heme exposure, but the order (i.e., cytochrome c_{551} < cytochrome c < cytochrome b_5), remains the same for reasonable values of the heme exposure. The parameters derived from this analysis were used to analyze studies of the bimolecular electron transfer between cytochromes c and c_{551}, the bimolecular electron transfer between cytochromes c and b_5, and the intracomplex electron transfer between cytochromes c and b_5. The correlation between the self-exchange parameters and those obtained from these studies is very good.

ELECTRON TRANSFER IS ONE OF THE MOST BASIC biological reactions, important in photosynthesis, oxidative phosphorylation, oxidation of endogenous and exogenous substrates, and maintenance of enzymes in active states (1–8). The rate constant is controlled by a variety of factors, including the driving force, distance, and reorganizational energy of the system (9–21).

The nature of the intervening residues and orientation of the prosthetic groups are considered important. In addition, conformational change is coming into focus as a controlling factor for biological electron transfer (22–25).

Current understanding of the factors that control electron transfer comes from a variety of types of studies. Measurement of electron-transfer rate constants between physiological partners is certainly important, particularly in an effort to understand how electron transfer occurs in vivo (1–9). However, until the advent of site-directed mutagenesis, these studies did not allow systematic variation of the properties of the proteins. Therefore, in the last two decades great effort has gone into measurement of electron-transfer rate constants between nonphysiological partners (15–21). Such studies include bimolecular electron transfer between two proteins or between a protein and an inorganic or organometallic complex. More recently, these studies have been expanded to include intramolecular electron transfer between two proteins within a complex or between the protein redox site and an organometallic species attached to the protein surface (16–18). Because diffusional considerations are nonexistent and protein–protein electrostatic interaction factors are minimal in intramolecular electron transfer, these studies allow particular focus on the dependence of electron transfer on distance, on the intervening residues, on the reorganizational energy of the two centers, and on the orientation of the two redox centers.

Still another approach to probing the factors that control electron transfer comes from measurement of electron self-exchange, electron transfer between the oxidized and reduced forms of the same molecule (20, 26): $\text{cyt}_a^{ox} + \text{cyt}_b^{red} \leftrightarrows \text{cyt}_a^{red} + \text{cyt}_b^{ox}$.

Electron self-exchange rate constants are of fundamental importance because the thermodynamic driving force (ΔG^0) is zero in this reaction, and hence there is no net driving force for the reaction. Differences in the electron self-exchange rate constants between similar proteins arise from differences in the electrostatic, geometric, and reorganizational characteristics of the proteins. Self-exchange rate constants for a variety of b- and c-type cytochromes, measured directly via NMR spectroscopic techniques, are given in Table I. These rate constants span at least five orders of magnitude, from 10^2 to 10^7 M^{-1} s^{-1}. An understanding of the factors that determine this wide range is important for a clear picture of biological electron transfer.

Our goal in this chapter is to delineate the factors that control electron transfer in three heme proteins: cytochrome c (48, 49), cytochrome c_{551} (50), and cytochrome b_5 (51–54). Self-exchange is a bimolecular process and as such can be expressed (20) as

$$k_{et} = SK_a\nu_n\kappa_{el} \exp\frac{-\Delta G_r^*}{RT} \tag{1}$$

where k_{et} is the observed rate constant for electron transfer; S is the steric factor, which reflects the hypothesis that electron transfer occurs primarily at the exposed heme edge in the cytochromes; K_a is the association constant for formation of the precursor state from the two separated electron-transfer partners; ν_n is the nuclear frequency factor; κ_{el} is the average probability of passing from the transition state to products; ΔG_r^* is the free energy of activation (the sum of both inner-sphere and outer-sphere reorganizational energies); R is the gas constant; and T is the temperature. We measured the self-exchange rate constants of these three proteins as a function of ionic strength and temperature; details of these measurements are reported elsewhere (42, 46, 55). The rate constants, in combination with protein dipole moments calculated from the crystal structures and molecular modeling of the electron-transfer complexes, allow us to estimate each of the parameters in eq 1 and thus provide a detailed picture of the factors that control electron transfer in cytochromes.

Steric Considerations

All three of the heme proteins discussed herein have part of the heme itself exposed to solvent (56, 57). Early suggestions that electron transfer occurs mainly through the exposed heme edge in bimolecular electron transfer (57) were followed by various experiments to establish this. Much of the work has involved derivatization of specific residues on the protein surface. In general, alterations close to the exposed heme edge have a larger effect on electron transfer than do derivatizations removed from the heme (58). More recently, a numerical analysis of the heme exposure, assuming exponential falloff of the electron-transfer rate constant, indicated that most of the electron transfer occurs though the exposed heme edge (55). Even if electron transfer is largely at or near the exposed heme edge, a number of different complexes may be involved, as has been discussed by Northrup et al. for the cytochrome c–cytochrome c peroxidase system (59), by Mauk et al. for the reaction between the dimethyl ester of cytochrome c and cytochrome b_5 (60), and by Slayton et al. for reactions of cytochrome b_5 with various partners (61).

If we assume that electron transfer occurs largely through the exposed heme edge, then it is necessary to know the fraction of the protein surface that is heme to analyze the data. To compare the electron-transfer rate constants of the three proteins discussed herein, we calculated this fraction by using a probe sphere of 1.5 Å and the Connolly algorithm (62), as implemented in the BIOGRAF molecular modeling program (63). The fraction of the surface area of the protein that is heme, ϕ, is 0.007, 0.012, and 0.038 for cytochrome c [we made appropriate substitutions in the X-ray structure of the protein from tuna (48, 49)], cytochrome c_{551} (50), and cytochrome b_5 (51), respectively.

The form of bovine cytochrome b_5 studied consists of 82 amino acid residues and differs from the lipase-solubilized form that has been charac-

Table I. Electron Self-Exchange Rate Constants of Cytochromes

Cytochrome	Amino Acids, No.	Charge on Fe(III) Protein[a]	pD[b]	Buffer	Temp., °C	k_{NMR},[c] $M^{-1}s^{-1}$	Ref.
Paracoccus denitrificans c_{550}	134	−6	7.5	10 mM phosphate + 100 mM NaCl	25	1.6×10^4	27
Rhodospirillum rubrum c_2	112	1	6.9		30	slow[d]	28
Saccharomyces cerevisiae c I	108	10	7.0	0.1 M phosphate	30	10^4	29
Crithidia oncopelti c_{557}	112	4	7.3		26	slow[d]	30
Candida krusei c	109	4	7	0.1–1 M salt	40	10^2–10^3	29, 31
Horse heart c	104	7.5	6.6	0.1 M phosphate + 0.1 M KCl	~25	5×10^4	31–34
			7	<0.05 M HEPES– 0.4 M salt	30	3×10^2– 1×10^4	
Horse heart CDNP-Lys-13	104	5.5	7.0	0.24 M cacodylate	25	6×10^3	35
Horse heart CDNP-Lys-72	104	5.5	7.0	0.24 M cacodylate	25	6×10^4	35
Euglena gracilis c_{552}	87	−8	7.0	50 mM phosphate	29	5×10^6	36
Chlorobium thiosulfatophilum c_{555}	86	7	7.0		13	slow[d,e]	37

8. DIXON & HONG   Electron Self-Exchange in Cytochromes

Protein	Residues	Charge	pH	Buffer	T (°C)	k (M^{-1} s^{-1})	Ref.
Desulfovibrio desulfuricans c_{553}	86	−2	7.7		8	slow^{d,e}	38
Alcaligenes faecalis c_{554}	~86		8	10 mM phosphate + 100 mM NaCl	40	~3×10^8	39
Rhodopseudomonas gelatinosa c_{551} I	85	5	7.4		8	10^5	40
Desulfovibrio vulgaris c_{553}	82	2	6.2		12	slow^{d,e}	38
Pseudomonas aeruginosa c_{551}	82	−2	7.0	50 mM phosphate	42	1.2×10^7	41, 42
Pseudomonas mendocina c_{551}	82	−1	6.2	CD$_3$OD–D$_2$O	−6	10^6–10^7	28
Pseudomonas stutzeri c_{551}	82	−2	6.8		30	4×10^6	28, 42
Cyt c undecapeptide	11		10.1	0.5 M cyanide	57	1.3×10^7	43
Cyt c octapeptide	8			>2 equiv of pyridine		>10^5	44
Cytochrome b_{562}	103		7.0	none	17	4×10^6	45
Cytochrome b_5	82	−7.5	7.0	0.12 M phosphate	25	2.6×10^3	46

^{a}Calculated from sequence except for *R. rubrum*, *C. krusei*, and horse (47).
^{b}Measured in D$_2$O.
^{c}From NMR saturation transfer, T_1, and line shape measurements.
^{d}Both the Fe(II) and Fe(III) oxidation states can be seen in a mixture of the two.
^{e}Subambient temperature was presumably used to slow electron exchange and thereby prevent excessive line broadening. The value of k_{NMR} is probably >10^5 M^{-1} s^{-1}.

terized crystallographically by removal of two residues from the amino terminus and nine residues from the carboxyl terminus. The sequence has recently been redetermined from the bovine liver cytochrome b_5 cDNA clone (52) and from gas-phase sequence analysis of tryptic peptides obtained from a tryptic hydrolysate of apocytochrome b_5 (46). The corrected sequence differs from that reported earlier (53, 54) in that residue 57 is now known to be aspargine and residues 11 and 13 are known to be glutamic acid and glutamine, respectively.

Cytochrome b_5 is somewhat different from cytochromes c and c_{551} in that one of the heme propionates is extended out into the solvent. To determine the contribution of this propionate, we removed the terminal CH_2CO_2H atoms from the data set. The heme fraction decreased to 0.027. We used the larger 0.038 in the calculations in this chapter; the smaller heme exposure would result in a smaller reorganizational energy.

If electron transfer were to occur only through the exposed heme edge, then the steric factor, S, would be simply the fraction of the surface of the protein that is heme squared or ϕ^2. This is, however, a minimum value because, as Marcus and Sutin pointed out (20), it is in general necessary to integrate over all mutual orientations and distances of the reacting pair. Marcus and Sutin used a value of S of 0.01 in their analysis of the electron self-exchange rate constant of cytochrome c (20). Given that the surface of cytochrome c is 0.7% heme, this value of S assumes that electron transfer is enhanced by additional factor of 15.

We made detailed numerical analyses of the X-ray structures of cytochromes c, c_{551}, and b_5, by using a model in which electron transfer falls off exponentially with distance (55). The model assumed that the orientation of the two hemes and the nature of the intervening residues had no effect on the electron-transfer rate constant and that electron transfer occurred only at the surface of the protein. The analysis showed that electron transfer at the heme edge accounts for 40% (cytochrome c) to 80% (cytochrome b_5) of the total electron transfer. These values would correspond to enhancements of 2.4 for cytochrome c and 1.3 for cytochrome b_5. For this chapter we chose an enhancement factor of 5, between the upper and lower estimates.

The heme exposures and heme–heme distances calculated from the crystal structures may not represent the values found in solution. In a study of the cytochrome c–cytochrome b_5 interaction, Wendoloski et al. (64) ran picosecond dynamics and found substantial motions of the residues in the interface between the two hemes. In particular, they observed that the side chain of phenylalanine-82 moved to a bridging position between the two hemes in the complex. They also found that the inter-iron distances in two simulations were 1.1–2.1 Å smaller than the 17.8-Å distance in the static model.

Protein motions in general are under increasing study (65–68). Motions in cytochromes have been measured with NMR spectroscopic techniques

(69, 70). Studies of amide NH exchange, including work on the cytochromes (71, 72), reveal that relatively large protein motions occur on a biological time scale (73). Myoglobin has also been investigated by molecular modeling (74). Its crystal structure shows no channel large enough to admit to ligands such as carbon monoxide. Therefore, protein fluctuations are necessary to accommodate the CO binding. The modeling indicates that more than one channel from the protein surface to the heme opens and closes dynamically and allows CO binding. The binding of xenon to myoglobin shows similar characteristics (75).

The Association Constant, K_a

The association constant of the two cytochromes in the proper geometry for electron transfer, K_a, can be estimated by calculating the effective volume over which the reaction occurs along the reaction coordinate multiplied by an electrostatic work term [exp $(-w_r/RT)$] (20),

$$K_a = 4\pi N r^2 \delta(r) \exp \frac{-w_r}{RT} \quad (2)$$

where N is Avogadro's number, r is the sum of the radii of the two electron-transfer proteins, $\delta(r)$ is the range of internuclear separations that contribute significantly to the reaction rate, and w_r is the work to bring the two proteins into the proper geometry for electron transfer. The value for $\delta(r)$ is usually taken to be that at which the electron-transfer rate constant falls to $1/e$ of its value at the distance of closest approach. In general, electron transfer is thought to fall off exponentially as exp $[-\beta(d - d_0)]$ (d is the distance between the two hemes in the electron-transfer complex); $\delta(r)$ may be estimated as β^{-1}. For cytochrome c, $r = 33.2$ Å (47). With $\beta = 0.9$ Å^{-1} and hence $\delta(r) = 1.1$ Å, we calculate $4\pi N r^2 \delta(r) = 9.3$ M^{-1}.

Values for cytochrome c_{551} and cytochrome b_5 are given in Table II. Multiplication of these values by the appropriate work terms (discussed in the next section) gives the K_a values. These K_a values are rough approximations because the specifics of the protein surface cannot be taken into account. They are not directly comparable with experimental measurement of protein association, because the latter includes geometries that are not productive for electron transfer. Our NMR spectroscopic studies have shown no evidence for dimerization of cytochromes c or c_{551} at concentrations up to approximately 10 mM. Cytochrome b_5 shows small changes in the chemical shifts and line widths of the hemin methyl resonances as a function of concentration. If these changes reflect protein dimerization, then the equilibrium constant is 25–30 M^{-1} (55).

Table II. Calculation of Reorganization Energies
for $k_{et} = SK_a\nu_n\kappa_{el} \exp(-G_r^*/RT)$

Factor	Cytochrome c	Cytochrome c_{551}	Cytochrome b_5
Heme fraction surface area	0.007	0.012	0.038
Steric factor	0.0012	0.0036	0.0361
Radius (Å)	16.6	14.4	15.9
$4\pi r^2 dr$	9.27	6.98	8.50
Work (kcal mol^{-1})	2.7	0.30	3.1
K_a (M^{-1})	0.097	4.2	0.045
Heme–heme distance	8.9	8.6	7.5
$\kappa_{el} = \exp(-\beta(d - d_o))$	4.9×10^{-3}	6.5×10^{-3}	1.7×10^{-2}
$SK_a\nu_n\kappa_{el}$	5.9×10^6	9.8×10^8	2.9×10^8
k_{et} (exptl.)	5.1×10^3	5.1×10^6	2.7×10^3
ΔG_r^* (kcal mol^{-1})	4.2	3.1	6.85
λ (eV)	0.72	0.54	1.2

NOTE: $\mu = 0.1$ M, 25 °C.

The Work Term and Electrostatic Considerations

Bimolecular electron transfer in biological systems occurs between species that usually have a net charge as well as a substantially asymmetric charge distribution. Given an appropriate model, one can use the dependence of the electron-transfer rate constant on ionic strength to calculate two related parameters: the energy needed to form the electron-transfer complex (i.e., the work term) at a given ionic strength and the rate constant extrapolated to infinite ionic strength.

Of the theoretical approaches presently available, we (55) and others (76, 77) find that of van Leeuwen (78) to be the best at the ionic strengths used in NMR experiments (0.1–1.5 M). This formalism treats each protein as both monopole and a dipole. The expression is:

$$\ln \frac{k_I}{k_{inf}} = -[Z_{ox}Z_{red} + (ZD)(1 + \kappa r) + (DD)(1 + \kappa r)^2] \frac{q^2}{4\pi\epsilon_0\epsilon kTr} f(\kappa) \quad (3a)$$

where k is the Boltzmann constant, q is the charge on the electron, ϵ is the dielectric constant, ϵ_0 is the permittivity of vacuum, and T is the temperature.

$$ZD = \frac{Z_{ox}D_{red}' + Z_{red}D_{ox}'}{qr} \quad (3b)$$

$$DD = \frac{D_{ox}'D_{red}'}{(qr)^2} \quad (3c)$$

$$f(\kappa) = \frac{1 - \exp(-\kappa r)}{\kappa r \left(1 + \frac{\kappa r}{2}\right)} \quad (3d)$$

where Z_{ox} and Z_{red} are the net charges of the oxidized and reduced protein, respectively; D_{ox}' and D_{red}' are the components of the dipole moments through the exposed heme edge, respectively; r is the sum of the radii of the two electron-transfer partners; κ depends on the ionic strength, μ, as $0.329\ \mu^{1/2}$; k_I is the rate constant at a given ionic strength; and k_{inf} is the rate constant at infinite ionic strength.

Calculation (79, 80) of the dipole moments from the X-ray structures gives the values in Table III. The dipole moment is origin-dependent for molecules with a net charge; the center of mass has been chosen as the origin for our calculations. Figure 1 shows the point where the positive end of the dipole extends through the surface of the protein for cytochrome b_5. The positive end of the dipole extends through the side of the protein away from the heme. As can be seen in Table III, the dipole moments for cytochromes c and b_5 are very similar in magnitude, but opposite in sign (i.e., the positive end of the dipole moment extends through the protein near the exposed heme edge in cytochrome c). The dipole moment of cytochrome c_{551} is similar to that of cytochrome c in orientation, but the magnitude of the dipole moment is much smaller for the former protein.

The dipole moments can also be estimated from the dependence of the observed self-exchange rate constant on the ionic strength. In these calculations, the data are fit to van Leeuwen's equations by using a nonlinear least-squares algorithm and three adjustable parameters: the dipole moments through the exposed heme edge of the oxidized and reduced proteins and the rate constant at infinite ionic strength. As can be seen in Table III, the dipole moments calculated in these two ways are in very good agreement. Thus, the van Leeuwen approach is a useful electrostatic formalism at this level of representation.

The monopole–monopole, monopole–dipole, and dipole–dipole contributions as a function of ionic strength for cytochrome b_5 are shown in Figure 2. The monopole–monopole term falls quickly to zero at these ionic

Table III. Net Charges and Dipole Moments Through the Exposed Heme Edge of Cytochromes

Protein	Oxidized	Reduced	Method
Cytochrome c	+7.5	+6.5	
$r_p^a = 16.5$ Å, $\Theta^b = 30°$	273	258	X-ray[c]
	300	275	fit to k_{ex}
Cytochrome c_{551}	−2	−3	
$r_p = 14.4$ Å, $\Theta = 17°$	150	120	X-ray
Cytochrome b_5	−7.5	−8.5	
$r_p = 15.9$ Å, $\Theta = 15°$	−250	−280	fit to k_{ex}
	−280	−330	X-ray

[a] Protein radius.
[b] Angle of the dipole moment with respect to the heme plane.
[c] Ref. 80.

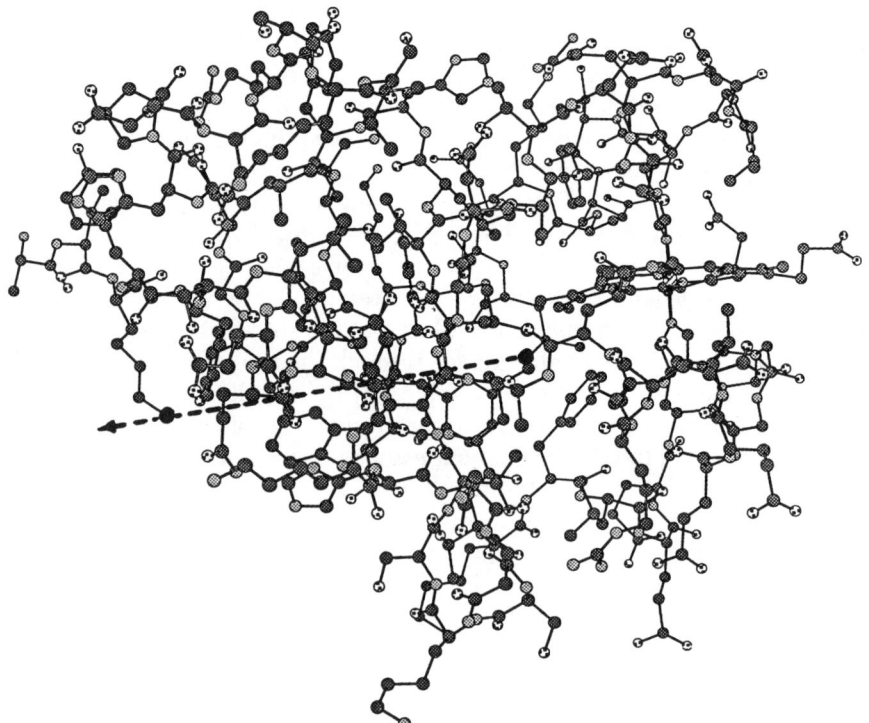

Figure 1. Cytochrome b_5 structure. The dashed line represents the dipole moment, which is from the center of mass (near Ile-23) through the surface of the protein (positive end of the dipole near the amino nitrogen of Lys-5).

strengths, as expected. The monopole–dipole term falls more slowly to zero, and the dipole–dipole term is almost constant between $\mu = 0.1$ and 1.5 M. Equation 3 also allows calculation of the contribution of the work term to K_a. Assuming that only electrostatic interactions are important in determining the rate constant as a function of ionic strength, the work needed to form the electron-transfer complex is $w_r = -RT \ln (k_I/k_{inf})$. The values of w_r are 2.7, 0.3, and 3.1 kcal mol^{-1} for the heme edge-to-heme edge complexes of cytochromes c, c_{551}, and b_5, respectively, at 0.1 M ionic strength.

Extrapolation of these interactions to infinite ionic strength allows estimation of the rate constant that ostensibly would be observed in the absence of electrostatic interactions. The k_{inf} for cytochrome c, c_{551}, and b_5 are 5.1 × 10^5, 2 × 10^7, and 6.9 × 10^5 M^{-1} s^{-1}, respectively. Thus, even when the electron self-exchange rate constants are extrapolated to infinite ionic strength, that of cytochrome c_{551} is still substantially larger than those of cytochrome c and cytochrome b_5. The origins of this condition are found in both steric and reorganizational energy differences.

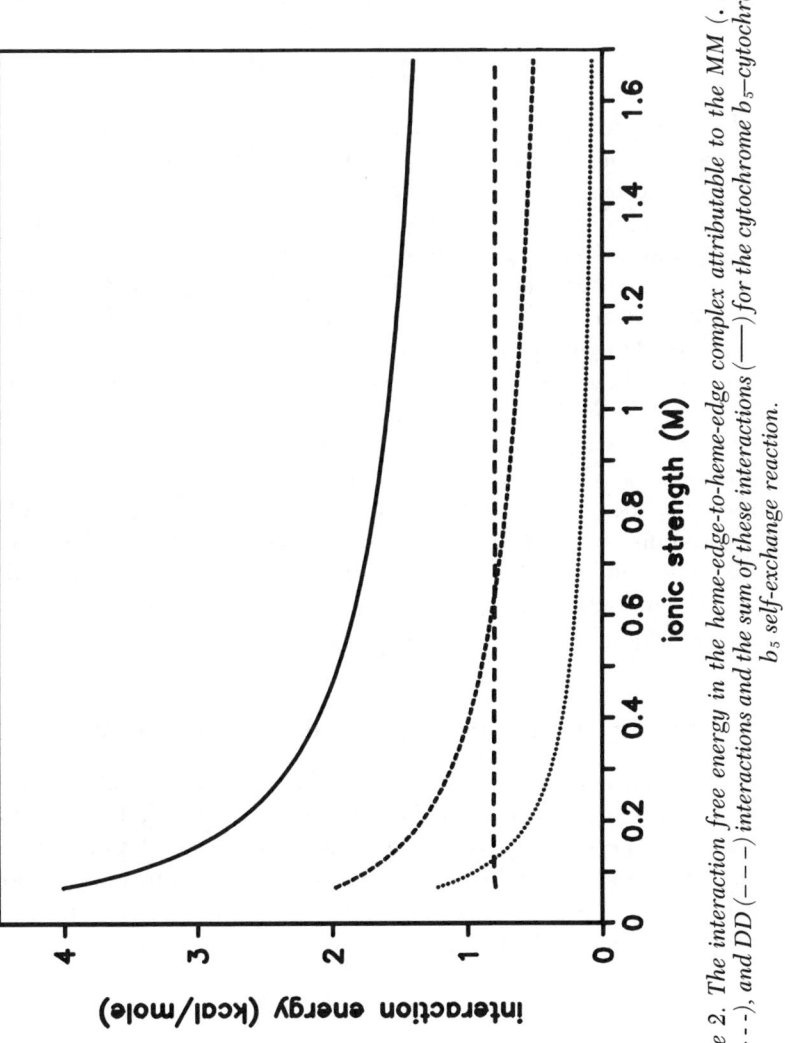

Figure 2. *The interaction free energy in the heme-edge-to-heme-edge complex attributable to the MM (· · ·), MD (– – –), and DD (– – –) interactions and the sum of these interactions (——) for the cytochrome b_5-cytochrome b_5 self-exchange reaction.*

Electron Transfer in the Heme-Edge-to-Heme-Edge Complex

Protein–protein electron transfer is nonadiabatic in most instances. That is, electron transfer does not occur each time the reactants achieve the transition-state geometry. It is generally thought that electron transfer falls off exponentially with distance. In this case, $\nu_n \kappa_{el}$ can be expressed as (20)

$$\nu_n \kappa_{el} = 10^{13} \exp[-\beta(d - d_0)] \tag{4}$$

where d_0 is the value of d (the distance between the two hemes in the electron-transfer complex) at which κ_{el} equals 1 and β is a system-dependent constant. On the basis of recent work of Gray and Malmström (81) in the ruthenated cytochrome c system, we chose a value for β of 0.9 Å^{-1} for our calculations. For d_0, we adopted the convention that $d_0 = 3$ Å (20, 81).

The value of d was calculated by modeling all three of the self-exchange complexes discussed in this chapter. In each case, a heme-edge-to-heme-edge geometry of the two proteins was created by using the crystal structures of the proteins and the BIOGRAF molecular modeling software (DREIDING force field). These models give distances of closest approach between the two hemes of 8.9 Å for cytochrome c, 7.5 Å for cytochrome b_5, and 8.6 Å for cytochrome c_{551}. Our results for cytochrome c are in very good agreement with those of Weber, who found a heme-edge-to-heme-edge distance of 9.4 Å with no minimization (20). These values of d allow calculation of the values of $\nu_n \kappa_{el}$ as shown in Table II. These values are, however, only approximate because it is generally necessary to integrate over all distances and mutual orientations of the electron-transfer partners.

Calculation of Self-Exchange Reorganizational Energies

Once the electron self-exchange rate constants are measured and values for the parameters S, K_a, and $\nu_n \kappa_{el}$ are estimated, one can calculate ΔG_r^* according to eq 1. These values are 4.2, 3.1, and 6.9 kcal mol^{-1} for cytochromes c, c_{551}, and b_5, respectively. The rate constant for electron transfer should depend on the free-energy change for the electron step ($\Delta G^{0\prime}$) and the reorganizational energy of the reaction, λ, as (26)

$$\Delta G_r^* = \left(\frac{\lambda}{4}\right)\left(\frac{1 + \Delta G^{0\prime}}{\lambda}\right)^2 \tag{5}$$

where $\Delta G^{0\prime}$ is $(\Delta G^0 + w_p - w_r)$, ΔG^0 is the free-energy change of the reaction, and w_p and w_r represent the work required to bring the reactants and products to the separation achieved in the electron-transfer complex. For self-exchange reactions, the terms w_r and w_p are equal to each other. When the thermodynamic driving force for the reaction, ΔG^0, is zero (as in the self-exchange reaction), $\lambda = 4\Delta G_r^*$. Therefore, the λ values for these three self-exchange reactions are 17, 12, and 28 kcal mol^{-1} or, in conventional energy units, 0.7, 0.5, and 1.2 eV for cytochromes c, c_{551}, and b_5, respectively.

Numerically, the reorganizational energies depend strongly on the values assumed for the steric factor. If one uses simply the fraction of the surface area that is heme squared as the steric factor, then the calculated λ values would be 0.4, 0.2, and 0.9 eV for cytochromes c, c_{551}, and b_5, respectively. On the other hand, if consideration of electron transfer over all distances and mutual orientations of the two proteins were to enhance the electron-transfer rate constant by a factor of 10 for both proteins, then the λ values for the three self-exchange reactions would be 0.9, 0.7, and 1.3 eV.

Cross Reactions

Cytochrome c and Cytochrome c_{551}.

Cytochrome c_{551} is a smaller protein than cytochrome c, with a more exposed heme and a closer heme-edge-to-heme-edge distance in the complex, as well as a smaller net charge and a smaller dipole moment. All of these points indicate that the electron self-exchange rate constant for cytochrome c_{551} should be larger than that of cytochrome c. We calculate the difference to be about 170-fold at 0.1 M ionic strength and 25 °C. Experimentally, a substantial difference is indeed observed. However, at μ = 0.1 M and 25 °C, the rate constant for cytochrome c_{551} (5.1 × 10^6 M^{-1} s^{-1}) is in fact three orders of magnitude greater than that of cytochrome c (5.1 × 10^3 M^{-1} s^{-1}). Given eq 1 as a model for electron transfer, the very fast rate of self-exchange for cytochrome c_{551} is explained as a smaller reorganizational energy for that protein.

One can also arrive at the conclusion that cytochrome c_{551} has a very low reorganizational energy by considering the absolute magnitude of the electron self-exchange rate constant, 10^7 M^{-1} s^{-1}. The diffusion-controlled interaction of two cytochromes c_{551} has a rate constant of about 5 × 10^9 M^{-1} s^{-1} (82). If only a small percentage of the surface area of the protein is reactive, then cytochrome c_{551} electron self-exchange is occurring essentially at diffusion control. This conclusion implies that there can be no other significant energy barrier (i.e., reorganizational energy) to the process.

The formalism expressed herein should be useful in analyzing the reaction between cytochrome c and cytochrome c_{551}. The reorganizational energy of electron transfer between two proteins can be expressed as (26)

$$\Delta G_r^* = \frac{\Delta G_{11}^* + \Delta G_{22}^* + \Delta G_{12}^{0'}}{2} + \frac{(\Delta G_{12}0')^2}{8(\Delta G_{11}^* + \Delta G_{22}^*)} \qquad (6)$$

where ΔG_r^* is the reorganizational free energy of the electron-transfer reaction, ΔG_{11}^* and ΔG_{22}^* are the reorganizational energies of the two self-exchange reactions, and $\Delta G_{12}^{0'}$ is the free-energy change for the electron-transfer step, calculated from the difference in redox potentials. The redox potentials of the two proteins are, fortuitously, similar (83–88). The reorganizational energy when $\Delta G_{12}^{0'} = 0$ should be the arithmetic mean of the reorganizational energies of the two partners (cf. eq 6), or approximately 3.8 kcal mol^{-1} in this case. We can use the net charges and dipole moments in Table III to calculate a work term of 2.0 kcal mol^{-1}. This result, in combination with data in Table II and a heme–heme distance of 8.7 Å, allows prediction of a rate constant of 3×10^4 M^{-1} s^{-1} ($\mu = 0.1$ M, 4 °C). The electron-transfer rate constant between cytochrome c and cytochrome c_{551} has been measured by two groups (89, 90). The rate constant for the bimolecular reaction at 0.1 M ionic strength and 4.5 °C is 1.6×10^4 M^{-1} s^{-1}. Thus, the predicted rate constant is in excellent agreement with that measured experimentally.

The electron-transfer reaction between cytochrome c and cytochrome c_{551} is known to be independent of ionic strength over the range $\mu = 0.04$–0.4 M (90). The net charges and dipole moments of the proteins allow one to calculate the expected dependence of the electron-transfer rate constant on ionic strength. In line with the experimental data, the calculated rate constant is independent of ionic strength over the range of interest. A fit to the data is shown in Figure 3. The only adjustable parameter, the rate constant for the electron-transfer reaction at infinite ionic strength, k_{inf}, is 3.6×10^4 M^{-1} s^{-1}.

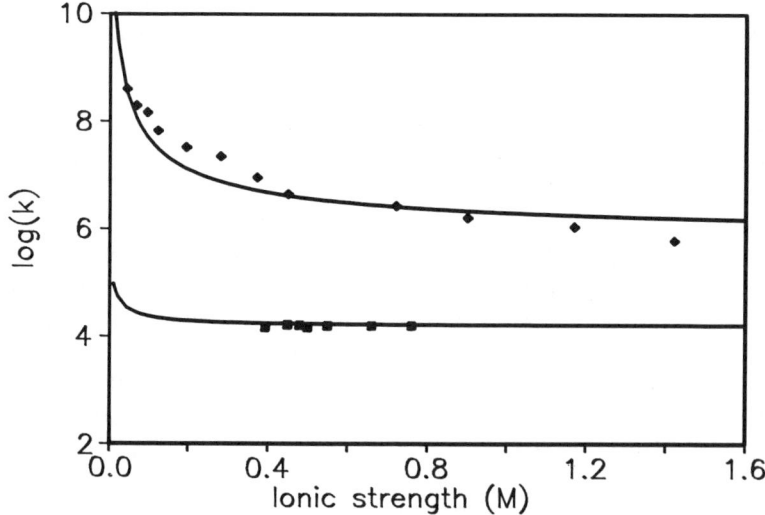

Figure 3. Electron-transfer rate constants for cytochrome–cytochrome reactions as a function of ionic strength. The solid lines are best fits to the data, given the dipole moments in Table III and the k_{inf} reported in the text. Key: ■, cytochromes c–c_{551} (90); ♦, cytochromes c–b_5 (94). (Reproduced with permission from ref. 55. Copyright 1989 Rockefeller University Press.)

Cytochromes c and b_5. Our analysis concludes that cytochrome b_5 has a larger reorganizational energy than cytochrome c. These two proteins are very similar in size and almost identical in net charge and net dipole moment. However, the heme is more exposed in cytochrome b_5. This difference leads to a larger S factor. The two hemes are able to get closer in the cytochrome b_5–cytochrome b_5 complex than in the cytochrome c–cytochrome c complex. The result is a smaller value of $\nu_n \kappa_{el}$. The changes in both S and $\nu_n \kappa_{el}$ predict a faster electron-transfer rate constant in the complex for cytochrome b_5.

The model predicts that the electron self-exchange rate constant should be 50 times larger for cytochrome b_5 than for cytochrome c. However, experiment shows that there is almost no difference. Given eq 1 as a model for electron transfer, the necessary conclusion is that the reorganizational energy for cytochrome b_5 is larger than that for cytochrome c. The heme in cytochrome b_5 is more exposed, and the greater exposure may lead to a larger reorganizational energy because the heme is calculated to have a larger reorganization in water than in the protein (*91*). Another possibility is that electrostatic repulsion between the extended heme propionates of the two cytochromes in the heme-edge-to-heme-edge complex is not adequately modeled by the van Leeuwen approach.

In the past 15 years, substantial attention has been paid to the electron-transfer reaction between cytochrome c and cytochrome b_5 (*92–94*). At low ionic strength, the two cytochromes form a complex that has been studied by using modeling (*64, 95*), optical (*96, 97*), and NMR spectroscopic techniques (*98, 99*). McLendon and Miller used pulse radiolysis to measure intramolecular electron transfer in the cytochrome c–cytochrome b_5 complex (*93*). Data as a function of the driving force of the reaction allowed them to calculate a λ of approximately 0.8 eV for the sum of the reorganizational energies of cytochromes c and b_5.

A value for λ can also be calculated as

$$\lambda = \frac{\Delta G_{11}{}^* + \Delta G_{22}{}^*}{2} \qquad (7)$$

The self-exchange reorganizational energies derived herein give a value for λ of 0.95 eV. The difference between the two estimates comes both from errors in the approximations for bimolecular electron transfer and from the assumption in the intracomplex electron-transfer experiments that the substitution of one metal for another at the active site results in no change beyond that of altering the redox potential. Overall, however, these two estimates of the reorganizational energy, one derived from intracomplex electron transfer as a function of free energy and the other derived from bimolecular self-exchange data, are in very good agreement.

At high ionic strengths, the reaction between cytochrome c and cytochrome b_5 becomes bimolecular. Data of Stonehuerner et al. (94) are given in Figure 3. The solid line is a fit to the data, including the charges and dipole moments given in Table III and a k_{inf} of $1.1 \times 10^5\,M^{-1}\,s^{-1}$. Considering that there is only one adjustable parameter, k_{inf}, the fit is very good.

Conclusion

In this chapter we have integrated experiment, calculations from crystal structure data, and molecular modeling to present a picture of electron transfer for cytochromes c, c_{551}, and b_5 that encompasses both bimolecular and unimolecular electron transfer. The two types of measurements are complementary and lead to similar conclusions for the reorganizational energy of the proteins in question. Electrostatic interactions appear to be quite well modeled by van Leeuwen theory. However, better models of both protein–protein electrostatic interactions and protein motions will be welcome.

Acknowledgments

We thank the National Institutes of Health for support of this work (DK 38826). D. W. Dixon thanks the National Science Foundation for a Career Advancement Award (NSF CHE 8707447). The molecular graphics facility was supported with funds from the Department of Education (2-4-01008).

References

1. *Antennas and Reactions Centers of Photosynthetic Bacteria;* Michel-Beyerle, M. E., Ed.; Springer-Verlag: New York, 1985.
2. Hatefi, Y. *Annu. Rev. Biochem.* **1985**, *54*, 1015–1069.
3. Pettigrew, G. W.; Moore, G. R. *Cytochromes c. Biological Aspects;* Springer–Verlag: New York, 1987.
4. Anraku, Y. *Annu. Rev. Biochem.* **1988**, *57*, 101.
5. Dixit, B. P. S. N.; Vanderkooi, J. M. *Curr. Top. Bioenerg.* **1984**, *13*, 159–202.
6. Poulos, T. L.; Finzel, B. C. In *Peptide and Protein Reviews;* Marcel Dekker: New York, 1984; Vol. IV, pp 115–171.
7. Ortiz de Montellano, P. R. *Acc. Chem. Res.* **1987**, *20*, 289–294.
8. Mathews, F. S. *Prog. Biophys. Molec. Biol.* **1985**, *45*, 1–56.
9. Moore, G. R.; Eley, C. G. S.; Williams, G. *Adv. Inorg. Bioinorg. Mech.* **1984**, *3*, 1–96.
10. Newton, M. D.; Sutin, N. *Annu. Rev. Phys. Chem.* **1984**, *35*, 437–480.
11. Sutin, N. *Prog. Inorg. Chem.* **1983**, *30*, 441–498.
12. Isied, S. S. *Prog. Inorg. Chem.* **1984**, *32*, 443–517.
13. Hush, N. S. *Coord. Chem. Rev.* **1985**, *64*, 135–157.
14. Haim, A. *Comments Inorg. Chem.* **1985**, *4*, 113–149.
15. Tollin, G.; Meyer, T. E.; Cusanovich, M. A. *Biochim. Biophys. Acta* **1986**, *853*, 29–41.
16. Mayo, S. L.; Ellis, W. R., Jr.; Crutchley, R. J.; Gray, H. B. *Science* **1986**, *233*, 948.

17. Scott, R. A.; Mauk, A. G.; Gray, H. B. *J. Chem. Educ.* **1985**, *62*, 932–938.
18. McLendon, G. *Acc. Chem. Res.* **1988**, *21*, 160–167.
19. Guarr, T.; McLendon, G. *Coord. Chem. Rev.* **1985**, *68*, 1–52.
20. Marcus, R. A.; Sutin, N. *Biochim. Biophys. Acta* **1985**, *811*, 265–322.
21. Dixon, D. W. In *Molecular Structure and Energetics: Principles of Enzyme Activity;* Liebman, J. F.; Greenberg, A., Eds.; Elsevier: New York, 1988; pp 169–225.
22. Bechtold, R.; Kuehn, C.; Lepre, C.; Isied, S. S. *Nature (London)* **1986**, *322*, 286–288.
23. Hoffman, B. M.; Ratner, M. A. *J. Am. Chem. Soc.* **1987**, *109*, 6237–6243.
24. McLendon, G. M.; Pardue, K.; Bak, P. *J. Am. Chem. Soc.* **1987**, *109*, 7540–7541.
25. Cheung, E.; English, A. M. *Inorg. Chem.* **1988**, *27*, 1078–1081.
26. Sutin, N.; Creutz, C.; Linck, R. G. In *Inorganic Reactions and Methods;* Zuckerman, J. J., Ed.; 1986; Vol. 15, pp 3–87.
27. Timkovich, R.; Cork, M. S.; Taylor, P. V. *Biochemistry* **1984**, *23*, 3526–3533.
28. Senn, H.; Wüthrich, K. *Biochim. Biophys. Acta* **1983**, *746*, 48–60.
29. Senn, H.; Eugster, A.; Wüthrich, K. *Biochim. Biophys. Acta* **1983**, *743*, 58–68.
30. Keller, R. M.; Picot, D.; Wüthrich, K. *Biochim. Biophys. Acta* **1979**, *580*, 259–265.
31. Gupta, R. K. *Biochim. Biophys. Acta* **1973**, *292*, 291–295.
32. Kowalsky, A. *Biochemistry* **1965**, *4*, 2382–2388.
33. Oldfield, E.; Allerhand, A. *Proc. Natl. Acad. Sci. U.S.A.* **1973**, *70*, 3531–3535.
34. Gupta, R. K.; Koenig, S. H.; Redfield, A. G. *J. Magn. Reson.* **1972**, *7*, 66–73.
35. Concar, D. W.; Hill, H. A. O.; Moore, G. R.; Whitford, D.; Williams, R. J. P. *FEBS Lett.* **1986**, *206*, 15–19.
36. Keller, R. M.; Wüthrich, K.; Schejter, A. *Biochim. Biophys. Acta* **1977**, *491*, 409–415.
37. Senn, H.; Cusanovich, M. A.; Wüthrich, K. *Biochim. Biophys. Acta* **1984**, *785*, 46–53.
38. Senn, H.; Guerlesquin, F.; Bruschi, M.; Wüthrich, K. *Biochim. Biophys. Acta* **1983**, *748*, 194–204.
39. Timkovich, R.; Cork, M. S. *Biochemistry* **1984**, *23*, 851–860.
40. Senn, H.; Wüthrich, K. *Biochim. Biophys. Acta* **1983**, *743*, 69–81.
41. Keller, R. M.; Wüthrich, K.; Pecht, I. *FEBS Lett.* **1976**, *70*, 180–184.
42. Timkovich, R.; Cai, M. L.; Dixon, D. W. *Biochem. Biophys. Res. Commun.* **1988**, *150*, 1044–1050.
43. Kimura, K.; Peterson, J.; Wilson, M.; Cookson, D. J.; Williams, R. J. P. *J. Inorg. Biochem.* **1981**, *15*, 11–25.
44. McLendon, G.; Smith, M. *Inorg. Chem.* **1982**, *21*, 847–850.
45. Moore, G. R.; Williams, R. J. P.; Peterson, J.; Thomson, A. J.; Mathews, F. S. *Biochim. Biophys. Acta* **1985**, *829*, 83–86.
46. Dixon, D. W.; Hong, X.; Woehler, S. E.; Mauk, A. G.; Sishta, B. *J. Am. Chem. Soc.*, in press.
47. Wherland, S.; Gray, H. In *Biological Aspects of Inorganic Chemistry [Symposium];* Addison, A. W.; Cullen, W. R.; Dolphin, D.; James, B. R., Eds.; Wiley: New York, 1978; pp 289–368.
48. Takano, T.; Dickerson, R. E. *J. Mol. Biol.* **1981**, *153*, 79–94.
49. Takano, T.; Dickerson, R. E. *J. Mol. Biol.* **1981**, *153*, 95–115.
50. Matsuura, Y.; Takano, T.; Dickerson, R. E. *J. Mol. Biol.* **1982**, *156*, 389–409.
51. Mathews, F. S.; Czerwinski, E. W. In *Enzymes in Biological Membranes;* 2nd ed.; Martonosi, A. N., Ed.; Plenum: New York, 1986; Vol. 4, pp 235–300, and references therein.
52. Christiano, A. B.; Steggles, A. W. *Nucl. Acids Res.* **1989**, *17*, 799.

53. Strittmatter, P.; Ozols, J. *J. Biol. Chem.* **1966**, *241*, 4787–4792.
54. Ozols, J.; Strittmatter, P. *J. Biol. Chem.* **1969**, *244*, 6617–6618.
55. Dixon, D. W.; Hong, X.; Woehler, S. E. *Biophys. J.* **1989**, *56*, 339–351.
56. Stellwagen, E. *Nature (London)* **1978**, *275*, 73–74.
57. Sutin, N. *Chem. Br.* **1972**, *8*, 148–151.
58. Dixon, D. W. In ref. 21; Liebman, J. F.; Greenberg, A., Eds.; pp 206–208.
59. Northrup, S. H.; Boles, J. O.; Reynolds, J. C. L. *Science* **1988**, *241*, 67–70, and references therein.
60. Mauk, M. R.; Mauk, A. G.; Weber, P. C.; Matthew, J. B. *Biochemistry* **1986**, *25*, 7085–7091; **1987**, *26*, 974.
61. Slayton, S. S.; Fisher, M. T.; Sligar, S. G. *J. Biol. Chem.* **1988**, *263*, 13544–13548.
62. Connolly, M. E. *J. Appl. Crystallogr.* **1983**, *16*, 548–558.
63. BIOGRAF Version 1.40; BioDesign, 199 S. Los Robles Ave., Suite 270, Pasadena, CA 91101.
64. Wendoloski, J. J.; Matthew, J. B.; Weber, P. C.; Salemme, F. R. *Science* **1987**, *238*, 794–796.
65. Huber, R. *Biochem. Soc. Trans.* **1987**, *15*, 1009–1020.
66. Hayashi, L. *Dynamic Analysis of Enzyme Systems: An Introduction;* Springer–Verlag: New York, 1986.
67. Welch, G. R. *The Fluctuating Enzyme;* Wiley: New York, 1986.
68. McCammon, J. A.; Harvey, S. C. *Dynamics of Proteins and Nucleic Acids;* Cambridge University Press: New York, 1987.
69. Williams, G.; Moore, G. R.; Williams, R. J. P. *Comments Inorg. Chem.* **1985**, *4*, 55–98.
70. Johnson, R. D.; La Mar, G. N.; Smith, K. M.; Parish, D. W.; Langry, K. C. *J. Am. Chem. Soc.* **1989**, *111*, 481–485.
71. Roder, H.; Elöve, G. A.; Englander, S. W. *Nature (London)* **1988**, *335*, 700–704.
72. Yu, L. P.; Smith, G. M. *Biochemistry* **1988**, *27*, 1949–1956.
73. Englander, S. W.; Kallenbach, N. R. *Q. Rev. Biophys.* **1984**, *16*, 521–655.
74. Kottalam, J.; Case, D. A. *J. Am. Chem. Soc.* **1988**, *110*, 7690–7697, and references therein.
75. Tilton, R. F.; Singh, U. C.; Kuntz, I. D.; Kollman, P. A. *J. Mol. Biol.* **1988**, *199*, 195–211.
76. Rush, J. D.; Koppenol, W. H. *Biochim. Biophys. Acta* **1988**, *936*, 187–198.
77. Rush, J. D.; Lan, J.; Koppenol, W. H. *J. Am. Chem. Soc.* **1987**, *109*, 2679–2682.
78. van Leeuwen, J. W. *Biochim. Biophys. Acta* **1983**, *743*, 408–421.
79. Jackson, J. D. *Classical Electrodynamics;* 2nd ed.; Wiley: New York, 1975; p 139.
80. Koppenol, W. H.; Margoliash, E. *J. Biol. Chem.* **1982**, *257*, 4426–4437.
81. Gray, H. B.; Malmström, B. G., submitted for publication.
82. Cantor, C. R.; Schimmel, P. R. *Biophysical Chemistry, Part II;* W. H. Freeman: San Francisco, CA, 1980; p 584.
83. Hwang, Y.-Y.; Kimura, T. *Biochemistry* **1984**, *23*, 2231–2236.
84. Taniguchi, V. T.; Sailasuta-Scott, N.; Anson, F. C.; Gray, H. B. *Pure Appl. Chem.* **1980**, *52*, 2275–2281.
85. Moore, G. R.; Pettigrew, G. W.; Pitt, R. C.; Williams, R. J. P. *Biochim. Biophys. Acta* **1980**, *590*, 261–271.
86. Leitch, F. A.; Moore, G. R.; Pettigrew, G. W. *Biochemistry* **1984**, *23*, 1831–1838.
87. Rogers, N. K.; Moore, G. R. *FEBS Lett.* **1988**, *228*, 69–73.
88. Rogers, N. K.; Moore, G. R.; Sternberg, M. J. E. *J. Mol. Biol.* **1985**, *182*, 613–616.
89. Greenwood, C.; Finazzi-Agro, A.; Guerrieri, P.; Avigliano, L.; Mondovi, B.; Antonini, E. *Eur. J. Biochem.* **1971**, *23*, 321–327.

90. Morton, R. A.; Overnell, J.; Harbury, H. A. *J. Biol. Chem.* **1970**, *245*, 4653–4657.
91. Churg, A. K.; Weiss, R. M.; Warshel, A.; Takano, T. *J. Phys. Chem.* **1983**, *87*, 1683–1694.
92. McLendon, G. L.; Winkler, J. R.; Nocera, D. G.; Mauk, M. R.; Mauk, A. G.; Gray, H. B. *J. Am. Chem. Soc.* **1985**, *107*, 739–740.
93. McLendon, G.; Miller, J. R. *J. Am. Chem. Soc.* **1985**, *107*, 7811–7816.
94. Stonehuerner, J.; Williams, J. B.; Millett, F. *Biochemistry* **1979**, *18*, 5422–5427.
95. Salemme, F. R. *J. Mol. Biol.* **1976**, *102*, 563–568.
96. Mauk, M. R.; Reid, L. S.; Mauk, A. G. *Biochemistry,* **1982**, *21*, 1843–1846.
97. Holloway, P. W.; Mantsch, H. H. *Biochemistry* **1988**, *27*, 7991–7993.
98. Eley, C. G. S.; Moore, G. R. *Biochem. J.* **1983**, *215*, 11–21.
99. Hartshorn, R. T.; Mauk, A. G.; Mauk, M. R.; Moore, G. R. *FEBS Lett.* **1987**, *213*, 391–395.

RECEIVED for review May 1, 1989. ACCEPTED revised manuscript September 7, 1989.

9

Electron-Transfer Kinetics of Singly Labeled Ruthenium(II) Polypyridine Cytochrome c Derivatives

Bill Durham, Lian Ping Pan, Seung Hahm, Joan Long, and Francis Millett

Department of Chemistry and Biochemistry, University of Arkansas, Fayetteville, AR 72701

> Cytochrome c was labeled at specific lysine groups with Ru(bipyridine)$_2$(dicarboxybipyridine) [Ru(bpy)$_2$(dcbpy)]. Singly labeled derivatives at lysines 7, 8, 13, 25, 72, 86, and 87 were isolated. Laser flash photolysis experiments showed that the excited state of the ruthenium complex undergoes an electron-transfer reaction with the heme in some of the derivatives. The first-order rate constants are 16×10^6 and 26×10^6 s^{-1} for the forward and backward electron-transfer reactions, respectively, in the derivative labeled at lysine 13. Derivatives also were prepared by allowing Ru(bpy)$_2$CO$_3$ to react with cytochrome c at pH 7, followed by reaction with imidazole. Purified derivatives containing Ru(bpy)$_2$(imidazole)$^{2+}$ linked to histidines 26 and 33 were obtained.

SMALL METAL COMPLEXES COVALENTLY LINKED TO METALLOPROTEINS have proven to be invaluable in the study of electron-transfer reactions. Much of the pioneering work in this area has been done by Gray, Isied, and co-workers, who used Ru(NH$_3$)$_5^{2+}$ bound to the naturally occurring histidine residues of cytochrome c (cyt c) (1–5), myoglobin (6), and other proteins (7–9). In these examples, the distances between the redox centers are well defined, as are the driving forces for these reactions. Armed with this information, these researchers have been able to compare the observed rate constants for electron transfer to those predicted by the theoretical treatments presented by Marcus (10–13).

0065-2393/90/0226-0181$06.00/0
© 1990 American Chemical Society

From these comparisons, it has become evident that the distance dependence for electron transfer through protein can be described as an exponential decrease with distance scaled by a term β (typically about 0.9–1.2 Å^{-1}). The free energy dependence of these reactions appears to be well described by the relations developed by Marcus. Few examples of reactions have driving forces of magnitude comparable to or larger than the reorganizational barriers, but such reactions are of considerable interest. Questions about the nature of the molecular material between redox centers and how it may affect the rate of electron transfer are currently being actively pursued.

Many clever strategies have been used to systematically vary the driving forces in these systems, including changing the metal in the heme portion (14) of the protein and permutation of the ligands on the ruthenium (15). The zinc-substituted proteins (6) provide long-lived excited states with which reactions with high driving forces can be explored.

We have developed a synthetic method (16) that allows the attachment of Ru(bpy)$_2$(dcbpy) to the amine end of lysine residues (bpy is bipyridine; dcbpy is the deprotonated form of 4,4'-dicarboxy-2,2'-bipyridine). Ru(bpy)$_2$(dcbpy) is essentially equivalent to the well-known complex Ru(bpy)$_3{}^{2+}$ (17). Ru(bpy)$_3{}^{2+}$ has a strongly oxidizing or reducing excited state with a lifetime of about 600 ns. It has been used in the study of electron-transfer reactions with many different reagents, including cyt c and several other proteins (13, 18, 19). The distance dependence question has been explored with esterified derivatives of Ru(bpy)$_3{}^{2+}$ in synthetic polymers (20).

The reactions of Ru(bpy)$_3{}^{2+}$ (i.e., the excited-state and ground-state 1+ and 3+ forms) are characterized by high driving forces and low reorganizational barriers. Thus, the (bpy)$_2$Ru(dcbpy–cyt c) derivatives may provide further opportunities to obtain information about electron transfer in the barrier-free regime. An added advantage is that the electron-transfer properties of the native heme iron can be explored in these studies. Proteins derivatized with Ru(bpy)$_2$(dcbpy) will have reasonably well-defined geometries from which the distances between electron-transfer centers can be obtained.

In the process of developing a practical synthetic route to the (bpy)$_2$Ru-(dcbpy–cyt c) derivatives, a valuable side reaction was discovered. This side reaction can ultimately be made to produce (bpy)$_2$(imid)Ru((His)cyt c) derivatives (imid is imidazole) in which the two coordination sites not taken up by bipyridine are occupied by a simple imidazole and the imidazole of a histidine residue. These derivatives show long-lived room-temperature emission that presumably will undergo photoredox reactions.

Synthesis of (bpy)$_2$Ru(dcbpy–cyt c) Derivatives

The general synthetic scheme used to prepare the (bpy)$_2$Ru(dcbpy–cyt c) derivatives is shown in Scheme I. The N-hydroxysuccinimide ester of dcbpy

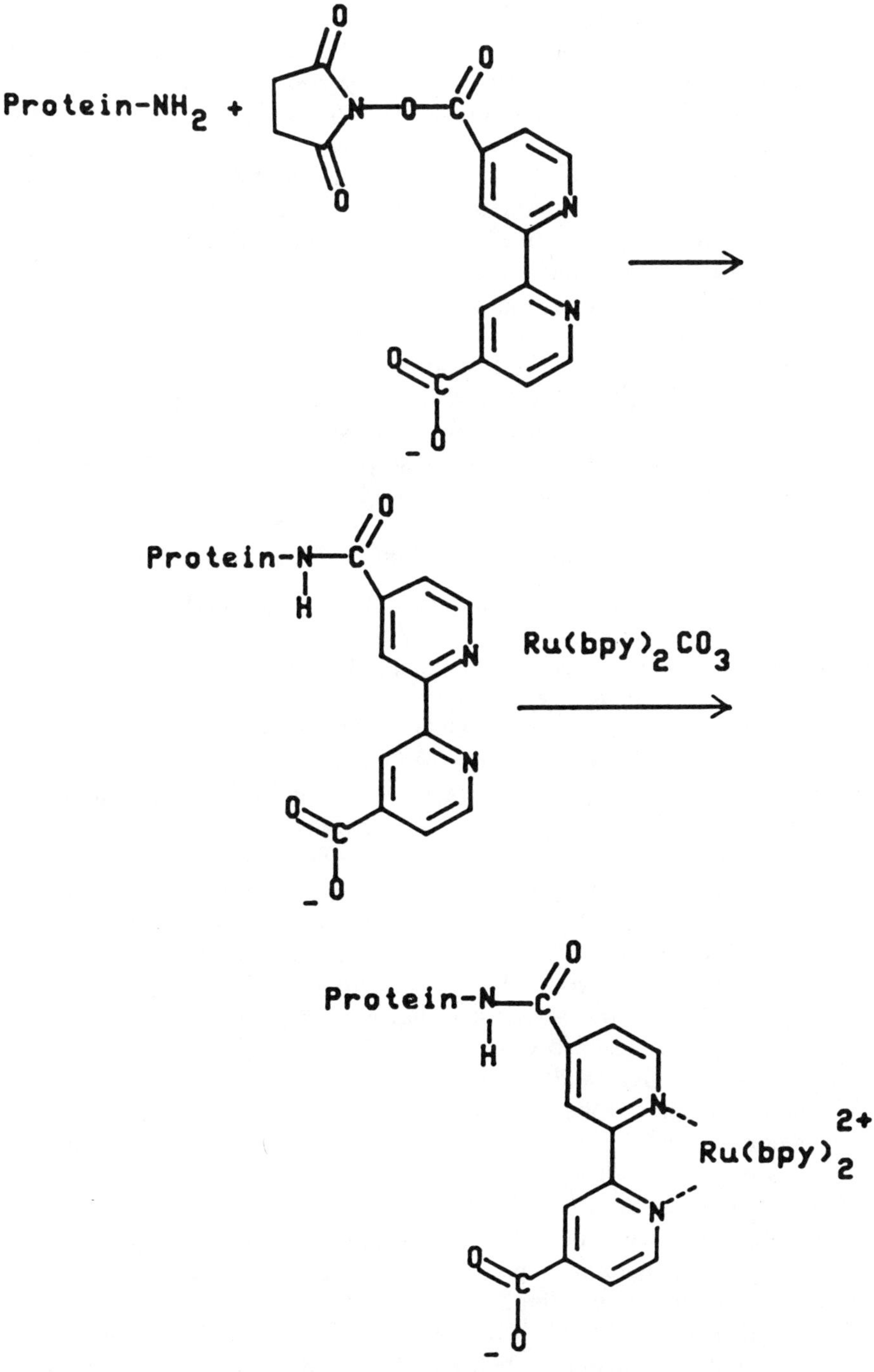

Scheme I. Preparation of (bpy)$_2$Ru(dcbpy–cyt c) derivatives. (Reproduced from ref. 16. Copyright 1988 American Chemical Society.)

was prepared by reaction with the partially deprotonated dcbpy in dimethylformamide (DMF) in the presence of dicyclohexylcarbodiimide. The ester was found to react with the lysine groups of cyt c in a few hours at room temperature. The major separation step performed at this stage is illustrated in Figure 1. The purified fractions were allowed to react with about a 10-fold excess of $Ru(bpy)_2CO_3$ (presumably $Ru(bpy)_2(H_2O)^{2+}$ in solution) and then chromatographed to remove unreacted starting material.

A few of the originally purified fractions contained mixtures of closely related derivatives that could be separated by further chromatography. High-performance liquid chromatographs (HPLCs) of tryptic digests of the ruthenium-containing derivatives and subsequent amino acid analysis of the appropriate fractions were used to identify the location of each label. A representative example is illustrated in Figure 2. The purified derivatives had heme redox potentials in the range of native cyt c, 250–260 mV. The derivatives also retained the 695-nm absorption band. This retention indicated that methionine 80 was not perturbed.

Synthesis of $(bpy)_2(imid)Ru((His)cyt\ c)$ Derivatives

It was discovered during the preparation of the $(bpy)_2Ru(dcbpy-cyt\ c)$ derivatives that $Ru(bpy)_2CO_3$ was able to specifically modify the histidine residues on cyt c upon prolonged incubation at pH 7. The reaction is reasonably efficient and does not require a large excess of reagent. It was possible to separate two singly labeled $(bpy)_2(H_2O)Ru((His)cyt\ c)$ derivatives by chromatography (CM–32) (data not shown). These derivatives, as well as the parent compound $Ru(bpy)_2(imid)(H_2O)^{2+}$, were not luminescent. However, one of us (Durham) had previously observed that solution-phase $Ru(bpy)_2(imid)_2^{2+}$ had an emission centered at 660 nm when excited at 450 nm. Therefore, the crude reaction mixture of $(bpy)_2(H_2O)Ru((His)cyt\ c)$ derivatives was incubated with 1 M imidazole to prepare $(bpy)_2(imid)Ru((His)cyt\ c)$ derivatives. The reaction, followed by monitoring the emission at 660 nm, was found to be complete after 18 h. Chromatography (CM–32) resulted in the separation shown in Figure 3. The UV–visible spectra of fractions 1 and 2 were equal to the sum of the spectra of one equivalent of $Ru(bpy)_2(imid)_2^{2+}$ and one equivalent of native cyt c (Figure 4); fraction 3 contained 2 equivalents of $Ru(bpy)_2(imid)_2^{2+}$.

Fractions 1 and 2 were rechromatographed (CM–32), and an aliquot of each was digested with trypsin and chromatographed on a reverse-phase HPLC column, as described by Pan et al. (16). In the chromatogram of fraction 2, the native peptide 28–38 was completely missing, and a $Ru(bpy)_2(imid)_2$-containing peptide that eluted a little later in the chromatogram had the same amino acid composition as 28–38. There were no other changes in the HPLC chromatogram relative to that of native cyt c. Therefore, fraction 2 is singly labeled at His 33. In a similar fashion, fraction 1 was found to be singly labeled at His 26.

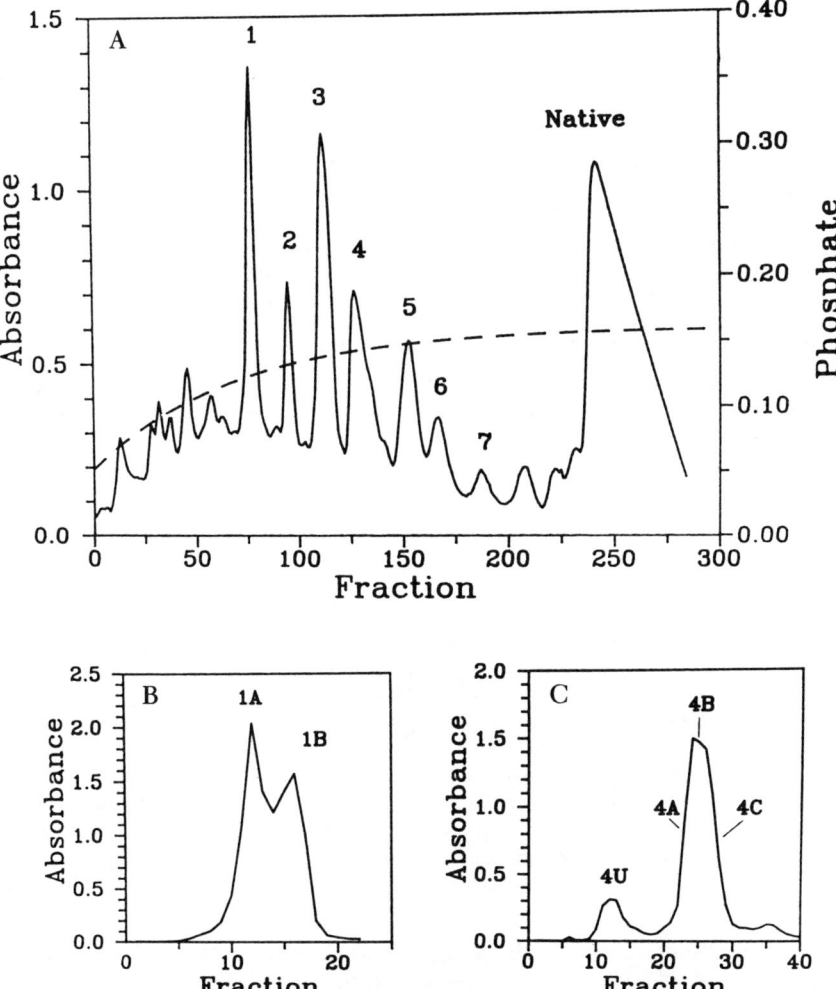

Figure 1. Purification of the dcbpy–cyt c and (bpy)$_2$Ru(dcbpy–cyt c) derivatives. A, The crude reaction mixture of dcbpy–cyt c (500 mg) was chromatographed on a 2.5- × 70-cm column (Bio-Rex 70) with an exponential gradient from 50 mM ammonium phosphate, pH 7.2, to 160 mM ammonium phosphate, pH 7.2. The flow rate was 25 mL/h and the fraction size was 3.8 mL. Absorbance was measured at 542 nm. B, Fraction 1 from A was rechromatographed on a 1.5- × 25-cm column (Whatman sulfopropyl SE–53) with an exponential gradient from 20 to 250 mM sodium phosphate, pH 6.0. The fraction size was 1 mL and the absorbance was measured at 542 nm. C, Repurified fraction 4 was treated with Ru(bpy)$_2$CO$_3$ and chromatographed on a 0.6- × 45-cm column (Whatman CM–32) with a gradient from 20 to 400 mM sodium phosphate, pH 6.0. The fraction size was 1 mL and absorbance was measured at 542 nm. The fraction marked 4U contained unmodified dcbpy–cyt c. (Reproduced from ref. 16. Copyright 1988 American Chemical Society.)

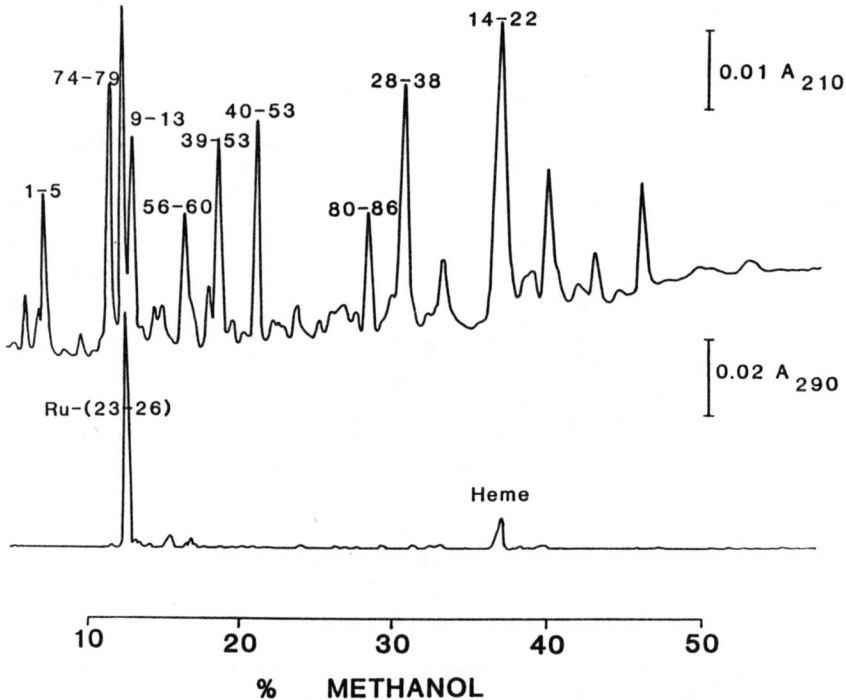

Figure 2. HPLC of the tryptic digest of (bpy)$_2$Ru(dcbpy–cyt c) fraction 4A. The tryptic digest (50 μg) was eluted on a column (Dynamax 300-A) with a linear gradient from 0.01% trifluoroacetic acid to 100% methanol. The native and ruthenium-containing peptides were identified by amino acid analysis and indicated on the figure. Small peptides such as 23–25 and 26–27 eluted in the void volume of the column. (Reproduced from ref. 16. Copyright 1988 American Chemical Society.)

The emission spectra of fractions 1 and 2 were very similar to that of Ru(bpy)$_2$(imid)$_2^{2+}$, with a maximum at 670 nm at 298 K (Figure 4) and 619 nm at 77 K.

Electron Transfer in (bpy)$_2$Ru(dcbpy–cyt c)

All experimental evidence supports the contention that the excited-state and ground-state parameters of Ru(bpy)$_2$(dcbpy) linked to cyt c are nearly identical to those of the free complex. Specifically, both have luminescence maxima at 662 nm at 298 K that shift to 606 nm at 77 K. The lifetime of the free complex and the lifetime of a derivative made from lysozyme, a protein with no heme iron, are also identical. The absorption spectra of the derivatives of cyt c are the sum of the cyt c and ruthenium complex spectra.

Cherry and Henderson (17) examined the excited-state decay of Ru(bpy)$_2$(dcbpy) and showed it to be similar to Ru(bpy)$_3^{2+}$. The addition of

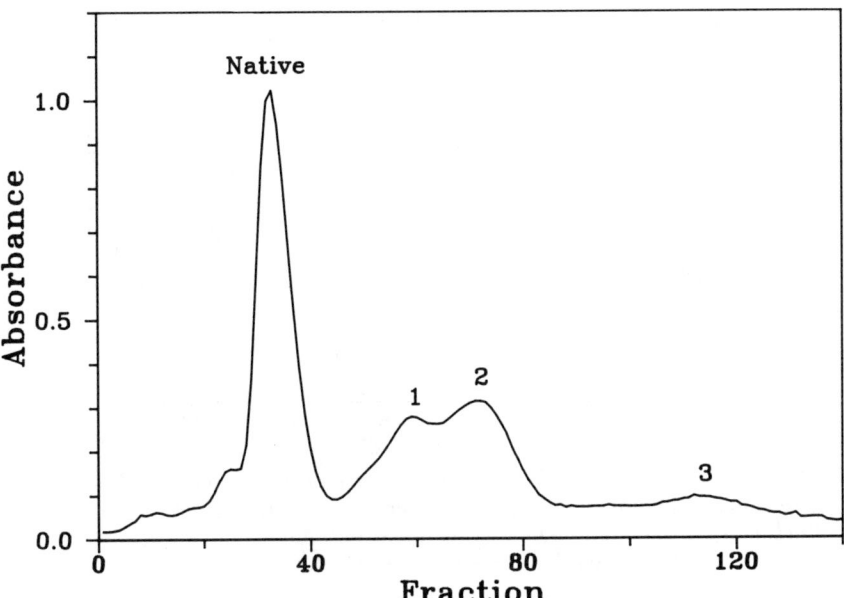

Figure 3. Purification of (bpy)$_2$(imid)Ru((His)cyt c) derivatives. The mixture was chromatographed on 1.5- × 30-cm column (Whatman CM–32) with an exponential gradient from 50 to 200 mM sodium phosphate, pH 6. The fraction size was 1 mL, and the absorbance was measured at 542 nm. The sample was prepurified by chromatography on a 1- × 10-cm column (Biogel P–2) with 20 mM sodium phosphate, pH 7.

an electron-transfer-quenching pathway complicates the decay of the excited state as indicated by Scheme II. In this scheme, all decay paths that do not involve electron transfer are represented by rate constant k_d, electron transfer from the excited state by k_1, and back electron transfer of the resulting ground-state molecules by k_2. The rate constant k_d contains radiative and nonradiative terms, in addition to possible contributions from energy-transfer quenching. Solution-phase Ru(bpy)$_3^{2+}$* has been shown to be quenched by electron transfer by cyt c (2). Experimentally, however, it is difficult to rule out the possibility that there is some contribution from energy transfer. It cannot be ruled out on spectroscopic grounds.

Our initial studies showed that (bpy)$_2$Ru(dcbpy–cyt c) derivatives exhibited quenching of both the luminescence intensity and lifetime. The magnitude of the quenching appeared to decrease with increasing ruthenium complex to heme distance.

In a subsequent investigation, transient absorption measurements made with laser flash photolysis equipment were carried out with the Ru(bpy)$_2$(dcbpy–cyt c) derivatives (22). The derivatives labeled at lysines 86, 87, and 8 showed no luminescence quenching, and no transients indicative

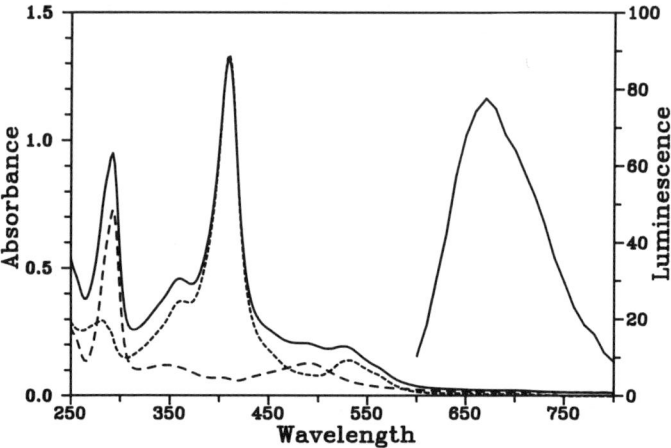

Figure 4. Absorbance and luminescence spectra of $(bpy)_2(imid)Ru((His)cyt\ c)$. Absorbance spectra shown for fraction 2 (—), native cyt c (----), and $Ru(bpy)_2(imid)_2^{2+}$ (- - -). The uncorrected luminescence spectrum of $Ru(bpy)_2(imid)_2^{2+}$ is shown on the right of the figure.

of electron transfer were found with these derivatives. The derivative with a modified lysine 13, however, showed a transient at 550 nm that clearly indicates the formation of an Fe(II) intermediate (Figure 5). Furthermore, the transient difference spectrum for the region around 550 nm shows minima at 542 and 556 nm that are isosbestic points in the conversion of oxidized to reduced cyt c. As expected, the transient is short-lived because of the back-reaction with ground-state Ru(III). The luminescence decay and transient absorbance measurements at 550 and 440 nm were simultaneously fit

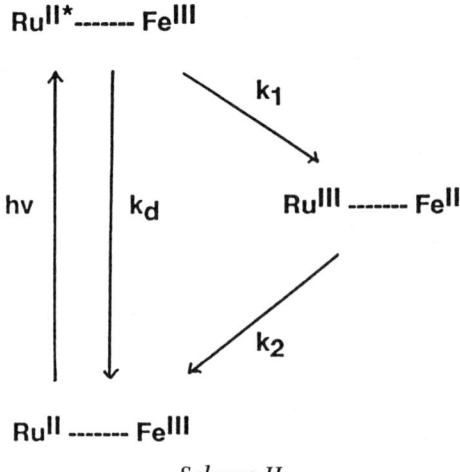

Scheme II.

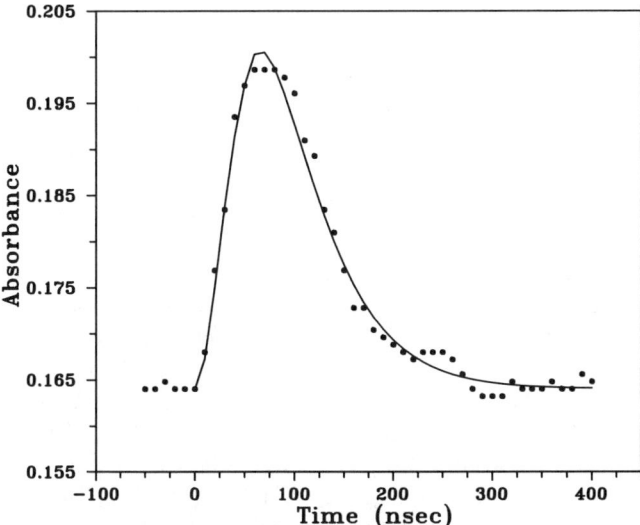

Figure 5. Transient absorbance of (bpy)$_2$Ru(dcbpy–(lysine 13)cyt c) recorded at 550 nm following a 25-ns pulse at 460 nm (Nd:YAG pumped coumarin 460 dye laser). Sample was purged with nitrogen.

to the following equations:

$$I(\text{Ru(II)}^*, \text{emission}) = I_0 e^{-(k_1+k_d)t} \qquad (1)$$

$$I(\text{Fe(II)}, 550 \text{ nm}) = A[e^{-(k_1+k_d)t} - e^{-k_2 t}] \qquad (2)$$

$$I(\text{Ru(II)}, 440\text{nm}) = \Delta\epsilon_{440} C_0 [1 - B e^{-(k_1+k_d)t} + C e^{-k_2 t}] \qquad (3)$$

where C_0 is the concentration of Ru(II) excited state produced by laser pulse, I is the signal intensity at any time, I_0 is the signal intensity at $t = 0$, $A = k_1 \Delta\epsilon_{550} C_0/(k_2-k_1-k_d)$, $B = (k_2-k_d)/(k_2-k_1-k_d)$, $C = k_1/(k_2-k_1-k_d)$, and the rate constants are as indicated in Scheme II. Corrections were also made for the finite rise and fall time of the laser pulse as described by Demas (23). The values for the first-order rate constants ($k_d = 8 \pm 3 \times 10^6$ s^{-1}, $k_1 = 16 \pm 3 \times 10^6$ s^{-1}, $k_2 = 26 \pm 5 \times 10^6$ s^{-1}) were obtained for the lysine 13 derivative (22). Derivatives modified at lysines 25, 27, 72, and 7 were also found to have transients associated with electron transfer. The measured first-order rate constants k_2 in these cases were $1.5 \pm 0.3 \times 10^6$, $30 \pm 5 \times 10^6$, $24 \pm 5 \times 10^6$, and $0.6 \pm 0.2 \times 10^6$ s^{-1}, respectively (22).

Because both cyt c and Ru(bpy)$_3^{2+}$ have been examined extensively, it is informative to compare the observed rate constants with those predicted by theory. In the context of the semiclassical treatment suggested by Marcus and Sutin (13), the following relation between free energy, distance, and rate is expected.

$$k_{el} = \frac{10^{13} \exp[-\beta(d-3)] \exp[-(\lambda + \Delta \mathring{G})^2]}{4\lambda RT} \qquad (4)$$

In this expression, ΔG is the free energy of reaction, k_{el} is the rate constant for electron transfer, d is separation distance, λ is the reorganizational barrier, and β is the scaler that describes how fast the rate decreases with separation distance. The term $(d-3)$, where d is the internuclear separation, is included to allow for a van der Waals contact distance of 3 Å, at which distance the first exponential term is assumed to be unity. Studies of the $Ru(bpy)_3^{2+}$ self-exchange rate suggest that contact is best measured with respect to the ligand atoms because of the delocalization of metal-electron density onto the bipyridines. Axup et al. (6) made use of a similar assumption in their analysis of electron transfer in $(NH_3)_5Ru(His)Mb$. In their case, the distance between the nearest ligand atoms was taken as a measure of d (i.e., heme or histidine ring atoms around Zn and histidine ring atoms or amine nitrogens on the ruthenium complex).

Estimates of the reorganizational barrier for the reaction of $Ru(bpy)_3^{3+}$ with heme Fe(II) can be made from self-exchange data for cyt c and $Ru(bpy)_3^{2+}$. Typical values given in the literature are 100 and 55 kJ, respectively (13, 21). The reorganizational barrier for electron transfer within a derivative based on these values is 78 kJ. The driving force for forward and backward electron transfer reactions using reduction potentials $E°(2+*/3+) = -0.72$ V and $E°(2+/3+) = 1.31$ V for $Ru(bpy)_2(dcbpy)$ are 0.98 and 1.05 V, respectively. Values of β are somewhat varied, but 0.9 Å^{-1} appears to be representative. In the case of the lysine 13 derivative, we estimate d to be in the range of 6–10 Å on the basis of constraints of the lysine tail, 9–16 Å for lysines 25 and 7, 6–12 Å for lysine 27, and 8–16 Å for lysine 7. Figure 6 shows the locations of the labeled lysines.

By using the parameters described, eq 4 can be used to predict a distance dependence for the reaction of $Ru(bpy)_2(dcbpy)^{3+}$ with the Fe(II) of the heme. The relation is illustrated in Figure 7. The best-fit line was obtained with $\beta = 0.9$ Å^{-1} and $\lambda = 43$ kJ.

Electron Transfer in (bpy)₂(imid)Ru((His)cyt c)

A quenching scheme identical to that describing the reactions of the $Ru(bpy)_2(dcbpy$-cyt c) derivatives can be used with the histidine derivatives.

The luminescence decay rates for fractions 1 and 2 were found to be 10 × 10^6 and 12 × 10^6 s^{-1}, respectively. These compared well with a rate constant of 14 × 10^6 s^{-1} measured with solution-phase $Ru(bpy)_2(imid)_2^{2+}$. No absorption transients indicative of electron transfer were observed. In

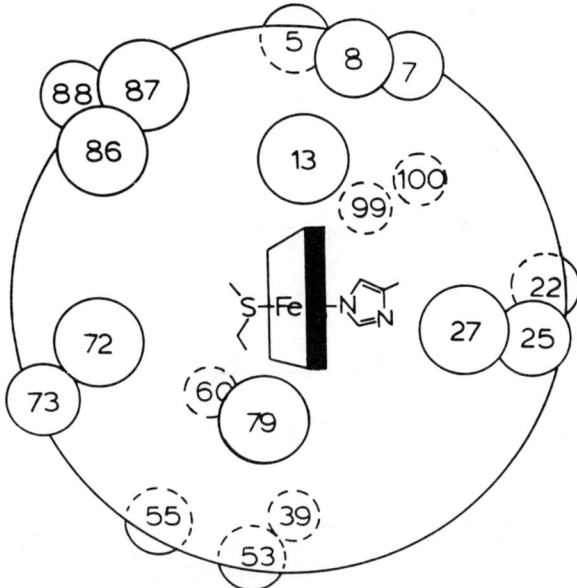

Figure 6. Schematic diagram of horse-heart cyt c viewed from the front side of the heme crevice. The approximate positions of the β carbon atoms at the lysine residues are indicated by closed circles and dashed circles for the residues located on the front and back of cyt c, respectively. (Reproduced from ref. 22. Copyright 1989 American Chemical Society.)

these derivatives, it appears that the rate of electron transfer from the excited state of ruthenium to the heme must be too small to compete with the natural luminescence decay rate.

Experimental Methods

Preparation of (bpy)$_2$(imid)Ru((His)cyt c) Derivatives. Horse-heart cyt c (2 mM Sigma type VI in 2 mL of 20 mM sodium phosphate, pH 7) was incubated with 4 mM Ru(bpy)$_2$CO$_3$ (22) for 16 h at 25 °C in the dark under anaerobic conditions. Imidazole, 1 M, was then added to the solution, and the incubation continued for an additional 18 h. The solution was oxidized with potassium ferricyanide and passed through a 1- × 10-cm column (Biogel P-2) to remove excess reagent and equilibrate the modified cyt c with 20 mM sodium phosphate, pH 7. The sample was eluted from a 1.5- × 30-cm column (Whatman CM-32) with an exponential gradient from 50 to 200 mM sodium phosphate, pH 6. Visible spectra of each fraction were recorded on a diode array spectrophotometer (Hewlett-Packard HP8452A). Fluorescence spectra were recorded on a spectrometer (Perkin-Elmer 650-40) using excitation at 450 nm. Luminescence lifetimes were measured as described by Pan et al. (16).

Identification of the Residue Modified. Each derivative was dialyzed into 0.1 M bicine, pH 8, at a concentration of 1 μg/μL and digested with 50 ng/μL TPCK-treated trypsin for 15 h. (TPCK is tosylamide-2-phenylethyl chloromethyl ketone.)

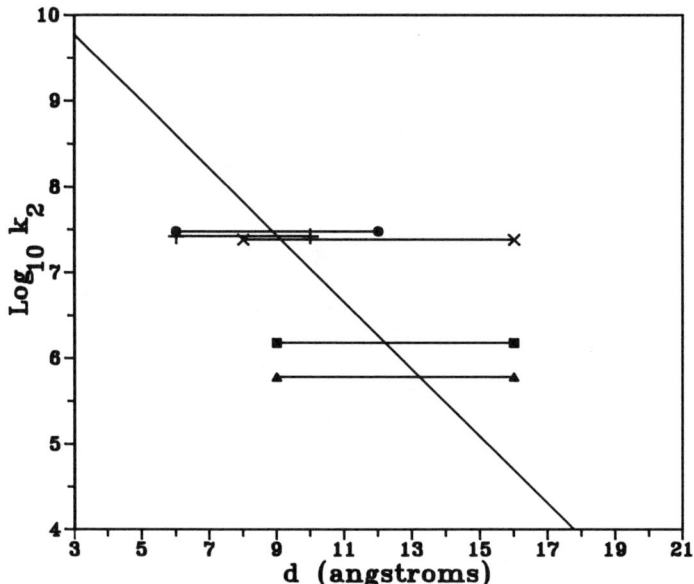

Figure 7. Plot of natural logarithm of the rate constant vs. separation distance d. Best-fit line calculated with β = 0.9 Å^{-1} and λ = 43 kJ. Data points for derivatives modified at lysines 7 (▲), 13 (+), 25 (■), 27 (●), and 72 (×) at the minimum and maximum estimated distance. (Reproduced from ref. 22. Copyright 1989 American Chemical Society.)

The tryptic digests were separated on a reverse-phase HPLC column (Dynamax 300 A) with a linear gradient from 0.01% trifluoroacetic acid to 100% methanol. The eluent was monitored at 210 and 290 nm by using two HPLC detectors in series. The amino acid composition of each peptide was determined as described by Pan et al. (16).

References

1. Yocom, K. M.; Shelton, J. B.; Shelton, J. R.; Schroeder, W. A.; Worosila, G.; Isied, S. S.; Bordignon, E.; Gray, H. B. *Proc. Natl. Acad. Sci. U.S.A.* **1982**, *79*, 7052–7055.
2. Winkler, J. R.; Nocera, D. G.; Yocom, K. M.; Bordignon, E.; Gray, H. B. *J. Am. Chem. Soc.* **1982**, *104*, 5798–5800.
3. Nocera, D. G.; Winkler, J. R.; Yocom, K. M.; Bordignon, E.; Gray, H. B. *J. Am. Chem. Soc.* **1984**, *106*, 5145–5150.
4. Isied, S. S.; Worosila, G.; Atherton, S. J. *J. Am. Chem. Soc.* **1982**, *104*, 7659–7661.
5. Isied, S. S.; Kuehn, C.; Worosila, G. *J. Am. Chem. Soc.* **1984**, *106*, 1722–1726.
6. Axup, A. W.; Albin, M.; Mayo, S. L.; Crutchley, R. J.; Gray, H. B. *J. Am. Chem. Soc.* **1988**, *110*, 435–439.
7. Kostic, N. M.; Margalit, R.; Che, C.-M.; Gray, H. B. *J. Am. Chem. Soc.* **1983**, *105*, 7765–7767.

8. Jackman, M. P.; McGinnis, J.; Powls, R.; Salmon, G. A.; Sykes, A. G. *J. Am. Chem. Soc.* **1988**, *110*, 5880–5887.
9. Che, C.-M.; Margalit, R.; Huey-Jenn, C.; Gray, H. B. *Inorg. Chim. Acta* **1987**, *135*, 33–35.
10. Marcus, R. A. *Disc. Faraday Soc.* **1960**, *29*, 21–31.
11. Marcus, R. A. *J. Phys. Chem.* **1963**, *67*, 853–857.
12. Marcus, R. A. *J. Chem. Phys.* **1965**, *43*, 679–701.
13. Marcus, R. A.; Sutin, N. *Biochim. Biophys. Acta.* **1985**, *811*, 265–322.
14. Karas, J. L.; Liebe, C. M.; Gray, H. B. *J. Am. Chem. Soc.* **1988**, *110*, 599–600.
15. Bechtold, R.; Kuehn, C.; Lepre, C.; Isied, S. S. *Nature* **1986**, *322*, 286–288.
16. Pan, L. P.; Durham, B.; Wolinska, J.; Millet, F. *Biochemistry* **1988**, *27*, 7180–7184.
17. Cherry, W. R.; Henderson, L. J., Jr. *Inorg. Chem.* **1984**, *23*, 983–986.
18. English, A. M.; Lum, V. R.; Delaive, P. J.; Gray, H. B. *J. Am. Chem. Soc.* **1982**, *104*, 870–871.
19. Brunschwig, B. S.; Delaive, P. J.; English, A. M.; Goldberg, M.; Gray, H. B.; Mayo, S. L.; Sutin, N. *Inorg. Chem.* **1985**, *24*, 3743–3749.
20. Guarr, T.; McGuire, M. E.; McLendon, G. *J. Am. Chem. Soc.* **1985**, *107*, 5104–5111.
21. Brown, G. M.; Sutin, N. *J. Am. Chem. Soc.* **1979**, *101*, 883–892.
22. Durham, B.; Pan, L. P.; Long, J.; Millett, F. *Biochemistry* **1989**, *28*, 8659–8665.
23. Demas, J. N. *Excited State Lifetime Measurements;* Academic: New York, 1983.

RECEIVED for review May 1, 1989. ACCEPTED revised manuscript August 10, 1989.

EXPERIMENTAL APPROACHES TO BIOLOGICAL ELECTRON TRANSFER:

INORGANIC COMPLEXES

10
High-Pressure Studies of Long-Range Electron-Transfer Reactions in Solution

Nita A. Lewis and Daniel V. Taveras

Department of Chemistry, University of Miami, Coral Gables, FL 33124

Electron transfer in the mixed-valent complex (μ-2,6-dithiaspiro-[3.3]heptane)decaamminediruthenium(II,III) hexafluorophosphate (I) was studied in D_2O under high pressure (1.5 kbar). On the basis of a tunneling theory, the volume of activation for this process was estimated to be -7.2 ± 0.5 cm^3/mol at several different concentrations and ionic strengths. This value results primarily from solvent rearrangement during the electron-transfer process. It may be approximated quite accurately by even the simplest electron-transfer theories. The advantages of using models based on ellipsoidal shapes and dielectric cavities are discussed, as well as the potential of high-pressure studies as a test of electron-transfer theories.

E<small>LECTRON-TRANSFER REACTIONS BETWEEN IONS AND MOLECULES</small> in solution, both intermolecular and intramolecular, have been the subject of intense experimental and theoretical investigation during the past three decades (1–3). Numerous experimental and theoretical approaches to studying these kinds of processes have been devised. One of the most recent approaches is the application of high pressure during the course of the investigation. Pressure as an experimental variable has been employed rather extensively in areas such as pressure-tuning spectroscopy (PTS) (4), kinetics (volumes of activation) (5), and synthesis (6). However, until recently its use in the study of electron-transfer reactions in solution was rather limited.

Theoretical Models

Effect of High Pressure. Particularly useful tools in the study of outer-sphere electron-transfer (OSET) reactions are the many theoretical models that sometimes are validated by experimental results or that may be used to explain such results. In 1974 Stranks (7) developed a theory, based on Marcus's and Hush's previous work as discussed by Cannon (1), that tried to predict the effect of high pressures on OSET processes in solution. The attempt was considered successful because almost all of the available experimental data were correctly reproduced by the theory. Unfortunately, it was discovered later that there was an algebraic error in one of Stranks's formulas. In fact, very few experimental volumes of activation are predicted by his calculations (8).

The specific reasons for the apparent failure of the Stranks–Marcus–Hush (SMH) theory are not yet fully understood. One possibility is that until very recently the experimental data were available only for intermolecular reactions, in which the dielectric continuum approximations may lead to erroneous estimates of electrostatic interactions. In those reactions the distance between the metal centers normally is not accurately known, but has to be estimated from the dimensions of the ligands surrounding the reacting centers.

Swaddle and co-workers (9–11) improved Stranks's model somewhat by including modern developments in electron-transfer theories, such as ellipsoidal dielectric-cavity models and nonadiabaticity contributions. They demonstrated (10) that the variation with pressure of the distance between reacting species was crucial in achieving reasonable agreement between volumes of activation determined experimentally and the corresponding theoretical estimates. Therefore, some insight may be obtained into the mechanics of the process by fixing the distance between two metal centers with a connecting ligand and studying the intramolecular electron transfer process under pressure.

Intramolecular Redox Reactions. Numerous studies of intramolecular redox reactions employing binuclear transition metal mixed-valence complexes have been reported (2) since the original paper on this subject by Creutz and Taube (12) appeared in 1973. Usually the energy of the intervalence transfer (IT) band for a symmetrical mixed-valence molecule (E_{op} or λ) is treated as arising from two contributions—an inner-sphere reorganizational energy, λ_{in}, which results from stretching and compressing internuclear bonds prior to electron transfer, and λ_{out}, which results from reorientation of solvent molecules.

$$\lambda = \lambda_{in} + \lambda_{out} \qquad (1)$$

The dielectric continuum model is usually employed to describe the solvent reorganization contribution; eq 2 results.

$$\lambda_{out} = (\Delta e)^2 \left(\frac{1}{2a_1} + \frac{1}{2a_2} - \frac{1}{d}\right)\left(\frac{1}{D_o} - \frac{1}{D_s}\right) \qquad (2)$$

where Δe is the charge transferred, a_1 and a_2 are the radii of the two reactants, d is the distance between their centers, D_o is the optical dielectric constant, and D_s is the static dielectric constant of the medium. Normally, a solid has a much smaller D_s than the corresponding liquid, typically by a factor of 10 or more. Thus, eq 2 predicts a large red shift of the IT band upon freezing a solution containing a mixed-valence complex. Hammack et al. (13) tested this idea by employing pressure-induced freezing of solutions containing the ferrocenium binuclear species and the complex $(bpy)_2ClRu(pyz)RuCl(bpy)_2{}^{3+}$, where bpy is 2,2'-bipyridine and pyz is pyrazine. Only very small shifts in IT band energy occurred. Hammack et al. concluded that the dielectric continuum model is inadequate in determining reorganizational energies in intervalence redox processes. Other systems were also tried, apparently without success (14).

The same authors later reported a PTS study (15) of the well-known Creutz–Taube ion, $(NH_3)_5Ru(pyz) Ru(NH_3)_5{}^{5+}$, in liquid D_2O up to 7 kbar. A total blue shift of only 35 cm^{-1} was observed. They concluded that the primary effect of pressure on the Creutz–Taube ion is to change the amount of electronic coupling between the metal centers by compression of the internuclear bonds. By comparison, the vibronic coupling is affected only very slightly. These results are in agreement with an already-high degree of electronic coupling suggested by other evidence.

Observations. We examined an electronically very weakly coupled complex, the dinuclear complex **I** (16), in D_2O at moderately high pressures (1.5 kbar) (17). We observed a shift of more than 200 cm^{-1} to a lower wavelength. This shift is interesting because the bulk modulus always increases with pressure, and therefore the energies of vibrational motions are normally

$$\left[(NH_3)_5Ru\ S\diagup\!\!\!\diagdown\!\!\!\diagup\!\!\!\diagdown S\ Ru(NH_3)_5 \right]^{5+}$$

I

expected to increase. That is, it is more difficult for a bond to stretch when the surrounding solvent molecules are compressed. Apparently, in the case of very weakly electronically coupled metal systems, the vibrational coupling of the dinuclear complex with the solvent molecules is enhanced by pressure, and this enhanced coupling leads to a decrease in the energy of the vibronic optical transition band (17).

The oscillator strength and hence the vibronic coupling was slightly enhanced by pressure, as shown in Figure 1. This enhanced coupling, com-

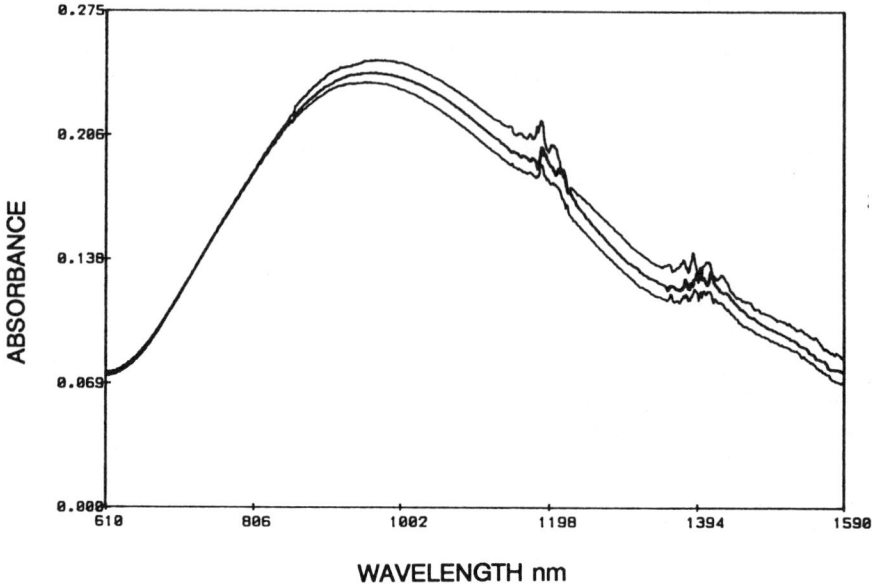

*Figure 1. IT band of complex **I** at different pressures. Key: top to bottom, the pressures are 1500, 750, and 1 bar.*

bined with the red shift, may be explained by assuming a flattening of the vibronic wells in the Marcus potential energy diagram caused by a pressure-induced softening of the spring constants that define the wells (17). An alternative explanation would be that a decrease occurs in the displacement between the two wells along the reaction coordinate. This explanation can be excluded, however, on the basis that it predicts a change in the energy of the optical transition, but not of the vibronic coupling.

It has been proposed (17) that, because of the limited electronic coupling between the metal atoms in this molecule, pressure effects on interactions with the medium predominate over those that would increase electronic coupling (i.e., compressions of internuclear bonds). This proposal accounts for the observation that complex **I** is sensitive to our pressure range (1–1500 bar), whereas the Creutz–Taube ion was virtually unaffected because this pressure range is insufficient to cause measurable changes in intermolecular distances. Small compressions of the Ru(II) and Ru(III) bond lengths of the Creutz–Taube ion leading to the observed stronger electronic coupling between the metal centers are possible at the higher pressures employed by Hammack et al. (15) (up to 7000 bar in solution).

Theoretical Calculations of Volume of Activation

It had been demonstrated previously that the primary mechanism for electron transfer in the spiro binuclear system is tunneling (18). Therefore, a nonadiabatic, multiphonon theory proposed by Hopfield (19, 20) was used to calculate the thermal rate of electron transfer at each pressure, k_{ab}, from the energy of the optical transition, E_{op}, and the extent of coupling, $|T_{ab}|$, namely,

$$k_{ab} = \left(\frac{2\pi}{\hbar}\right) |T_{ab}|^2 (2\pi \sigma^2)^{-1/2} \exp\left(\frac{-(E_a - E_b - \Delta)^2}{2\sigma^2}\right) \quad (3)$$

$$|T_{ab}|^2 = \frac{(n^2 + 1)(3)(2300)E_{op}\hbar c\epsilon}{(2n)(2\pi^2)(N)a^2 e^2 (2\pi \sigma^2)^{1/2}} \exp\left(\frac{-(E_a - E_b - \Delta)^2}{2\sigma^2}\right) \quad (4)$$

where n is the refractive index of the connecting ligand, $\hbar$ is Planck's constant divided by 2π, c is the speed of light, N is Avogadro's number, a is the distance between metal centers, e is the charge of the electron, σ is the half-bandwidth at 0.61 of the IT band at the peak maximum, E_a and E_b are the redox energies, Δ is the vibronic coupling parameter, and ϵ is the extinction coefficient of the IT band at the peak maximum (18). With these data and eq 5 (5),

$$\left(\frac{\delta \ln k_{ab}}{\delta P}\right)_T = -\left(\frac{\Delta V^\ddagger}{RT}\right) \quad (5)$$

(where P is pressure, T is absolute temperature, and R is the gas constant), $\Delta V^\ddagger$, the volume of activation for the intramolecular electron exchange, was estimated to be -7.5 ± 0.2 cm^3/mol at an ionic strength of 0.08 M.

To clarify the significance of this quantity, expressions for the volume of activation have been derived from several different electron-transfer theories. The results suggest that the experimental value for complex **I** arises almost exclusively from solvent rearrangement and that it may be predicted accurately by even the simplest models.

The expression for the total volume of activation is derived by application of eq 6 to all the contributions to the electron-transfer free energy of activation, $\Delta G^\ddagger$, producing eq 8, which was originally postulated by Stranks (7).

$$\left(\frac{\delta \Delta G^\ddagger}{\delta P}\right)_T = -\Delta V^\ddagger \quad (6)$$

$$\Delta G^\ddagger = \Delta G_{coul}^\ddagger + \Delta G_{DH}^\ddagger + \Delta G_{ir}^\ddagger + \Delta G_{solv}^\ddagger \quad (7)$$

$$\Delta V^\ddagger = \Delta V_{coul}^\ddagger + \Delta V_{DH}^\ddagger + \Delta V_{ir}^\ddagger + \Delta V_{solv}^\ddagger \quad (8)$$

where the subscripts denote the coulombic work to bring the reactants together at zero ionic strength (coul), the Debye–Hückel correction at a nonzero ionic strength (DH), the internal rearrangement of bond lengths and angles (ir), and the solvent rearrangement during the redox process (solv). In a compound such as **I** the reactants need not be brought together, and it follows that the first two terms in eq 7 disappear. Furthermore, the $\Delta V_{ir}^{\ddagger}$ term is negligible because the pressures necessary for significant bond length contraction are well beyond our available range (7, 10, 11). Therefore, eq 8 is greatly simplified, being apparently limited to the term for solvent rearrangement, that is,

$$\Delta V^{\ddagger} = \Delta V_{solv}^{\ddagger} = \left(\frac{\delta \Delta G_{solv}^{\ddagger}}{\delta P}\right)_T = \left(\frac{1}{4}\right)\left(\frac{\delta \lambda_{out}}{\delta P}\right)_T \quad (9)$$

$$\lambda_{out} = \left(\frac{\Delta e^2}{8\pi}\right)\left(\frac{1}{D_o} - \frac{1}{D_s}\right)\int (D_A - D_B)^2 \, d\tau \quad (10)$$

where D_A and D_B are the dielectric displacement vectors created in vacuo by the precursor and successor complexes, respectively, and $d\tau$ is a volume differential (2). This is quite advantageous. Several expressions are reported in the literature to calculate λ_{out}, whose differences stem primarily from the method used to solve the integral in eq 10 (21).

Hard-Spheres Model. The model used by Stranks (7) to calculate the contribution of solvent rearrangement around the reactants in the transition state was the classical conducting hard spheres model. In this model the electrostatic influence of the spheres upon each other is neglected, but the electronic polarization of the solvent, as the instantaneous charge distribution on each ion changes, is considered. The resulting equation is (2):

$$\Delta V^{\ddagger} = \left(\frac{e^2}{4}\right)\left(\frac{1}{2a_1} + \frac{1}{2a_2} - \frac{1}{d}\right)\frac{\delta}{\delta P}\left(\frac{1}{D_o} - \frac{1}{D_s}\right)_T \quad (11)$$

where e is the charge of the electron, a_1 and a_2 are the radii of the two spheres, and d is the intermetal distance.

For the calculation of $\Delta V^{\ddagger}$, a_1 and a_2 were considered to be 3.5 Å (pentaammineruthenium moiety radius) (17), whereas d was taken as the through-space length of the connecting spiro ligand (9.2 Å). The pressure-dependent term was calculated to be $(1.167 \pm 0.015) \times 10^{-5}$ atm^{-1} from the data of Reisler and Eisenberg (22) for the refractive index and Srinivasan and Kay (23) for the dielectric constant of D_2O at high pressures. The resulting value is -7.1 ± 0.1 cm^3/mol (Table I). A refinement to the hard-spheres model advanced by Kharkats (24), which holds for any d, also was used.

$$\Delta V^{\ddagger} = \left(\frac{e^2}{4}\right)\left[\frac{1}{a} - \frac{1}{d} - \left(\frac{2a}{d^2 - a^2}\right) - \frac{1}{d}\ln\frac{(d+a)^2}{d^2 - a^2}\right]\frac{\delta}{\delta P}\left(\frac{1}{D_o} - \frac{1}{D_s}\right)_T \quad (12)$$

where $a_1 = a_2 = a$. The $\Delta V^{\ddagger}$ calculated is -7.0 ± 0.1 cm^3/mol.

Table I. Calculated Volumes of Activation in D_2O for Intramolecular Electron Transfer Process in Complex I

Method	Equation	$\Delta V^{\ddagger a}$	$\Delta V^{\ddagger b}$
Hard spheres	11	−7.1 (0.1)	—
Modified hard spheres	12	−7.0 (0.1)	—
Separated spheres	13	−7.0 (0.1)	—
Perfect ellipse	14	−11.7 (0.2)	−7.2 (0.1)
Normal ellipse	16	−10.3 (0.1)	−7.0 (0.1)
BES ellipse	18	−6.3 (0.2)	−4.3 (0.1)
BES simplified	20	−8.4 (0.1)	−7.4 (0.1)

NOTE: $\Delta V^{\ddagger}$ values are given in cubic centimeters per mole. The values in parentheses are standard errors.
^{a}Calculated using $(d/2 + a)$ as semimajor axis and a as semiminor axis (dotted-line ellipse in Figure 2).
^{b}Calculated using the dimensions determined from the minimum enclosing volume convention (full line ellipse in Figure 2).

Dielectric-Cavity Models. These models are more appropriate for mixed-valence dinuclear complexes like **I**, because they account for the difference in permittivity of the space between the metals (connecting ligand) compared to the bulk solvent. However, the expressions obtained are much more elaborate because of the addition of imaging effects and exclusion of the space occupied by the cavity during the integration of the dielectric displacements in the medium.

The separated-dielectric-spheres model also was developed by Kharkats (25). This model is based on two separated spheres of permittivity D_{in} embedded in a medium of permittivity D_s, with the charge being transferred from one center to the other. The result is approximate because the inhomogeneity of the external field and the redistribution of the charges on the reactants caused by interaction with the induced dipoles are neglected (21). The expression for $\Delta V^{\ddagger}$ is

$$\Delta V^{\ddagger} = \left(\frac{e^2}{4}\right)\left(\frac{1}{a} - \frac{1}{d} + \frac{a^3}{d^4}\right)\frac{\delta}{\delta P}\left(\frac{1}{D_o} - \frac{1}{D_s}\right)_T +$$
$$\left(\frac{2a^3}{d^4}\right)\left[\left(\frac{D_s^2 - D_s D_o}{(2D_s + D_{in})^2}\right)\frac{\delta}{\delta P}\left(\frac{1}{D_o}\right)_T + \right.$$
$$\left.\left(\frac{2D_s D_{in} - D_o D_{in} + 2D_s D_o}{D_o(2D_s + D_{in})^3}\right)\left(\frac{\delta D_s}{\delta P}\right)_T\right] \quad (13)$$

For this calculation, D_{in} was assumed to be the square of the refractive index of the connecting ligand. Because eq 13 requires the use of invariant values of D_s and D_o, the calculation was performed with the values at 250, 750, and 1250 bar, giving a resulting average $\Delta V^{\ddagger} = -7.0 \pm 0.1$ cm^3/mol.

Ellipsoidal Cavity Model. Cannon (26) was the first to apply the ellipsoidal cavity model of Kirkwood and Westheimer (27) to mixed-valence dinuclear complexes. In this case, the cavity is formed by revolution of an ellipse around its major axis. The expression for $\Delta V^{\ddagger}$ was first derived by Swaddle (9), namely

$$\Delta V^{\ddagger} = \left(\frac{e^2}{4}\right)\left(\frac{d^2}{2AB^2}\right) S(l_o) \frac{\delta}{\delta P}\left(\frac{1}{D_o} - \frac{1}{D_s}\right)_T \tag{14}$$

$$S(l_o) = \sum_{n=0}^{\infty} \frac{\frac{1}{2}[1 - (-1)^n](2n + 1)l_o\ (l_o^2 - 1)Q_n(l_o)}{P_n(l_o)} \tag{15}$$

where A and B are the semimajor and semiminor axes, respectively; l_o is the reciprocal of the eccentricity of the ellipse (i.e., $l_o = A/(A^2 - B^2)^{1/2}$; $S(l_o)$ is the so-called "shape" factor; and $P_n(l_o)$ and $Q_n(l_o)$ are associated Legendre polynomials of degree n. These polynomials, determined in the present work by a computer calculation using standard recursive formulas, were checked against published tabulations (28).

Two sets of dimensions A and B were used for the calculations with ellipsoidal cavity models. The first one used the ordinary assumption that $A = (d/2 + a)$ and $B = a$. In contrast, the second set of dimensions followed the convention recently suggested by Brunschwig, Ehrenson, and Sutin (29) of the minimum enclosing volume. The latter method was applied graphically and produced the values $A = 9.3$ and $B = 4.3$. Both results are listed in Table I and depicted in Figure 2.

To achieve his result, Cannon (26) assumed that the change in dipole moment from precursor to successor complex could be replaced with little error by another dipole of equal moment, but with charges located at the foci of the ellipsoid. German (30) solved this problem by generalizing the Kirkwood–Westheimer theory to the case in which charges are located any-

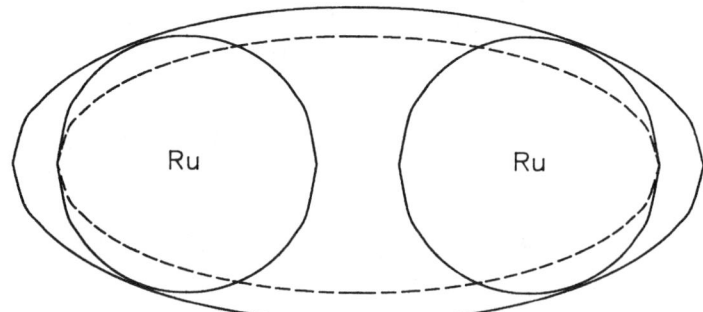

Figure 2. Diagram of the ellipses whose dimensions were used in the calculations. The full line represents the minimum enclosing volume, whereas the dotted line results from using the $(NH_3)_5Ru$ radius as semiminor axis.

where in the ellipsoidal cavity. For a symmetrical dinuclear compound with $D_{in} <<< D_s$, the corresponding expression for $\Delta V^{\ddagger}$ would be

$$\Delta V^{\ddagger} = \left(\frac{e^2}{4}\right)\left(\frac{1}{R}\right) S(l_o) \frac{\delta}{\delta P}\left(\frac{1}{D_o} - \frac{1}{D_s}\right)_T \quad (16)$$

$$S(l_o) = \sum_{n=0}^{\infty} \frac{[1-(-1)^n]^2 \left[\frac{2n+1}{n+1}\right][P_n(\mu)]^2}{[P_n(l_o)]^2 \left[l_o - \frac{Q_{n+1}(l_o)}{Q_n(l_o)}\right]} \quad (17)$$

where R is the interfocal distance, $2(A^2 - B^2)^{1/2}$, and μ is a prolate spheroidal coordinate equal to d/R.

The direct integration approach of Cannon (26) and German (30) avoids consideration of certain features like the differential reaction field. This was pointed out by Brunschwig et al. (29), who used a thermodynamic cycle to calculate the energy required to recharge the ellipsoid from its initial charge distribution to the final nonequilibrium charge distribution. The adjustment of D_o, as part of the total solvent permittivity, occurs simultaneously with the actual charge transfer. Other authors interpret the solvent reorganization energy as the difference between the free energy changes for charge transfer in media of permittivity D_s and D_o (29).

Two pressure-dependent expressions were obtained from the Brunschwig–Ehrenson–Sutin (BES) model. In the first expression, the imaging effects were assumed to be unresponsive to the changes in pressure, hence resulting in equations similar to eqs 16 and 17, namely,

$$\Delta V^{\ddagger} = \left(\frac{e^2}{4}\right)\left(\frac{1}{R}\right) S(l_o) \frac{\delta}{\delta P}\left(\frac{1}{D_o} - \frac{1}{D_s}\right)_T \quad (18)$$

$$S(l_o) = \sum_{n=0}^{\infty} \frac{2[1-(-1)^n](2n+1)[P_n(\mu)]^2 Q_n(l_o)}{P_n(l_o)} \quad (19)$$

Nevertheless, the form of the expressions given by BES also allows evaluation of the effect of pressure on the imaging effects, thus resulting in eq 20.

$$\Delta V^{\ddagger} = \left(\frac{e^2}{4R}\right)\left[\sum_{n=0}^{\infty} \frac{X_n(1-Y_n)}{(D_s - D_{in}Y_n)^2}\left(\frac{\delta D_s}{\delta P}\right)_T - \sum_{n=0}^{\infty} \frac{X_n(1-Y_n)}{(D_o - D_{in}Y_n)^2}\left(\frac{\delta D_o}{\delta P}\right)_T\right] \quad (20)$$

$$X_n = \frac{2[1 - (-1)^n](2n + 1)[P_n(\mu)]^2 Q_n(l_o)}{P_n(l_o)} \tag{21}$$

$$Y_n = \frac{\dfrac{l_o - P_{n+1}(l_o)}{P_n(l_o)}}{\dfrac{l_o - Q_{n+1}(l_o)}{Q_n(l_o)}} \tag{22}$$

Whenever absolute values for D_s and D_o were needed, the calculation was performed at a fixed pressure and then averaged with the results obtained at other pressures.

The fact that the results obtained depended heavily on the evaluation of the Legendre polynomials did not seem to be very critical. Even though all the calculations were performed up to $n = 10$, stepwise tabulations showed that most of the expressions converged rather rapidly. As a further check on the possibility of mathematical artifacts, the value of activation volume exerted by a change in the eccentricity l_o of the ellipse was calculated while keeping a constant original volume. This approach can be envisioned as squeezing out a sealed balloon of elliptical shape. The $\Delta V^\ddagger$ found through the BES model did not vary, as shown in Figure 3.

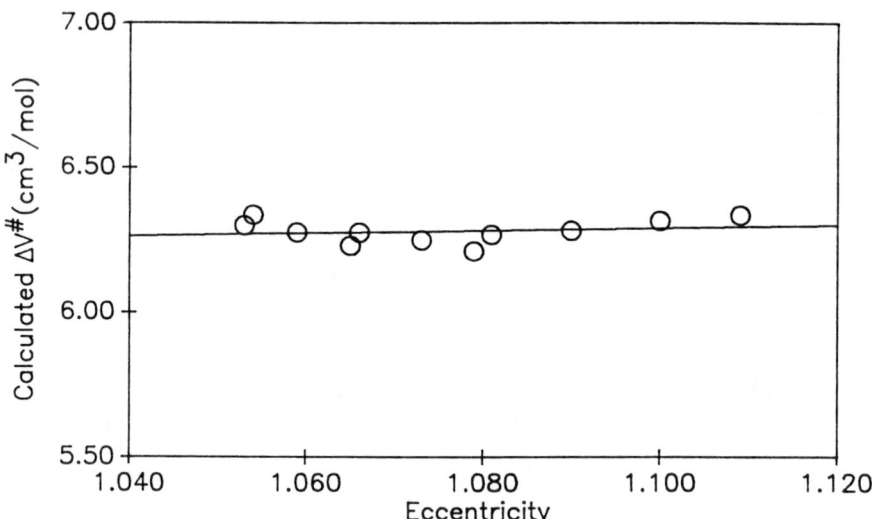

Figure 3. Change in the $\Delta V^\ddagger$ calculated with the BES approach as a function of the eccentricity of the ellipse. The absolute geometric volume of the ellipse was held constant for all the points.

Medium Effects on Volume of Activation Value

Evidence reported by our group (*31, 32*) demonstrated that intramolecular electron transfer in very weakly coupled systems like **I** are especially sensitive to medium effects. In highly polar solvents such as D_2O, ionic strength and ion-pair formation seem to dominate, whereas in solvents of low dielectric constant, concentration effects may be important (*33*).

The variation of the IT band frequency with pressure was measured at six different concentrations over a threefold concentration range (1.5–4.5 mM) by using $Ce(NH_4)_2(NO_3)_6$; no systematic changes were observed. The average for the volume of activation was -7.2 ± 0.5 cm^3/mol. An ionic strength dependence study was also tried, in which fixed concentrations (1.5 mM) of both complex **I** and Ce(IV) were subjected to different amounts of added NH_4NO_3 in the concentration range 0.05–0.30 M. No variation of $\Delta V^{\ddagger}$ could be discerned.

Discussion

Table I lists the values for the volume of activation of the intramolecular electron transfer process in complex **I** calculated from several theoretical models. The agreement with the activation volume estimated from the IT band piezochromism is quite satisfactory. Along with Swaddle's previous work, this agreement lends support to the validity of the SMH model. A judicious choice of experimental system seems to be important in achieving successful results. This probably reflects the importance of the system's geometry on the calculations. The most important property of complex **I** is obviously the fixed distance between the metal centers. The mathematical derivation of the final expressions was greatly simplified because only the dielectric properties of the solvent were pressure-dependent. Likewise, the choice of D_2O as the solvent enabled us to use permittivity data reported in the literature that was measured directly under pressure.

The numbers in Table I also indicate that the results of the calculations were consistent with their respective physical meaning. For example, the ellipsoidal-cavity models did not yield good values unless the minimum enclosing volume condition was used. Figure 2 shows that using the $(NH_3)_5Ru$ radius as the semiminor axis generates a smaller space that might not include the whole complex (including the solvent molecules directly coupled to it). In the cavity models, the medium inside the cavity is excluded during the integration of eq 10. The term $\Delta V_{solv}^{\ddagger}$ essentially accounts for the electrostriction of solvent molecules outside this cavity. If $\Delta V_{solv}^{\ddagger}$ is too small, some molecules that should be inside the cavity are considered as rearranging. The minimum enclosing volume condition apparently corrects this situation.

On the other hand, the very good values obtained with the simpler spherical models were somewhat surprising. It has been suggested (29) that the interconnecting ligand sometimes has a refractive index n not too different from the solvent used. Because D_{in} normally is assumed to be the square of n for the ligand, there is not much difference between D_{in} and D_o. (In our case, D_o is approximately 25% higher than D_{in}.) The consequence is a numerically correct result that is not easily interpreted. Table I also indicates that, for the BES model, the full expression is less accurate than the simplified version. This finding implies either that the imaging effects are not affected by pressure or that the complicated nature of the full pressure-dependent expression introduces numerical errors into the calculation.

Although many quantum mechanical models (34) in principle allow the calculation of thermal rate constants from optical data, the approach of Hopfield (19, 20) is simple enough to be experimentally useful without the need for additional approximations. Any systematic errors in the rates originating from fundamental failures in the theory should cancel out because the volume of activation was estimated from the slope of a linear plot. This condition means that employing high-pressure measurements of intramolecular electron-transfer rates may be an excellent method of assessing the validity of any given theoretical model.

Conclusions

The SMH model of electron transfer in solution under high pressures may represent an adequate approach to the interpretation of experimental high-pressure data, provided that some of the variables are either eliminated or simplified through rational selection of experimental probes. Because of the apparent importance of an accurate assessment of the distance between the reactants, the study of mixed-valence dinuclear complexes featuring intramolecular electron transfer may be most advantageous. It seems likely that only complexes with a medium-to-high component of vibrational coupling in the vibronic band exhibit enough sensitivity to pressure changes to be experimentally useful.

High-pressure studies may be employed to test electron-transfer theories. Any systematic errors introduced by assumptions and approximations of any given theory are likely to originate in such a way that those errors will cancel out in the pressure-dependent formulae. The results reported in the present work are only our first report of the use of high-pressure measurements to distinguish features of the various electron-transfer theories. Experiments in a variety of media, direct measurement of intramolecular electron-transfer rates, and study of a series of complexes with varying lengths of the interconnecting ligand under pressure are presently under investigation in our laboratories.

A detailed description of the equipment and experimental procedures that we used can be found elsewhere (16).

Acknowledgments

We are very thankful to the University of Miami for a fellowship to D. V. Taveras and for providing funds to purchase the equipment. Partial support from the National Institutes of Health (Grant No. 36857) is greatly appreciated. We also acknowledge enlightening discussions with Y. S. Obeng.

References

1. Cannon, R. D. *Electron Transfer Reactions*; Butterworths: London, 1980.
2. *Progress in Inorganic Chemistry*; Lippard, S. J., Ed.; Wiley: New York, 1983; Vol. 30.
3. Zewail, A. H.; Connor, J. N. L.; Sutin, N., Eds.; *J. Phys. Chem.* **1986**, *90*, Number 16.
4. Drickamer, H. G. *Acc. Chem. Res.* **1986**, *19*, 355.
5. Van Eldik, R. In *Inorganic High Pressure Chemistry: Kinetics and Mechanisms*; Van Eldik, R., Ed.; Elsevier: Amsterdam, 1986; p 1.
6. Le Noble, W. J. In *High Pressure Chemistry and Biochemistry*; Van Eldik, R.; Jonas, J., Eds.; Reidel: Dordrecht, Holland, 1987; p 295.
7. Stranks, D. R. *Pure Appl. Chem.* **1974**, *38*, 303.
8. Wherland, S. *Inorg. Chem.* **1983**, *22*, 2349.
9. Swaddle, T. W. In *Inorganic High Pressure Chemistry: Kinetics and Mechanisms*; Van Eldik, R., Ed.; Elsevier: Amsterdam, 1986; p 273.
10. Spiccia, L.; Swaddle, T. W. *Inorg. Chem.* **1987**, *26*, 2265.
11. Doine, H.; Swaddle, T. W. *Inorg. Chem.* **1988**, *27*, 665.
12. Creutz, C.; Taube, H. *J. Am. Chem. Soc.* **1973**, *95*, 1086.
13. Hammack, W. S.; Drickamer, H. G.; Lowery, M. D.; Hendrickson, D. N. *Chem. Phys. Lett.* **1986**, *132*, 231.
14. Hammack, W. S.; Drickamer, H. G.; Lowery, M. D.; Hendrickson, D. N. *Inorg. Chem.* **1988**, *27*, 1307.
15. Hammack, W. S.; Lowery, M. D.; Hendrickson, D. N.; Drickamer, H. G. *J. Phys. Chem.* **1988**, *92*, 1771.
16. Stein, C. A.; Taube, H. *J. Am. Chem. Soc.* **1981**, *103*, 693.
17. Lewis, N. A.; Obeng, Y. S.; Taveras, D. V.; Van Eldik, R. *J. Am. Chem. Soc.* **1989**, *111*, 924.
18. Stein, C. A.; Lewis, N. A.; Seitz, G. *J. Am. Chem. Soc.* **1982**, *104*, 2596.
19. Hopfield, J. J. *Proc. Natl. Acad. Sci. U.S.A.* **1974**, *71*, 3640.
20. Hopfield, J. J. *Biophys. J.* **1977**, *18*, 311.
21. German, E. D.; Kuznetsov, A. M. *Electrochim. Acta* **1981**, *26*, 1595.
22. Reisler, E.; Eisenberg, H. *J. Chem. Phys.* **1965**, *43*, 3875.
23. Srinivasan, K. R.; Kay, R. L. *J. Chem. Phys.* **1974**, *60*, 3645.
24. Kharkats, Y. I. *Electrokhimiya* **1973**, *9*, 881.
25. Kharkats, Y. I. *Electrokhimiya* **1976**, *12*, 592.
26. Cannon, R. D. *Chem. Phys. Lett.* **1977**, *49*, 299.
27. Kirkwood, J. G.; Westheimer, F. H. *J. Am. Chem. Soc.* **1938**, *6*, 506.

28. *National Bureau of Standards Tables of Associated Legendre Functions;* Columbia University Press: New York, 1945.
29. Brunschwig, B. S; Ehrenson, S.; Sutin, N. *J. Phys. Chem.* **1986,** *90,* 3657.
30. German, E. D. *Chem. Phys. Lett.* **1979,** *64,* 295.
31. Lewis, N. A.; Obeng, Y. S. *J. Am. Chem. Soc.* **1988,** *110,* 2306.
32. Lewis, N. A.; Obeng, Y. S. *J. Am. Chem. Soc.* **1989,** *111,* 7624.
33. Sinha, U.; Lowery, M. D.; Hammack, W. S.; Hendrickson, D. N.; Drickamer, H. G. *J. Am. Chem. Soc.* **1987,** *109,* 7340.
34. Mikkelsen, K. V.; Ulstrup, J.; Zakaraya, M. G. *J. Am. Chem. Soc.* **1989,** *111,* 1315.

RECEIVED for review May 1, 1989. ACCEPTED revised manuscript August 22, 1989.

11

Intramolecular Electron Transfer from Photoexcited Ru(II) Diimine Complexes to N,N'-Diquaternarized Bipyridines

Russell H. Schmehl[1], Chong Kul Ryu[1], C. Michael Elliott[2],
C. L. E. Headford[2], and S. Ferrere[2]

[1]Department of Chemistry, Tulane University, New Orleans, LA 70118
[2]Department of Chemistry, Colorado State University, Fort Collins, CO 80523

A series of complexes of the type $[L_2Ru(II)(4.x.3\text{-}DQ^{2+})]^{4+}$ was prepared where L is either 2,2'-bipyridine or 4,4',5,5'-tetramethyl-2,2'-bipyridine and $4.x.3\text{-}DQ^{2+}$ is a ligand in which a 4,4'-dimethyl-2,2'-bipyridine links to a diquaternary 2,2'-bipyridine through a methylene chain (x). Rate constants for intramolecular electron transfer from the excited Ru(II) complex to the diquaternary 2,2'-bipyridine decrease as the length of the bridging chain increases from x = 2 to 12. The observed electron-transfer rate exhibits an even–odd chain length alternation for x = 2 to 6. A large reorganizational barrier is obtained (1.6 V) from the temperature dependence of the electron-transfer rate for the x = 6 complex. Rate constants for the back reaction were estimated from yields for trapping the Ru(III) of the intermediate by triethylamine. A preliminary account is given of environmental effects on the intramolecular electron transfer.

THE DEPENDENCE OF ELECTRON-TRANSFER RATES on donor–acceptor separation is of fundamental interest in charge-transfer chemistry (1–4). This dependence is particularly true for electron-transfer reactions in biological systems (2–4), where the distance can be large (>15 Å).
Model systems have been prepared that link an electron donor and acceptor through a sterically rigid spacer framework for which the donor–acceptor distance is clearly defined (5–13). Several elegant porphyrin–quinone complexes have been prepared (14–18) in which photo-

induced electron transfer occurs from the excited porphyrin to a quinone linked to the porphyrin through a rigid bridge. Closs and co-workers (1, 5–8) examined electron transfer from the biphenyl anion through steroid and decalin links to naphthalene. An exponential dependence of the electron-transfer rate on the number of bonds between the donor and acceptor is observed for linked derivatives in which the stereochemistry of the link to the bridging alkyl framework is fixed (i.e., both donor and acceptor are equatorial). The results indicate that through-bond electronic coupling is an important component in long-distance (thus nonadiabatic) electron transfer in this system.

Far more work has been done on donor–acceptor systems linked by flexible chains or bridging groups with several slowly interconverting conformers (19). Much of this work has been directed toward understanding intramolecular chain dynamics in flexible polymers (20, 21). Studies of the dependence of the rates of intramolecular excimer formation or electron transfer on the number of methylene units in the bridging chain have yielded a mix of results (22–28). For instance, the rate of intramolecular electron transfer in alkyl-linked anthracene–alkylamine systems in polar solvents is only weakly dependent on the chain length and shows a minimum for the three-methylene-bridged complex (28).

Mataga and co-workers (29, 30) reported a systematic decrease of the rate constant for electron transfer (k_{et}) with increasing chain length (two, four, and six methylenes) for linked octaethylporphyrin–benzoquinone complexes. In the more complex bridged tetraphenylporphyrin–benzoquinone systems studied by Connolly and others (31–33), evidence is obtained for ground-state Π complex formation in complexes connected by two or three linking methylenes. Intramolecular self-exchange electron transfer in alkyl-linked α-naphthyl and N-phthalimide moieties was studied by Szwarc and co-workers (20). An exponential decrease with increasing chain length up to six methylenes was observed in proprionitrile. For longer linkages, the rate was found to be essentially invariant with solvent.

In all of these alkyl-linked systems, at least one of the two reactive species is uncharged and ground-state complex formation is frequently observed. A much smaller body of information exists for flexible chain-linked donor–acceptor complexes in which both reactive sites have like charges. Intervalence-transfer (IT) absorption in mixed-valence complexes (34) (eq 1) has shown that electronic coupling between the metal centers decreases rapidly as the number of methylenes in the bridging ligand (1,n-bis(4-pyridyl)alkanes) increases. In fact, no IT band is observed for the 1,2-bis(4-pyridyl)ethane bridged dimer (35, 36)

$$[(bpy)_2ClRu(II)(py\text{-}(CH_2)_n\text{-}py)Ru(III)(NH_3)_5]^{5+} \xrightarrow{h\nu}$$

$$[(bpy)_2ClRu(III)(py\text{-}(CH_2)_n\text{-}py)Ru(II)(NH_3)_5]^{5+} \quad (1)$$

where bpy is 2,2'-bipyridine and py is pyridine.

Hurst and co-workers (37, 38) examined photoinduced electron transfer from Cu(I) olefin complexes linked to Co(III) pentammine through alkenoic acids, aminoalkenes, and pyridylalkenes. For the aminoalkene series, a steady decrease in the quantum yield for photoreduction of the Co(III) center was observed with increasing alkyl chain length up to eight methylene bridging carbons, where the quantum yield was below the limit of measurement (38). They attributed the chain-length dependence to both the effect of coulombic repulsion of the two centers on the distribution of conformers in solution and the rapid relaxation of the Cu(I)-to-olefin(π^*) metal-to-ligand charge-transfer (MLCT) state (<10 ns), which is faster than chain folding (10–100 ns) (21).

Elliott, Kelley, and co-workers (39–41) reported the dynamics of intramolecular electron transfer from photoexcited Ru(II) diimine chromophores to several diquaternized species diquaternary 2,2'-bipyridine linked through the bridging ligand in structure 1. The ligand designation 4.x.n-DQ^{2+} corresponds to a polymethylene bridge with a variable number of methylenes, x, linked through the 4 position of the coordinating 2,2'-bipyridine to a diquaternized 2,2'-bipyridine with n = two, three, or four methylenes linking the nitrogen centers of the bipyridine.

1

The rate constants for the photoinduced process (eq 2) are dependent upon both the exergonicity of the process and a rapid intramolecular redox equilibrium between the bridging bipyridine and the other ligands in the coordination sphere of the Ru(II) (L is bpy, 4,4'-dimethyl-2,2'-bipyridine (DMB), or 4,4',5,5'-tetramethyl-2,2'-bipyridine (TMB)) in the photoexcited complex (39).

$$[(L)_2Ru(II)(4.x.3\text{-}DQ^{2+})]^{4+} \xrightarrow{h\nu,\ k_{et}} [(L)_2Ru(III)(4.x.3\text{-}DQ^{+})]^{4+} \quad (2)$$

This chapter describes the effect of changing the number of bridging methylene carbons, x, on the rate of photoinduced electron transfer in complexes with a fixed exergonicity (L is bpy or TMB and n = 3). The effect of environmental perturbation on the electron-transfer dynamics is also presented. A preliminary account is given of measurement of the rapid thermal back electron transfer (eq 3) by trapping the transient Ru(III) complex with triethylamine.

$$[(L)_2Ru(III)(4.x.3\text{-}DQ^+)]^{4+} \xrightarrow{k_b} [(L)_2Ru(II)(4.x.3\text{-}DQ^{2+})]^{4+} \qquad (3)$$

Results for the forward electron transfer (eq 2) show that the electron-transfer rate decreases with increases in the number of methylene carbons and that rates for electron transfer depend on whether the bridge has an odd or even number of carbon atoms.

Redox and Spectroscopic Properties of $[(L)_2Ru(4.x.3\text{-}DQ^{2+})]^{4+}$

The bridging ligands were prepared as reported earlier (39, 40, 42) and diquaternarized with 1,3-dibromopropane to yield $4.x.3\text{-}DQ^{2+}$ ($x = 2, 3, 4, 5, 6,$ or 12). The bridging ligand ($4.x.3\text{-}DQ^{2+}$) was then heated (175 °C) with an excess of $[Ru(L)_2Cl_2]$ (L is bpy or TMB) in ethylene glycol under N_2 for 1 h to produce the resulting complex, $[(L)_2Ru(4.x.3\text{-}DQ^{2+})]^{4+}$. Purification of the complex was achieved by repeated chromatography on silica gel with 5:4:1 acetonitrile:water:saturated aqueous KNO_3 (39, 40).

The redox behavior for the series of complexes with fixed L was identical; one-electron oxidation of the complex (eq 4) was reversible by cyclic voltammetry and corresponds to the metal-localized, Ru(III/II), couple (40).

$$[(L)_2Ru(III)(4.x.3\text{-}DQ^{2+})]^{5+} \xleftrightarrow{+e^-} [(L)_2Ru(II)(4.x.3\text{-}DQ^{2+})]^{4+} \qquad (4)$$

Five reversible reductive waves were observed by cyclic voltammetry (40). The first two occur (eq 5) at -0.64 and -0.92 V vs. the sodium-saturated calomel electrode (SSCE) and correspond to two sequential reductions of the $4.x.3\text{-}DQ^{2+}$.

$$[(L)_2Ru(II)(4.x.3\text{-}DQ^{2+})]^{4+} \xleftrightarrow{+e^-} [(L)_2Ru(II)(4.x.3\text{-}DQ^+)]^{3+} \qquad (5a)$$

$$[(L)_2Ru(II)(4.x.3\text{-}DQ^+)]^{3+} \xleftrightarrow{+e^-} [(L)_2Ru(II)(4.x.3\text{-}DQ^0)]^{2+} \qquad (5b)$$

The remaining reductions correspond to sequential reduction of the diimine ligands coordinated to the Ru(II) center. The first of these reductions is at -1.36 V for $[(bpy)_2Ru(4.2.3\text{-}DQ^0)]^{2+}$, virtually identical to the first reduction potential of $[(bpy)_2Ru(DMB)]^{2+}$, which is bpy localized. Table I summarizes redox data for complexes in which L is either bpy or TMB.

The absorption and emission properties of Ru(II) diimine complexes have been widely studied (43). The spectroscopic characteristics of the $[(L)_2Ru(4.x.3\text{-}DQ^{2+})]^{4+}$ series closely parallel those of the parent complex, $[(L)_2Ru(DMB)]^{2+}$, except that the emission quantum yields are smaller and

Table I. Physical Properties of Complexes with the Ligand 4.6.3-DQ^{2+}

Property	[(bpy)$_2$Ru(L)]$^{4+}$	[(TMB)$_2$Ru(L)]$^{4+}$
$E°$ [Ru(III/II)]^a	1.24	1.06
$E°$ [DQ$^{(2+/+)}$]^a	−0.64	−0.64
$E°$ (L'/L'−)^b	−1.36	−1.51
E_{em} (298 K), nm^c	624	634
Φ_{em} (298 K)^c	0.008	0.003
k_r, s^{-1}	8.1 × 10^4	8.3 × 10^4
ΔE, V^d	0.36	0.46

^{a}Potentials, $E°$, vs. SSCE in CH$_3$CN.
^{b}L' is bpy for [(bpy)$_2$Ru(L)]$^{4+}$ and DMB of the 4.6.3-DQ^{2+} for [(TMB)$_2$Ru(L)]$^{4+}$.
^{c}In deaerated CH$_3$CN. Emission maxima, E_{em}, and quantum yields, Φ_{em}, are uncorrected for detector response.
^{d}Determined from the approximate E_{00}, $E°$ [DQ$^{(2+/+)}$], and $E°$ [Ru(III/II)] using $\Delta E = E_{00} − E°$ [Ru(III/II)] + $E°$ [DQ$^{(2+/+)}$].

the luminescence decay rates are much faster than those of the parent complex. The absorption is Ru(II) to L(π*) MLCT. Emission arises from the ^{3}(MLCT) state that is formed following intersystem crossing from the ^{1}(MLCT) state. For the parent complex, the intersystem crossing efficiency is unity (44).

The ^{3}(MLCT) state of [(bpy)$_2$Ru(DMB)]$^{2+}$ is quenched efficiently by the diquaternary 2,2'-bipyridine. Flash photolysis and steady-state trapping (of transient Ru(III)) studies indicate that the quenching occurs by electron transfer from the excited complex to the diquaternary 2,2'-bipyridine (43, 45, 46). The luminescence and redox characteristics of the Ru(II) complex–diquaternary bipyridine systems studied here are given in Table I. The exergonicity of excited-state electron transfer in [(L)$_2$Ru(4.6.3-DQ^{2+})]$^{4+}$ is estimated from the Ru(III/II) and DQ(2+/+) potentials and the zero–zero emission energy, E_{00}, of [(L)$_2$Ru(DMB)]$^{2+}$ (39–41). Exergonicities in the range of −0.36 to −0.46 V indicate that the reactions are in the normal free energy region (45, 46, 47, 48).

Chain-Length Dependence of Photoinduced Electron Transfer

The luminescence lifetimes of the complexes in the series [(bpy)$_2$Ru(4.x.3-DQ^{2+})]$^{4+}$ in CH$_3$CN at 298 K are given in Table II. The rate of intramolecular electron transfer from the ^{3}(MLCT) state of the Ru(II)complex to the linked diquaternary bipyridine (eq 2) is obtained from the difference in luminescence decay rates of the diquaternary bipyridine-containing complex (k_{obs}) and the parent complex of the series, [(bpy)$_2$Ru(DMB)]$^{2+}$, k_0 (eq 6).

$$k_{et} = k_{obs} − k_0 \qquad (6)$$

Table II. Luminescence Lifetimes and Intramolecular Electron-Transfer Rate Constants for [(bpy)$_2$Ru(L)] Complexes in CH$_3$CN at 298 K

Ligand L	τ, ns	$k_{et} \times 10^6$, s^{-1}	Number of C–C Bonds
DMB	690 ± 5	—	—
4.2.3-DQ^{2+}	1.7 ± 0.5	590	3
4.3.3-DQ^{2+}	42 ± 3	22	4
4.4.3-DQ^{2+}	18 ± 3	54	5
4.5.3-DQ^{2+}	128 ± 3	6.4	6
4.6.3-DQ^{2+}	99 ± 2	8.7	7
4.12.3-DQ^{2+}	373 ± 6	1.2	13

This rate assumes that electron transfer occurs exclusively from the ^{3}(MLCT) state and that the intersystem crossing efficiency, η_{isc}, is unity for the diquaternary bipyridine-containing complexes. If reduction of the diquaternary bipyridine occurs from the ^{1}(MLCT) state, η_{isc} must be less than 1. Because the luminescence lifetime (τ_{em}), emission quantum yield (Φ_{em}), and η_{isc} are related (eq 7), relative changes in the intersystem crossing efficiency can be determined from quantum yield and lifetime data if it is assumed that the radiative decay rates, k_r, of the parent complex and the diquaternized-species diquaternary 2,2'-bipyridine-containing complex are the same.

$$\Phi_{em} = \eta_{isc} k_r \tau_{em} \qquad (7)$$

For the complex [(bpy)$_2$Ru(4.6.3-DQ^{2+})]$^{4+}$, the emission quantum yield in CH$_3$CN is 0.008 ± 0.001 and the luminescence lifetime is 99 ns, giving a value of 8×10^4 s^{-1} for $\eta_{isc} k_r$. This value is close to the [(bpy)$_2$Ru(DMB)]$^{2+}$ radiative decay rate of 8.3×10^4. The similarity implies that electron transfer from the ^{1}(MLCT) state of [(bpy)$_2$Ru(4.6.3-DQ^{2+})]$^{4+}$ is not a major decay path. Bimolecular quenching can also be excluded as a decay path for the ^{3}MLCT state under the experimental conditions used. The complex concentration was low (<2×10^{-5} M) and the bimolecular quenching rate constant is less than diffusion-controlled (2×10^8 M^{-1} s^{-1}).

Figure 1 illustrates the dependence of ln (k_{et}) on the number of carbon–carbon bonds between the two redox centers. Although the data are limited, it is clear that, for odd numbers of bonds up to 7, an exponential decrease in k_{et} occurs. The exponential falloff is expected if the predominant conformation of the complexes in solution is fully extended and conformational folding is slow on the time scale of the experiment. The observed alternation of k_{et} with the number of bonds is also consistent with electron transfer from fully extended conformers. The observed exponential decrease for the series bridged by an odd number of bonds suggests that the electron-

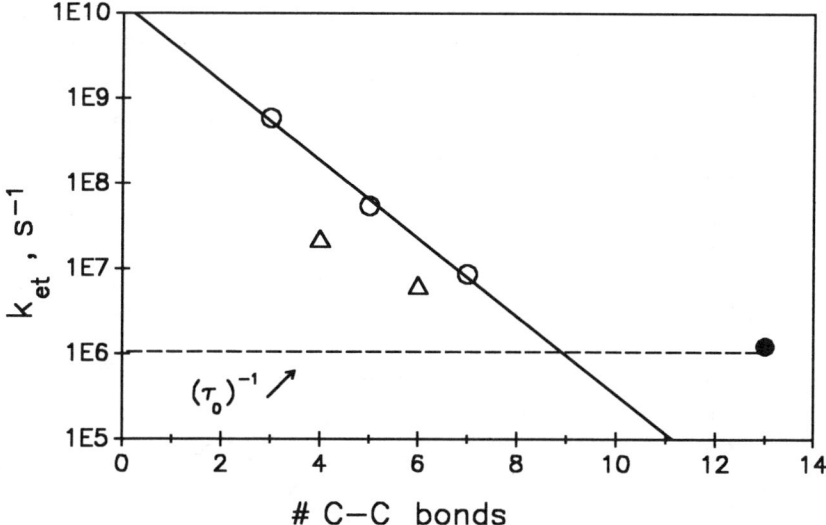

Figure 1. Plot of ln (k_{et}) vs. the number of C–C bonds between the Ru(II)-coordinated bipyridine and the diquaternized species diquaternary 2,2'-bipyridine. The solid line represents the least-squares fit to data from complexes with an odd number of C–C bonds (○, odd; △, even). The dashed line represents the decay rate of [(bpy)$_2$Ru(DMB)]$^{2+}$.

transfer rate will be slower than the decay rate of the unquenched complex (dashed line) when more than nine bonds separate the Ru(II) complex from the diquaternized species diquaternary 2,2'-bipyridine. For the 13-bond bridged complex, the measured k_{et} is greater than $(k_0)^{-1}$, an indication that conformational changes following excitation of the complex result in electron transfer.

Studies of Winnik (21) show that the dynamics of conformational changes in linear alkyl polymers are on the 10–100-ns time scale. The luminescence lifetimes of the [(bpy)$_2$Ru(4.x.3-DQ^{2+})]$^{4+}$ series are in the range of 1–100 ns (except for the 4.12.3 complex). Therefore, the observed rates are likely to be a function of both the rate of conformational change and electronic tunneling from the dominant solution conformers. Flexible chain-bridged donor–acceptor complexes are poor models for examining distance effects on electron-transfer reactions. However, such effects may be seen in systems that have electron-transfer rates that are faster than conformational rearrangement rates ($k_{et} > 10^7$ s^{-1}) or have steric or electrostatic factors that hinder conformational motions.

The excited-state electron transfer in these complexes exhibits temperature dependence between 180 and 300 K. Luminescence lifetimes of [(TMB)$_2$Ru(4.6.3-DQ^{2+})]$^{4+}$ and [(TMB)$_2$Ru(DMB)]$^{2+}$ in 4:1 ethanol:methanol above the liquid-to-glass transition temperature are shown in Figure 2. Both complexes show temperature-dependent luminescence decays. The behavior of [(TMB)$_2$Ru(DMB)]$^{2+}$ is similar to that of other diimine Ru(II) complexes that exhibit thermally activated internal conversion from the ^{3}(MLCT) state to a metal-centered, ^{3}(MC), state (43).

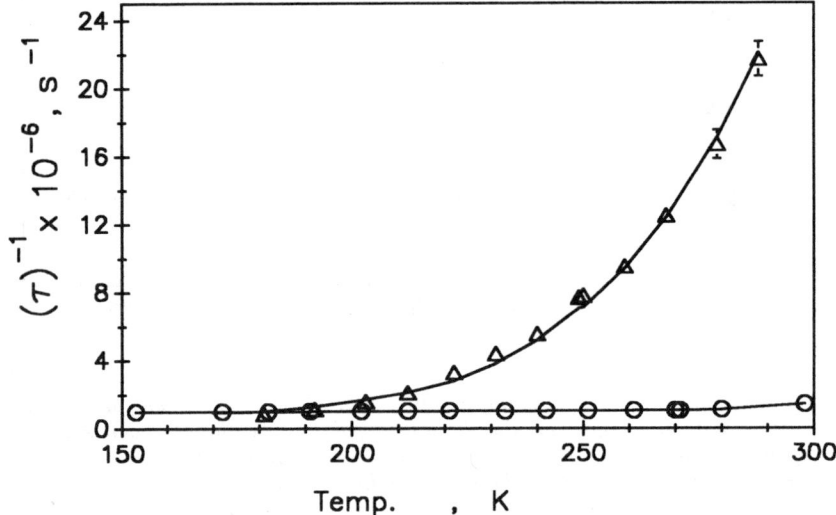

Figure 2. Temperature-dependent luminescence-decay-rate constants for [(TMB)$_2$Ru(4.6.3-DQ^{2+})]$^{4+}$ (△) and [(TMB)$_2$Ru(DMB)]$^{2+}$ (○). Solid lines represent fits to the data (see text).

The temperature dependence of intramolecular electron transfer (eq 2) can be examined by using eq 6 to determine k_{et} at each temperature. Activation parameters obtained from Eyring plots yield an activation enthalpy ($\Delta H^{\ddagger}$) of 4.7 kcal/mol and an activation entropy ($\Delta S^{\ddagger}$) of −8.9 eu. For intramolecular electron transfer in polyproline-linked Os–Ru ammines, the activation enthalpy, and hence the reorganizational energy, has been shown to depend on the distance between the two redox active centers (49).

Further, it was shown that the distance dependence of the nuclear factor is greater than that of the electronic factor in these systems. If it is assumed that $\Delta S = 0$ for the reaction and that there is no temperature dependence of solvent dielectric properties, then the reorganizational energy, λ, can be determined from $\Delta H^{\ddagger}$; in [(TMB)$_2$Ru(4.6.3-DQ^{2+})]$^{4+}$, $\lambda = 1.6$ V. Estimates of λ of approximately 0.8 V have been obtained from studies of bimolecular excited-state electron transfer between Ru(II) bipyridyl complexes and a series of diquaternary bipyridines and viologens (45–48). In this complex, the larger λ may result from the increased distance or from conformational changes required prior to electron transfer.

In principle, the rate of back electron transfer from the reduced diquaternary bipyridines to the Ru(III) (eq 3) can be determined either from flash photolysis or steady-state photolysis, in which one of the transient species is chemically trapped. Subnanosecond flash photolysis studies (39) of $[(L)_2Ru(4.2.3-DQ^{2+})]^{4+}$ showed that the back electron transfer is faster than the rise time of the apparatus used. Thus, a lower limit for the rate constant was estimated to be 3×10^{10} s^{-1} (39).

One approach to chemically trapping one of the intermediate ions formed in eq 2 is reduction of the Ru(III) by a mild reductant such as triethylamine or triethanolamine. This approach has been widely used in schemes to generate reduced viologens or diquaternary bipyridines for studies of hydrogen evolution from water using colloidal catalysts (50–52). The scheme for trapping Ru(III) by triethylamine in the series $[(L)_2Ru(4.x.3-DQ^{2+})^{4+}]$ is shown in eqs 8–12.

$$[(L)_2Ru(II)(4.x.3-DQ^{2+})]^{4+} \xrightarrow{h\nu,\, k_{et}} [(L)_2Ru(III)(4.x.3-DQ^+)]^{4+} \quad (8)$$

$$[(L)_2Ru(III)(4.x.3-DQ^+)]^{4+} \xrightarrow{k_b} [(L)_2Ru(II)(4.x.3-DQ^{2+})]^{4+} \quad (9)$$

$$[(L)_2Ru(III)(4.x.3-DQ^+)]^{4+} + (C_2H_5)_3N \xrightarrow{k_t}$$
$$[(L)_2Ru(II)(4.x.3-DQ^+)]^{3+} + (C_2H_5)_3N^{+\cdot} \quad (10)$$

$$(C_2H_5)_3N^{+\cdot} + (C_2H_5)_3N \rightarrow (C_2H_5)_3NH^+ + (C_2H_5)_2N(\dot{C}HCH_3) \quad (11)$$

$$[(L)_2Ru(II)(4.x.3-DQ^{2+})]^{4+} + (C_2H_5)_2N(\dot{C}HCH_3) \rightarrow$$
$$[(L)_2Ru(II)(4.x.3-DQ^+)]^{3+} + (C_2H_5)_2N=CHCH_3 \quad (12)$$

The radical cation of triethylamine formed in eq 10 reacts with a second mole of $(C_2H_5)_3N$ to produce a radical that can reduce a second diquaternary bipyridine (eq 12). Thus, a single photon will produce two reduced diquaternary bipyridines. The overall quantum yield (Φ_{obs}) for the process is given by eq 13.

$$\Phi_{obs} = 2\eta_{isc}[k_{et}/(k_r + k_n + k_{et})]\{k_t[(C_2H_5)_3]/(k_b + k_t[(C_2H_5)_3N])\} \quad (13)$$

The intersystem crossing efficiency, η_{isc}, is assumed to be unity, and the electron-transfer efficiency, $k_{et}/(k_r + k_n + k_{et})$, η_q, is obtained from the emission lifetimes of $[(L)_2Ru(4.x.3-DQ^{2+})]^{4+}$, τ_c, and $[(L)_2Ru(DMB)]^{2+}$, τ_0, as $1 - (\tau_c/\tau_0)$. An estimate of the intramolecular back-electron-transfer rate may be obtained from the efficiency of trapping, $k_t[(C_2H_5)_3N]/(k_b + k_t[(C_2H_5)_3N])$, and knowledge of the trapping rate, k_t. Photolysis of the complexes $[(bpy)_2Ru(4.x.3-DQ^{2+})]^{4+}$ in CH_3CN containing 0.5 M $(C_2H_5)_3N$ leads to the production of the reduced diquaternary bipyridine, $[(bpy)_2Ru(4.x.3-DQ^+)]^{3+}$, after 10–30 min of photolysis for the 5-, 6-, and 12-carbon bridged species.

Table III lists observed quantum yields for DQ$^+$ production and ap-

Table III. Quantum Yields for Trapping of [(bpy)$_2$Ru(4.x.3-DQ$^+$)]$^{4+}$ with (C$_2$H$_5$)$_3$N

Complex	$\eta_q{}^a$	$\Phi_{obs}{}^b$	k_b, s^{-1}
[(tmb)$_2$Ru(4.6.3-DQ^{2+})]	0.95	0.003	1.3 × 10^{10}
[(bpy)$_2$Ru(4.5.3-DQ^{2+})]	0.87	0.003	1.0 × 10^{10}
[(bpy)$_2$Ru(4.6.3-DQ^{2+})]	0.84	0.002	2.0 × 10^{10}
[(bpy)$_2$Ru(4.12.3-DQ^{2+})]	0.45	0.007	3.0 × 10^9

NOTE: These values assume that η_{isc} = 1 and k_t = 2 × 10^7 M^{-1}s^{-1}. See text eq 13.
^{a}Fraction of excited states quenched by electron transfer.
^{b}Observed quantum yield for diquaternized species formation. Relative error is ±50%.

proximate intramolecular back-electron-transfer rates with an assumed trapping rate constant of 2 × 10^7 M^{-1} s^{-1} (46). The margins of error are large because of the uncertainty in the luminescence lifetimes of the complexes, coupled with the fact that the reduced diquaternary bipyridine slowly decomposes. As a result, the calculated back-electron-transfer rate constants represent upper limiting values. Results from bimolecular trapping of photoproducts from photolysis of [Ru(bpy)$_3$]$^{2+}$ and diquaternary bipyridine in aqueous solution (46) yield values of 1–4 × 10^{10} M^{-1} s^{-1} for the geminate back-electron-transfer rate.

Although limited data are available here, the results are qualitatively comparable with distance-dependence effects observed for the forward electron transfer (eq 2). Back-electron-transfer rates that are much faster than quenching rates are expected because the reorganizational energies of the forward and back electron transfer should be comparable. Given the activation parameters obtained for the forward process, the back reaction should be close to the activationless limit (λ = 1.6 V and ΔG = −1.88 V).

Environmental Effects on Intramolecular Electron Transfer

A few simple experiments illustrate the relative importance of coulombic and dynamic conformational changes on excited-state electron transfer in [(bpy)$_2$Ru(4.x.3-DQ^{2+})]$^{4+}$. Addition of tetraethylammonium perchlorate (0.1 M) to acetonitrile solutions of both the 6- and 12-carbon bridged complexes results in an increase in k_{et}, as shown in Table IV. The observed effect is consistent with a strong influence of electrostatic factors on the distribution of conformers in solution prior to electron transfer.

Viscosity effects on electron transfer in these complexes depend on the particular medium. For example, as shown in Table IV, there is only a factor of 2.2 decrease in the luminescence lifetime of [(TMB)$_2$Ru(4.6.3-DQ^{2+})]$^{4+}$ in changing the viscosity, η, of the medium from glycerol at 273 K (η = 220,000 cp) to 4:1 ethanol:methanol at the same temperature (η = 1.2 cp). However, in poly(methyl methacrylate) (PMMA) glasses at room temperature, the luminescence lifetime of [(bpy)$_2$Ru(4.6.3-DQ^{2+})]$^{4+}$ becomes bi-

Table IV. Effects of Medium on Intramolecular Excited-State Electron Transfer

Complex	Medium	Temperature, K	τ^a, ns
$[(bpy)_2Ru(4.12.3\text{-}DQ^{2+})]^{4+}$	CH_3CN	298	373 ± 6
	CH_3CN, 0.1 M TEAP[b]	298	93 ± 3
$[(bpy)_2Ru(4.6.3\text{-}DQ^{2+})]^{4+}$	CH_3CN	298	99 ± 2
	CH_3CN, 0.1 M TEAP[b]	298	57 ± 4
	PMMA[c]	298	608 ± 5 (33%)
			1460 ± 8 (67%)
$[(bpy)_2Ru(4.6.3\text{-}DQ^{2+})]^{4+}$	$C_2H_5OH\text{-}CH_3OH$	298	36 ± 3
	glycerol	298	99 ± 4
	$C_2H_5OH\text{-}CH_3OH$	276	63 ± 3
	glycerol	273	141 ± 6

[a] τ is luminescence lifetime of ^{3}MLCT emission.
[b] TEAP is tetraethylammonium perchlorate.
[c] PMMA is poly(methyl methacrylate) glass.

exponential and the longer of the two decays approaches that of $[(bpy)_2Ru(DMB)]^{2+}$ in PMMA. The nearly complete disappearance of intramolecular electron transfer in PMMA may result from slow dielectric relaxation of the medium (53–55). These effects are currently being investigated in greater detail.

Summary

This work demonstrates that excited-state electron transfer in flexible chain-linked donor–acceptor complexes with a large coulombic repulsion between the donor and acceptor can show effects similar to those observed in rigid bridged systems. For the series $[(bpy)_2Ru(4.x.3\text{-}DQ^{2+})]^{4+}$, both electronic and nuclear factors are important in determining the rate of intramolecular electron transfer, as reflected by the large distance dependence and observable temperature dependence. Ion recombination in this complex series is rapid ($>10^{10}$ s^{-1}), even in complexes with long alkyl bridges. Further, examination of excited-state electron transfer of these complexes in viscous and rigid media illustrates that dielectric relaxation effects can significantly affect the electron-transfer rate, particularly when the medium dielectric relaxation rate is slower than the excited-state relaxation rate.

Acknowledgments

R. H. Schmehl acknowledges the support of the Petroleum Research Fund, administered by the American Chemical Society, and the Louisiana Educational Quality Support Fund, administered by the Louisiana State Board of Regents. C. M. Elliott thanks the National Science Foundation (Grant No. CHE 8516904) for support.

References

1. Closs, G. L.; Miller, J. R. *Science* **1988**, *240*, 440, and references therein.
2. Marcus, R. A.; Sutin, N. *Biochem. Biophys. Acta* **1985**, *811*, 265.
3. Guarr, T.; McLendon, G. *Coord. Chem. Rev.* **1985**, *68*, 1.
4. Hush, N. S. *Coord. Chem. Rev.* **1985**, *64*, 135.
5. Closs, G. L.; Piotrowiak, P.; MacInnis, J. M.; Fleming, G. R. *J. Am. Chem. Soc.* **1988**, *110*, 2652.
6. Closs, G. L.; Calcaterra, L. T.; Green, N. J.; Penfield, K. W.; Miller, J. R. *J. Phys. Chem.* **1986**, *90*, 3673.
7. Calcaterra, L. T.; Closs, G. L.; Miller, J. R. *J. Am. Chem. Soc.* **1983**, *105*, 670.
8. Miller, J. R.; Calcaterra, L. T.; Closs, G. L. *J. Am. Chem. Soc.* **1984**, *106*, 3047.
9. Oliver, A. M.; Craig, D. C.; Paddon-Row, M. N.; Kroon, J.; Verhoeven, J. W. *Chem. Phys. Lett.* **1988**, *150*, 366.
10. Oevering, H.; Paddon-Row, M. N.; Heppener, M.; Oliver, A. M.; Cotsaris, E.; Verhoeven, J. W.; Hush, N. S. *J. Am. Chem. Soc.* **1987**, *109*, 3258.
11. Kroon, J.; Oliver, A. M.; Paddon-Row, M. N.; Verhoeven, J. W. *Recl. Trav. Chim. Pays-Bas* **1988**, *107*, 509.
12. Wegewijs, B.; Hermant, R. M.; Verhoeven, J. W.; Kunst, A. G. M.; Rettschnick, R. P. H. *Chem. Phys. Lett.* **1987**, *140*, 587.
13. Verhoeven, J. W.; Paddon-Row, M. N.; Hush, N. S.; Oevering, M.; Heppener, M. *Pure Appl. Chem.* **1986**, *58*, 1285.
14. Wasielewski, M. R.; Johnson, D. G.; Svec, W. A.; Kersey, K. M.; Minsek, D. W. *J. Am. Chem. Soc.* **1988**, *110*, 7219.
15. Wasielewski, M. R.; Niemczyk, M. P.; Svec, W. B.; Pewitt, E. B. *J. Am. Chem. Soc.* **1985**, *107*, 5562.
16. Gust, D.; Moore, T. A.; Moore, A. L.; Makings, L. R.; Seely, G. R.; Ma, X.; Trier, T. T.; Gao, F. *J. Am. Chem. Soc.* **1988**, *110*, 7567.
17. Gust, D.; Moore, T. A.; Moore, A. L.; Barrett, D.; Harding, L. O.; Makings, L. R.; Liddell, P. A.; De Schryver, F. C.; van der Auweraer, M.; Bensasson, R. V.; Rougee, M. *J. Am. Chem. Soc.* **1988**, *110*, 321.
18. Gust, D.; Moore, T. A.; Liddell, P. A.; Nemeth, G. A.; Makings, L. R.; Moore, A. L.; Barrett, D.; Pessiki, P. J.; Bensasson, R. V.; Rougee, M.; Chachaty, C.; DeSchryver, F. C.; van der Auweraer, M.; Holzwarth, A. R.; Connolly, J. S. *J. Am. Chem. Soc.* **1987**, *109*, 846.
19. Kavarnos, G. J.; Turro, N. J. *Chem. Rev.* **1986**, *86*, 401.
20. Shimada, K.; Shimozato, Y.; Szwarc, M. *J. Am. Chem. Soc.* **1975**, *97*, 5834.
21. Winnik, M. A. *Acc. Chem. Res.* **1977**, *10*, 173.
22. Brimage, B. R. G.; Davidson, R. S. *J. Chem. Soc., Chem. Commun.* **1976**, 827.
23. Crawford, M. K.; Eisenthal, K. B. *Chem. Phys. Lett.* **1981**, *79*, 529.
24. Crawford, M. K.; Eisenthal, K. B. *J. Am. Chem. Soc.* **1982**, *104*, 5874.
25. Okada, T.; Masahito, M.; Mataga, N.; Sakata, Y.; Misumi, S. *J. Am. Chem. Soc.* **1981**, *103*, 4715.
26. Swinnen, A. M.; van der Auweraer, M.; De Schryver, F. C.; Nakatani, K.; Okada, T.; Mataga, N. *J. Am. Chem. Soc.* **1987**, *109*, 321.
27. Vannikov, A. V.; Grishina, A. D. *Russ. Chem. Rev.* **1987**, *56*, 633.
28. Winnik, M.; Zachariasse, K. *J. Phys. Chem.* **1984**, *88*, 2964.
29. Nishitani, S.; Kurata, N.; Sakata, Y.; Misumi, S.; Migita, M.; Okada, T.; Mataga, N. *Tetrahedron Lett.* **1981**, *22*, 2099.
30. Migita, M.; Okada, T.; Mataga, N.; Nishitani, S.; Kurata, N.; Sakata, Y.; Misumi, S. *Chem. Phys. Lett.* **1981**, *84*, 263.
31. Schmidt, J. A.; McIntosh, A. R.; Weedon, A. C.; Bolton, J. R.; Connolly, J. R.; Hurley, J. K.; Wasielewski, M. R. *J. Am. Chem. Soc.* **1988**, *110*, 1733.

32. Connolly, J. S. In *Photochemical Conversion and Storage of Solar Energy—1982;* Rabani, J., Ed.; Weizmann Scientific Press of Israel: Jerusalem, Israel, 1982; Part A, p 175.
33. McIntosh, A. R.; Siemiarczuk, A.; Bolton, J. R.; Stillman, M. J.; Ho, T.-F.; Weedon, A. C. *J. Am. Chem. Soc.* **1983**, *105*, 7215.
34. Creutz, C. *Progr. Inorg. Chem.* **1983**, *30*, 1, and references therein.
35. Callahan, R. W.; Brown, G. M.; Meyer, T. J. *J. Am. Chem. Soc.* **1974**, *96*, 7829.
36. Callahan, R. W.; Brown, G. M.; Meyer, T. J. *Inorg. Chem.* **1975**, *14*, 1443.
37. Farr, J. K.; Hulett, L. G.; Lane, R. H.; Hurst, J. K. *J. Am. Chem. Soc.* **1975**, *97*, 2654.
38. Norton, K. A.; Hurst, J. K. *J. Am. Chem. Soc.* **1982**, *104*, 5960.
39. Cooley, L. F.; Headford, C. E. L.; Elliott, C. M.; Kelly, D. F. *J. Am. Chem. Soc.* **1988**, *110*, 6673.
40. Elliott, C. M.; Freitag, R. A.; Blaney, D. D. *J. Am. Chem. Soc.* **1985**, *107*, 4647.
41. Elliott, C. M.; Freitag, R. A. *J. Chem. Soc., Chem. Commun.* **1985**, 156.
42. Schmehl, R. H.; Auerbach, R. A.; Wacholtz, W. F.; Elliott, C. M.; Freitag, R. F. *Inorg. Chem.* **1986**, *25*, 2440.
43. Juris, A.; Balzani, V.; Barigelletti, F.; Campagna, S.; Belser, P.; von Zelewsky, A. *Coord. Chem. Rev.* **1988**, *52*, 85.
44. Wacholtz, W. F.; Auerbach, R. A.; Schmehl, R. H. *Inorg. Chem.* **1986**, *25*, 227.
45. Chan, S.-F.; Chou, M.; Creutz, C.; Matsubara, T.; Sutin, N. *J. Am. Chem. Soc.* **1981**, *103*, 369.
46. Creutz, C.; Keller, A. D.; Sutin, N.; Zipp, A. P. *J. Am. Chem. Soc.* **1982**, *104*, 3618.
47. Creutz, C.; Sutin, N. *Pure Appl. Chem.* **1980**, *52*, 2717.
48. Bock, C. R.; Connor, J. A.; Gutierrez, A. R.; Meyer, T. J.; Whitten, D. G.; Sullivan, B. P.; Nagle, J. K. *J. Am. Chem. Soc.* **1979**, *101*, 4815.
49. Isied, S. S.; Vassilian, A.; Wishart, J. F.; Creutz, C.; Schwarz, H. A.; Sutin, N. *J. Am. Chem. Soc.* **1988**, *110*, 635.
50. *Energy Resources through Photochemistry and Catalysis;* Gratzel, M., Ed.; Academic: New York, 1983.
51. *Photogeneration of Hydrogen;* Harrriman, A.; West, M. A., Eds.; Academic: London, 1982.
52. Kalyanasundaram, K. *Coord. Chem. Rev.* **1982**, *46*, 159.
53. Nielson, R. M.; McManis, G. E.; Golovin, M. N.; Weaver, M. J. *J. Phys. Chem.* **1988**, *92*, 3441.
54. McCrum, N. G.; Read, B. E.; Williams, G. *Anelastic and Dielectric Effects in Polymeric Solids;* Wiley: New York, 1967; p 255.
55. Miller, J. R.; Beitz, J. V.; Huddleston, R. K. *J. Am. Chem. Soc.* **1984**, *106*, 5057.

RECEIVED for review May 1, 1989. ACCEPTED revised manuscript July 17, 1989.

12

Bridged Mixed-Valence Systems

How Polarizable Bridging Ligands Can Lead to Interesting Spectroscopic and Conductive Properties

Mary Jo Ondrechen, Saeed Gozashti, Li-Tai Zhang, and Feimeng Zhou

Department of Chemistry, Northeastern University, Boston, MA 02115

> *In bridged mixed-valence dimers of the form M–L–M (where M is a metal ion and L is a bridging ligand), two remote metal ions sometimes can be strongly coupled to each other via certain bridging ligands, even if they are separated by several angstroms or more and have essentially zero d-orbital overlap. In discrete bridged dimers of this type, the frequency maximum and the shape of the intervalence transfer band are strongly dependent on the bridging species. This type of interaction could lead to interesting conductive properties in the extended-chain linear polymeric system ... –M–L–M–L–M–L–M–L–M– .... More importantly, the system properties are controllable by chemical substitution on the bridging ligand L. In both the discrete systems and the extended systems, properties are dependent upon the electron occupation. Model Hamiltonians for these systems, which contain essential one- and two-electron terms, are presented. The important interactions in the discrete systems, and their implications for extended systems, will be discussed.*

A HOST OF BRIDGED MIXED-VALENT OR BINUCLEAR COMPLEXES have been synthesized in the past 20–30 years (1–5). In a bridged dimer, two metal atoms or ions are held together by some bridging ligand. The successful efforts to synthesize such compounds were motivated by a desire to make prototype model compounds that could offer new understanding of some very fundamental and intriguing questions: In the inner-sphere mechanism for electron transfer, how does the electron-transfer step occur (6)? How does electron transfer occur in biological systems (7–9)? Are two or more classical structures of a particular species in equilibrium or in resonance (10–14)?

Periodic one-dimensional analogues of the discrete bridged dimeric complex also may be synthesized. Quasi-one-dimensional solids have attracted substantial interest in chemistry, physics, materials science, and engineering (15). Some of these one-dimensional chains consist of transition metal ions joined together by a bridging ligand (16, 17).

One of the most exciting and challenging new developments in all of science in the past couple of years is the discovery of new superconducting materials with higher critical temperatures (T_c) for superconductivity (18, 19). Many of these new superconductors may be thought of as mixed-valence compounds of copper, in which copper ions are joined together in a two-dimensional net by oxide bridging ligands.

In this chapter, we discuss some of the advantages of coupling metal atoms or ions together by using bridging ligands. The effects of the electronic structure of the bridge on the properties of the whole system, and how these effects might be exploited, are discussed. Special emphasis is placed on complexes of ruthenium, in which the strength of the metal–ligand π backbonding interaction is strongly dependent on electron occupation. In the next section, particular systems are discussed. Then model Hamiltonians containing one- and two-electron terms to describe the systems of interest are presented.

Systems

Discrete Bridged Dimers. In a discrete bridged dimeric complex M–L–M, two metal atoms or ions M are joined together by some bridging ligand L. The degree of (indirect) M–M coupling and all of the properties that arise from this coupling are changed if the bridging ligand L is altered. In all of the discrete examples discussed in this section, the metal–metal distance is long enough so that the direct metal–metal interaction is weak.

In a bridged mixed-valence complex, the frequency maximum, width, and shape of the intervalence transfer (IT) band in the optical absorption (OA) spectrum have been shown to vary considerably as the bridging ligand is changed (20, 21). For example, in a series of bridged mixed-valence dimers of ruthenium, the frequency maximum is shifted by a factor of 2 and the width increases by a factor of 10 from the most strongly coupled bridging ligand to the most weakly coupled bridging ligand (20, 21).

In this section we show how the bridging ligands influence the spectra of the discrete bridged dimer. Some expressions for transition energies and a criterion for delocalization of the odd electron are obtained, in terms of the bridging-ligand properties.

We showed in earlier papers (22–26) that the ground- and excited-state potential surfaces (which correspond to initial and final states in the intervalence transfer transition) are changed completely when the coupled electronic state or states on the bridging ligand are included explicitly in the model Hamiltonian. In a two-state model (27–29) in which the two metal ions are assumed to be directly coupled, the only nuclear degree of freedom coupled to the IT transition is the antisymmetric (or vibrational difference) coordinate, represented in structure 1.

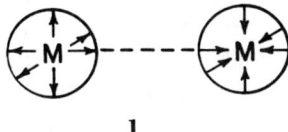

1

Consider, on the other hand, a three-site model Hamiltonian, where a single electronic state on the bridge (state 2) is coupled to one electronic state on each of the metal ions (states 1 and 3) as:

$$\hat{H} = \hat{H}_e + \hat{H}_v + \hat{H}_{e-v} \tag{1}$$

$$\hat{H}_e = J(a_1^\dagger a_2 + a_2^\dagger a_3 + a_2^\dagger a_1 + a_3^\dagger a_2) + \alpha a_2^\dagger a_2 \tag{2}$$

$$\hat{H}_v = \sum_{i=1}^{3} \left(\frac{p_i^2}{2m_i} + \frac{1}{2} m_i \omega_i^2 q_i^2 \right) \tag{3}$$

$$\hat{H}_{e-v} = \sum_{i=1}^{3} A_i q_i a_i^\dagger a_i \tag{4}$$

Here $a_i^\dagger$ and a_i are, respectively, the creation and annihilation operators for the ith electronic state. The momentum and the coordinate for the ith nuclear degree of freedom are given by p_i and q_i, respectively. These vibrational modes are assumed to be harmonic oscillators with reduced mass, m_i, and frequency, ω_i. The parent metal basis states 1 and 3 are assumed to be degenerate, and the energy gap between the bridge basis state 2 and the metal basis states is given by α. J is the electronic coupling (or resonance integral) between the bridge and each metal state. A_i is the vibronic coupling between the electronic state and the vibrational mode on the ith site. With the explicit inclusion of an electronic state on the bridge, the allowed electronic transitions are now coupled to the totally symmetric (or vibrational sum) coordinate on the metal atoms (22), represented in structure 2.

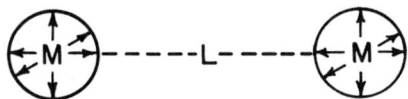

2

In addition, the vibrational coordinate on the bridging species, q_2, may also be coupled to the electronic transitions. In the case of very strong resonance interaction between the metal and bridging species (i.e., J is large and negative), the totally symmetric vibrational modes dominate the width and shape of the OA bands (25). If the bridge state is omitted from the model Hamiltonian, these important totally symmetric vibrational modes are artificially decoupled from the electronic degrees of freedom (26, 30).

Whether the bridged mixed-valence complex M–L–M has the odd electron localized [Robin–Day (31) Class I or II] or delocalized [Robin–Day Class III], of course, depends on the nature of the bridging species L. Equations 1–4 may be solved in the adiabatic (Born–Oppenheimer) approximation to obtain an approximate criterion for delocalization of the odd electron. The ground-state potential-energy surface has single-minimum form whenever the inequality condition

$$A^2(U + \alpha)[KU(U - \alpha)]^{-1} < 1 \tag{5}$$

is satisfied, where

$$K \equiv m_1\omega_1^2 = m_3\omega_3^2 \tag{6}$$

and where

$$U = (\alpha^2 + 8J^2)^{1/2} \tag{7}$$

This condition depends on the relative energy (α) of the coupled bridge orbital and the metal–bridge electronic coupling (J).

If inequality 5 is satisfied, then the electron transfer is faster than the vibrational time scale ($\sim 10^{-13}$ s), the two classical structures are in resonance, and the complex is average-valent (or Robin–Day Class III). In this case, the three molecular orbitals (MO) of the complex have bonding, nonbonding, and antibonding (B, N, and A) character (32). These orbitals may be represented schematically as in structure 3. These three MOs resemble the Hückel MOs of the pi system of the allyl radical.

The bridged system with the MOs depicted in structure 3 will have allowed transitions at energies (E) given approximately by

$$E_A - E_N = \frac{U + \alpha}{2} \tag{8}$$

and

$$E_N - E_B = \frac{U - \alpha}{2} \tag{9}$$

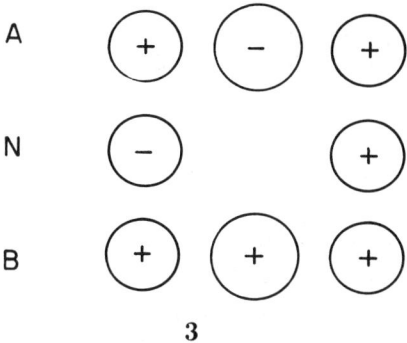

3

(The B-to-A transition is symmetry-forbidden.) Here we have ignored the small shifts in frequency maxima caused by the difference in zero-point energy between the ground- and excited-state surfaces. These transition energies are governed by the bridge-dependent quantities α and J.

In references 33 and 34, we argued that the Creutz–Taube ion (35), a pyrazine-bridged mixed-valence dimer of ruthenium, is such an average-valent structure, with MOs resembling those shown schematically in structure 3. In this complex, the metal–bridge electronic coupling is large enough ($J = -0.8$ eV) to cause complete delocalization of the unpaired electron. This conclusion is consistent with a number of experimental observations (36–38). The model given by eqs 1–4 successfully predicts the position and shape of the intervalence bands.

On the basis of this model, it is possible to "tailor" a bridging ligand, perhaps by chemical substitution (39), to create a discrete dimer that absorbs light at a particular frequency. For instance, if a methoxy group is placed on the pyrazine ring of the Creutz–Taube ion, the width at half-maximum of the IT (intervalence transfer) band doubles and the frequency maximum shifts toward the blue (39). Similarly, we can control the extent of bridge-mediated (indirect) metal–metal coupling and, therefore, the shape of the potential surfaces and the degree of electron delocalization. This capability suggests some potentially very exciting possibilities for one- and two-dimensional periodic bridged systems.

One-Dimensional Bridged Chains. Extended chains of metal ions M and bridging ligands L with the linear structure . . .–M–L–M–L–M–L–M–L–M–. . . are potentially very important because of the possibility of conductivity along the M–L axis and also of control of the conductive properties through synthetic alteration of the bridging ligand. Some of the "redox polymers" reported recently (40, 41) have this type of structure. Linear, low-polymer analogues of the Creutz–Taube ion were reported earlier by Tom (42). A number of metal phthalocyanine (PcM) and metal porphyrin (PrM) bridged chains (see structure 4), where

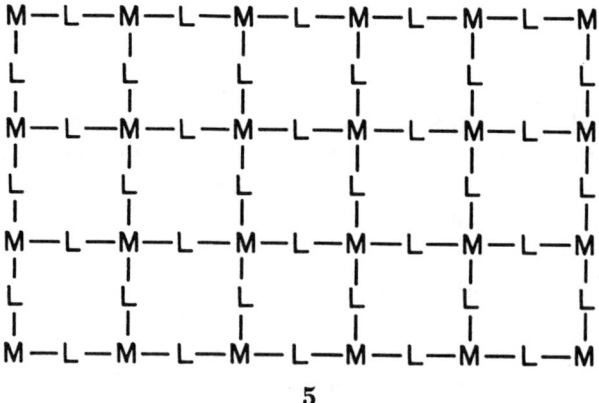

the four N atoms represent the macrocyclic species and where L is the bridging ligand, have been studied to date and are known to exhibit semiconducting properties (16, 17, 43–47). Under some conditions, these systems have been found to become conductors (45).

There is already evidence that the conductive properties of these chains are dependent upon the bridging species. For example, $[PcFe(dib)]_n$ has an electrical conductivity of 2×10^{-5} S/cm (where dib is p-diisocyanobenzene); if 2,3,5,6-tetramethyl-1,4-diisocyanobenzene is substituted for the dib bridging ligand, the conductivity drops to 1×10^{-7} S/cm (17). $[PcRu(pyz)]_n$ has a conductivity of 1×10^{-7} S/cm, but $[PcRu(tz)]_n$ has a conductivity of 1×10^{-2} S/cm (17). (Here pyz is pyrazine and tz is 1,2,4,5-tetrazine.).

In systems of this type, the conductivity is strongly dependent upon the electron occupation. For example, if the chain is partially oxidized, the conductivity generally increases (17, 45).

Two-Dimensional Superconducting Systems. Most of the new copper oxide superconducting materials (18, 19, 48, 49) may be thought of as mixed-valence compounds of Cu(II) and Cu(III), in which the copper ions are joined together by the polarizable oxide bridging ligand to form the two-dimensional structure represented in structure 5.

For electron-accounting purposes only, we regard the holes as located on the copper atoms. However, the holes may occur entirely or partially on the oxygen atoms (SO) (50). The system could also be thought of as a mixed-valence compound of O^{2-} and O^{1-}, with Cu(II) acting as a bridging ligand.

The critical temperature for superconductivity, T_c, is dependent upon the stoichiometry of the material, and hence upon the electron occupation. For example, in the system $YBa_2Cu_3O_{6+x}$, where $0 < x < 1$, T_c depends upon x and reaches a maximum at a value less than $x = 1$ (*51*).

One- and Two-Electron Contributions to the Total Energy

One-Electron Terms. The Hückel (or tight-binding) model Hamiltonian contains only one-electron terms and has the form:

$$H_{Huc} = \sum_i E_i a_i^\dagger a_i + \sum_{i,j}{}' J_{i,j}(a_i^\dagger a_j + a_j^\dagger a_i) \qquad (10)$$

where the prime on the summation indicates the restrictions that i and j are nearest-neighbors and that $i > j$. Equation 2 is such a model. In the systems considered here, only one basis orbital per atom is important. E_i is the energy of the isolated basis orbital on atom i and J_{ij} is the resonance integral for the interaction between nearest-neighbor atoms i and j.

A significant feature of the Hückel model is that the MO energy-level diagram that results from diagonalization of eq 10 is independent of the electron occupation of the MOs. This model might be reasonable for conjugated pi systems of carbon, for example, but it is clearly not correct for transition metal systems where electron correlation is significant. Thus, when a Hückel model is applied to a transition metal complex, the values one finds for the parameters E_i and J_{ij} apply only to that complex with fixed charge and fixed spin state. These one-electron models may have some utility in the prediction of the properties of bridged complexes as functions of the electronic structure of the bridging ligand, but they are obviously inadequate in the description of properties (such as transition energies and conductivities) as functions of electron occupation.

Electron–Electron Repulsion. The description of properties as functions of electron occupation are particularly important in polymeric systems. If the (undoped) linear chain $(ML)_N$ contains N metal ions and each metal ion M has two possible oxidation states, n and $n + 1$, then there are $N + 1$ possible values for the total charge on the chain. With dopants that partially oxidize or partially reduce the infinite-membered linear chain, a continuum of values for the electron occupation of the chain is possible in principle.

The Hubbard model (*52, 53*), which has been used widely as a better alternative to the Hückel model, incorporates electron–electron repulsion between electrons of opposite spin on the same atom. (We again confine ourselves, for purposes of discussion, to cases where only one basis orbital per atom is important.) The Hamiltonian may be written as

$$H_{\text{Hub}} = H_{\text{Huc}} + \sum_i U_i a_{i\uparrow}^\dagger a_{i\uparrow} a_{i\downarrow}^\dagger a_{i\downarrow} \tag{11}$$

where U_i is the Coulomb repulsion between two electrons of opposite spin in the same spatial orbital. These terms lead to the raising and lowering of orbital energies as the electron occupation is changed. In a transition metal complex, the ionization potential may change by several electronvolts if the complex is oxidized or reduced by one electron. These models have been applied extensively in solid-state physics. The applications of such a model to discrete mixed-valence dimers of transition metals is discussed in reference 53. Exact solutions to eq 11 for the linear chain have been obtained by Lieb and Wu (54).

To date, no Hubbard model has been found to exhibit superconductivity. However, several groups have looked at extended Hubbard (55) models for the copper oxide systems, seeking regions of parameter space where two particles (or two holes) exhibit an effective attractive pairing interaction (56). An extended Hubbard Hamiltonian may be written as

$$H_{\text{EH}} = H_{\text{Hub}} + {\sum_{i,j}}' V_{i,j} a_{i\uparrow}^\dagger a_{i\uparrow} a_{j\downarrow}^\dagger a_{j\downarrow} \tag{12}$$

The prime indicates that i and j are nearest neighbors. Thus, the extended Hubbard model is like the Hubbard model, except that electron–electron repulsion between electrons on nearest-neighbor atoms is also included and is given the symbol V_{ij}. We can indeed find values for the parameters such that there is apparent attractive pairing between electrons (or between holes) (56). However, an artificially large value for the nearest-neighbor repulsion V will tend to force the pairing of electrons. In that case, the extended Hubbard model does not give much physical insight into the possible mechanism for effective attraction (and hence for superconductivity) in that part of parameter space.

Occupation-Dependent Resonance Energy. In our studies of the electronic structure of some discrete systems, such as the back-bonding and pi-bonding monomeric complexes $[Ru(NH_3)_5pyz]^{2+}$ and $[Ru(NH_3)_5pyz]^{3+}$, we find that a simple Hubbard model does not contain sufficient flexibility to account properly for the observed ionization potentials and transition energies and for our calculated energy differences (57, 58). For example, the Hubbard model predicts too large shifts in the energy gap between the

Ru 4d orbitals and the π^* pyz orbital upon oxidation of the 2+ species to the 3+ species.

Experimental evidence (59–61) shows that Ru(II) back-bonds strongly, but Ru(III) does not. Thus, the metal–ligand electronic coupling (or effective resonance integral) is strongly dependent on the electron occupation. To introduce greater flexibility and detail (for the description of back-bonding complexes) into the model given by eq 11, we include an explicit, linear dependence of the resonance integral J_{ij} upon the electron occupation on the metal n_M. With $J(n_M)$, an additional two-electron integral is effectively introduced into the model. Like the extended Hubbard model, this model (which includes occupation-dependent resonance energy) can have an effective attractive interaction between electrons for certain values of the parameters (57, 62). Within the framework of the new model, this interaction arises from increased resonance stabilization upon the addition of an electron.

Extended-chain bridged complexes in which the metal ions back-bond to the bridging ligand should have interesting occupation-dependent conductive properties. For example, the fully reduced bridged chain of $[Ru(II)-L]_n$ may show increased conductivity as the system is partially oxidized. However, it should show decreased conductivity as it approaches full oxidation to Ru(III), because the metal–ligand resonance integral becomes small for low electron occupation on the metal ion.

Conclusions

The spectroscopic and conductive properties of bridged mixed-valence compounds may be controlled by synthetic alteration of the bridging ligand. The term that is probably most readily controlled by chemical substitution on the bridge is the energy of the coupled bridge orbital, given by α in eq 2 and by E_i in eq 10.

The occupation dependence of the properties of these systems is also of interest. Chains of bridged mixed-valent species may exhibit conductivity of electrons, holes, or both electrons and holes, depending on the occupation of the system.

The apparent occupation dependence of the resonance integral in the back-bonded complexes is intriguing. Under some conditions it can lead to an effective attractive pairing interaction between electrons (62). The mechanism for superconductivity in the new higher T_c materials is not yet clear, although many have been proposed. It is not even certain at this time whether the mechanism is purely electronic or phonon-mediated. Nevertheless, the search for plausible electronic mechanisms for effective attractive interactions between electrons is important.

The dependence of the properties of bridged mixed-valence compounds upon the bridging ligand and upon the electron occupation has not yet been exploited fully.

Acknowledgments

Acknowledgment is made to the National Science Foundation for financial support for this research under grant number CHE–8820340. Parts of this work were made possible by the Chemistry Department Computer Facility at Northeastern University, which was partially funded by the National Science Foundation under grant number CHE–8700787. We thank Jan Linderberg, George McLendon, and Mark Ratner for valuable discussions. We thank the Alfred P. Sloan Foundation for a fellowship awarded to M. J. Ondrechen.

References

1. *Mixed Valence Compounds;* Brown, D. B., Ed.; Reidel: Dordrecht, Netherlands, 1980.
2. *Mechanistic Aspects of Inorganic Reactions;* Rorabacher, D. B.; Endicott, J. F., Eds.; ACS Symposium Series 198; American Chemical Society: Washington, DC, 1982.
3. Taube, H. *Angew. Chem., Int. Ed. Engl.* **1984**, *23*, 329.
4. Richardson, D. E.; Taube, H. *Coord. Chem. Rev.* **1984**, *60*, 107.
5. Mikkelsen, K. V.; Ratner, M. A. *Chem. Rev.* **1987**, *87*, 113.
6. Cotton, F. A.; Wilkinson, G. *Advanced Inorganic Chemistry*, fourth ed.; Wiley: New York, 1980; p 1206.
7. McLendon, G.; Guarr, T.; McGuire, M.; Simolo, K.; Strauch, S.; Taylor, K. *Coord. Chem. Rev.* **1985**, *64*, 113.
8. Marcus, R. A.; Sutin, N. *Biochim. Biophys. Acta* **1985**, *811*, 265.
9. McLendon, G. *Acc. Chem. Res.* **1988**, *21*, 160.
10. von Doering, W. E.; Roth, W. R. *Angew. Chem., Int. Ed. Engl.* **1963**, *2*, 115.
11. Grob, C. A. *Acc. Chem. Res.* **1983**, *16*, 426.
12. Brown, H. C. *Acc. Chem. Res.* **1983**, *16*, 432.
13. Olah, G. A.; Prakash, G. K. S.; Saunders, M. *Acc. Chem. Res.* **1983**, *16*, 440.
14. Walling, C. *Acc. Chem. Res.* **1983**, *16*, 448.
15. *Theoretical Aspects of Band Structures and Electronic Properties of Pseudo-One-Dimensional Solids;* Kamimura, H., Ed.; Reidel: Dordrecht, Netherlands, 1985.
16. Hanack, M. *Israel J. Chem.* **1985**, *25*, 205.
17. Hanack, M. *Coord. Chem. Rev.* **1988**, *83*, 115.
18. Bednorz, J. G.; Muller, K. A. *Z. Phys. B* **1986**, *64*, 189.
19. *Chemistry of High-Temperature Superconductors;* Nelson, D. L.; Whittingham, M. S.; George, T. F., Eds.; ACS Symposium Series 168; American Chemical Society: Washington, DC, 1987.
20. Creutz, C. *Prog. Inorg. Chem.* **1983**, *30*, 1.
21. Tanner, M.; Ludi, A. *Inorg. Chem.* **1981**, *20*, 2348.
22. Root, L. J.; Ondrechen, M. J. *Chem. Phys. Lett.* **1982**, *93*, 421.
23. Ondrechen, M. J.; Ko, J.; Root, L. J. *J. Phys. Chem.* **1984**, *88*, 5919.
24. Ko, J.; Ondrechen, M. J. *Chem. Phys. Lett.* **1984**, *112*, 507.
25. Ko, J.; Ondrechen, M. J. *J. Am. Chem. Soc.* **1985**, *107*, 6161.
26. Ondrechen, M. J.; Ko, J.; Zhang, L. T. *Int. J. Quantum Chem., Quantum Chem. Symp.* **1986**, *19*, 393.
27. Fulton, R. L.; Gouterman, M. *J. Chem. Phys.* **1964**, *41*, 2280.

28. Kudinov, E. K.; Firsov, Y. A. *Sov. Phys. Solid State (Engl. Transl.)* **1965**, *7*, 435.
29. Piepho, S. B.; Krausz, E. R.; Schatz, P. N. *J. Am. Chem. Soc.* **1978**, *100*, 2296.
30. Ko, J., Ph.D. Thesis, Northeastern University, Boston, MA, 1986.
31. Robin, M. B.; Day, P. *Adv. Inorg. Chem. Radiochem.* **1967**, *10*, 247.
32. Ondrechen, M. J.; Ellis, D. E.; Ratner, M. A. *Chem. Phys. Lett.* **1984**, *109*, 50.
33. Zhang, L. T.; Ko, J.; Ondrechen, M. J. *J. Am. Chem. Soc.* **1987**, *109*, 1666.
34. Ondrechen, M. J.; Ko, J.; Zhang, L. T. *J. Am. Chem. Soc.* **1987**, *109*, 1672.
35. Creutz, C.; Taube, H. *J. Am. Chem. Soc.* **1969**, *91*, 3988.
36. Beattie, J. K.; Hush, N. S.; Taylor, P. R.; Raston, C. L.; White, A. H. *J. Chem. Soc. Dalton Trans.* **1977**, *1121*.
37. Hush, N. S.; Edgar, A.; Beattie, J. K. *Chem. Phys. Lett.* **1980**, *69*, 128.
38. Fürholz, U.; Bürgi, H. B.; Wager, F. E.; Stebler, A.; Ammeter, J. H.; Krausz, E.; Clark, R. J. H.; Ludi, A. *J. Am. Chem. Soc.* **1984**, *106*, 121.
39. Hush, N. S. In ref 2; Rorabacher, D. B.; Endicott, J. F., Eds.; p 301.
40. Calvert, J. M.; Schmehl, R. H.; Sullivan, B. P.; Facci, J. S.; Meyer, T. J.; Murray, R. W. *Inorg. Chem.* **1983**, *22*, 2151.
41. Jernigan, J. C.; Murray, R. W. *J. Am. Chem. Soc.* **1987**, *109*, 1738.
42. Tom, G. M., Ph.D. Thesis, Stanford University, Stanford, CA, 1975.
43. Collman, J. P.; McDevitt, J. T.; Leidner, C. R.; Yee, C. T.; Torrance, J.; Little, W. A. *J. Am. Chem. Soc.* **1987**, *109*, 4606.
44. Dirk, C. W.; Inabe, T.; Schoch, K. F.; Marks, T. J. *J. Am. Chem. Soc.* **1983**, *105*, 1539.
45. Canadell, E.; Alvarez, S. *Inorg. Chem.* **1984**, *23*, 573, and references cited therein.
46. Hanack, M.; Deger, S.; Lange, A.; Zipplies, T. *Synth. Met.* **1986**, *15*, 207.
47. Hanack, M.; Deger, S.; Keppler, U.; Lange, A.; Leverenz, A.; Rein, M. *Synth. Met.* **1987**, *19*, 739.
48. *Novel Superconductivity;* Wolf, S.; Kresin, V., Eds.; Plenum: New York, 1987.
49. Sleight, A. W. *Science (Washington, D.C.)* **1988**, *242*, 1519.
50. Chen, G.; Goddard, W. A. *Science (Washington, D.C.)* **1988**, *239*, 899.
51. Cava, R. J.; Batlogg, B.; Chen, C. H.; Rietman, E. A.; Zahurak, S. M.; Werder, D. *Nature (London)* **1987**, *329*, 423.
52. Hubbard, J. *Proc. Roy. Soc. London Ser. A.* **1964**, *281*, 401.
53. Hale, P. D.; Ratner, M. A.; Hofacker, G. L. *Chem. Phys. Lett.* **1985**, *119*, 264.
54. Lieb, E. H.; Wu, F. Y. *Phys. Rev. Lett.* **1968**, *20*, 1445.
55. Varma, C. M.; Schmitt-Rink, S.; Abrahams, E. *Solid State Commun.* **1987**, *62*, 681.
56. Hirsch, J. E.; Tang, S.; Loh, E.; Scalapino, D. J. *Phys. Rev. Lett.* **1988**, *60*, 1668.
57. Zhang, L. T., Ph.D. Thesis, Northeastern University, Boston, MA, 1989.
58. Zhang, L. T.; Ondrechen, M. J. *Inorg. Chem.*, submitted for publication.
59. Ford, P. C.; Rudd, De F. P.; Gaunder, P.; Taube, H. *J. Am. Chem. Soc.* **1969**, *90*, 1187.
60. Gress, M. E.; Creutz, C.; Quicksall, C. O. *Inorg. Chem.* **1981**, *20*, 1522.
61. Creutz, C.; Chou, M. H. *Inorg. Chem.* **1987**, *26*, 2995.
62. Gozashti, S.; Zhang, L.-T.; Wu, X. M.; Zhou, F.; Ondrechen, M. J.; *Chem. Phys. Lett.*, in press.

RECEIVED for review May 1, 1989. ACCEPTED revised manuscript July 28, 1989.

13

Chiral Recognition by Metal–Ion Complexes in Electron-Transfer Reactions

Rosemary A. Marusak, Thomas P. Shields, and A. Graham Lappin[1]

Department of Chemistry, University of Notre Dame, Notre Dame, IN 46556

Chiral induction in the reaction between horse cytochrome c(II) and $[Co(ox)_3]^{3-}$ (ox^{2-} is oxalate(2−)) was investigated. The usefulness of $[\Lambda\text{-}Co(ox)_3]^{3-}$ as a stereoselective probe is established in the oxidation of $[Co(en)_3]^{2+}$ (en is 1,2-diaminoethane), in which two products are formed: $[Co(en)_3]^{3+}$ by an outer-sphere pathway with an enantiomeric excess of 8% Δ and $[Co(en)_2(ox)]^+$ by an inner-sphere pathway with an enantiomeric excess of 1.5% Λ. Stereoselectivity in the reaction with cytochrome c(II) averages 9%, with a preference for the Λ form of the oxidant. Equilibrium dialysis experiments with cytochrome c(III) indicate that $[\Lambda\text{-}Co(ox)_3]^{3-}$ binds preferentially to the protein and that this ion pair is a good model for an active electron-transfer precursor complex. The mechanism is discussed in terms of known anion binding sites on the protein surface.

R<small>EPORTS OF CHIRAL INDUCTION IN ELECTRON-TRANSFER REACTIONS</small> involving metal–ion complexes have been published over the past decade (1–5). The induction is a direct measure of the relative reactivities of an optically active reagent, Δ-A^{ox} (ox is oxidized) with the enantiomeric forms of the electron-transfer reaction partners Δ-B^{red} and Λ-B^{red} (red is reduced), expressed as $k_{\Delta\Delta}/k_{\Delta\Lambda}$, eqs 1 and 2.

$$\Delta\text{-A}^{ox} + \Delta\text{-B}^{red} \xrightarrow{k_{\Delta\Delta}} \Delta\text{-A}^{red} + \Delta\text{-B}^{ox} \quad (1)$$

$$\Delta\text{-A}^{ox} + \Lambda\text{-B}^{red} \xrightarrow{k_{\Delta\Lambda}} \Delta\text{-A}^{red} + \Delta\text{-B}^{ox} \quad (2)$$

[1]Address correspondence to this author.

0065-2393/90/0226-0237$06.00/0
© 1990 American Chemical Society

The reaction between [Co(edta)]⁻ [edta⁴⁻ is 1,2-diaminoethane-N,N,N',N'-tetraacetate(4–)] and [Co(en)₃]²⁺ (en is 1,2-diaminoethane) is perhaps the most extensively studied system from the point of view of stereoselectivity (4, 5). It is an outer-sphere process, and when [Δ-Co(edta)]⁻ is used, the product, [Co(en)₃]³⁺, shows a 10% enantiomeric excess of the Λ isomer, a ΔΛ process where $k_{\Delta\Lambda}/k_{\Delta\Delta}$ is 1.22. These stereoselectivity studies are possible because of the substitution inertness of the cobalt(III) complexes and the slow electronic self-exchange between cobalt(III) and cobalt(II), which prevents racemization by a self-exchange mechanism.

Comparisons with ion-pairing stereoselectivities for inert isostructural analogues (5, 6) suggest a dominant role for precursor complex stereoselectivity in the chiral induction. A model of the precursor complex has been derived from studies with structurally related derivatives. In the model, the orientation of the [Co(edta)]⁻ oxidant is well defined, hydrogen-bonding through its carboxylate face to the amine protons of the [Co(en)₃]²⁺ reductant. This latter reagent is indiscriminate in its interactions because of its relatively high symmetry and its conformational lability.

The importance of hydrogen bonding through the carboxylate face of the oxidant in this reaction has prompted studies with the related complex [Co(ox)₃]³⁻, in which carboxylate faces are also available, as oxidant (7). The complex is readily resolved (8); it has a reduction potential of 0.57 V (vs. the normal hydrogen electrode) (9) and a low self-exchange rate (10), which are ideal for stereoselectivity studies. However, it is heat- and light-sensitive and is prone to rapid racemization, even in the solid state.

One reason for investigating the use of [Co(ox)₃]³⁻ as a stereoselective oxidant lies in its value as a probe of the mechanisms of electron transfer with metalloproteins below their isoelectric points. Chiral recognition in reactions of metalloproteins has proved to be rather difficult to detect, despite the obvious chirality of the reagents involved. Early workers in the field failed to detect rate differences in the reduction of Δ- and Λ-[Co(en)₃]³⁺ by parsley ferredoxin (11) and the reduction of horse cytochrome c by Δ- and Λ-[Co(sep)]²⁺ (sep is sepulchrate, 1,3,6,8,10,13,16,19-octaazabicyclo[6,6,6]eicosane) (12). However, rate measurements are rather insensitive and cannot distinguish stereoselectivities below the order of 10% or so. More recently, Bernauer and Sauvain (13) reported stereoselectivity in the reaction of chiral iron(II) complexes with plastocyanin. The intimate mechanism involved is not known, as the iron reagents have been shown to favor inner-sphere mechanisms (14).

Stereoselectivity in the reduction of [Co(ox)₃]³⁻ by cobalt(II) in 1,2-diaminoethane solutions is reported in this chapter to demonstrate the usefulness of this reagent as a stereoselective oxidant. Because of the utility of this reagent, [Co(ox)₃]³⁻ is employed as a chiral probe in the oxidation of horse cytochrome c(II). This particular reaction is chosen for several reasons. First, the protein carries a +6 charge at pH 7; the −3 charge on the oxidant enhances the electrostatic attraction between the reagents, a factor important in influencing the stereoselectivity. Second, the reaction is relatively well characterized, and two previous kinetic studies are reported (15, 16). Finally, electron transfer at cytochrome c takes place through the exposed heme edge, around which there are well-defined anion-binding sites. NMR spectroscopic studies (17, 18) with the paramagnetic analogue $[Cr(ox)_3]^{3-}$ have pinpointed three areas on the protein surface at which binding takes place, a weak site distant from the heme edge and two stronger binding sites in the vicinity of the reaction center. The role of these binding sites in the electron-transfer process will be discussed.

Experimental Details

The preparation of $K_3[Co(ox)_3] \cdot 3.5H_2O$ (molar absorptivity, $\epsilon_{605} = 165$ M^{-1} cm^{-1}) and resolution of the complex were carried out by literature methods (7, 8). The absolute configuration is taken as $[\Lambda-(-)_{589}-Co(ox)_3]^{3-}$ ($\Delta\epsilon_{622} = 3.80$ M^{-1} cm^{-1}) (19). The complex is light-sensitive, and all manipulations were carried out in the dark. Racemization of the complex amounted to 10–12% over a 2-h period; this value was applied as a correction factor in the studies. The reaction stoichiometry and products were determined by high-performance liquid chromatography (HPLC) with an ion-exchange column (Waters Protein Pak SP–5PW Sephadex).

Typical reaction conditions involved addition of $[Co(ox)_3]^{3-}$ (10^{-3} M) to a solution containing cobalt(II) (10^{-2} M) in excess 1,2-diaminoethane (2.0 × 10^{-2}–10^{-1} M) and appropriate supporting electrolyte, with rapid stirring and under an atmosphere of argon to prevent aerial oxidation of the cobalt(II) complex. In some experiments, oxalate ion and 1,2-^{13}C-oxalate ion (3 × 10^{-3} M) were added to the cobalt(II) solution. After completion of the reaction, the mixture was cooled and ice-cold 6 M HCl added so that the resultant pH was less than 1. Aliquots of this mixture were then subject to analysis. The two products $[Co(en)_3]^{3+}$ and $[Co(en)_2(ox)]^+$ were isolated on a 1- × 10-cm column (Sephadex SP C–25) and a 1- × 20-cm column (Dowex 50X2–400), washed with water and dilute acid, and eluted with 1.0 M HCl and 0.01 M HCl, respectively.

The electron-transfer stereoselectivity was determined by examining the optical purity of the reaction products. Absolute configurations are $[\Lambda-(+)-Co(en)_2(ox)]^+$ ($\epsilon_{500} = 103$ M^{-1} cm^{-1}, $\Delta\epsilon_{520} = 2.65$ M^{-1} cm^{-1}) (20) and $[\Lambda-(+)-Co(en)_3]^{3+}$ ($\epsilon_{467} = 88$ M^{-1} cm^{-1}, $\Delta\epsilon_{493} = 1.90$ M^{-1} cm^{-1}) (21). Kinetic measurements were made anaerobically at 25.0 °C and 0.1 M ionic strength (NaClO$_4$), under pseudo-first-order conditions with an excess of reductant and with an excess of the 1,2-diaminoethane ligand as a buffer. The decomposition of $[Co(ox)_3]^{3-}$ was monitored at 605 nm.

Horse cytochrome c (Sigma, Type VI) was used without further purification. Samples of the oxidized protein were dialyzed in appropriate buffer solutions for at least 2 h before use. The reduced protein was obtained by the addition of a few crystals of sodium dithionite to the oxidized protein, followed by dialysis with argon-saturated buffer under an atmosphere of argon. In a typical equilibrium dialysis experiment, 5 mL of 10^{-3} M cytochrome c(III) (10^{-2} M Tris buffer, 0.1 M ionic strength (KCl)) was dialyzed against 5 mL of 3 × 10^{-3} M racemic $[Co(ox)_3]^{3-}$ in the same buffer for 1 h at 0 °C. The $[Co(ox)_3]^{3-}$ solution was then examined for optical activity.

Electron-transfer stereoselectivities were determined by mixing 2 mL of 5 × 10^{-4} M cytochrome c(II) in appropriate buffer at 0.1 M ionic strength (KCl) with equal volumes containing a two- to fivefold excess of $[Co(ox)_3]^{3-}$, also in buffer at 23 °C. After the reaction proceeded to completion, the protein was removed on an ion-exchange column (Sephadex SP C–25) and the optical activity in the resulting $[Co(ox)_3]^{3-}$ was determined. Kinetic measurements were made at 550 nm under pseudo-first-order conditions, with an excess of $[Co(ox)_3]^{3-}$ in 5×10^{-3} M buffer and ionic strength 0.1 M (KCl).

Results and Discussion

Oxidation of Cobalt(II) in 1,2-Diaminoethane Solutions by $[Co(ox)_3]^{3-}$. The stoichiometric reduction of $[Co(ox)_3]^{3-}$ by cobalt(II) in aqueous solutions of 1,2-diaminoethane (en) results in the formation of two products that can be separated by HPLC or conventional chromatography (Figure 1). The products are identified spectroscopically as $[Co(en)_3]^{3+}$ and $[Co(en)_2(ox)]^+$, and the relative amount of each is dependent on the en concentration. At low [en], the dominant product is $[Co(en)_2(ox)]^+$. The proportion of $[Co(en)_3]^{3+}$ increases with increasing [en] (Figure 2), a result

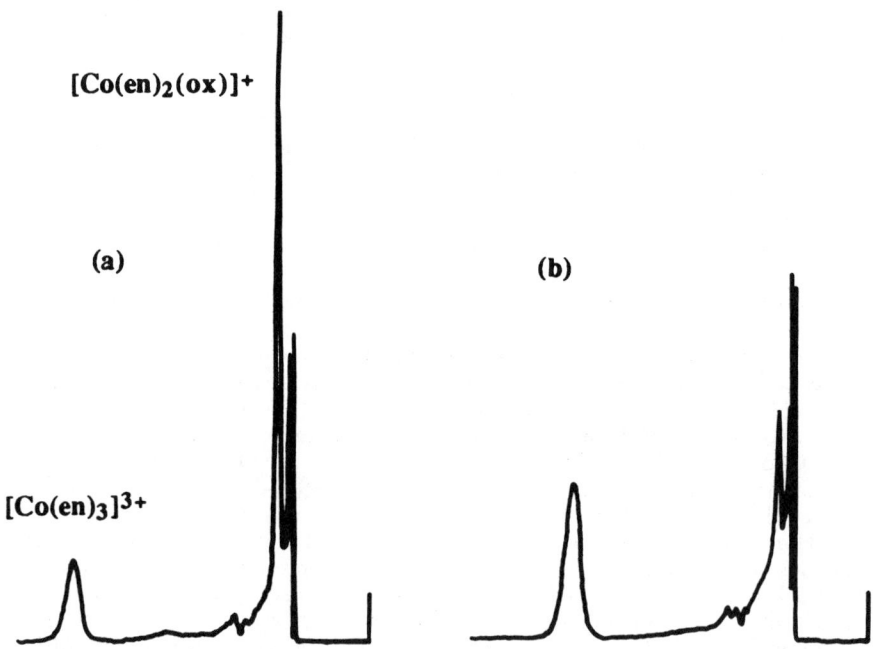

Figure 1. *HPLC analysis (SP–5PW Sephadex) of the products of the reaction of $[Co(ox)_3]^{3-}$ with cobalt(II) in 1,2-diaminoethane solutions. (a), [en] = 0.122 M; (b), [en] = 0.183 M. The wavelength is 346 nm, where excess [cobalt(II)] has little absorbance. There is an initial Schlieren gradient caused by the passage of excess 1,2-diaminoethane, but the cation peaks are readily identified.*

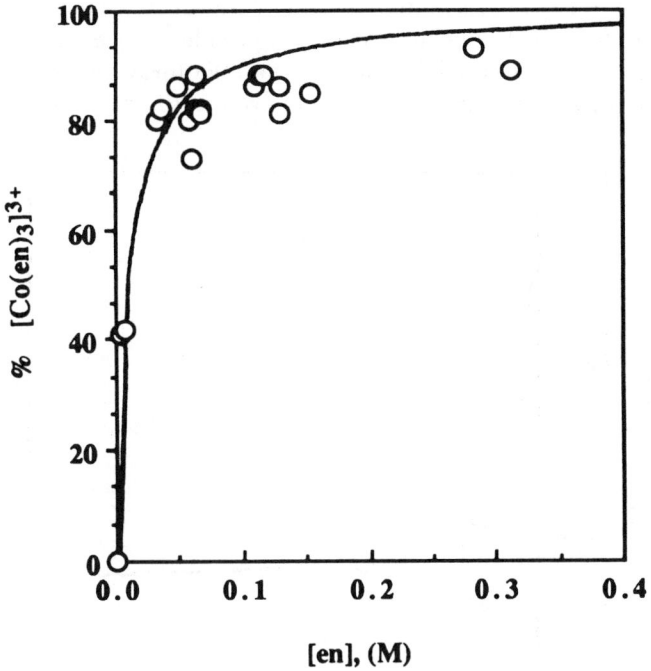

Figure 2. Plot of percent $[Co(en)_3]^{3+}$ product as a function of [en] for the oxidation of [cobalt(II)] in 1,2-diaminoethane solutions by $[Co(ox)_3]^{3-}$ at 25.0 °C and 0.10 M ionic strength.

suggesting that the reaction involves two parallel pathways differing in their dependence on [en].

The kinetics of the reaction are also consistent with this observation. The reaction is first-order in $[Co(ox)_3]^{3-}$ and [Co(II)] concentrations. The second-order rate constant, k_{so}, shows a strong dependence on [en] (Figure 3), which is explained by the mechanism in eqs 3–5, with the resulting rate law, eq 6. Best-fit parameters are $k_o = 390 \pm 20$ M^{-1} s^{-1} and $k_i = 3300 \pm 300$ M^{-1} s^{-1}, with a value for $K_3 = 2000$ M^{-1} (22). These parameters show reasonable agreement with values for k_i/k_o estimated from the stoichiometry results.

$$[Co(en)_2]^{2+} + en \rightleftharpoons [Co(en)_3]^{2+} \qquad K_3 = 2000 \text{ M}^{-1} \qquad (3)$$

$$[Co(en)_2]^{2+} \; [Co(ox)_3]^{3-} \xrightarrow{k_i} [Co(en)_2(ox)]^+ + \text{``}[Co(ox)_2]^{2-}\text{''} \qquad (4)$$

$$[Co(en)_3]^{2+} + [Co(ox)_3]^{3-} \xrightarrow{k_o} [Co(en)_3]^{3+} + \text{``}[Co(ox)_3]^{4-}\text{''} \qquad (5)$$

$$k_{so} = \frac{k_i + K_3 k_o [en]}{1 + K_3 [en]} \qquad (6)$$

where k_{obs} is the observed rate constant. Pathway k_o, which leads to the formation of $[Co(en)_3]^{3+}$, is most likely outer-sphere in nature because no intermediates are observed in the reaction. Pathway k_i, which leads to $[Co(en)_2(ox)]^+$, is shown from experiments run in the presence of free ^{13}C-

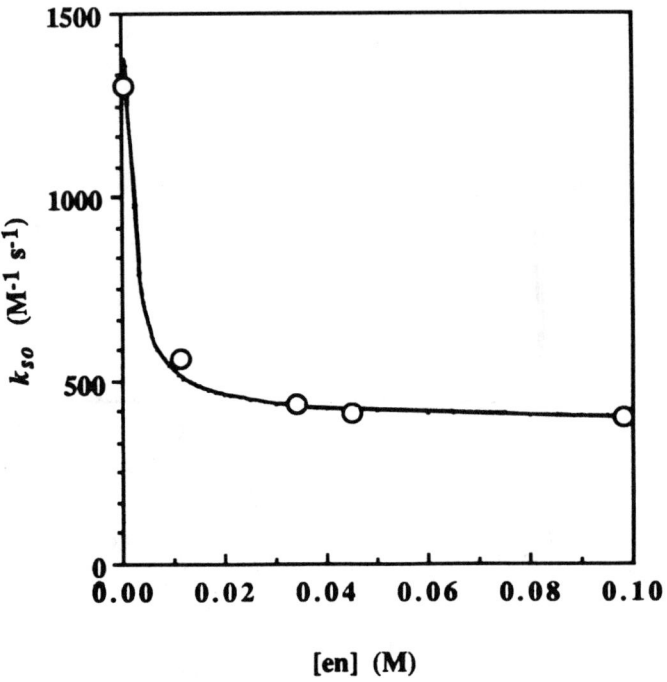

Figure 3. Plot of the second-order rate constant, k_{so}, as a function of [en] for the oxidation of [cobalt(II)] in 1,2-diaminoethane solutions by $[Co(ox)_3]^{3-}$ at 25.0 °C and 0.10 M ionic strength.

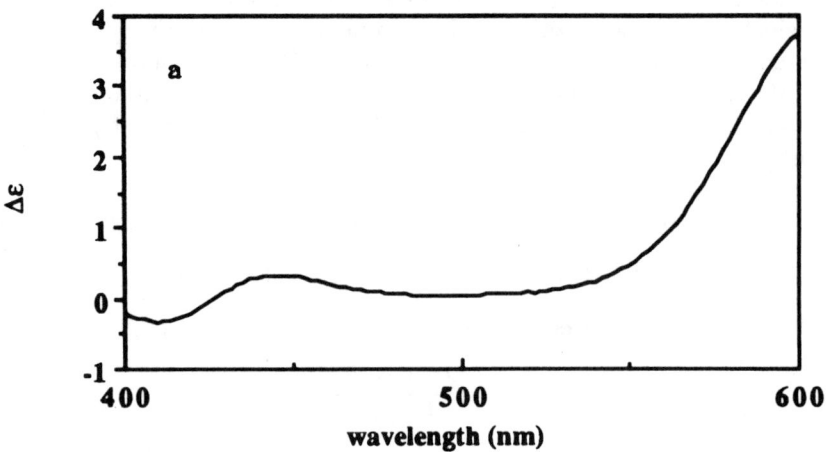

Figure 4. Circular dichroism spectra of (a), $[Co(ox)_3]^{3-}$. Continued on next page.

labeled oxalate ion to incorporate none of the added label. Hence, the oxalate in $[Co(en)_2(ox)]^+$ is derived from the $[Co(ox)_3]^{3-}$ as a result of an inner-sphere electron transfer in which a doubly bridged oxalate ion is transferred.

Stereoselectivity in this reaction can be investigated by the use of optically active $[\Lambda\text{-}Co(ox)_3]^{3-}$ as oxidant. The reaction products are optically stable and not prone to racemization by a self-exchange mechanism. The circular dichroism spectra of the isolated products under well-defined conditions are shown in Figure 4. These results indicate that for the $[Co(en)_3]^{3+}$ there is an 8% enantiomeric excess of the Δ isomer; for the $[Co(en)_2(ox)]^+$ there is a 1.5% enantiomeric excess of the Λ isomer. These stereoselectivities

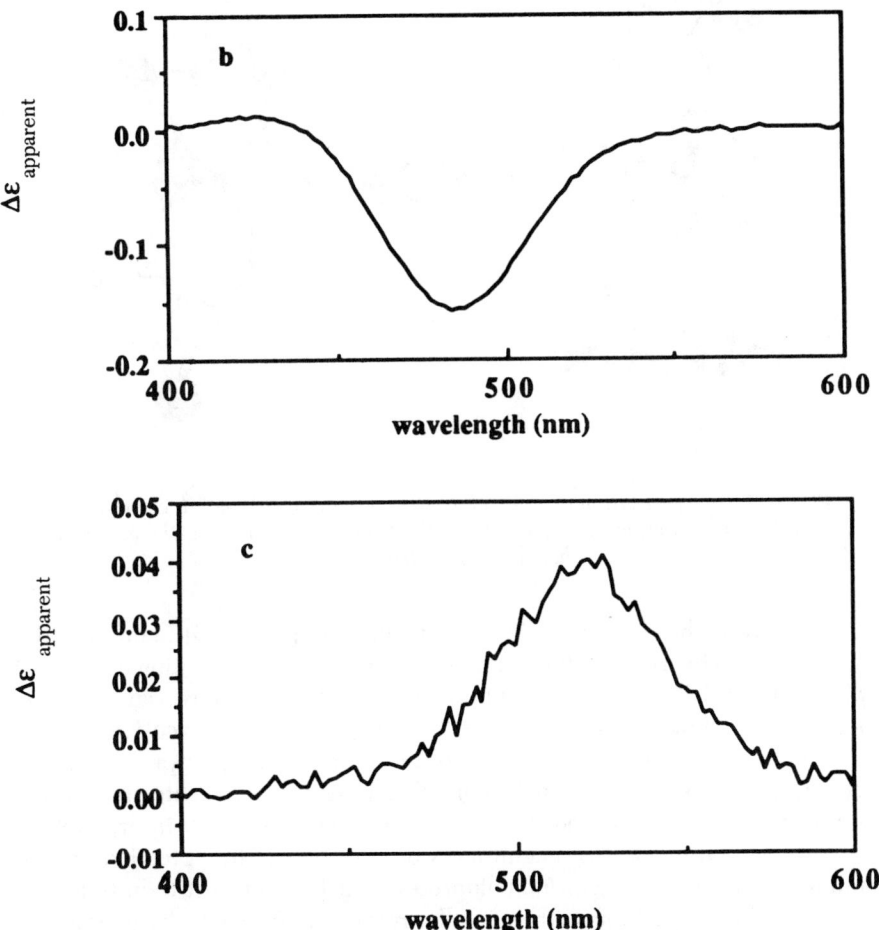

Figure 4. (b), $[Co(en)_3]^{3+}$ produced in the reaction of $[Co(ox)_3]^{3-}$ with [cobalt(II)] in 1,2-diaminoethane solutions; and (c), $[Co(en)_2(ox)]^+$ produced in the reaction of $[Co(ox)_3]^{3-}$ with [cobalt(II)] in 1,2-diaminoethane solutions.

are modest but nevertheless demonstrate the utility of $[Co(ox)_3]^{3-}$ as a stereoselective oxidant. Somewhat surprisingly, chiral induction in the outer-sphere reaction is greater than that in the inner-sphere reaction. The doubly bridged intermediate in the inner-sphere reaction holds the chiral centers around 5 Å apart (Figure 5), so that there is little intimate contact. Hence, transfer of chirality is difficult. On the other hand, for the outer-sphere reaction, intimate contact between the coordination spheres is possible, and indeed likely.

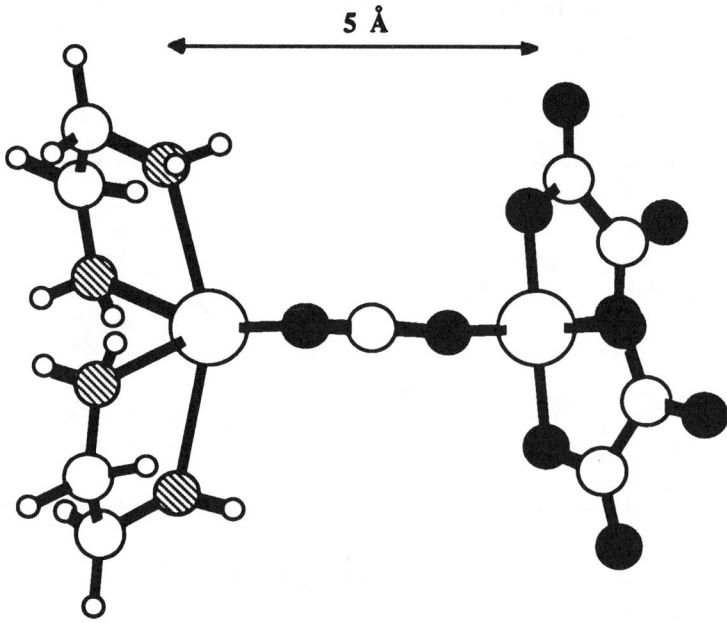

Figure 5. Representation of the oxalate-bridged inner-sphere intermediate proposed in the reaction of $[Co(ox)_3]^{3-}$ with $[Co(en)_2]^{2+}$, showing the separation between the chiral centers.

As with the corresponding oxidation by $[Co(edta)]^-$, outer-sphere stereoselectivity reflects the high symmetry and conformational flexibility of the $[Co(en)_3]^{2+}$ reductant. Where the reductant is more rigid and sterically demanding, as with $[Co((\pm)\text{-chxn})_3]^{2+}$ (($\pm$)-chxn is rac-1,2-diaminocyclohexane), stereoselectivities can be much larger, approaching 70% enantiomeric excess (23). Such high stereoselectivity indicates an intimate interaction in which the coordination spheres of the reactants interpenetrate. Stereoselectivities of this magnitude would facilitate the detection of chiral induction in reactions with metalloproteins. It is of interest then to identify whether the interactions with a metalloprotein will provide a large or modest stereoselectivity, and thus to determine the specificity of the interaction.

Oxidation of Horse Cytochrome c(II) by $[Co(ox)_3]^{3-}$. The reaction between horse cytochrome c(II) and $[Co(ox)_3]^{3-}$ is a well-characterized single-electron transfer that has been the subject of two previous kinetic investigations (15, 16). Under pseudo-first-order conditions with an excess of oxidant and at pH 7.0, oxidation of cytochrome c(II) by $[Co(ox)_3]^{3-}$ is a first-order process for greater than three half-lives. Pseudo-first-order rate constants are presented in Table I. A plot of the first-order rate constant, k_{obs}, against [Co(III)] is linear, with no evidence for rate saturation to indicate

Table I. Rate Constants for the Reaction of $[Co(ox)_3]^{3-}$ with Horse Cytochrome c(II)

pH	10^3 [Co(III)], M	10^3 k_{obs}, s^{-1}	k_{so}, $M^{-1}s^{-1}$
4.10[a]	1.50	13.3	8.9
4.44[a]	1.44	11.3	7.9
4.98[a]	1.43	10.4	7.3
5.28[a]	1.37	10.7	7.8
5.99[a]	1.39	11.3	8.1
5.97[b]	1.42	8.47	6.0
6.46[b]	1.40	8.33	6.0
6.81[b]	1.61	9.40	5.9
6.88[b]	1.41	8.47	6.0
7.90[b]	1.40	8.07	5.8
5.55[c]	1.43	10.1	7.1
6.11[c]	1.44	10.8	7.5
6.58[c]	1.51	11.3	7.5
6.70[d]	1.61	11.1	6.9
6.72[d]	1.45	10.2	7.0
7.17[d]	1.45	8.92	6.1
7.68[d]	1.60	10.4	6.5
7.99[e]	1.77	10.5	5.9
8.57[e]	1.78	10.4	5.8
9.02[e]	1.78	9.33	5.2
7.00[f]	0.36	2.74	7.7
7.00[f]	0.64	3.76	5.9
7.00[f]	1.13	6.66	5.9
7.00[f]	1.68	9.33	5.6
7.00[f]	2.39	14.3	6.0
6.78[g]	1.61	10.1	6.3
6.80[h]	1.61	9.70	6.0
6.80[i]	1.61	9.40	5.8

NOTE: Reaction conditions were 25.0 °C, 0.10 M ionic strength (KCl), and [cyt c(II)] = 5×10^{-6} M.
[a] 5×10^{-3} M acetate.
[b] 1.7×10^{-3} M phosphate.
[c] 5×10^{-3} M MES.
[d] 5×10^{-3} M HEPES.
[e] 1.7×10^{-3} M borate.
[f] 3.3×10^{-3} M phosphate.
[g] 5.5×10^{-4} M phosphate and 5×10^{-3} M HEPES.
[h] 8.3×10^{-4} M phosphate and 5×10^{-3} M HEPES.
[i] 1.11×10^{-3} M phosphate and 5×10^{-3} M HEPES.

the presence of kinetically important ion-pair formation. Rate saturation has not been detected in previous work on this reaction or in reactions with other negatively charged reagents such as $[Fe(CN)_6]^{3-}$ (18, 24).

In weakly coordinating buffers, the second-order rate constant for the electron transfer is almost independent of pH over the range 4.5–7.5 (Figure 6), with a value of 7.1 ± 0.4 M^{-1} s^{-1}, compared to the previously reported

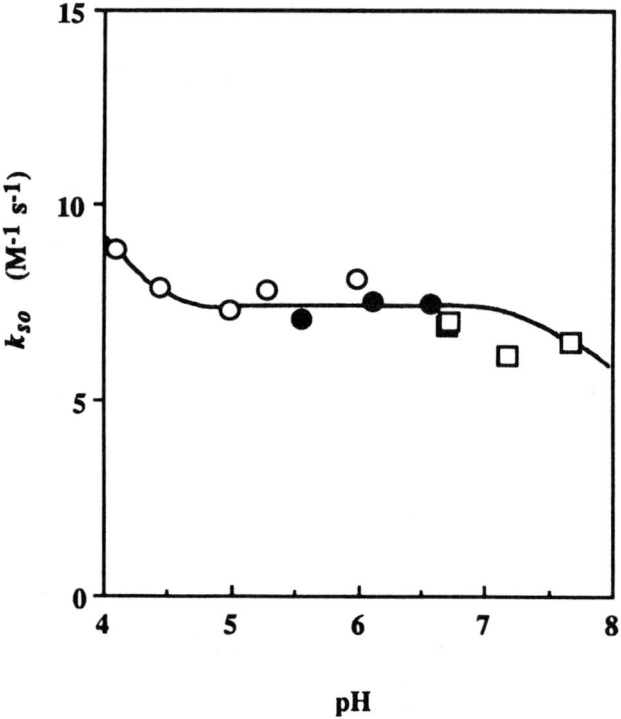

Figure 6. Dependence of k_{so} on pH for the reaction of horse cytochrome c(II) with $[Co(ox)_3]^{3-}$ at 25.0 °C and 0.10 M ionic strength (KCl). Key: ○, acetate buffer; ●, MES (2-[N-morpholino]ethanesulfonic acid); □, HEPES (N-2-hydroxyethylpiperazine-N'-2-ethanesulfonic acid).

value of 5.5 M^{-1} s^{-1} in 0.5 M phosphate at pH 7.0 (15). Below pH 4.5, the reaction shows a slight trend, increasing with decreasing pH, perhaps responding to changes in the protein charge. Above pH 8, a more marked rate reduction is expected on thermodynamic grounds. In the presence of chloride ion, the reaction shows inhibition by phosphate to the extent of about 20% of the reaction rate (Figure 7), a feature noted previously (16). Phosphate ion is known to bind at sites on the protein surface that are used by all anions (25). Apparently $[Co(ox)_3]^{3-}$ is able to oxidize both chloride- and phosphate-bound forms of the protein at approximately the same rate.

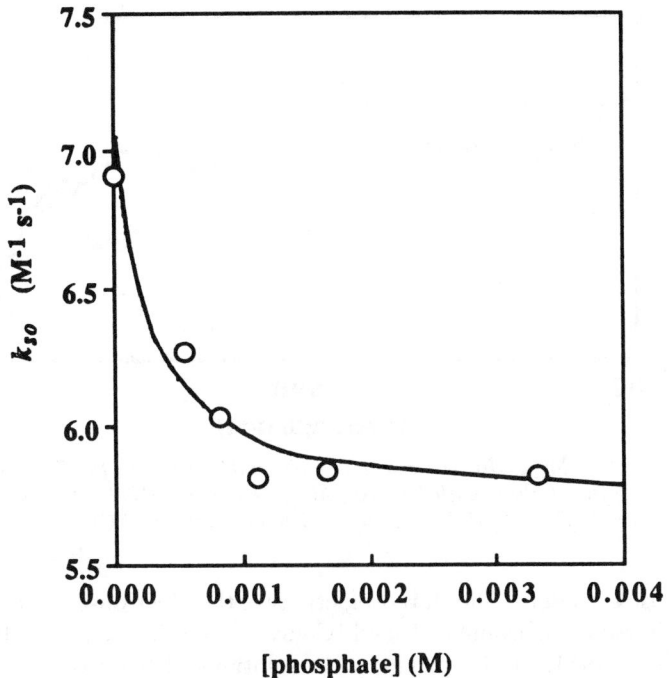

Figure 7. Dependence of k_{so} *on the total phosphate concentration at pH 6.8 for the reaction of horse cytochrome c(II) with* $[Co(ox)_3]^{3-}$ *at 25.0 °C and 0.10 M ionic strength (KCl).*

Equilibrium. When horse cytochrome c(III) is dialyzed against a solution containing $[Co(ox)_3]^{3-}$, equilibration of the binding of the complex ion with the protein takes place. One isomeric form of the ion binds preferentially to the protein, and its concentration is depleted in the bulk solution. Detection of stereoselectivity is achieved by examining the circular dichroism spectrum of the bulk solution. It is also possible to examine the enantiomeric enhancement of the protein binding by dialysis of the equilibrated protein solution with buffer and examination of the complex released. The two methods are complementary in that the enantiomer enhanced in the first experiment is the opposite of that enhanced in the second experiment.

True equilibrium conditions cannot be established because $[Co(ox)_3]^{3-}$ racemizes within a few hours in solution, even at 0 °C. Figure 8 shows the circular dichroism spectrum of the bulk $[Co(ox)_3]^{3-}$ after dialysis. The circular dichroism obtained by dialyzing the equilibrated dialysis sack with buffer has been subtracted. Quite clearly, this spectrum shows that the presence of the Δ isomer is enhanced in the bulk solution and indicates preferential binding to the protein by the Λ isomer. With an estimate of the equilibrium constant for association of racemic $[Co(ox)_3]^{3-}$ with horse cytochrome c(III) of 200 M^{-1} (18), the ratio K_Λ/K_Δ is calculated to be in excess of 1.04.

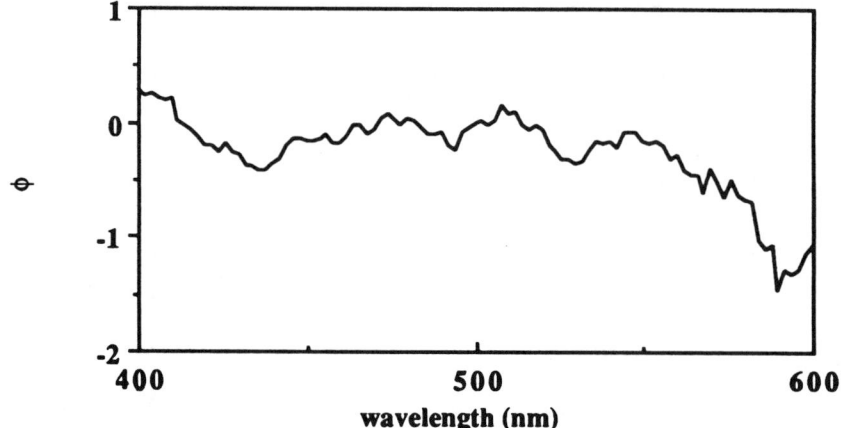

Figure 8. Circular dichroism spectrum (ɸ, millidegrees) of 1.5×10^{-3} M $[Co(ox)_3]^{3-}$ after dialysis with 1×10^{-3} M cytochrome c(III) in 10^{-2} M HEPES buffer at pH 7.0 and 0.10 M ionic strength (KCl).

Stereoselectivity. The stereoselectivity of the electron-transfer reaction between horse cytochrome c(II) and $[Co(ox)_3]^{3-}$ can be examined by determining the optical activity produced in solutions of $[Co(ox)_3]^{3-}$ when it is reacted with a stoichiometric deficiency of the protein. A typical spectrum obtained in this manner is shown in Figure 9. The spectrum, that of the Δ isomer, indicates that the $[\Lambda\text{-}Co(ox)_3]^{3-}$ reacts preferentially with the protein. The stereoselectivity, little affected by pH over the range 4.2–7.6, is rela-

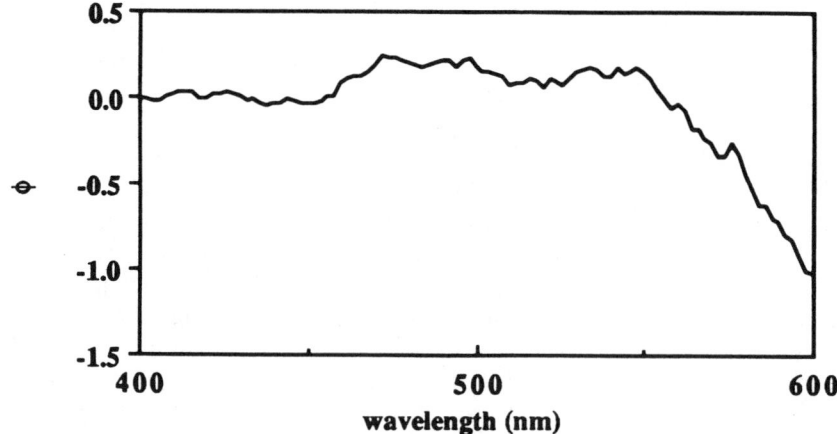

Figure 9. Circular dichroism spectrum (ɸ, millidegrees) of 3.85×10^{-4} M $[Co(ox)_3]^{3-}$ obtained in the reaction of cytochrome c(II) (2.3×10^{-4} M) with excess $[Co(ox)_3]^{3-}$ (1.53×10^{-3} M) in HEPES buffer, pH 6.97, at 23 °C and 0.10 M ionic strength (KCl).

tively insensitive to the presence of phosphate ion and the supporting electrolyte, with k_Λ/k_Δ averaging 1.19 (Table II). This insensitivity implies that the binding of phosphate is not important in stabilizing the ion pair with $[Co(ox)_3]^{3-}$, a feature that is in general agreement with the small effect of phosphate on the reaction rate.

Table II. Stereoselectivity in the Oxidation of Horse Cytochrome c(II) by $[Co(ox)_3]^{3-}$

pH	10^4 [cyt c(II)], M	10^4 [[Co(ox)$_3$]$^{3-}$], M	ee[a], %	k_Λ/k_Δ
4.43[b]	2.2	14.3	7 ± 2	1.15
6.97[c]	2.4	15.3	6 ± 2	1.13
7.01[c]	3.0	11.3	11 ± 1	1.25
6.73[d]	1.6	15.3	10 ± 1	1.22
6.87[d]	1.2	15.0	10 ± 1	1.22
4.70[e]	2.4	14.2	4 ± 2	1.08

NOTE: Reaction conditions were 23 °C and 0.10 M ionic strength (KCl).
[a] ee is enantiomeric excess; Λ form in all cases.
[b] 5×10^{-3} M acetate.
[c] 5×10^{-3} M HEPES.
[d] 1.7×10^{-3} M phosphate.
[e] No supporting electrolyte.

Stereoselectivity in the electron-transfer reaction between cytochrome c(II) and $[Co(ox)_3]^{3-}$ is small, a fact suggesting that the interactions involved are not particularly specific. However, a qualitative interpretation of the data is not uninformative. Although there is no kinetic evidence for an intermediate in the reaction, the overall process can be separated into two kinetically distinct steps: formation of an electron-transfer precursor complex between the protein and the oxidant, followed by electron transfer within this assembly.

Both of these steps can provide a source of stereoselectivity, but experience with reactions between metal–ion complexes suggests that precursor complex formation assumes a dominant role (5). Although it is not possible in this instance to examine stereoselectivity directly in the precursor complex, the process can be modeled by the equilibrium dialysis experiments in which the only difference is the oxidation state of the protein. Some caution is recommended in this interpretation because the protein does undergo a small conformational change during the redox process. This change is known to affect the conformation at the heme edge, the presumed site of electron transfer (18).

Binding Sites. NMR spectroscopic investigations of the binding of the paramagnetic analogue $[Cr(ox)_3]^{3-}$ to cytochrome c(III) reveal the location of at least three anion-binding sites on the protein surface (Figure 10). Two

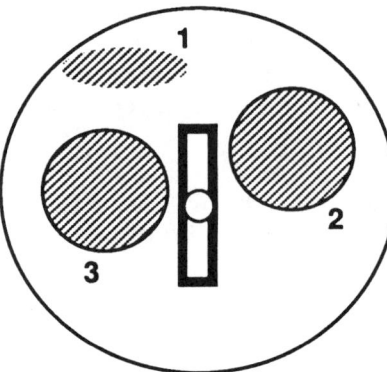

Figure 10. Representation of cytochrome c, showing the heme edge and anion-binding sites as indicated in the text. Site 1 is on the top side of the protein, away from the heme edge; sites 2 and 3 are close to the heme edge.

of these sites are close to the heme edge. However, a third (weak) binding site on the opposite side of the protein, distant from the heme edge, can be ignored in the present argument. Of the two sites close to the heme edge, site 3 has the highest affinity for $[Cr(ox)_3]^{3-}$ and has been shown to be kinetically important in the oxidation by $[Fe(CN)_6]^{3-}$ (25). It seems likely that it is the primary binding site for $[Co(ox)_3]^{3-}$ in the equilibration and electron-transfer studies. Phosphate is known to bind most strongly at sites 1 and 2 (26), so that the interaction of $[Co(ox)_3]^{3-}$ at site 3 might be expected to be relatively insensitive to the presence of phosphate, as has been found.

The detailed information available about the intimate electron-transfer mechanism makes this particular reaction ideal for exploring the interpretation of stereoselectivity data. In general, when metal–ion complexes are capable of binding to different sites on the surface of metalloproteins, an important question in determining the detailed mechanism is the role of the precursor in the overall reaction. There are two kinetically indistinguishable possibilities: mechanism A (eqs 7 and 8), where the precursor is active and forms an intermediate on the reaction profile, and mechanism B (eqs 9 and 10), where the precursor is inactive, the so-called "dead-end" mechanism.

$$\text{A} \quad \text{cyt c(II)} + [Co(ox)_3]^{3-} \rightleftarrows \{\text{cyt c(II)},[Co(ox)_3]^{3-}\} \quad (7)$$

$$\{\text{cyt c(II)},[Co(ox)_3]^{3-}\} \rightarrow \text{cyt c(III)} + \text{``}[Co(ox)_3]^{4-}\text{''} \quad (8)$$

$$\text{B} \quad \text{cyt c(II)} + [Co(ox)_3]^{3-} \rightleftarrows \{\text{cyt c(II)},[Co(ox)_3]^{3-}\} \quad (9)$$

$$\text{cyt c(II)} + [Co(ox)_3]^{3-} \rightarrow \text{cyt c(III)} + \text{``}[Co(ox)_3]^{4-}\text{''} \quad (10)$$

In mechanism A, the ion-pairing and electron-transfer steps involve an identical arrangement of reagents and should reflect similar stereoselectivities.

However, this is not necessarily the case in mechanism B. If, for example, strong ion pairing with a Λ stereoselectivity is demonstrated, the interaction of the metalloprotein with two equivalents of the oxidant will produce an excess of the Δ isomer in solution. In the case of the inactive (dead-end) complex, the Δ isomer will be preferentially reduced (unless stereoselectivity in step 10 exceeds that in step 9). With the active complex, the Λ isomer will be preferentially reduced. This distinct preference provides a readily available means for differentiating between the two mechanisms.

The results obtained for the oxidation of cytochrome c(II) by $[Co(ox)_3]^{3-}$, in which ion-pairing and electron-transfer stereoselectivities are comparable in both magnitude and sense, strongly suggest that mechanism A is operating. The result is not particularly surprising. However, it demonstrates the mechanistic potential of stereoselectivity studies and points the way to experiments in which the role of the ion pair is more controversial.

Acknowledgments

The authors acknowledge the generous support of this work by the National Science Foundation (Grant No. CHE 87–02012).

References

1. Geselowitz, D. A.; Taube, H. *J. Am. Chem. Soc.* **1980**, *102*, 4525–4526.
2. Kondo, S.; Sasaki, Y.; Saito, K. *Inorg. Chem.* **1981**, *20*, 429–433.
3. Lappin, A. G.; Laranjeira, M. C. M.; Peacock, R. D. *Inorg. Chem.* **1983**, *22*, 786–791.
4. Geselowitz, D. A.; Hammershøi, A.; Taube, H. *Inorg. Chem.* **1987**, *26*, 1842–1845.
5. Osvath, P.; Lappin, A. G. *Inorg. Chem.* **1987**, *26*, 195–202.
6. Sakaguchi, U.; Yamamoto, I.; Izumoto, S.; Yoneda, H. *Bull Chem. Soc. Jpn.* **1983**, *56*, 153–156.
7. Bailar, J. C.; Jones, E. M. *Inorg. Synth.* **1939**, *1*, 37.
8. Kaufman, G. B.; Takahashi, L. T.; Sugisaka, N. *Inorg. Synth.* **1966**, *8*, 208–209.
9. Hin-Fat, L.; Higginson, W. C. E. *J. Chem. Soc. A* **1967**, 298–301.
10. Ehighaokhuo, J. O.; Ojo, J. F.; Olubuyide, O. *J. Chem. Soc., Dalton Trans.* **1985**, 1665–1667
11. Armstrong. F. A.; Sykes, A. G. *J. Am. Chem. Soc.* **1978**, *100*, 7710–7715.
12. Toma, H. E.; Murakami, R. A. *Inorg. Chim. Acta* **1984**, *93*, L33–L36.
13. Bernauer, K.; Sauvain, J. *J. Chem. Soc., Chem. Commun.* **1988**, 353–354.
14. Bernauer, K.; Puosaz, P.; Porret, J.; Jeanguenat, A. *Helv. Chim. Acta* **1988**, *71*, 1339–1344.
15. Holwerda, R. A.; Knaff, D. B.; Gray, H. B.; Clemmer, J. D.; Crowley, R.; Smith, J. M.; Mauk, A. G. *J. Am. Chem. Soc.* **1980**, *102*, 1142–1146.
16. Rush, J. D.; Lan, J.; Koppenol, W. H. *J. Am. Chem. Soc.* **1987**, *109*, 2679–2682.
17. Eley, C. G. S.; Moore, G. R.; Williams, G.; Williams, R. J. P. *Eur. J. Biochem.* **1982**, *124*, 295–303.
18. Moore, G. R.; Eley, C. G. S.; Williams, G. In *Advances in Inorganic and Bioinorganic Mechanisms*; Sykes, A. G., Ed.; Academic: New York, 1984; Vol. 3, pp 1–96.

19. Okazaki, H.; Kushi, Y.; Yoneda, H. *J. Am. Chem. Soc.* **1985,** *107,* 4183–4189.
20. Jordan, W. T.; Brennan, B. J.; Froebe, L. R.; Douglas, B. E. *Inorg. Chem.* **1973,** *12,* 1827–1831.
21. McCaffery, A. J.; Mason, S. F.; Norman, B. J.; Sargeson, A. M. *J. Chem. Soc. A* **1968,** 1304–1310.
22. Smith, R. M.; Martell, A. E. *Critical Stability Constants;* Plenum: New York, 1975; Vol. 2, p 36.
23. Marusak, R. A.; Osvath, P.; Kemper, M.; Lappin, A. G. *Inorg. Chem.* **1989,** *28,* 1542–1548.
24. Butler, J.; Davies, D. M.; Sykes, A. G. *J. Inorg. Biochem.* **1981,** *15,* 41–53.
25. Butler, J.; Davies, D. M.; Sykes, A. G.; Koppenol, W. H.; Osheroff, N.; Margoliash, E. *J. Am. Chem. Soc.* **1981,** *103,* 469–471.
26. Osheroff, N.; Brautigan, D. L.; Margoliash, E. *Proc. Natl. Acad. Sci. U.S.A.* **1980,** *77,* 4439–4443.

RECEIVED for review May 1, 1989. ACCEPTED revised manuscript July 24, 1989.

14

The Role of Free Energy in Interligand Electron Transfer

L. K. Orman, D. R. Anderson, T. Yabe, and J. B. Hopkins[1]

Department of Chemistry, Louisiana State University, Baton Rouge, LA 70803

Electron transfer between ligands in the metal-to-ligand charge transfer (MLCT) excited states of Ru(II) polypyridine complexes was investigated by using a powerful two-color picosecond Raman technique. Photoexcitation into the MLCT absorption band results in the promotion of an electron to one of three possible ligand acceptors. In mixed-ligand complexes it is unambiguously established that excitation initially produces an ensemble of complexes with electrons localized on dissimilar ligands. Transient Raman spectroscopy is used to measure the rates of interligand electron transfer as a function of the free energy separation between ligands (ΔG_0). The apparent rate increases from $\leq 10^6$ to $\geq 10^{11}$ s^{-1} with a change in ΔG_0 of only 0.12 eV. The free energy dependence of the electron-transfer rate is clearly much steeper than predicted by simple Marcus theory. The implications of these results are discussed with respect to the possible mechanisms that enhance electron transfer in metal complexes.

THE PHOTOCHEMISTRY OF POLYPYRIDYL COMPLEXES OF RU(II) has been an object of long-term interest (1–4). In recent years, research has continued into the utilization of these complexes as light-active antennae (5, 6) that pump electrons into subsequent reactions. Interligand electron transfer (ILET) in the excited metal-to-ligand charge-transfer (MLCT) states is a crucial aspect of excited-state photochemistry that must be understood to optimize the efficiency of these complexes as a source of photoelectrons. This chapter discusses the rate of ILET in the excited MLCT states of mixed-ligand polypyridyl complexes of Ru(II).

[1]Address correspondence to this author.

0065–2393/90/0226–0253$06.00/0
© 1990 American Chemical Society

Previous investigations of the excited MLCT states were performed by using fluorescence (7, 8) and nanosecond Raman spectroscopy (9–12). In mixed-ligand complexes, there is limited experimental evidence (7, 8) for dual luminescence originating from dissimilar ligands in the same complex. In support of the fluorescence experiment, transient Raman spectroscopy produced similar results (11, 12). This led to the conclusion that the ligand-localized MLCT states are coupled strongly enough to permit facile electron transfer. For many substituted bipyridine complexes, the fluorescence lifetime at 298 K is ~500 ns. From this it is generally concluded that the rate of ILET is $>>1/500$ ns.

This chapter discusses picosecond transient Raman experiments that investigate the free energy dependence of ILET. Raman spectroscopy is a direct measurement of this process because resolved vibrational bands can be identified unambiguously with the MLCT state localized on each type of ligand in the complex. The free energy gap between ligands (ΔG_0) is tuned by chemical substitution of bipyridine ligands. In all cases, substitution takes place on the opposite side of the ring from the Ru–nitrogen bond to keep the electronic coupling between ligands constant. Contrary to previous results, the rate of ILET varies from $\leq 10^6$ to $\geq 10^{11}$ s^{-1} and is very strongly dependent on free energy. This conclusion assumes knowledge about the energy gap between ligands, although the energy gap is not directly measured. The implications of this uncertainty are discussed later.

Experimental Details

The details of the experimental apparatus are described in several recent publications (13, 14). The laser system used in these studies is mode-locked continuous-wave NdYAG laser coupled with a high repetition rate chirped pulse regenerative amplifier that was pioneered (15, 16) at Louisiana State University.

The 1-kHz repetition rate and high peak power of this laser are ideal for transient Raman investigations. One of the unique features of this laser is that the pulse width can be varied between 10 and 150 ps by controlling the magnitude of the frequency chirp. In the data presented, the pulse width and laser power for each particular experiment are stated in the figure caption.

The two-color experiment involved excitation of the sample by pump and probe pulses at two different wavelengths: pump pulses at 532 nm, with an energy of 0.3 mJ, and probe pulses at 354.7 nm, with an energy of 0.03 mJ, were typically used. The laser power at 354.7 nm was limited by using an NRC laser power attenuator. The probe pulses were optically delayed from the pump pulses to produce transient Raman spectra at various temporal delays. A chopper was used to allow 354.7 nm only or both colors to excite the sample. No background signals correlated solely to the 532-nm laser were observed. Because the chopping action takes place on a fast time scale compared to the frequency of the major noise components, spectra with very high signal-to-noise ratios can be obtained. A computer sorts the Raman signal from each laser pulse into two channels according to the phase of the chopper. In this way, two separate spectra are simultaneously obtained. The first channel corresponds to excitation with both 354.7 and 532 nm, and the second channel to 354.7 nm alone. The pure two-color spectrum is generated by subtracting the one-color background components from the raw two-laser signal.

Detection consists of a scanning double monochromator, photomultiplier tube, and gated integration. The sample is a free-flowing jet of optical quality that flows at a speed sufficient to ensure that no two laser shots interrogate the same region of solution. The cold experiments were performed in a cylindrically shaped Dewar flask oriented such that the principle symmetry axis was vertical. An aluminum sample cell 0.375 in. thick makes up the bottom of the Dewar flask. When the Dewar flask is filled with a refrigerant, this cell is in direct contact with the cryogenic fluid above it with an area of 7 in^2. Both dry ice–acetone and liquid nitrogen were used as refrigerants.

The aluminum cell contained a 5-mL sample reservoir optically accessed from the bottom through a quartz window. Because the sample cell was physically part of the Dewar flask, the entire cell–Dewar flask assembly was rotated to keep successive laser pulses from probing the same region of sample. The possibility of laser heating was monitored in situ by measuring the ratio of the Stokes–anti-Stokes solvent bands. No evidence of laser heating was observed (13).

The compounds discussed in this chapter are $[Ru(bpy)_3]^{2+}$, $[Ru(Me_2bpy)_3]^{2+}$, $[Ru(5,5'-Me_2bpy)_3]^{2+}$, $[Ru(Ph_2bpy)_3]^{2+}$, $[Ru((COOH)_2bpy)_3]^{2+}$, $[Ru(bpym)_3]^{2+}$, and the mixed-ligand analogs. Abbreviations used are as follows: bpy is 2,2'-bipyridine, Me$_2$bpy is 4,4'-dimethyl-2,2'-bipyridine, Ph$_2$bpy is 4,4'-diphenyl-2,2'-bipyridine, (COOH)$_2$bpy is 4,4'-dicarboxy-2,2'-bipyridine, and bpym is 2,2'-bipyrimidine. The compounds were prepared by previously published (13) methods and purified chromatographically (Sephadex LH–20).

The mixed-ligand complexes were checked for impurities consisting of the symmetric trisubstituted complexes by using a spectroscope (Bruker AM400 high-resolution NMR). The ring protons corresponding to the bpy and substituted bpy ligands can easily be resolved. Assignments are straightforward by comparison to the symmetric complexes. Integration of the peaks corresponding to the protons on the respective ligands indicates that an impurity of tris complex, if it is present at all, must be less than 1% of the composition of the sample. In fluorescence studies, even a small amount of impurity can lead to erroneous results because the detection efficiency is very sensitive to the quantum yield of fluorescence. Raman spectroscopy is less sensitive to the lifetime of the excited state, as long as the lifetime exceeds the 10-ps laser pulse width. In some of our results the relative intensities of the Raman bands from two dissimilar ligands in the same complex are roughly equal. This result cannot be explained by impurities.

Spectroscopic Results

Two-color transient Raman spectroscopy is a powerful method for studying the dynamics of excited states. The excited-state dynamics can be obtained by introducing an optical delay between the pump and probe lasers. In addition, because the probe laser is maintained at a relatively low energy density, multiphoton artifacts that might arise from stimulated Raman are eliminated. Furthermore, the assignments of vibrational bands to the ground and excited state are relatively straightforward because the low-energy density-probe laser only weakly produces the excited state. Figure 1 illustrates this point. In this figure the Raman spectra obtained at 354.7 nm for $[Ru(Me_2bpy)_2(bpy)]^{2+}$ appear under conditions where only the probe laser interrogates the sample (Figure 1B) and where simultaneous excitation to the MLCT state occurs (Figure 1A). The two spectra are subtracted and the

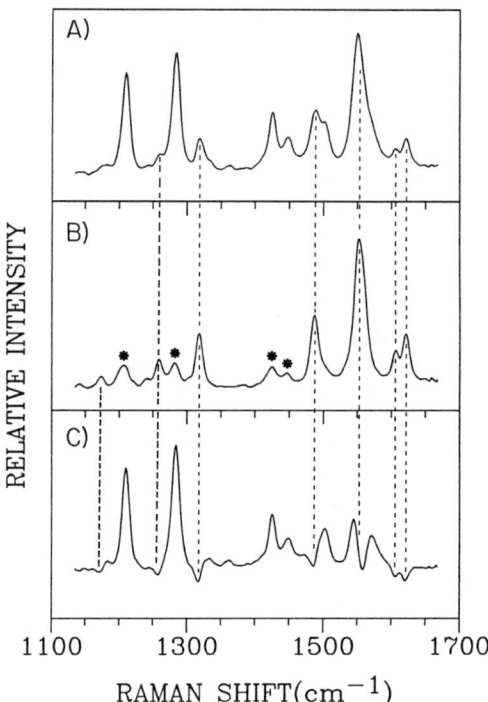

Figure 1. Transient Raman spectrum of the MLCT state of $[Ru(Me_2bpy)_2(bpy)]^{2+}$ in H_2O with two-color excitation at 532 nm and probe at 354.7 nm. Time delay between pump and probe lasers is 0 ps. Dashed lines connect ground-state bands in each figure. Concentration was 2 mM. Key: A, two-color spectrum including one-color background components obtained by simultaneous excitation at 354.7 and 532 nm; pulse energy at 354.7 nm was 30 µJ in a 3-mm beam waist; pulse energy at 532 nm was 200 µJ in a 1-mm beam waist; B, one-color background spectrum obtained with 354.7-nm excitation alone; asterisks indicate excited-state bands; and C, pure two-color spectrum obtained by subtracting the one-color background components from the experimental two-color spectrum. (Reproduced with permission from reference 13. Copyright 1989 American Institute of Physics.)

result is shown in Figure 1C. In the latter, positive-going peaks represent excited-state features. Ground-state bands appear to be bleached, as indicated by negative-going peaks. The excited-state bands correspond to those previously assigned as the ligand-localized MLCT state.

The apparent bleaching effect occurs for two reasons. First, the pump pulse depletes the ground-state population by exciting the MLCT transition. Second, the optical density of the solution increases at the probe wavelength because of the presence of the transient MLCT absorption. This increased optical density reduces the path length of the probe laser in the sample, and thereby causes the intensity of ground-state bands to be diminished.

Transient Raman spectra of $[Ru(Me_2bpy)_2(bpy)]^{2+}$ in methanol (not shown) exhibit (14) strong depletion of solvent bands, in addition to depletion of bands corresponding to the ground state of the complex. The fact that solvent bands appear to deplete indicates that a change in optical density is the dominant effect leading to the apparent depletion of ground-state bands in the spectrum shown in Figure 1C. In fact, the solvent bands provide an internal reference for correcting the spectra for depletion, as described in a separate publication (14).

The assignments of vibrational bands with respect to ligand parentage must also be considered. Fortunately, the coupling between ligands is so small in the compounds studied that vibrational bands of one particular ligand do not change in frequency when a different type of ligand is substituted into the complex. The ligand parentage in mixed complexes can therefore be assigned by direct comparison to the tris-substituted complexes. An example is shown in Figure 2 for the pure excited-state spectra of A, $[Ru(bpy)_3]^{2+}$; B, $[Ru(Me_2bpy)_2(bpy)]^{2+}$; and C, $[Ru(Me_2bpy)_3]^{2+}$. This

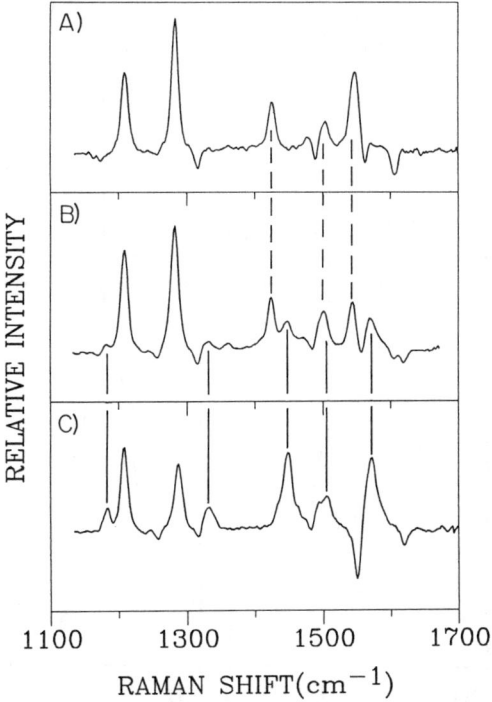

Figure 2. Pure two-color spectra obtained by subtracting the one-color signal shown in Figure 6 from the raw two-color signal similar to that in Figure 5A. Dashed lines connect excited-state bpy bands. Solid lines connect excited-state Me_2bpy bands. Concentration was 2 mM in H_2O. Key: A, $[Ru(bpy)_3]^{2+}$; B, $[Ru(Me_2bpy)_2(bpy)]^{2+}$; and C, $[Ru(Me_2bpy)_3]^{2+}$. (Reproduced with permission from reference 13. Copyright 1989 American Institute of Physics.)

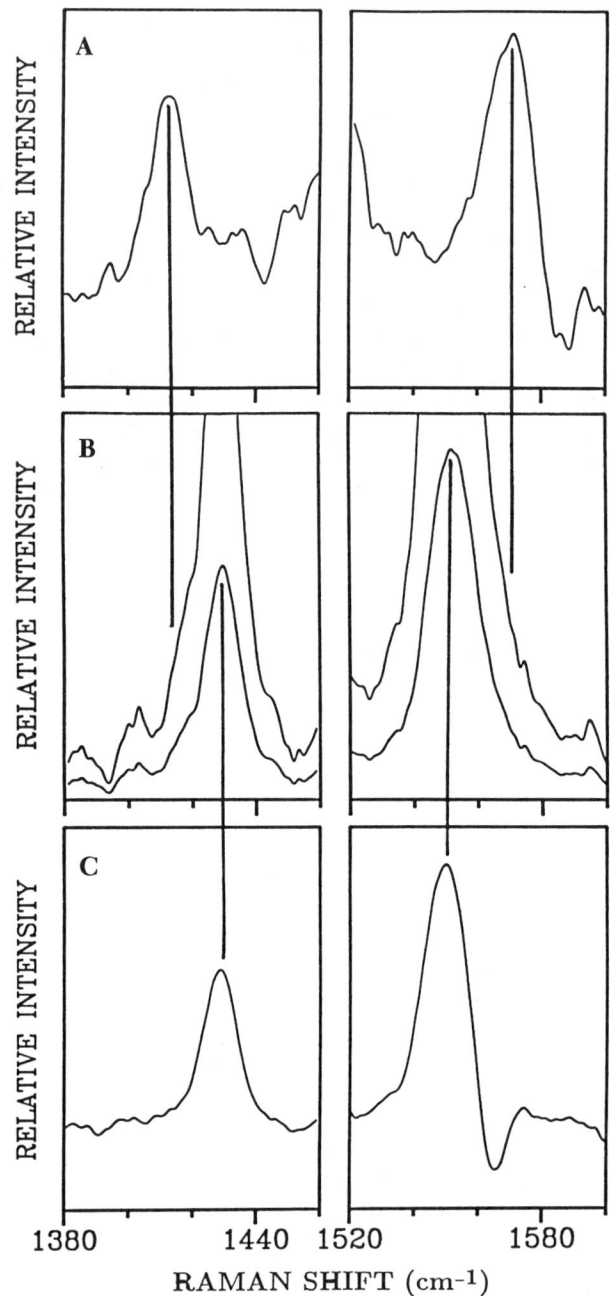

Figure 3. Pure two-color transient Raman spectra. Concentration was 2 mM in H₂O. Key: A, [Ru(5,5'-Me₂bpy)₃]²⁺; B, [Ru(5,5'-Me₂bpy)₂(bpy)]²⁺ with an expanded view as an insert; and C, [Ru(bpy)₃]²⁺

method is straightforward and leaves little doubt as to the nature of the assignments. Figure 2C clearly shows excited-state features corresponding to the ligand-localized MLCT states of both bpy and Me_2bpy.

Spectra for other complexes are shown in Figures 3–6. There are two important features in this data. The first point is shown in Figures 2 and 3, where the pure excited-state spectra of the mixed complex are shown in the middle spectrum. The top and bottom views show the respective symmetric trisubstituted complexes. In each figure, the mixed-ligand spectra exhibit features corresponding to an excited MLCT state where the electron is localized on both types of ligand present in the complex.

The second important feature in the transient Raman spectra is presented in Figures 4–6. The $[Ru(Ph_2bpy)(bpy)_2]^{2+}$, $[Ru(bpym)(bpy)_2]^{2+}$, and $[Ru((COOH)_2bpy)(bpy)_2]^{2+}$ spectra exhibit excited-state bands characteristic of only one ligand in the mixed-ligand complex. For $[Ru(bpym)(bpy)_2]^{2+}$ the spectral congestion is minimal, in that there are no bpym bands at frequencies anticipated for bands corresponding to the bpy ligand. This result implies

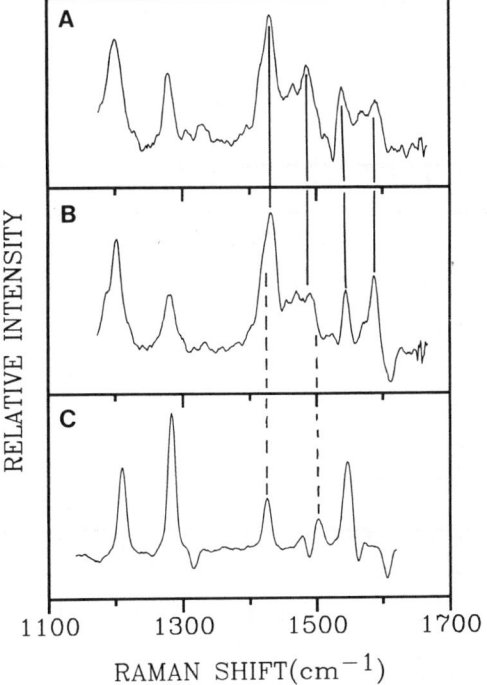

Figure 4. Pure two-color transient Raman spectra. Solid lines connect bands assigned to the MLCT state of the Ph_2bpy ligand. Dashed lines indicate the expected (but absent) position of bands assigned to the MLCT state of the bpy ligand. Concentration was 2 mM in H_2O. Key: A, $[Ru(Ph_2bpy)_3]^{2+}$; B, $[Ru(Ph_2bpy)(bpy)_2]^{2+}$; and C, $[Ru(bpy)_3]^{2+}$.

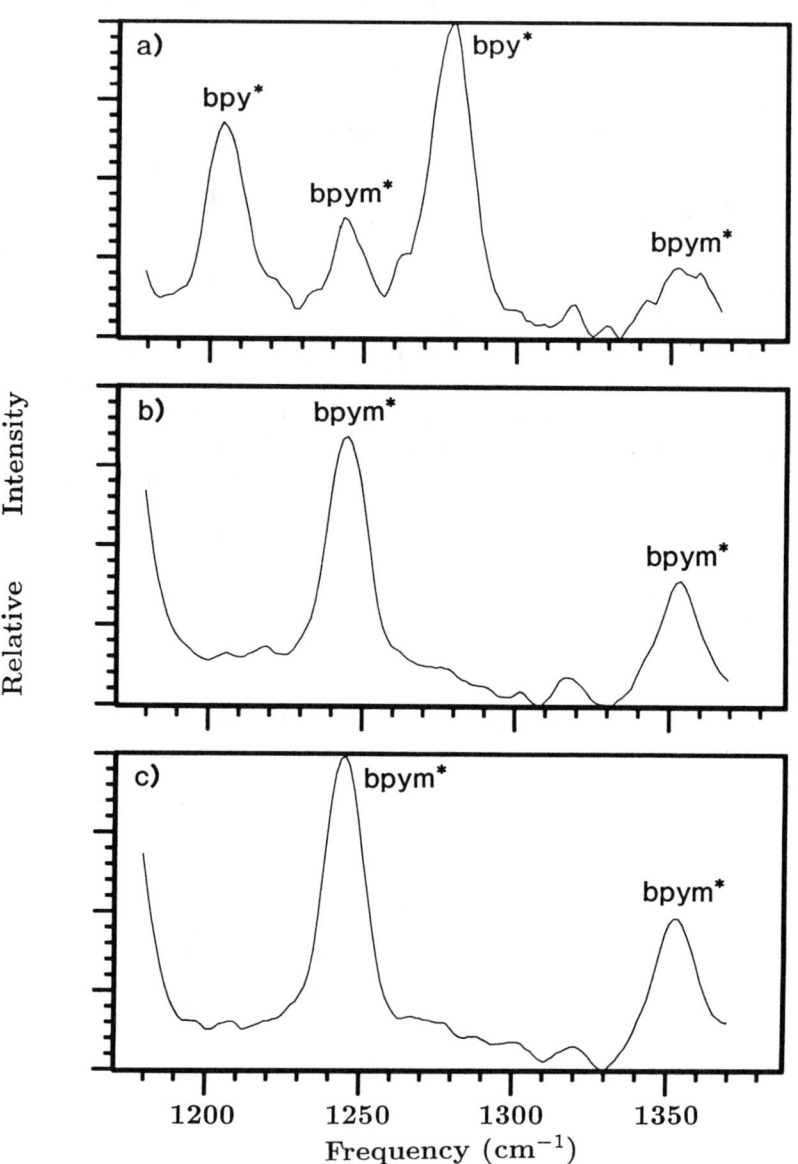

Figure 5. Pure two-color picosecond Raman spectra of $[Ru(bpy)_2(bpym)]^{2+}$ obtained by excitation at 532 nm and probing at 354.7 nm. Concentration is 2 mM. Key: a, expanded spectrum of 50:50 mixture of $[Ru(bpy)_3]^{2+}$ and $[Ru(bpym)_3]^{2+}$; b, expanded spectrum of $[Ru(bpy)_2(bpym)]^{2+}$ with a time delay between the pump and probe of 0 ps; and c, expanded spectrum of $[Ru(bpy)_2(bpym)]^{2+}$ with the time delay between the pump and probe pulses at 4000 ps.

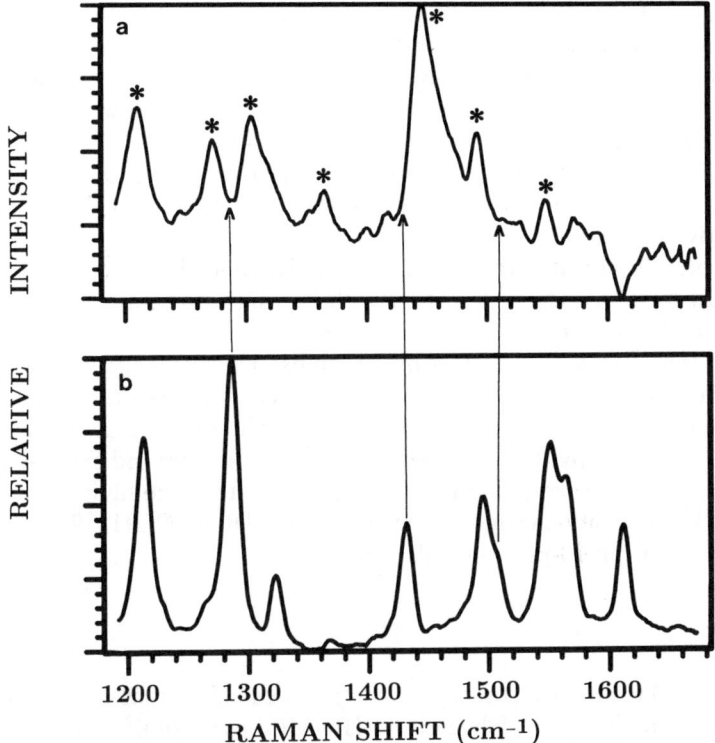

Figure 6. a, Pure two-color picosecond Raman spectra in H_2O of $[Ru((COOH)_2bpy)(bpy)_2]^{2+}$; asterisks indicate excited-state bands assigned to the $(COOH)_2bpy$ ligand; arrows indicate the expected (but absent) position of the bands corresponding to the MLCT state localized on the bpy ligand. b, One-color picosecond Raman spectrum of $[Ru(bpy)_3]^{2+}$.

that any excitation to the higher energy bpy ligand has decayed on a time scale fast compared to the 10-ps probe pulse. The fact that the bpy ligand has initially undergone excitation can be seen in the ground-state Raman spectrum. Resonance Raman bands can be observed and assigned to both the bpym and bpy ground-state ligands. The bpy bands are resonantly enhanced in the spectrum to approximately the same degree as in $[Ru(bpy)_3]^{2+}$. This fact indicates that the laser is in resonance with the excited MLCT state of bpy. This specific enhancement is a consequence of the importance of the Franck–Condon factor in the resonance Raman enhancement (17). The effect of the Franck–Condon factor is to selectively enhance those bands along which there is a distortion in the potential between the ground and excited state in resonance.

For the ground state of $[Ru(bpym)(bpy)_2]^{2+}$, this distortion appears to be the same as that occurring in $[Ru(bpy)_3]^{2+}$ because the observed ground-state bpy bands in $[Ru(bpym)(bpy)_2]^{2+}$ have the same frequencies and relative intensities as those in [Ru(bpy)$_3$]$^{2+}$. Given that the MLCT states are assumed to be localized, a distortion of potential along the vibrational coordinates in bpy can occur only with MLCT excitation to the bpy ligand. There is no change in the magnitude of the resonance Raman enhancement of the ground-state bpy bands in [Ru(bpym)(bpy)$_2$]$^{2+}$ compared to [Ru(bpy)$_3$]$^{2+}$. Therefore, it is expected that the laser produces an initial population of roughly one-third to two-thirds on the bpym and bpy ligand, respectively.

Figure 5 indicates that the bpy ligand is not observed in the excited state. In fact, to escape detection with the signal-to-noise ratio shown in Figure 5, the bpy bands must be present at an intensity at least 75 times smaller than the corresponding bpym bands. This is taken as evidence that the initial excitation to the bpy-localized MLCT state has decayed on a time scale faster than the 10-ps observation time. The same reasoning is applied to [Ru((COOH)$_2$bpy)(bpy)$_2$]$^{2+}$. Here again, the spectra indicate an absence of Raman bands corresponding to the excited state of the higher-energy bpy ligand. The indication is that the rate of ILET in [Ru((COOH)$_2$-(bpy)(bpy)$_2$]$^{2+}$ is faster than the 10-ps laser pulse width.

Dynamics

Two-color Raman spectroscopy was used to establish that electrons are localized on both types of ligands in [Ru(Me$_2$bpy)$_2$(bpy)]$^{2+}$. The two-color excited-state spectrum for [Ru(Me$_2$bpy)$_2$(bpy)]$^{2+}$ is shown in Figure 7 for various time delays between the pump and probe lasers. No dynamics are observed on a picosecond time scale. In fact, Figure 7 shows that there are no dynamics occurring all the way out to 500 ns. The implication from this result is that electron transfer from the Me$_2$bpy ligand to the bpy ligand occurs at a rate slower than 2×10^6 s^{-1}.

The energy gap between these two ligands is estimated (13) at 890 cm^{-1} from the reduction potentials of [Ru(bpy)$_3$]$^{2+}$ and [Ru(Me$_2$bpy)$_3$]$^{2+}$. The difference in the 2+/1+ reduction potentials of the respective tris-substituted complexes is assumed to approximate the MLCT excited-state energy gap. The reduction is believed to be that of the complexed ligand, thereby mimicking excited-state behavior. If this estimate is in error, there is an alternative explanation for the lack of dynamics observed in the [Ru(Me$_2$bpy)$_2$(bpy)]$^{2+}$ excited state. The possibility exists that fast electron transfer establishes thermal equilibrium between the two dissimilar ligands. It should be possible to determine if the bpy and Me$_2$bpy ligands are in thermal equilibrium by investigating the temperature dependence of the Raman spectrum. In other words, if fast ILET is occurring, it should be possible to collapse the population on the higher-energy ligand into that of the lower by cooling the sample.

Low-temperature Raman spectra are shown in Figure 8. No change in

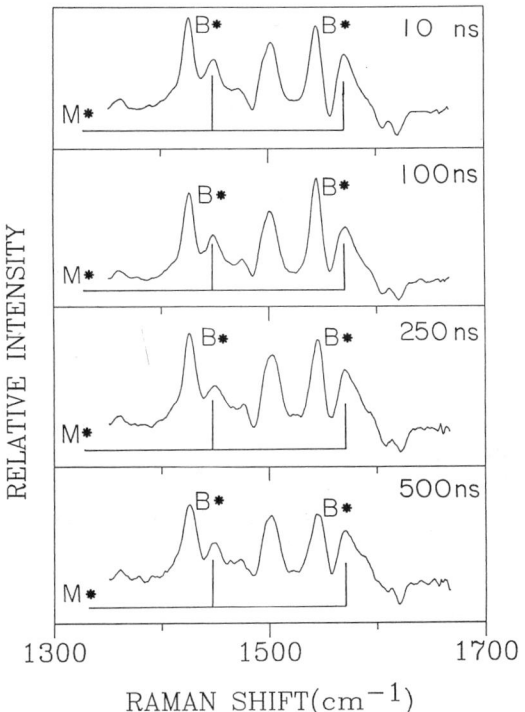

Figure 7. Pure two-color spectrum of the MLCT state of [Ru(Me₂bpy)₂(bpy)]²⁺ produced by excitation at 532 nm (200 µJ, 1-mm beam waist) and probed at 354.7 nm (30 µJ, 1-mm beam waist). The temporal delay between the pump and probe lasers is given in the figure. Solid lines labeled M* indicate bands assigned to the excited state of Me₂bpy. Bands labeled B* indicate bands assigned to the excited state of bpy. Data shown here were obtained by subtracting the one-color spectrum from the raw two-color spectrum. Concentration was 2 mM in H₂O. (Reproduced with permission from reference 13. Copyright 1989 American Institute of Physics.)

the relative intensities of the bands assigned to the Me$_2$bpy and bpy ligand are observed over a temperature range of 90–295 K. Between these extremes in temperature, simple statistical mechanics predicts that even an energy gap of 60 cm^{-1} would exhibit a factor of 2 change in population if thermal equilibrium applied. It is difficult (although not impossible) to believe that the estimated energy gap is high by more than a factor of 15, leading to an actual energy gap of ≤60 cm^{-1} between the Me$_2$bpy and bpy ligand. The more likely conclusion is that thermal equilibrium does not apply and that the lack of dynamics results from slow electron transfer at a rate $k_{\text{ILET}} \leq 2 \times 10^6$ s^{-1}.

The temporal dynamics were also investigated for the [Ru(5,5'-Me$_2$bpy)$_2$-(bpy)]$^{2+}$ complex. The result is the same as for [Ru(Me$_2$bpy)$_2$(bpy)]$^{2+}$. No

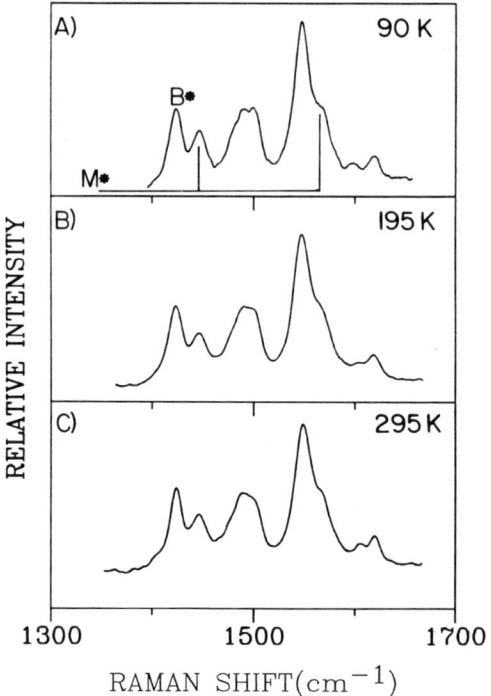

Figure 8. One-color spectrum of the MLCT state of $[Ru(Me_2bpy)_2(bpy)]^{2+}$ produced by excitation at 354.7 nm (30 µJ, 250-µm beam waist). The temperatures are given in the figure. Concentration was 2 mM in CD_3OD for B and C. A was obtained at 2 mM in a 60:40 $CD_3OD:D_2O$ glass. Solid lines labeled M* indicate excited-state bands of Me_2bpy. Bands labeled B* correspond to the excited state of bpy. (Reproduced with permission from reference 13. Copyright 1989 American Institute of Physics.)

dynamics were observed, out to the longest delay (50 ns) investigated. The data are summarized in Table I. The rate constants quoted for the first two entries are lower limits of the rate estimated from the longest delay investigated in each case.

Discussion

One of the major conclusions of this chapter is illustrated by the data summarized in Table I. The transient Raman data provide a direct means of clearly demonstrating that electron populations are established on the lower-energy ligand in less than 10 ps for energy gaps ≥ 1130 cm^{-1}. The conclusions are not as straightforward for energy gaps less than 1130 cm^{-1}. The Raman spectra for the mixed-ligand complexes are extensively overlapped. Even so, the two-color transient Raman spectra unambiguously establish that there are isolated bands that can be assigned to the MLCT state localized on each

Table I. Apparent Rate of Interligand Electron Transfer as a Function of Free-Energy Gap Between Donor and Acceptor Ligands

Complex	Reaction Coordinate	ΔG_o, cm^{-1}	k_{ILET}, s^{-1}
[Ru(Me$_2$bpy)$_2$(bpy)]$^{2+}$	(Me$_2$bpy) → (bpy)	890	$\leq 2 \times 10^6$ [a]
[Ru(5,5'-Me$_2$bpy)$_2$(bpy)]$^{2+}$	(5,5'-Me$_2$bpy) → (bpy)	1130	$\leq 2 \times 10^7$ [a]
[Ru(Ph$_2$bpy)(bpy)$_2$]$^{2+}$	(bpy) → (Ph$_2$bpy)	1300	$\geq 2 \times 10^{11}$
[Ru((COOH)$_2$bpy)(bpy)$_2$]$^{2+}$	(bpy) → (COOH)$_2$bpy)	3080	$\geq 2 \times 10^{11}$
[Ru(bpym)(bpy)$_2$]$^{2+}$	(bpy) → (bpym)	3240	$\geq 2 \times 10^{11}$

[a] These rates are lower limits representing the maximum delays investigated (500 ns for Me$_2$bpy and 50 ns for 5,5'-Me$_2$bpy).

distinct ligand. The populations do not appear to evolve in time between 10 and 500 ns. In addition, the population of electrons localized on the higher-energy ligand do not cool into the lower-energy ligand over a temperature range of 90–298 K.

The present data cannot distinguish between the following two possibilities. Either the estimated energy gap separating the lowest-lying triplet MLCT states is grossly in error (that is, the estimated energy gap is high by more than a factor of 15) or the lowest-lying triplet MLCT states are weakly coupled (such that electron transfer does not occur within the observation time of the experiment). It might seem at first glance that the latter possibility is inconsistent with the observation of fast electron transfer for energy gaps greater than 1130 cm^{-1}. However, this is not necessarily the case.

Light at 354.7 and 532 nm initially excites into the singlet MLCT state, which decays into the triplet state in less than 10 ps. The term "singlet", which refers to the spin configuration of the state before spin orbit coupling, does not imply that the electronic state is a pure singlet. It is possible that electron transfer is fast in the initially excited singlet electronic state, but slow in the lowest lying triplet states. When ILET on the singlet surface is not complete before intersystem crossing to triplet occurs, some population is left on the higher energy ligand. Once in the triplet state, ILET might in fact be very slow. This mechanism would account for the lack of dynamics observed in the triplet state. For energy gaps ≥1130 cm^{-1}, ILET is apparently complete on the singlet surface within the time in which intersystem crossing occurs. In this case, because the picosecond Raman experiment probes only the triplet surface, no population is observed on the higher energy ligand.

There are many reasons why the rate of ILET is expected to be fast on the singlet potential surface. At higher energies, mixing with additional electronic states can enhance the overall electronic coupling between different ligands. One such possibility is coupling to metal-centered d–d states. Caspar and Meyer (19) showed the importance of metal-centered states in the energy-gap dependence of luminescence from the triplet states in analogous complexes. It was determined (19) that the energy transfer from triplet MLCT to metal-centered d–d states is thermally activated. For the higher energy singlet states, thermal activation is not a necessary prerequisite to energy transfer. In fact, it is energetically possible for complete mixing to occur between the singlet and d–d states. The end result of mixing the singlet state with all of the electronic states available at the higher energies will ultimately be to enhance the probability for radiationless decay. Further work is necessary to clarify the mechanism of electron transfer in these complexes. In particular, direct measurement of the excited-state energy gaps is now in progress.

References

1. Maverick, A; Gray, H. B. *Pure Appl. Chem.* **1980**, *52*, 2339.
2. Meyer, T. J. *Acc. Chem. Res.* **1978**, *11*, 94.
3. Creutz, C.; Sutin, N. *Proc. Natl. Acad. Sci. U.S.A.* **1975**, *72*, 2858.
4. Crosby, G. A. *Acc. Chem. Res.* **1975**, *8*, 231.
5. Nocera, D. G.; Winkler, J. R.; Yocom, K. M.; Bordignon, E.; Gray, H. B. *J. Am. Chem. Soc.* **1984**, *106*, 5145.
6. Krueger, J. S.; Mayer, J. E.; Mallouk, T. E. *J. Am. Chem. Soc.* **1988**, *110*, 8232.
7. DeArmond, M. K.; Carlin, C. M. *Coord. Chem. Rev.* **1981**, *36*, 325.
8. Juris, A.; Barigelletti, F.; Balzani, V.; Belser, P.; von Zelewsky, A. *Inorg. Chem.* **1985**, *24*, 202.
9. McClanahan, S. F.; Dallinger, R. F.; Holler, F. J.; Kincaid, J. R. *J. Am. Chem. Soc.* **1985**, *107*, 4853.
10. Mabrouk, P. A.; Wrighton, M. S. *Inorg. Chem.* **1986**, *25*, 526.
11. Caspar, J. V.; Westmoreland, T. D.; Allen, G. H.; Bradley, P. G.; Meyer, T. J.; Woodruff, W. H. *J. Am. Chem. Soc.* **1984**, *106*, 3492.
12. Chung, Y. C.; Leventis, N.; Wagner, J.; Leroi, G. E. *Inorg. Chem.* **1985**, *24*, 1966.
13. Orman, L. K.; Chang, Y. J.; Anderson, D. R.; Yabe, T.; Xu, X.; Yu, S.-C.; Hopkins, J. B. *J. Chem. Phys.* **1989**, *90*, 1469.
14. Chang, Y. J.; Xu, X.; Yabe, T.; Yu, S.-C.; Anderson, D. R.; Orman, L. K.; Hopkins, J. B. *J. Phys. Chem.*, accepted for publication.
15. Chang, Y. J.; Veas, C.; Hopkins, J. B. *Appl. Phys. Lett.* **1986**, *49*, 1758.
16. Chang, Y. J.; Anderson, D. R.; Hopkins, J. B. *International Conference on Lasers, 1986;* McMillan, R. W., Ed.; STS Press: McLean, VA, 1987; p 169.
17. Heller, E. J. *Acc. Chem. Res.* **1981**, *14*, 368.
18. Marcus, R. A. *J. Chem. Phys.* **1965**, *43*, 679.
19. Caspar, J. V.; Meyer, T. J. *Inorg. Chem.* **1983**, *22*, 2444.

RECEIVED for review May 1, 1989. ACCEPTED revised manuscript September 29, 1989.

THEORETICAL ASPECTS
OF SOLID-STATE SYSTEMS

15

Band Orbital Mixing and Electronic Instability of Low-Dimensional Metals

Myung-Hwan Whangbo

Department of Chemistry, North Carolina State University, Raleigh, NC 27695–8204

Electronic instability of a low-dimensional metal leading to its metal–insulator or metal–superconductor transition is discussed from the viewpoint of orbital mixing between filled and empty levels near the Fermi level of a normal metallic state. This chapter examines the concept of Fermi surface nesting, its relationship to a metal–insulator transition, the correlation between real and reciprocal space properties, and the difference in orbital mixing between insulating and superconducting states.

SOLID-STATE MATERIALS ARE OFTEN CLASSIFIED according to how their resistivities (ρ) vary as a function of temperature (T). Thus, metals and semiconductors are characterized by positive and negative slopes in their ρ vs. T plots. The stability of a metallic state depends on several factors, such as temperature and pressure. When the temperature is lowered, a metal may become a semiconductor (**1**) or a superconductor (**2**).

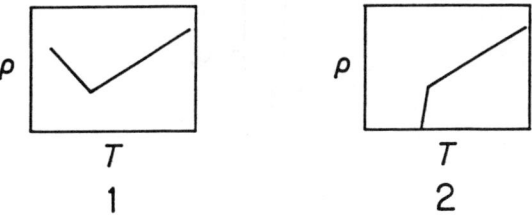

The electronic structure of a solid is described by energy bands as shown in **3**, where rectangular boxes represent allowed regions of energy (i.e.,

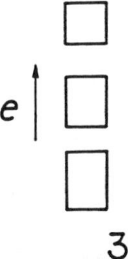

3

energy bands). A given band consists of N discrete levels, where N is the total number of unit cells in a solid. Because $N \to \infty$ for all practical purposes, all energy levels falling within a band are allowed. In a one-electron band picture, electron–electron repulsion is neglected so that each band level can be filled with two electrons. In this picture, a semiconductor (or an insulator) contains only completely filled and completely empty bands, so that an energy gap (i.e., band gap E_g) exists between the highest occupied and the lowest unoccupied band levels (*see* **4a**). (An insulator is a semiconductor

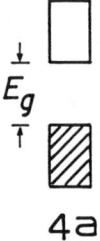

4a

with a large band gap.) On the other hand, a metal has at least one partially filled band (*see* **4b**) so that no energy gap exists between the highest occupied

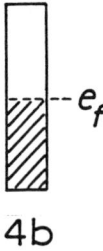

4b

level (i.e., the Fermi level e_f) and the lowest unoccupied band level.

The electronic instability of a metal that leads to either a metal–insulator or a metal–superconductor transition may be described on the basis of orbital mixing between the occupied and unoccupied levels (*1–3*). A new electronic state derived from the orbital mixing may become more stable than the metallic state when the energy gain resulting from the interactions between the occupied and unoccupied levels outweighs the inherent energy increase caused by introducing higher-lying, unoccupied levels. Because the energy difference between the occupied and unoccupied levels around the Fermi level can be very small, the extent of this energy increase can be made very small. Consequently, orbital mixing among the band levels in the vicinity of the Fermi level is crucial for a metal–insulator or a metal–superconductor transition.

From the viewpoint of one-electron band theory, a metal–insulator transition occurs when the Fermi surface of a partially filled band is well nested. A metal may have a chance to become superconducting only when it is free from electronic instability toward a metal–insulator transition. This chapter considers how the concept of Fermi surface nesting comes about, why Fermi surface nesting leads to an electronic instability toward a metal–insulator transition, how real and reciprocal space properties are related, and finally how the orbital mixing leading to a superconducting state differs from that leading to an insulating state.

Fermi Surface Nesting

For simplicity, let us consider a two-dimensional (2D) rectangular lattice 5, with repeat distances a and b. The coordinate of a lattice site is given by (ma, nb), where m and n are integers. If one orbital represents each lattice site, the orbital located at the site (ma, nb) may denoted by χ_{mn}. In this notation, the orbital at the coordinate origin is given by χ_{00} and those at its nearest-neighbor sites along the a- and b- directions by χ_{10} and χ_{01}, respectively. The allowed energy levels of lattice 5 are described by the band

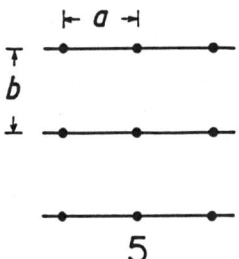

5

orbitals $\phi(k_a, k_b)$,

$$\phi(k_a, k_b) = N^{-1/2} \sum_m \sum_n \exp(ik_a ma) \cdot \exp(ik_b nb) \cdot \chi_{mn} \tag{1}$$

where k_a and k_b are wave vectors along the a and b directions, respectively, and the terms $\exp(ik_a ma)$ and $\exp(ik_b nb)$ are coefficients for the site orbital χ_{mn}. For convenience, the wave vectors k_a and k_b may be confined to the following values: $-\pi/a \leq k_a \leq \pi/a$ and $-\pi/b \leq k_b \leq \pi/b$. These values define the first Brillouin zone (shown in Figure 1) in wave vector (i.e., reciprocal) space. Those wave vectors define all possible values for the coefficients $\exp(ik_a ma)$ and $\exp(ik_b nb)$ in eq 1.

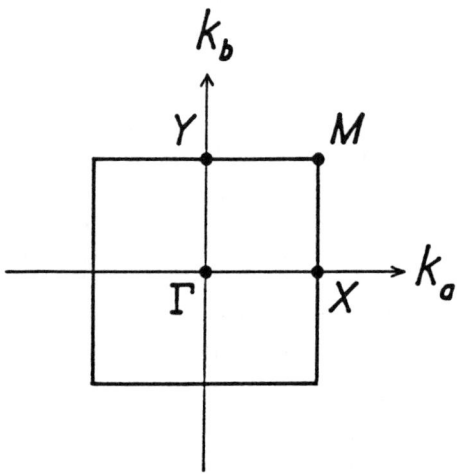

Figure 1. First Brillouin zone of the 2D rectangular lattice 5, where $\Gamma = (0,0)$, $X = (\pi/a, 0)$, $Y = (0, \pi/b)$, and $M = (\pi/a, \pi/b)$.

The band energies $E(k_a, k_b)$ associated with the orbitals $\phi(k_a, k_b)$ are expressed as

$$E(k_a, k_b) = \alpha + 2\beta_a \cos k_a a + 2\beta_b \cos k_b b \tag{2}$$

when the overlap between site orbitals is neglected (e.g., $\langle \chi_{00}|\chi_{10}\rangle = \langle \chi_{00}|\chi_{01}\rangle = 0$). Here α is the Coulomb integral, and β_a and β_b are the nearest-neighbor resonance (i.e., hopping) integrals along the a and b directions, respectively. Given H^{eff} as an effective Hamiltonian, α, β_a, and β_b are defined as

$$\alpha = \langle \chi_{00}|H^{\text{eff}}|\chi_{00}\rangle \tag{3a}$$

$$\beta_a = \langle \chi_{00}|H^{\text{eff}}|\chi_{10}\rangle \tag{3b}$$

$$\beta_b = \langle \chi_{00}|H^{\text{eff}}|\chi_{01}\rangle \tag{3c}$$

The nature of the band energies $E(k_a, k_b)$ is usually examined by plotting $E(k_a, k_b)$ as a function of wave vectors (i.e., band dispersions) along certain lines of the first Brillouin zone (e.g., $\Gamma \rightarrow X \rightarrow M \rightarrow Y \rightarrow \Gamma$ in Figure 1). Figure 2 shows three examples of band dispersions that illustrate how the ratio of the resonance integrals, β_b/β_a, affects the shape of band dispersion. Figures

2a, 2b, and 2c represent the cases of $\beta_b/\beta_a = 0$, $\beta_b/\beta_a < 1$, and $\beta_b/\beta_a = 1$, respectively. In Figure 2, each dashed line refers to the Fermi level for the case in which the band is half-filled, which occurs if there is one electron per site to contribute to the band.

Figure 2 shows that wave vectors in a certain region of the first Brillouin zone lead to occupied band orbitals, and those in the remaining region to

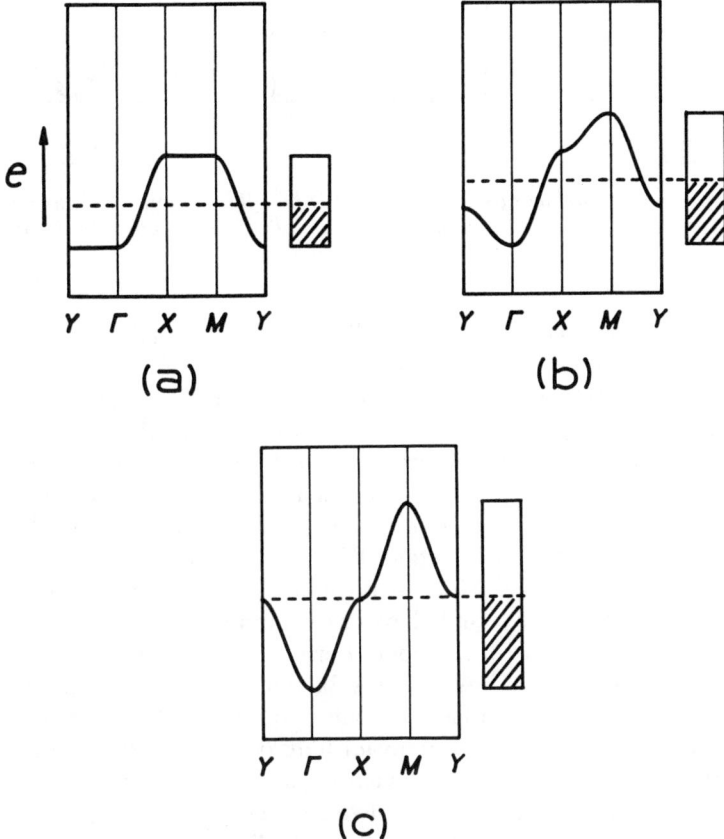

Figure 2. Dispersion relation of eq 2 for (a) $\beta_b/\beta_a = 0$, (b) $\beta_b/\beta_a < 1$, and (c) $\beta_b/\beta_a = 1$.

unoccupied band levels. This relationship is shown in Figures 3a, 3b, and 3c, which correspond to the band dispersions of Figures 2a, 2b, and 2c, respectively. In Figure 3, the wave vectors of the shaded regions are occupied (i.e., they lead to occupied band levels), and those of the unshaded region are unoccupied. Because all wave vectors are equally probable, the size of the occupied wave vector region is proportional to the band filling. Thus, for a half-filled band, one-half of the first Brillouin zone is occupied.

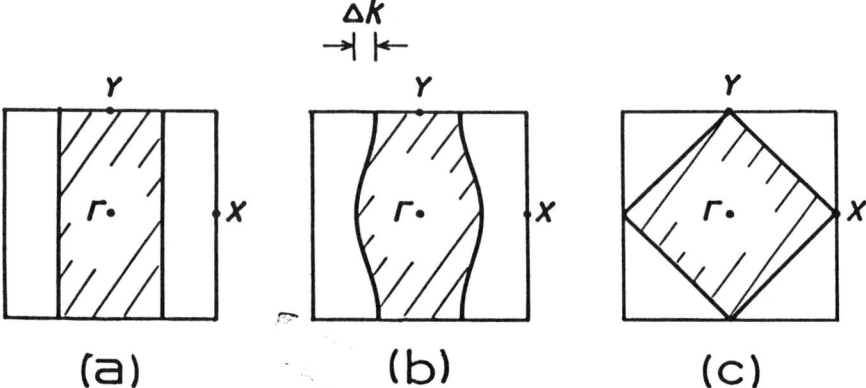

Figure 3. Fermi surfaces associated with the half-filled bands of Figure 2. (a) $\beta_b/\beta_a = 0$, (b) $\beta_b/\beta_a < 1$, and (c) $\beta_b/\beta_a = 1$. Filled wave vectors are indicated by hatching.

The Fermi surface of a partially filled band is defined as the boundary that separates the occupied wave vector region from the unoccupied wave vector region (4). The Fermi surface of Figure 3a or 3b consists of separated lines, and hence is said to be open. For the electrons with the crystal momenta $hk/2\pi$ along a certain wave vector direction, there exists no band energy gap if the wave vector line crosses a Fermi surface. In such a case, a partially filled band system under consideration exhibits a metallic character along the wave vector direction. When a straight wave vector line does not cross a Fermi surface, which is possible if the Fermi surface is open, the partially filled band system behaves as a nonmetal along that wave vector direction. The presence of an open Fermi surface characterizes a one-dimensional (1D) metal. Thus, Figures 3a and 3b both represent 1D metals. The wave-like Fermi surface of Figure 3b reflects the fact that interactions along the b direction are nonvanishing (i.e., $\beta_b \neq 0$). The extent of how strongly the Fermi surface deviates from the ideal 1D Fermi surface (i.e., the straight lines of Figure 3a) is measured by the wave vector ratio, $\Delta k/(\pi/a)$. Note that $\Delta k = 0$ for $\beta_b/\beta_a = 0$ (Figure 3a), and $\Delta k = 1$ for $\beta_b/\beta_a = 1$ (Figure 3c). Thus, the $\Delta k/(\pi/a)$ ratio scales with the β_b/β_a ratio.

The Fermi surface of Figure 3c, obtained for the case of $\beta_b/\beta_a = 1$, has a rectangular shape. A Fermi surface with a closed loop is said to be closed. A 2D metal is characterized by the presence of a closed Fermi surface. Figures 4a and 4b show the Fermi surfaces associated with the band of Figure 2c for the cases in which the band is less and more than half-filled, respectively. The Fermi surface of Figure 4a is a "distorted circle" centered at Γ. In contrast, the Fermi surface of Figure 4b features "distorted circles" centered at the four corners of the first Brillouin zone. (This becomes clearer

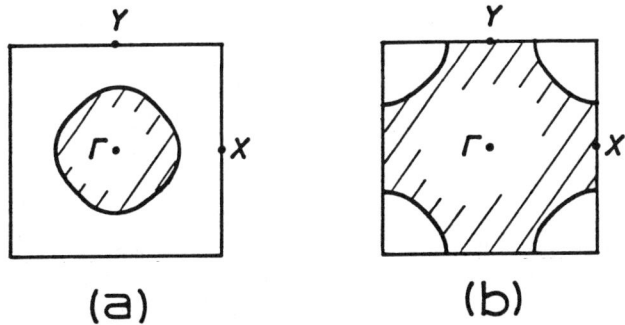

Figure 4. Fermi surfaces associated with the band of Figure 2c. (a) Less than half filled, (b) more than half filled.

when the pattern of Figure 4b is repeated by translating it in the k_a and k_b directions.) The Fermi surface of Figure 4a becomes more circular as the band filling is further reduced, and so does that of Figure 4b as the band filling increases further. The Fermi surfaces of Figures 4a and 4b both represent 2D metals.

A piece of a Fermi surface may be translated by a single wave vector **q** and superimposed on another piece of the Fermi surface. In such a case, the Fermi surface is said to be nested by the wave vector **q**. For example, the Fermi surfaces of Figure 4 are not nested, but those of Figures 3a, 3b, and 3c have the nesting vectors shown in Figures 5a, 5b, and 5c, respectively.

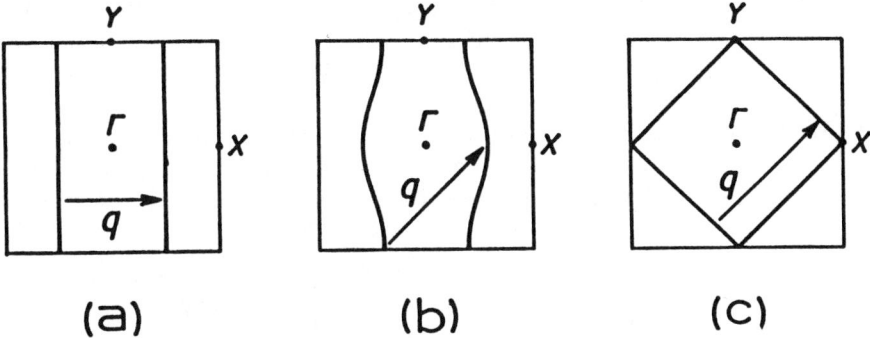

Figure 5. Nesting vectors associated with the Fermi surfaces of Figure 3. (a) $\beta_b/\beta_a = 0$, (b) $\beta_b/\beta_a < 1$, and (c) $\beta_b/\beta_a = 1$.

In Figure 5b it is not immediately obvious to see what the nesting vector does, because Figure 5b shows only the portion of the Fermi surface belonging to the first Brillouin zone. With an extended zone representation of the Fermi surface, which is obtained by repeating the pattern of Figure 5b in the two wave-vector directions, it becomes clear that the nesting vector q superimposes the left-hand-side piece of the Fermi surface onto the right-hand-side piece.

Electronic States Derived from a Normal Metallic State by Orbital Mixing

A metallic state predicted by one-electron band theory (i.e., a normal metallic state) is not stable when its Fermi surface is nested, and it becomes susceptible to a metal–insulator transition under a suitable perturbation. We now examine the nature of the nonmetallic electronic states that are derived from a normal metallic state upon mixing its occupied and unoccupied band energy levels (1–3). For simplicity, we will discuss this orbital mixing on the basis of the ideal 1D band structure shown in Figure 2a. This band is dispersive only along the direction parallel to $\Gamma \rightarrow X$ (i.e., the a direction in real space). In this situation, the energy $E(k_a, k_b)$ is independent of k_b. Thus, by dropping our reference to the wave vector k_b, eqs 1 and 2 are simplified as

$$\phi(k_a) = N^{-1/2} \sum_m \exp(ik_a ma) \cdot \chi_m \quad (4a)$$

$$E(k_a) = \alpha + 2\beta_a \cos k_a a \quad (4b)$$

The band dispersion of eq 4b for the first Brillouin zone, $-\pi/a \leq k_a \leq \pi/a$, is shown in Figure 6, where k_f and $-k_f$ are the Fermi wave vectors. Thus, the wave vectors in the $-k_f < k_a \leq k_f$ region are occupied, and those in the remaining region are unoccupied. In this 1D representation, the Fermi surfaces are given by the two points k_f and $-k_f$, and the nesting vector is $2k_f$. (By definition, a Fermi surface occurs in the form of points in 1D representation, lines in 2D representation, and surfaces in three-dimensional (3D) representation, in which three independent wave vectors are used to describe intersite interactions along three directions. Thus, in 3D repre-

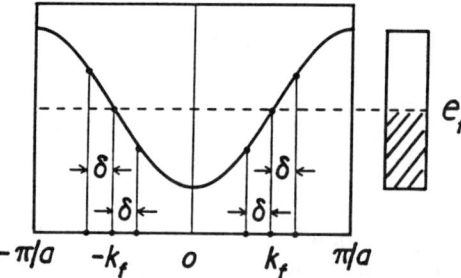

Figure 6. *Dispersion relation of eq 4b, where* δ *is a small positive number in reciprocal space.*

sentation, the Fermi surface of a 1D metal is given by two separate planelike surfaces, that of a 2D metal by a cylinderlike surface, and that of a 3D metal by a spherelike surface.) Given δ as a small positive number, the occupied band orbitals near the Fermi level are given by $\phi(-k_f + \delta)$ and $\phi(k_f - \delta)$, and the unoccupied orbitals by $\phi(k_f + \delta)$ and $\phi(-k_f - \delta)$. The occupied orbital $\phi(-k_f + \delta)$ differs from the unoccupied orbital $\phi(k_f + \delta)$ by the wave vector $2k_f$, and $\phi(k_f - \delta)$ differs similarly from $\phi(-k_f - \delta)$. Mixing the orbitals $\phi(-k_f + \delta)$ and $\phi(k_f + \delta)$ leads to new orbitals $\psi(-k_f + \delta)$ and $\psi(k_f + \delta)$, defined as

$$\psi(-k_f + \delta) = (1 + \gamma^2)^{-1/2} [\phi(-k_f + \delta) + \gamma\phi(k_f + \delta)] \tag{5a}$$

$$\psi(k_f + \delta) = (1 + \gamma^2)^{-1/2} [-\gamma\phi(-k_f + \delta) + \phi(k_f + \delta)] \tag{5b}$$

where γ is a mixing coefficient.

It is important to analyze spin and charge-density changes induced by the orbital mixing shown in eq 5. For simplicity, we assume that the overlap between site orbitals vanishes (e.g., $\langle\chi_0 | \chi_1\rangle = 0$). Then the charge distribution resulting from the unmodified orbital $\phi(k)$ (i.e., $\phi^*(k)\phi(k)$) is given by

$$\phi^*(k)\phi(k) = N^{-1} \sum_m \chi_m^* \chi_m \tag{6}$$

Thus, all site orbitals have the same coefficient, $1/N$, so that all sites have an identical density. The new orbitals, $\psi(-k_f + \delta)$ and $\psi(k_f + \delta)$, lead to the following density distributions:

$$\psi^*(-k_f + \delta)\psi(-k_f + \delta) = N^{-1} \sum_m [1 + 2\gamma(1 + \gamma^2)^{-1} \cos 2k_f ma] \cdot \chi_m^* \chi_m \tag{7a}$$

$$\psi^*(k_f + \delta)\psi(k_f + \delta) = N^{-1} \sum_m [1 - 2\gamma(1 + \gamma^2)^{-1} \cos 2k_f ma] \cdot \chi_m^* \chi_m \tag{7b}$$

Equation 7 is independent of the parameter δ, which measures how far away the orbitals $\phi(-k_f + \delta)$ and $\phi(k_f + \delta)$ are from the Fermi level. The coefficients of $\chi_m^* \chi_m$ in eq 7 are not uniform, but their magnitudes vary in a periodic manner that is described by the nesting vector $2k_f$ as $2\gamma(1 + \gamma^2)^{-1} \cos 2k_f ma$. Consequently, the orbital mixings defined by eq 5 introduce density waves with respect to the metallic state. If the absence of a density wave in a chain is represented by a straight line, **6a**, then the presence of it can be represented by a wavy line, **6b** or **6c**. In **6b** and **6c**, electron gain

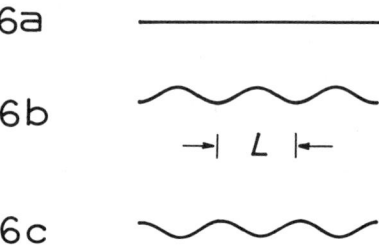

and loss with respect to the density distribution of **6a** are represented by the crests and troughs of the waves, respectively. Line **6b** shows density accumulation in the region where **6c** shows density depletion. Thus, the density waves **6b** and **6c** are out of phase. The repeat distance, L, of **6b** or **6c** is increased by a factor of $(2\pi/a)/2k_f$ with respect to that of **6a**, so that $L = \pi/k_f$. For the 1D band given by eq 4b, the k_f value for the band occupancy $1/n$ (e.g., $n = 2$ for a half-filled band) is given by $k_f = \pi/na$, so that $L = na$.

The density waves given by the orbitals $\psi(-k_f + \delta)$ and $\psi(k_f + \delta)$ in eq 7 are out of phase because $\psi(-k_f + \delta)$ accumulates electron density on the sites where $\psi(k_f + \delta)$ depletes electron density, and vice versa. When we fill the orbitals $\psi(-k_f + \delta)$ with two electrons, leaving the orbitals $\psi(k_f + \delta)$ empty, a charge-density wave state (CDW) results. In this state, each site has an equal density of up-spin and down-spin electrons, but the total electron density at each site varies in a periodic manner. An example of CDW derived from a half-filled metallic band of a 1D chain is shown in **7** (2),

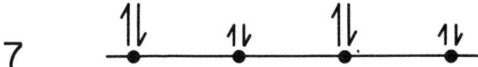

where the length of an upward or downward arrow at each lattice site represents the magnitude of the up-spin or down-spin electron density, respectively. If we fill the orbitals $\psi(-k_f + \delta)$ with an up-spin electron and the orbitals $\psi(k_f + \delta)$ with a down-spin electron, there occurs a spin-density wave state (SDW). In this state, the total electron density at each site is identical, but each site has unequal densities of up-spin and down-spin electrons that vary in a periodic manner. An example of an SDW derived from a half-filled band of a 1D chain is depicted in **8** (2), where the net spins

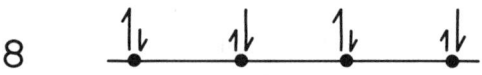

of the lattice sites have an antiferromagnetic ordering. Though not shown, we obtain results identical to 8 by mixing the unoccupied orbitals $\psi(-k_f - \delta)$ with the occupied orbitals $\psi(k_f - \delta)$.

This orbital mixing cannot occur unless a certain perturbation, H', is introduced to a metallic system so that a new effective Hamiltonian becomes $H^{\text{eff}} + H'$. The orbitals $\phi(-k_f + \delta)$ and $\phi(k_f + \delta)$ are not eigenfunctions of $H^{\text{eff}} + H'$, so the interaction between $\phi(-k_f + \delta)$ and $\phi(k_f + \delta)$ under the perturbation H' (i.e., $<\phi(-k_f + \delta) \mid H' \mid \phi(k_f + \delta)>$) does not necessarily vanish. The perturbation causing the CDW state associated with a Fermi surface nesting vector, $2k_f$, is a lattice vibration (i.e., phonon) with the wave vector $2k_f$. The CDW is referred to as commensurate when the ratio $(\pi/a)/2k_f$ is an integer (e.g., 2, 3, . . .) and incommensurate otherwise. A commensurate CDW state is susceptible to a permanent lattice distortion as a result of electron–phonon interactions. The perturbation causing an SDW state is an on-site repulsion, U (2). Given that the total density at each site is identical, the total energy of a system can be lowered by introducing spin polarization on each site because it reduces the extent of on-site repulsion.

Fermi Surface Nesting and Electronic Instability

We now consider why Fermi surface nesting plays a prominent role in inducing a CDW or SDW state from a normal metallic state. To simplify our notation, an occupied wave vector (k_a, k_b) may be represented by $\mathbf{k}$, and an unoccupied wave vector (k_a', k_b') by $\mathbf{k}'$. Orbital mixing between an occupied orbital $\phi(\mathbf{k})$ and an unoccupied orbital $\phi(\mathbf{k}')$ leads to new orbitals $\psi(\mathbf{k}) \propto \phi(\mathbf{k}) + \gamma\phi(\mathbf{k}')$ and $\psi(\mathbf{k}') \propto -\gamma\phi(\mathbf{k}) + \phi(\mathbf{k}')$. These new orbitals introduce electron-density waves described by the wave vector $\mathbf{q} = \mathbf{k} - \mathbf{k}'$, which are out of phase in their density distributions. When a Fermi surface has a nesting vector $\mathbf{q}$, this orbital mixing can be performed for all the wave vectors in the nested region of the first Brillouin zone, thereby leading to the sets of new orbitals $\{\psi(\mathbf{k})\}$ and $\{\psi(\mathbf{k}')\}$ differing in their wave vectors by $\mathbf{q} = \mathbf{k} - \mathbf{k}'$. A CDW state is obtained when the orbitals $\psi(\mathbf{k})$ are each doubly occupied. An SDW state is obtained when the orbitals $\psi(\mathbf{k})$ and $\psi(\mathbf{k}')$ are each singly occupied by up-spin and down-spin electrons, respectively.

With a nested Fermi surface, the sets of orbitals $\{\psi(\mathbf{k})\}$ and $\{\psi(\mathbf{k}')\}$ include those derived from mixing the occupied and unoccupied levels [$\phi(\mathbf{k})$ and $\phi(\mathbf{k}')$, respectively] in the vicinity of the Fermi level. The energy difference between such orbitals $\phi(\mathbf{k})$ and $\phi(\mathbf{k}')$ is small, so that the orbital mixing between them is significant, and so is the interaction energy $<\phi(\mathbf{k}) \mid H' \mid \phi(\mathbf{k}')>$. In addition, the energy lowering that results from such an interaction matrix element can be gained from all the wave vectors $\mathbf{k}$ and $\mathbf{k}'$ related by the nesting vector $\mathbf{q} = \mathbf{k} - \mathbf{k}'$ in the nested region of the first Brillouin zone.

This is why the electronic instability toward a CDW or an SDW state is strong when a normal metallic state possesses a complete Fermi surface nesting. As the extent of Fermi surface nesting diminishes, that of the electronic instability decreases.

Both CDW and SDW states are insulating in nature because a band energy gap is created at the Fermi level as a result of orbital mixing (2). An insulating state has no Fermi surface, by definition, because its highest occupied and lowest unoccupied levels are not degenerate. Thus, for a metal–insulator phase transition arising from a CDW or an SDW formation, the orbital mixing induced by the Fermi surface nesting is said to destroy the Fermi surface. A slightly incomplete Fermi surface nesting occurs when the pieces of the Fermi surface involved have slightly different curvatures in certain wave vector regions. For such a case as well, there exists an electronic instability resulting from the nested portions of the Fermi surface. The latter are destroyed upon the associated orbital mixing. After a phase transition resulting from this orbital mixing, the small unnested portions of the Fermi surface give rise to small pocketlike new Fermi surfaces around them, so that the resulting new electronic state is still metallic (5). That is, an incomplete Fermi surface nesting leads to an incomplete destruction of the Fermi surface, which is responsible for metal–metal transitions.

So far we have considered that a band level below e_f is completely filled (i.e., occupancy of 1); that above e_f is completely empty (i.e., occupancy of 0). This picture is valid for all levels when $T = 0$, but only for the levels lying outside the vicinity of the Fermi level (e.g., $e < e_f - 4k_BT$ and $e > e_f + 4k_BT$, where k_B is the Boltzmann constant) when $T > 0$. For the levels lying close to the Fermi level (e.g., $e_f - 4k_BT < e < e_f + 4k_BT$), whose orbital mixing plays a crucial role in lowering the energy of a 1D metal, thereby leading to a metal–insulator transition, their orbital occupancy $f(e)$ at nonzero temperature is given by the Fermi–Dirac distribution function, $f(e) = \{1 + \exp[(e - e_f)/k_BT]\}^{-1}$. Thus, $f(e) < 1$ for $e < e_f$, and $f(e) > 0$ for $e > e_f$. For example, in Figure 6, the occupancy of $\phi(-k_f + \delta)$ is less than 1, and that of $\phi(k_f + \delta)$ is larger than 0 (at $T > 0$). Consequently, the energy gain resulting from the orbital mixing between $\phi(-k_f + \delta)$ and $\phi(k_f + \delta)$ is maximum at $T = 0$ and decreases as T is raised. Thus, only when T is lowered below a certain temperature does the energy gain associated with the orbital mixing become substantial enough to cause a metal–insulator transition.

CDW Instability and Real vs. Reciprocal Space Correlations

Metal–insulator transition arising from a CDW instability is not abrupt, but typically undergoes a series of steps (6). This process can be illustrated by considering a 1D metal as composed of weakly interacting chains. At a high temperature, each chain has no tendency for CDW formation, so all chains have uniform density distributions, as illustrated in Figure 7a. Below a

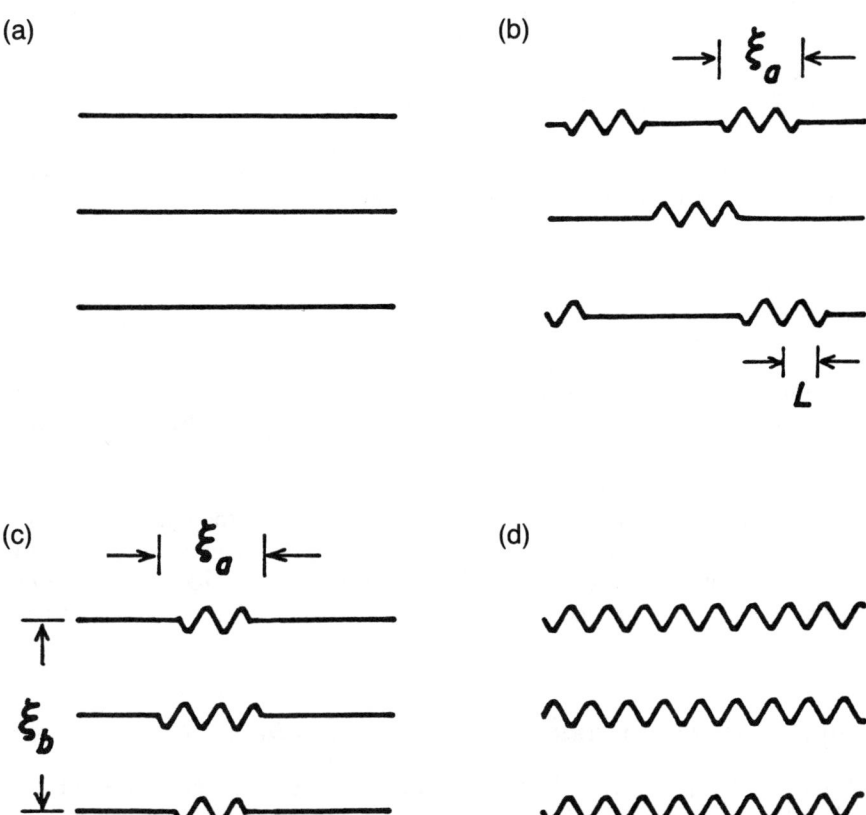

Figure 7. Schematic representation of what happens to a 1D metal with CDW instability as a function of temperature T. (a) Uniform electron-density distribution at $T > T_{1D}$, (b) 1D fluctuation at $T_x < T < T_{1D}$, (c) lateral ordering at $T_c < T < T_x$, and (d) long-range order at $T < T_c$.

certain temperature T_{1D}, each chain has a tendency for CDW formation. As shown in Figure 7b, a CDW is formed and destroyed dynamically at many parts of each chain, and CDW formation in one chain is independent of those in other chains. The average length of a CDW segment (i.e., coherent length) is ξ_a. Below a certain temperature T_x ($<T_{1D}$), CDW segments among different chains begin to order along the interchain direction as shown in Figure 7c, where ξ_b is the coherent length along the interchain direction. Below a certain temperature T_c ($<T_x$), CDW formation in each chain is complete and CDWs among different chains are ordered, as shown in Figure 7d. Thus, a long-range order sets in. The coherent lengths ξ_a and ξ_b increase gradually upon lowering the temperature as the extents of long-range order

along the intra- and interchain directions increase. These coherent lengths become infinite when a long-range order is complete in both directions.

Crystal structures are largely determined by single-crystal X-ray diffraction measurements, whose diffraction patterns are recorded in reciprocal space. The temperature dependence of the diffraction pattern of a 1D metal is closely related to what happens to it in real space, as discussed in connection with Figure 7. At $T > T_{1D}$, the diffraction pattern of a 1D metal shows only Bragg peaks, as shown in Figure 8a. At $T_x < T < T_{1D}$, the diffraction pattern shows diffuse lines (or sheets in 3D representation) perpendicular to the k_a direction, which are located at $\pm 2k_f$ from the rows of the Bragg peaks, as shown in Figure 8b. The thickness of the diffuse line is given by ξ_a^{-1}. At $T_c < T < T_x$, the diffuse lines are transformed into diffuse spots (or rods in 3D representation) centered at $(2k_f, q_b)$ and its equivalent positions, as shown in Figure 8c. The thickness of a diffuse spot along the k_a and k_b directions is given by ξ_a^{-1} and ξ_b^{-1}, respectively. Below T_c, a long-range order sets in so that the diffuse spots are converted into superlattice spots, as shown in Figure 8d. The diffuse spot thicknesses along the intra- and interchain directions are inversely proportional to the coherent lengths ξ_a and ξ_b, respectively. Therefore, the diffuse spots become smaller as the extent of long-range order along the two directions increases. They eventually become superlattice spots after a long-range order sets in along both directions.

Superconductivity and Electronic Instability

When the temperature is lowered, a metal may become susceptible to another type of electronic instability, formation of a superconducting state. For a metal to be superconducting, it should avoid the electronic instability toward a metal–insulator transition leading to a CDW or an SDW state associated with a good Fermi surface nesting. In general, the Fermi surface of a 1D metal is well-nested, so a 1D metal rarely undergoes a metal–superconductor transition. As seen from Figures 3c and 4, the Fermi surface nesting of a certain 2D metallic system can be removed by partial oxidation or partial reduction. For example, when half-filled, the CuO_2 layer $x^2 - y^2$ band of a high-temperature copper oxide superconductor has a well-nested Fermi surface (as in Figure 3c) (7). In such a case, the CuO_2 layer is no longer a metal but exhibits an antiferromagnetic state (i.e., an example of an SDW state). The CuO_2 layer shows superconductivity only when some electrons are removed from (8, 9) or added to (10, 11) the $x^2 - y^2$ band so that its Fermi surface loses nesting character (as in Figure 4).

From the viewpoint of one-electron band theory, a superconducting state also involves orbital mixing among the band levels above and below the Fermi level. However, the way this orbital mixing comes about differs considerably from that discussed for CDW and SDW states. Charge carriers

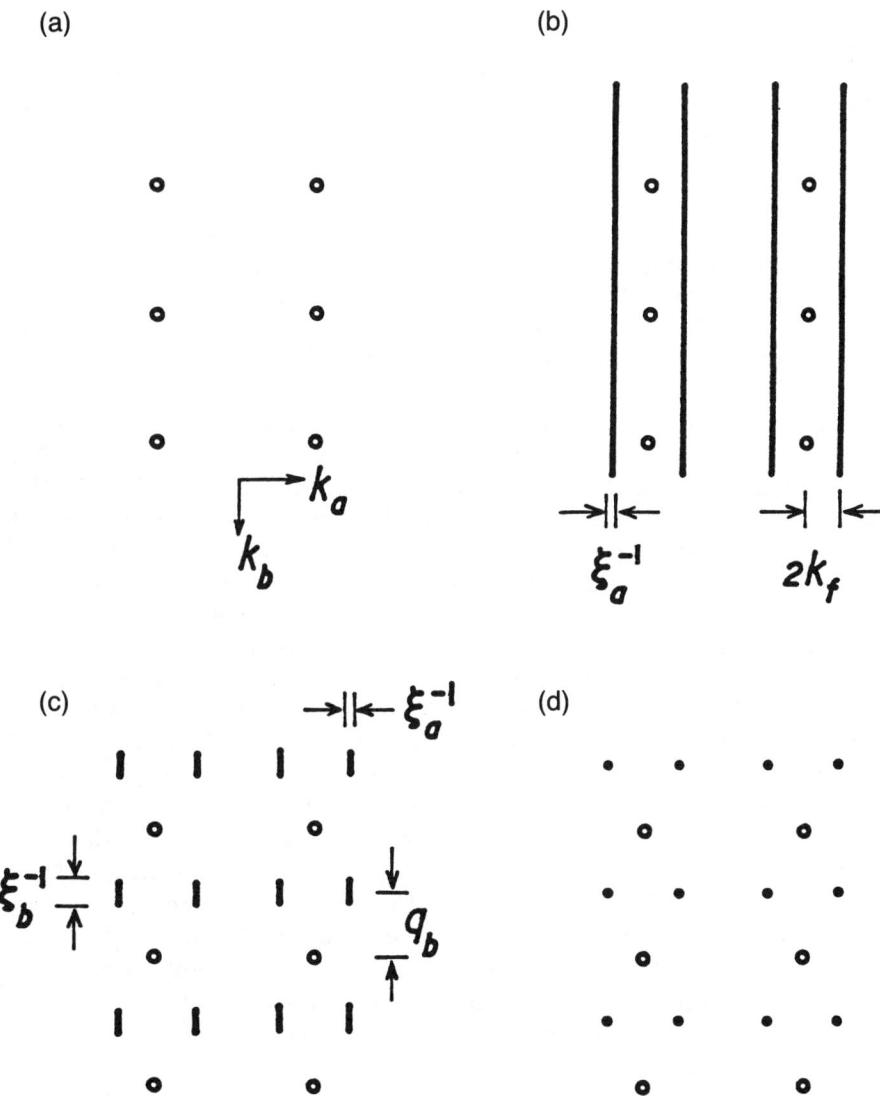

Figure 8. Schematic representation of diffraction patterns of a 1D metal with CDW. (a) Only Bragg peaks are present at $T > T_{1D}$, (b) diffuse lines occur at $T_x < T < T_{1D}$, (c) diffuse lines become diffuse spots at $T_c < T < T_x$, and (d) diffuse spots become superlattice spots at $T < T_c$.

of a superconducting state are not individual electrons, but pairs of electrons (called Cooper pairs) having opposite momenta (i.e., opposite wave vectors) (12–15). Thus, Cooper pairs are not described by individual band orbitals $\phi(k)$ and $\phi(k')$, but by product functions $\phi(k)\phi(-k)$ and $\phi(k')\phi(-k')$, where k and k' refer to occupied and unoccupied wave vectors of a normal metallic state, respectively. The energy lowering that gives rise to superconductivity is induced by the interaction of an occupied pair function $\phi(\mathbf{k})\phi(-\mathbf{k})$ with an unoccupied pair function $\phi(\mathbf{k}')\phi(-\mathbf{k}')$ (i.e., $<\phi(\mathbf{k})\phi(-\mathbf{k})\,|H'|\,\phi(\mathbf{k}')\phi(-\mathbf{k}')>$). The perturbation H' causing this mixing is electron–phonon interaction in traditional superconductors.

As a consequence of the interaction between pair functions, the character of the unoccupied pair function is mixed into that of the occupied pair function. In this indirect way, a superconducting state incorporates unoccupied orbital character into occupied orbital character. Interactions between pair functions $\phi(\mathbf{k})\phi(\mathbf{k})$ and $\phi(\mathbf{k}')\phi(\mathbf{k}')$ introduce an energy gap at the Fermi level. This energy gap is a few multiples of $k_B T_c$ at absolute zero temperature, where T_c refers to the superconducting transition temperature, and gradually decreases to zero at T_c (12–14).

For traditional superconductors described by the BCS theory (12–15), Cooper pair formation is induced by electron–phonon interaction. A moving electron causes a slight, momentary lattice deformation around itself. The deformation affects the motion of a second electron in the wake of the first in such a way that, effectively, the two electrons move as an entity, as if bound together by an attractive force. The extent of electron–phonon coupling is measured by the electron–phonon coupling constant, λ.

Among the several factors affecting the magnitude of T_c (15), the most important one is the electron–phonon coupling constant, λ. In general, T_c increases with increasing λ. For a lattice with atoms of mass M and with a phonon spectrum effective for the electron–phonon coupling, the constant λ varies as $\lambda \propto 1/M<\omega^2>$, where $<\omega^2>$ is the square of the phonon frequencies averaged over the phonon band (14), and the $M<\omega^2>$ term has the dimension of a force constant. Therefore, a large λ results when the lattice has a low-frequency phonon spectrum (i.e., soft phonons arising from vibrations with shallow potential wells). Thus, when the lattice is soft toward the low-frequency phonons crucial for superconductivity, the electron–phonon coupling constant, λ, is large, thereby raising the T_c (16).

For the high-temperature copper oxide superconductors (8–11), no satisfactory theory has been developed yet. Nevertheless, an interesting structure–property relationship is found for the copper oxide hole-superconductors (i.e., those superconductors whose CuO_2 layers have less than half-filled $x^2 - y^2$ bands) by plotting the T_c values against the in-plane Cu–O bond lengths, r_{Cu-O} (17). For every class of the T_c vs. r_{Cu-O} correlations, there exists an optimum in-plane Cu–O bond length, r_{opt}, for which the T_c is a maximum, such that the T_c decreases gradually as r_{Cu-O} either increases or decreases from r_{opt}.

Analysis of the T_c vs. r_{Cu-O} correlations suggests (17) that the strength of the in-plane Cu–O bond, which can be altered either by changing the extent of partial oxidation of the CuO_2 layers or by changing the size of the cations occupying the 9-coordinate sites adjacent to the CuO_2 layers, is an important parameter governing the magnitude of T_c. Namely, there exists an optimum in-plane Cu–O bond strength for which the T_c is a maximum, and the T_c decreases as the in-plane Cu–O bond strength either increases or decreases from the optimum value. This conclusion also applies to the copper oxide electron-superconductors (i.e., those superconductors whose CuO_2 layers have more than half-filled $x^2 - y^2$ bands) (18). For instance, the T_c vs. r_{Cu-O} plot obtained for the various $Eu_{1.85-x}La_xCe_{0.15}CuO_4$ samples with varying x shows a maximum. Substituting La^{3+} for Eu^{3+} does not change the extent of the partial reduction in the CuO_2 layer, but it increases the average r_{Cu-O}, thereby weakening the in-plane Cu–O bond on average, because La^{3+} is greater in ionic size than Eu^{3+}.

Certainly, for a superconducting state to occur, the energy increase associated with the introduction of unoccupied orbital character should be smaller than the energy gain resulting from interaction among pair functions. Depending upon the nature and strength of the perturbations causing orbital mixing, a normal metallic state with nested Fermi surface may in principle lead to a superconducting state when the temperature is lowered. Fermi surface nesting gives rise to a metal–insulator transition such as CDW or SDW formation in most cases, but to a superconducting state when the interaction matrix elements $<\phi(\mathbf{k})|H'|\phi(\mathbf{k}')>$ responsible for CDW or SDW formation are small compared with the interaction matrix elements $<\phi(\mathbf{k})\phi(\mathbf{k})|H'|\phi(\mathbf{k}')\phi(\mathbf{k}')>$ that cause a superconducting state (5, 19). When the relative stabilities of CDW, SDW, and superconducting states are similar, preference of one state over the other is delicately balanced by a change in temperature and pressure.

Concluding Remarks

The electronic structure of a CDW, an SDW, or a superconducting state may be described in terms of orbital mixing between filled and empty levels of a normal metallic state. In contrast to the case of CDW and SDW states, orbital mixing in a superconducting state occurs indirectly via interactions between filled and empty pair functions. Because incorporation of unoccupied orbital character raises energy, only those levels in the vicinity of the Fermi level are crucial for a metal–insulator or metal–superconductor transition. The perturbations causing CDW and SDW states are phonon and on-site repulsion, respectively, and occurrence of these two states require Fermi surface nesting. Our discussion is limited to those electronic states that originate from a normal metallic state, and is appropriate when the on-site repulsion U is small compared with the band width W (e.g., $W = 4\beta_a$ in the 1D band shown in Figure 6) (1, 2, 20–22). When U is greater than W, electrons are localized on lattice sites. It is difficult to describe the low-lying excited states of such a localized state by a band electronic structure theory because the latter is based upon the assumption that electrons are delocalized throughout the lattice. Localized electronic systems are typically examined in terms of model Hamiltonians (e.g., spin and Hubbard Hamiltonians) designed to study their low-lying excited states (21).

Acknowledgment

This work was supported by the U.S. Department of Energy, Office of Basic Energy Sciences, Division of Materials Sciences under Grant DE–FG05–86ER45259.

References

1. Whangbo, M.-H. *J. Chem. Phys.* **1979**, *70*, 4963.
2. Whangbo, M.-H. *J. Chem. Phys.* **1980**, *73*, 3854.
3. Whangbo, M.-H. *J. Chem. Phys.* **1981**, *75*, 4983.
4. Jones, H. *Theory of Brillouin Zones and Electronic States in Crystals;* 2nd ed.; North-Holland: Amsterdam, Netherlands, 1975; pp 43–49.
5. Whangbo, M.-H.; Canadell, E. *Acc. Chem. Res.* **1989**, *22*, 375.
6. Pouget, J. P. In *Low Dimensional Electronic Properties of Molybdenum Bronzes and Oxides;* Schlenker, C., Ed.; Reidel: Dordrecht, Netherlands, 1989; Chapter 3.
7. Whangbo, M.-H.; Evain, M.; Beno, M. A.; Williams, J. M. *Inorg. Chem.* **1987**, *26*, 1829, 1831, 1832.
8. Sleight, A. W. *Science* **1988**, *242*, 1519.
9. Sleight, A. W.; Subramanian, M. A.; Torardi, C. C. *Mater. Res. Soc. Bull.* **1989**, *XIV*, 45.
10. Tokura, Y.; Takagi, H.; Uchida, S. *Nature* **1989**, *337*, 345.
11. James, A. C. W. P.; Zahurak, S. M.; Murphy, D. W. *Nature* **1989**, *338*, 240.
12. Bardeen, J.; Cooper, L. N.; Schrieffer, J. R. *Phys. Rev.* **1957**, *108*, 1175.
13. Grassie, A. D. C. *The Superconducting State;* Sussex University Press: London, 1975; Chapter 2.
14. Solymar, L.; Walsh, D. *Lectures on the Electrical Properties of Materials;* 4th ed.; Oxford University Press: Oxford, England, 1988; Chapter 14.
15. McMillan, W. L. *Phys. Rev.* **1968**, *167*, 331.
16. Whangbo, M.-H.; Williams, J. M.; Schultz, A. J.; Emge, T. J.; Beno, M. A. *J. Am. Chem. Soc.* **1987**, *109*, 90.
17. Whangbo, M.-H.; Kang, D. B.; Torardi, C. C. *Physica C*, **1989**, *158*, 371.
18. Wang, E.; Tarascon, J.-M.; Greene, L. H.; Hull, G. W. *Phys. Rev. B.*, in press.
19. Whangbo, M.-H.; Canadell, E. *J. Am. Chem. Soc.* **1988**, *110*, 358.
20. Whangbo, M.-H. *Acc. Chem. Res.* **1983**, *16*, 95.
21. Brandow, B. H. *Adv. Phys.* **1977**, *26*, 651.
22. Mott, N. F. *Metal Insulator Transitions;* Barnes and Noble: New York, 1977.

RECEIVED for review June 6, 1989. ACCEPTED revised manuscript August 7, 1989.

16

Ceramic Superconductors

Single-Valent versus Mixed-Valent Oxides

John B. Goodenough

Center for Materials Science and Engineering, University of Texas at Austin, Austin, TX 78712-1084

The problem of electronic transport in oxides is reviewed briefly. The situation in the main-group oxides is contrasted with that in the rare-earth oxides. In $Ba_{1-x}K_xBiO_3$, for example, introduction of mixed valence allows superconductivity to compete with the charge-density wave stable in the "single-valent" stoichiometry; and in ferromagnetic EuO, oxidation leads to small-polaron conduction, but reduction gives a metal–semiconductor transition below the Curie temperature. The d-block transition-metal oxides are of intermediate character; in these oxides, either localized magnetic moments or charge-density waves may compete with superconductivity. However, an abrupt transition from antiferromagnetic order to high-critical-temperature (T_c) superconductivity, such as that found in some copper oxides, is not observed in oxides where a single-band model is clearly applicable. The special features that allow for a two-band model in the high-T_c oxide superconductors are emphasized. In addition, attention is called to a thermal-expansion mismatch between A–O and B–O bonds in structures related to the ABO_3 perovskites, and the relationship of the electronic and structural features is pointed out. These special features provide the necessary framework for a strong-coupling theory of superconductivity.

CERAMISTS HAVE TRADITIONALLY BEEN MORE INTERESTED in refractory insulators than in metallic oxides. Consequently, the 1986 report of high-T_c (critical temperature) superconductivity in an oxide appeared to many to be an especially extraordinary finding, even though several oxides were already known to be superconductors.

0065-2393/90/0226-0287$09.75/0
© 1990 American Chemical Society

To understand metallic conductivity in oxides, it is useful to consider separately three different types of atomic outer electrons: the valence s and p electrons active in bonding; the rare-earth $4f^n$ manifolds that everywhere remain localized, interacting only weakly with neighboring atoms; and the transition-metal d electrons. The d electrons are of intermediate character; in some oxides they act like valence electrons, in others like localized electrons, and in a few they exhibit a transitional character. The latter situation occurs in the high-T_c copper oxide superconductors; they exhibit a transition from an antiferromagnetic semiconductor in the single-valent state to superconductivity in a mixed-valent state. However, superconductivity gives way to normal metallic behavior after a relatively narrow compositional range. Any model of the superconductive properties of these copper oxides must address the question of why superconductivity occurs in only a narrow compositional range.

Main-Group Oxides

Most main-group oxides remain refractory insulators; attempts to render them electronic conductors by chemical doping tend to be frustrated by the spontaneous incorporation of compensating native defects. The problem is well illustrated by the binary compound MgO, which has the rock-salt structure.

Construction of the electronic energies on the basis of an ionic model is illustrated in Figure 1. The $O^{-/2-}$ redox energy for the free ion lies above the energy of vacuum (E_{vac}); the formation of a free O^{2-} ion requires overcoming a negative electron affinity. The $Mg^{2+/+}$ redox energy lies below E_{vac} by the second ionization energy of magnesium. E_I is the energy required to remove the remaining 3s electron from Mg^+ to an O^- ion at infinite distance to create free Mg^{2+} and O^{2-} ions. Assembly of these ions into the rock-salt structure of MgO leads to a gain of electrostatic Madelung energy $E_M > E_I$, which stabilizes the compound. Conservation of energy lowers the $O^{-/2-}$ and raises the $Mg^{2+/+}$ redox energies by the same amount, and an $E_M > E_I$ produces a crossover of the two energies. Introduction of covalent hybridization lowers the effective charge on the ions from their value in a point-charge model, which lowers the splitting $E_M - E_I$. However, this lowering is compensated by the repulsion between bonding and antibonding states introduced by this hybridization. Therefore the point-charge model gives a good zero-order approximation for the binding energy of the crystal if appropriate core–core short-range repulsive energies are introduced. Finally, the translational symmetry of the crystal introduces a broadening of the energy levels into energy bands, but the bandwidths remain narrow enough to have a large energy gap (E_g) between a filled valence band of primarily O-2p character and an empty conduction band of primarily

16. GOODENOUGH *Ceramic Superconductors* 289

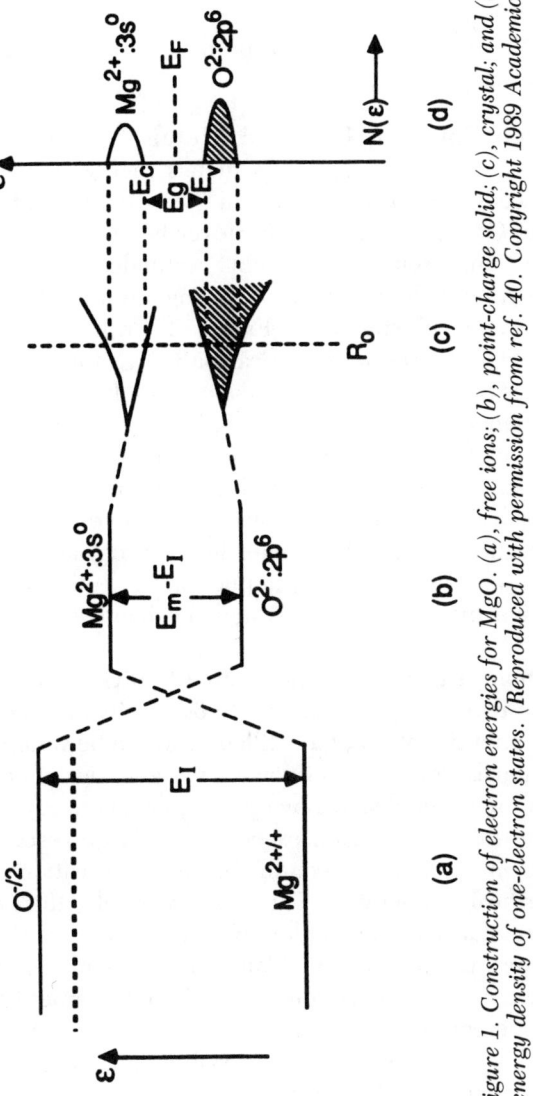

Figure 1. Construction of electron energies for MgO. (a), free ions; (b), point-charge solid; (c), crystal; and (d), energy density of one-electron states. (Reproduced with permission from ref. 40. Copyright 1989 Academic.)

Mg-3s character. These bands are therefore given the formal identifications O^{2-}:$2p^6$ and Mg^{2+}:$3s$.

In MgO, the energy gap E_g is sufficiently large and the band edges are so positioned that the Mg^{2+}:$3s$ conduction band is energetically inaccessible to electrons and the O^{2-}:$2p^6$ band remains inaccessible to holes. Chemical doping that would raise the Fermi energy (E_F) into the conduction band or would lower it into the valence band produces native defects that maintain E_F within the gap.

Conduction Bands of Group B Metals. On the other hand, the conduction bands of heavy Group B metals in higher valence states are energetically accessible. Although the Sn^{4+}:$5s$ conduction band of SnO_2 lies 3 eV above the O^{2-}:$2p^6$ valence band, nevertheless it can be doped to give a metal that is transparent, and the PbO_2 cathode of a lead–acid battery is metallic even without the injection of hydrogen on discharge.

The situation is illustrated in Figure 2 for three cubic perovskites: $BaSnO_3$, $BaPbO_3$, and the hypothetical $BaBiO_3$. Stoichiometric $BaSnO_3$ is a semiconductor because the Sn^{4+}:$5s$ conduction band lies discretely above the O^{2-}:$2p^6$ valence band, as in SnO_2; $BaPbO_3$ is a semimetal because the Pb^{4+}:$6s$ and O^{2-}:$2p^6$ bands overlap, as in PbO_2. Cubic $BaBiO_3$ would have a greater overlap of the "Bi^{4+}:$6s$" and O^{2-}:$2p^6$ bands, but the conduction band would be half-filled, thereby raising E_F above the top of the valence band. Clearly, an ionic model becomes less appropriate for these heavy B-metal oxides where a small $E_M - E_I$ implies strong covalent mixing between the metal-6s and oxygen-2p orbitals and the bandwidths cause E_g to disappear.

The situation in the hypothetical cubic $BaBiO_3$ must be distinguished from the other two cases because the Bi:6s conduction band is half-filled. The electronic energies of a partially filled band can be stabilized by a change in the translational symmetry that lowers the energies of occupied states at the expense of unoccupied states by opening up an energy gap at the Fermi surface. Where this energy gain exceeds the elastic restoring energy that favors higher symmetry, a semiconductor–metal transition can be expected below some critical temperature T_t (1). Alternatively, the electron–phonon interactions can stabilize the condensation of Cooper pairs below a superconductive critical temperature T_c. Thus, the lattice instabilities responsible for semiconductor–metal transitions below T_t are competitive with the stabilization of superconductivity.

Chemists know that Bi^{4+} is not a stable valence state; it is unstable against the disproportionation reaction

$$2\ Bi^{4+} \rightarrow Bi^{3+} + Bi^{5+} \qquad (1)$$

Thus, it comes as no surprise that cubic $BaBiO_3$ does not exist at room temperature. As illustrated in Figure 3, the oxygen atoms are cooperatively

16. GOODENOUGH *Ceramic Superconductors*

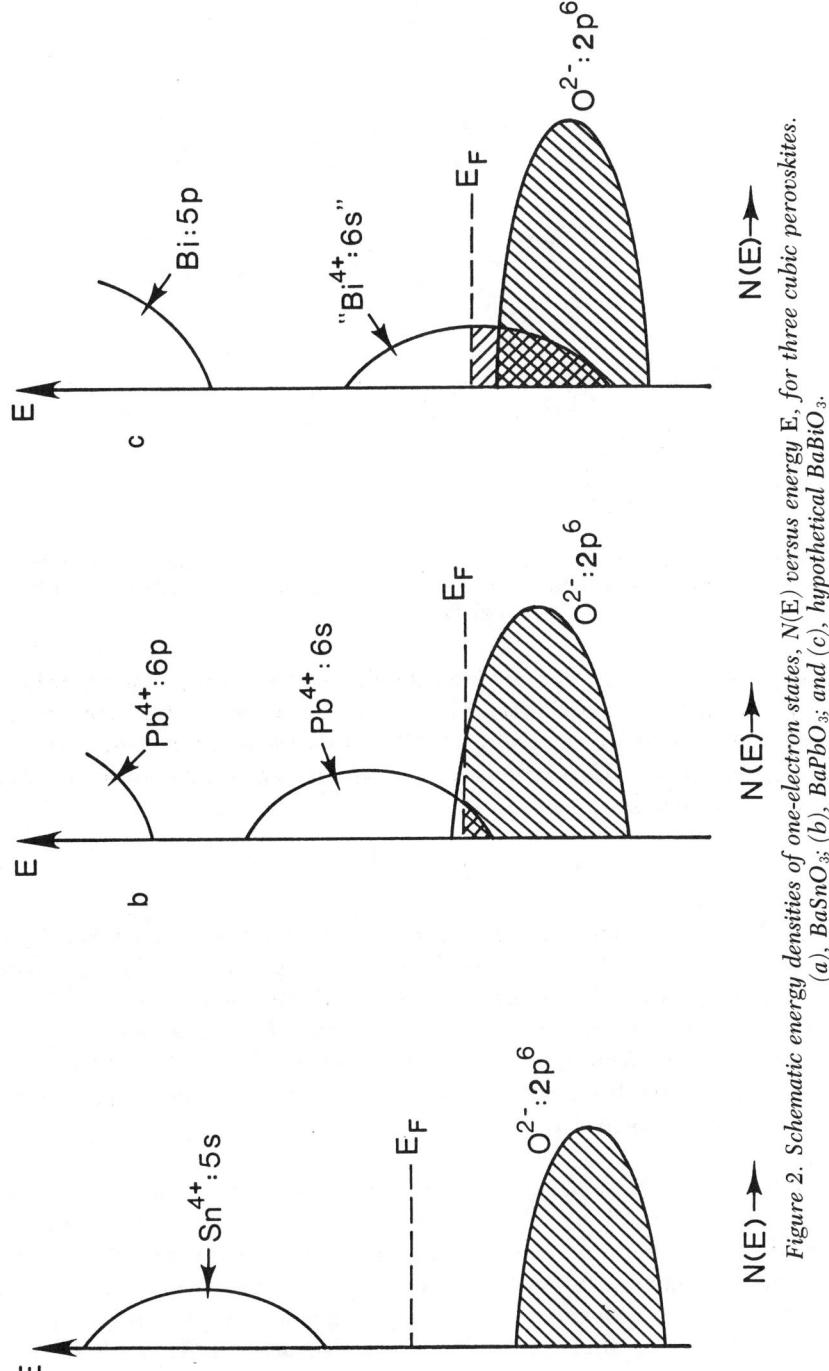

Figure 2. *Schematic energy densities of one-electron states, N(E) versus energy E, for three cubic perovskites. (a), $BaSnO_3$; (b), $BaPbO_3$; and (c), hypothetical $BaBiO_3$.*

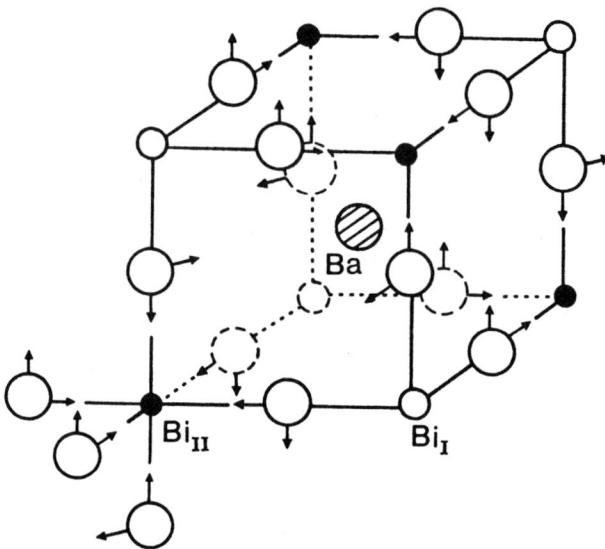

Figure 3. Oxygen-atom displacements from ideal-perovskite positions in monoclinic $BaBiO_3$. (Reproduced with permission from ref. 41. Copyright 1989 Materials Resource Society.)

displaced so as to create two distinguishable Bi atoms, a Bi_I with larger Bi–O distances and a Bi_{II} with shorter Bi–O distances. There is also a cooperative tilting of the octahedra relative to a [100] axis that lowers the symmetry from cubic to monoclinic. These displacements clearly signal a disproportionation reaction that may be described as

$$2\ Bi^{4+} \rightarrow Bi_I^{(4-\delta)+} + Bi_{II}^{(4+\delta)+} \qquad (2)$$

where $\delta = 1$ would correspond to a classic disproportionation into Bi^{3+} and Bi^{5+}. In fact, a $\delta \sim 0.5$ is found (2); a $\delta < 1$ represents a charge-density wave (CDW). No matter what the value of δ, the creation of two distinguishable Bi sites by the freezing out of a vibrational breathing mode changes the translational symmetry of the crystal so as to split the conduction band in two, leaving the Fermi energy E_F in an energy gap. $BaBiO_3$ is thereby rendered a semiconductor.

Polar-State Stabilization. Stabilization of a polar state on the Bi-atom array is possible only because the electrostatic correlation energy U between Bi-6s electrons, which is the energy cost to create a nonuniform distribution of charge, is smaller than the gain Δ in covalent-hybridization energy associated with the motion of the oxygen atoms toward one Bi atom and away from the other (1). In the physics literature, a $\Delta > U$ is referred to as a "negative U" (i.e., a $U_{\text{eff}} = U - \Delta < 0$). This condition can be stable only where the free-atom U is relatively small, as for the Bi-6s electrons, and even there only for a range of bandwidths, as is illustrated schematically in Figure 4.

If K is substituted for Ba in $Ba_{1-x}K_xBiO_3$, the BiO_3 framework becomes oxidized. As the Bi-6s conduction band becomes less than half-filled, the wavelength q of the CDW that opens up a gap at E_F increases and the stabilization energy of the CDW decreases, lowering the semiconductor–metal transition temperature T_t (1). In fact, q need not remain commensurate with the periodicity of the crystal structure, and the $Ba_{1-x}K_xBiO_3$ system exhibits a monoclinic–tetragonal transition with increasing x that probably represents a transition from a commensurate to an incommensurate CDW (3–5). As the K-substituted phases are difficult to synthesize, the phase diagram of Figure 5 must be considered tentative. A single crystal with $x = 0.13$ was found to be cubic at room temperature and to exhibit no magnetic anomaly characteristic of superconductivity down to lowest temperature (6), but it was inadvertently destroyed before a T_t could be measured in the sample. However, near $x = 0.25$ there is an abrupt onset of superconductivity with a T_c that decreases monotonically with increasing x over the narrow compositional interval $0.25 < x < 0.5$. A two-phase region probably separates the CDW and superconductor compositions, but this feature has not yet been established.

Anomalous Superconductivity. Although the $Ba_{1-x}K_xBiO_3$ superconductors appear to be described by the Bardeen–Cooper–Schrieffer (BCS) weak-coupling theory (7), they exhibit a T_c that is higher than expected from that theory for the measured density $N(E_F)$ of one-electron states at the Fermi energy. Therefore, this system is classified with the high-T_c copper oxides as representative of anomalous superconductivity. In this respect, it differs from the other known oxide superconductors such as TiO and $Li[T_2]O_4$. Moreover, confinement of superconductivity to a narrow compositional range (i.e., a narrow electron–atom ratio) is a striking feature that has not yet been addressed theoretically.

Examination of Figure 2(c) is instructive. Oxidation of the BiO_3 array lowers E_F toward the top of the O-$2p_\pi$ band; experimental evidence that E_F lies within the O-$2p_\pi$ band in the superconductor composition $x = 0.4$ is now available (8). A distinguishing feature of a high-T_c oxide superconductor is an E_F that cuts—or nearly cuts—two bands, a π or π^* band of primarily O-$2p_\pi$ character and a σ^* band of primarily cationic (Bi-6s or Cu-3d) character. In the other oxide superconductors, such as TiO or $Li[Ti_2]O_4$, E_F cuts a single cationic d band. The possible significance of this distinguishing feature will be discussed in a final speculative comment.

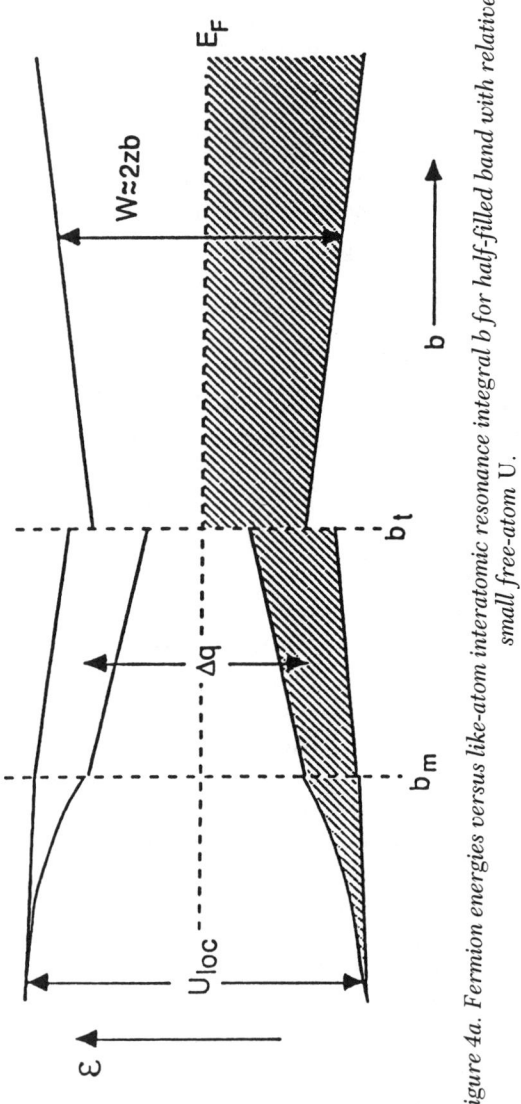

Figure 4a. Fermion energies versus like-atom interatomic resonance integral b for half-filled band with relatively small free-atom U.

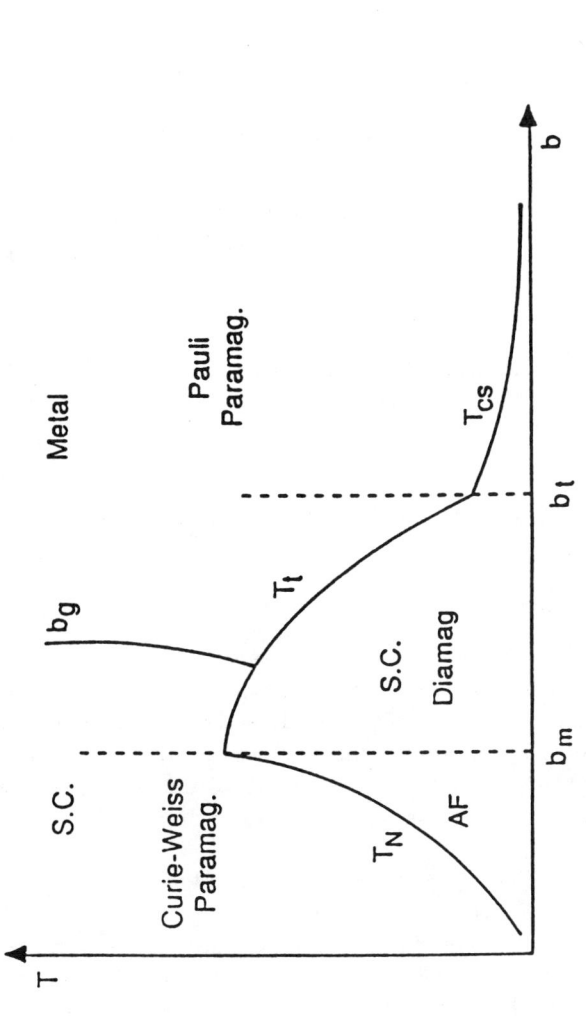

Figure 4b. *Transition temperatures versus like-atom interatomic resonance integral b for half-filled band with relatively small free-atom U. (Reproduced with permission from ref. 42. Copyright 1987 Pennsylvania State University.)*

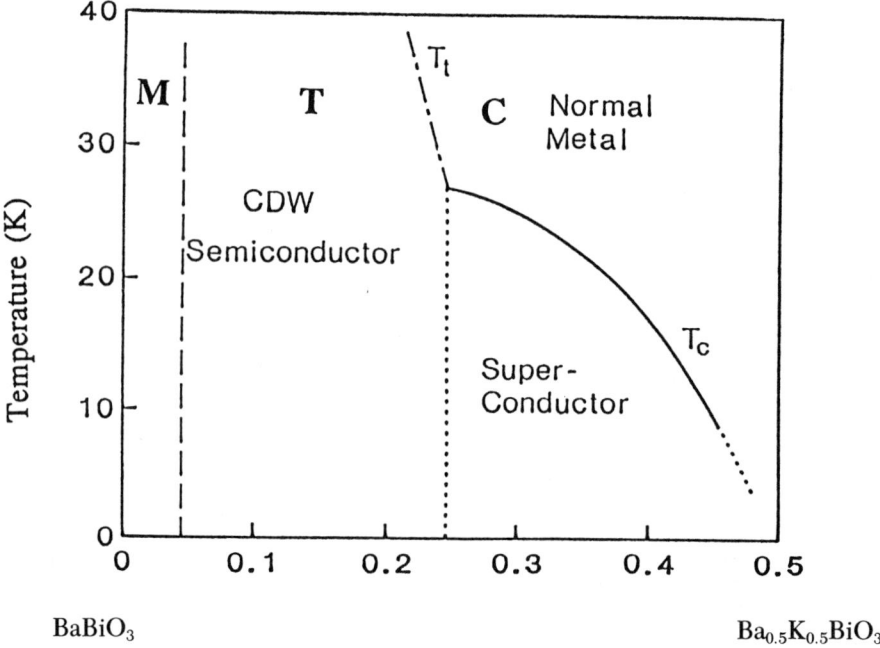

Figure 5. *Tentative phase diagram for the system* $Ba_{1-x}K_xBiO_3$.

Rare-Earth Oxides

Electronic Structure. The 4f electrons of the rare-earth ions are tightly bound to their atomic nucleus and are screened from interactions with neighboring atoms by the $5s^2 5p^6$ closed-shell core electrons. As a result, the intraatomic interactions among the 4f electrons of the $4f^n$ configuration are much larger than the interatomic 4f–4f interactions. Consequently, the Hamiltonian describing a $4f^n$ configuration has the form

$$H = H_o + V_{el} + \Delta_{LS} + \Delta_{cf} + H_z \tag{3}$$

where H_o is the spherical approximation and the successive terms represent successive perturbation corrections to this approximation. The first three terms, which include the electrostatic covalent interactions between the 4f electrons (V_{el}) and the spin-orbit coupling ($\Delta_{LS} = \lambda \mathbf{L} \cdot \mathbf{S}$) responsible for multiplet splittings, are present in the free atom and give rise to an atomic magnetic moment

$$\mu_J = gJ\mu_B \tag{4}$$

where g is the Landé spectroscopic splitting factor, J is the total angular momentum quantum number, and μ_B is the Bohr magneton. The energy V_{el} gives rise to the splittings U_n between successive $4f^n$ configurations (i.e., $4f^n$ and $4f^{n+1}$). Because the radial extension of the 4f electrons is relatively small, the energies U_n are large; $U_n > E_g$, where E_g is the energy gap between the top of the O-2p valence band and the bottom of the Ln-5d conduction band. (Stronger covalent mixing of the O-2p orbitals with Ln-6s versus Ln-5d orbitals raises the bottom of the Ln-6s band above the bottom of the Ln-5d band.) A $U_n > E_g$ restricts the number of energetically available $4f^n$ configurations to one or, at most, two; two are accessible only if a $4f^n$ configuration has its energy falling within E_g.

The last two terms in eq 3 are the crystal-field splitting Δ_{cf}, which primarily results from Ln-4f interactions with the nearest-neighbor oxide ions, and the Zeeman splitting H_z attributed to the internal magnetic fields produced by interatomic magnetic-exchange interactions with nearest-neighbor magnetic atoms. Because Δ_{cf} and H_z are both weaker than Δ_{LS}, the effective atomic magnetic moment μ_{eff} obtained from magnetic-susceptibility measurements provides an excellent measure of the $4f^n$ configuration at a rare-earth ion. Any 5d electrons present act like valence electrons; therefore, the "valence" of a rare-earth ion is generally defined as the total number of outer electrons on the atom minus the number n of the ground-state $4f^n$ configuration.

Characteristics. Some consequences of these considerations are illustrated in Figure 6 for GdO and EuO, each with the same rock-salt structure as MgO. The valence $O^{2-}:2p^6$ and conduction Ln:5d bands are constructed in a manner analogous to the construction of the valence and conduction bands of MgO, but with differential covalent mixing of O-2p with Ln-6s and 5d orbitals rendering the bottom of the conduction band Ln:5d in character. The splitting U between $4f^8$ and $4f^7$ energy levels is particularly large because the $4f^7$ configuration just completes the half-shell. The $4f^7$ level for Gd falls well below the top of the valence band, and the $4f^8$ energy lies well above the bottom of the conduction band. Therefore, only the $4f^7$ level can be occupied, and the Gd has the formal valence Gd^{3+} according to the magnetic criterion. However, each Gd can transfer at most two electrons to the oxygen, so the third valence electron occupies the Gd-5d conduction band, and GdO is a metallic antiferromagnet, the interatomic antiferromagnetic exchange being indirect through the itinerant 5d electrons. In EuO, on the other hand, the $Eu^{2+}:4f^7$ configuration lies in an energy gap about 1.1 eV below the bottom of the Eu-5d conduction band. Therefore, EuO is a semiconductor that can be doped either n-type or p-type (negative or positive charge charriers) because both the Eu-5d and Eu-$4f^7$ energies are accessible (9). Moreover, the dominant interatomic magnetic exchange is ferromagnetic via a Eu:$4f^7$–Eu:5d superexchange interaction, and EuO is a ferromagnet at low temperature.

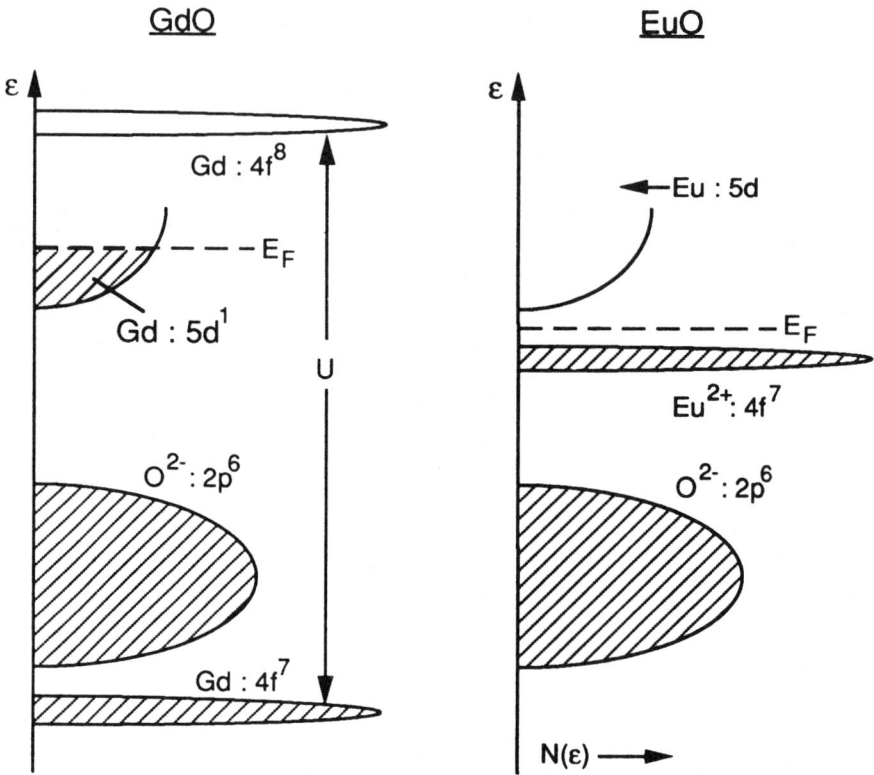

Figure 6. Schematic of $4f^n$ manifold energies relative to the energies of the conduction and valence bands of GdO and EuO.

Oxidation of EuO to $Eu_{1-\delta}O$ introduces cation vacancies and lowers E_F into the narrow $Eu^{2+}:4f^7$ level, which represents the $Eu^{3+/2+}$ redox energy. Because the 4f electrons are localized, the electron-transfer time τ_h for the reaction

$$Eu^{3+} + Eu^{2+} \leftrightarrows Eu^{2+} + Eu^{3+} \quad (5)$$

is slow relative to the period ω_R^{-1} of an optical-mode vibration ($\tau_h > \omega_R^{-1}$), which means that the empty $4f^7$ energy at a Eu^{3+} ion is raised above the occupied $4f^7$ energy at a Eu^{2+} ion by a local reorganization energy. The situation is completely analogous to the redox energies of an ion in solution, except that the periodicity of the crystal structure requires a configuration mobility that averages the electron occupancy at a given Eu atom over time. The mobile electron (or hole in this case) and its local deformation is called a "small polaron". Because it has a diffusive motion, its drift mobility is given

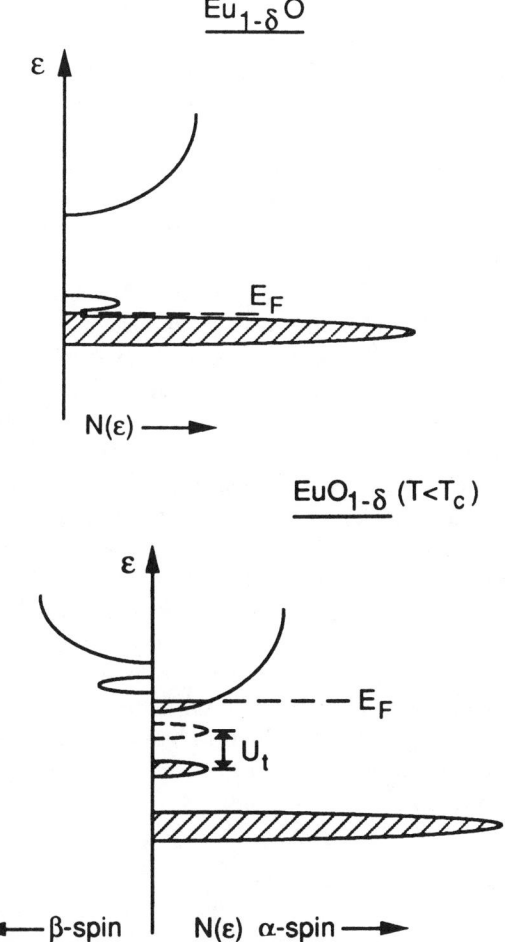

Figure 6. Continued. Schematic of $4f^n$ manifold energies relative to the energies of the conduction and valence bands of $Eu_{1-\delta}O$ and $EuO_{1-\delta}$.

by the Einstein relation

$$\mu = \frac{e_o D}{kT} = \left(\frac{e_o D_o}{kT}\right) \exp\left(\frac{-\Delta G_m}{kT}\right) \qquad (6)$$

where e_o is the magnitude of the electronic charge and $\Delta G_m = \Delta H_m - T \Delta S_m$ is the motional free energy. The motional enthalpy ΔH_m is a measure of the magnitude of the local reorganization energy $T \Delta S_m$. The electronic conductivity is related to the electron mobility via the relation.

$$\sigma = p e_o \mu \qquad (7)$$

where p is the concentration of mobile holes. It follows that the conductivity of $Eu_{1-\delta}O$ retains an activation energy through μ, even where p is independent of temperature, so the mixed-valent system also exhibits semiconductor behavior.

Reduction of EuO to $EuO_{1-\delta}$ produces oxygen vacancies V_O that stabilize a two-electron trap state from out of the conduction band (10). This situation differs from phosphor-doped silicon, where the donor state is a single-electron trap. Because the trap states are condensed out of the conduction band, their energies ride with the bottom of the conduction band. On cooling through the ferromagnetic Curie temperature T_c, states with spins parallel to the spin of the $4f^7$ configuration (α spin direction) have their energies shifted relative to the states of opposite spin (β spin direction) by $2\mu_B H_i$, where H_i is the internal molecular field resulting from the interatomic magnetic-exchange interactions. The two electrons trapped at an oxygen vacancy have opposite spins, so one rides with the bottom of the α-spin band and the other with the bottom of the β-spin band. The energy of the α-spin donor state is lowered by U_t in a single-electron trap. As the spontaneous magnetization M_s increases below T_c, H_i increases; at a critical value of M_s the β-spin donor energy crosses the bottom of the α-spin conduction band. As the temperature continues to decrease, the β-spin donor electrons are emptied into the α-spin conduction band. These conduction electrons increase T_c, thereby increasing M_s, which produces a positive feedback that results in a sharp semiconductor-to-metal transition with decreasing temperature. The electrons in the conduction band are itinerant ($\tau_h < \omega_R^{-1}$), so there is no local reorganization to trap them. Therefore, their mobility is given by the expression

$$\mu = \frac{e_o \tau_s}{m^*} \tag{8}$$

where m^* is the electron effective mass and τ_s is the mean free time between scattering events. With a fixed mobile-electron concentration and a mobility that decreases with increasing temperature because scattering from phonons increases in probability with temperature, the electronic conductivity has a metallic character.

Transition-Metal (d-Block) Oxides

A description of the outer d electrons at transition-metal cations in an oxide begins with the recognition that the d orbitals interact much more strongly with neighboring atoms than the 4f orbitals, but not as strongly as the outer s and p orbitals. Whereas the interatomic interactions give a bandwidth $W > U$ for the s and p electrons, but a $(\Delta_{cf} + H_z) < \Delta_{LS}$ for the 4f electrons, the d electrons are of intermediate character; they have a cubic component

Δ_c of the crystalline-field splitting Δ_{cf} and a bandwidth W resulting from interactions with other transition-metal atoms that may be comparable to or even greater than the V_{el} term responsible for the interatomic correlation splitting U between d^n and d^{n+1} manifolds. It is therefore generally useful to begin with the influence of the cubic component of the crystalline field Δ_c attributable to interactions with the nearest-neighbor oxide ions as expressed by the Hamiltonian

$$H = H_o + (V_{el} + \Delta_c) \qquad (9)$$

By symmetry, the cubic component of the crystalline field splits the fivefold degeneracy of the d orbitals into two accidentally degenerate groups, the twofold-degenerate e orbitals having the angular dependences

$$f_e \sim \frac{1}{\sqrt{2}} \frac{(z^2 - x^2) + (z^2 - y^2)}{r^2} \quad \text{and} \quad \frac{(x^2 - y^2)}{r^2} \qquad (10)$$

and the threefold-degenerate t_2 orbitals having the angular dependences

$$f_t \sim \frac{yz \pm izx}{r^2} \quad \text{and} \quad \frac{xy}{r^2} \qquad (11)$$

This splitting quenches the azimuthal orbital angular momentum $L_z = -i\hbar \, \partial/\partial\phi$ associated with the atomic d orbitals $[(x^2 - y^2) \pm ixy]/r^2$. Covalent hybridization is introduced by allowing a virtual electron transfer from an O^{2-} ion to an empty d orbital; from second-order perturbation theory, the covalent-mixing parameters are given by

$$\lambda \approx \frac{b^{ca}}{\Delta E} \qquad (12)$$

where b^{ca} is the cation–anion resonance integral (electron-transfer-energy matrix element) and ΔE is the energy separation of the acceptor and donor orbitals of the charge transfer. The result of this covalent hybridization for octahedral-site coordination is (3)

$$\Delta_c \approx \Delta_M + \tfrac{1}{2}(\lambda_\sigma^2 - \lambda_\pi^2)\Delta E_1 + \tfrac{1}{2}\lambda_s^2 \Delta E_2 \qquad (13)$$

where Δ_M is a relatively small electrostatic component, ΔE_1 and ΔE_2 are the energy separations of the empty cation-d^{n+1} configuration and the O-2p donor and the O-2s donor orbitals. The subscripts σ, π, and s refer to mixing with O-2p$_\sigma$, O-2p$_\pi$, and O-2s orbitals, respectively. The crystal-field wave functions having e-orbital and t_2-orbital symmetry in octahedral-site coordination are

$$\psi_e = N_\sigma(f_e - \lambda_s\phi_s - \lambda_\sigma\phi_\sigma) \tag{14a}$$

$$\psi_t = N_\pi(f_t - \lambda_\pi\phi_\pi) \tag{14b}$$

where the ϕ_i are the appropriately symmetrized O-2s, O-2p_σ, and O-2p_π orbitals; N_σ and N_π are normalization constants. Extension of the d wave functions out over the anions via the covalent mixing accomplishes two important things: it reduces the parameter U from its value for the free ion to that more nearly given by successive redox potentials for the hydrated cation in aqueous solutions, and it allows important interatomic cation–anion–cation interactions to occur.

Intraatomic Energy Considerations. The next step in the problem is to consider the relative magnitudes of Δ_c and the intraatomic-exchange component Δ_{ex} in V_{el} for high-spin configurations. A high-spin configuration ($\Delta_c < \Delta_{ex}$) is found only where the d orbitals are sufficiently localized to impart a localized magnetic moment to the transition-metal cation.

Once these intraatomic energy considerations have been obtained, it is then necessary to consider the interatomic interactions between like atoms, which introduce a bandwidth W to the configuration d^n. If the condition $W < U$ holds, then the d^n configuration is localized and imparts a localized magnetic moment to the transition-metal atoms. In the limit $W < \Delta_{LS} < U$, the interatomic exchange interactions in a single-valent compound are treated by second-order perturbation theory; this treatment gives rise to the superexchange spin–spin interaction. For half-filled orbitals, this latter interaction is antiferromagnetic; the corresponding magnetic-ordering (Néel) temperature is

$$T_N \sim \frac{b^2}{U_{eff}} \tag{15}$$

where $b \sim \epsilon\lambda^2$ for cation–anion–cation interactions and $b \sim \epsilon \exp(-R/\rho)$ for cation–cation interactions. In these expressions, ϵ is a one-electron energy, R is the interatomic separation, and ρ is a parameter in units of R. The energy

$$U_{eff} = U' + \begin{cases} 0 \\ \Delta_{ex} \\ \Delta_c \end{cases} \tag{16}$$

includes any contribution from Δ_c or Δ_{ex} that must be added for a particular value of n in the d^n configuration (*11*).

Localized-Electron Configuration. This localized-electron picture applies to the $3d^5$ configuration of a Mn^{2+} ion in the antiferromagnetic insulator MnO, and construction of an energy diagram proceeds as outlined in Figure 7. In this case, the Madelung stabilization raises the $Mn^{2+}:3d^5$ configuration, as well as the $Mn^{2+}:4s$ energies, above the top of the $O^{2-}:2p^6$ band. The energy separation U of the $3d^6$ and $3d^5$ configuration is large because $U_{eff} = U' + \Delta_{ex}$ applies on adding an electron to a set of half-filled orbitals, as is found in the high-spin $d^5 = t_2^3e^2$ configuration. On the other hand, the energy separation between the $3d^5$ and $3d^4$ configurations is only $U_{eff} = U'$, which is relatively small for the crystal-field ψ_e orbitals. Thus, both the $Mn^{3+/2+}$ and $Mn^{4+/3+}$ redox energies lie above the top of the $O^{2-}:2p^6$ valence band, so the three valence states Mn^{2+}, Mn^{3+}, and Mn^{4+} are chemically accessible in oxides (*see also* Figure 8). However, a $U_{eff} = U' + \Delta_c$ lowers the $3d^3$ configuration below the top of the $O^{2-}:2p^6$ band, so it is not possible to oxidize manganese beyond Mn^{4+} in an octahedral interstice of an oxide. The relatively small U' separating the $3d^4$ and $3d^5$ configurations makes it possible for a "negative-U" disproportionation to occur.

$$2 Mn^{3+} \rightarrow Mn^{2+} + Mn^{4+} \qquad (17)$$

However, in practice this reaction is only found at the surface of a manganese oxide; for example, in strong acidic media it allows $Li[Mn_2]O_4$ to be transformed to λ-MnO_2 (*12*).

The superconductor TiO corresponds to a single-valent compound having $W > U'$, where $U_{eff} = U'$ for a $3d^2$ configuration in an octahedral interstice. For a $W > U'$, the 3d electrons are itinerant, and tight-binding band theory must be applied to the electrons in the crystal-field ψ_t orbitals. The resulting density-of-states curve $N(E)$ for TiO (we neglect the spontaneous introduction of atomic vacancies in the structure) is compared to that of MnO in Figure 8. By contrast, VO contains a half-filled set of t_2 orbitals ($V^{2+}:t_2^3e^0$), so $U_{eff} = U' + \Delta_c$ is larger. Nevertheless, the bandwidths of the $V^{3+/2+}$ and $V^{4+/3+}$ redox energies overlap ($W \geq U$), so the compound is semimetallic with a deep minimum in the $N(E)$ versus E curve at E_F. In this case, the correlation splitting U is large enough to introduce an important enhancement of the Pauli paramagnetic susceptibility, but the compound remains metallic with no vanadium magnetic moments, and hence no antiferromagnetic order, at low temperatures.

For a half-filled band, these considerations lead to the phase diagram of Figure 9, which shows a transition from an antiferromagnetic semiconductor for $b < b_g$ (i.e., $W < U$), where there are localized magnetic moments on the cations, to a Pauli-paramagnetic metal for $b > b_g$ (i.e., $W > U$), where the existence of a Fermi surface at E_F allows for a transition to superconductivity below a T_{cs} given by the weak-coupling BCS theory. It is immediately apparent from this figure that, with a single-band model, the

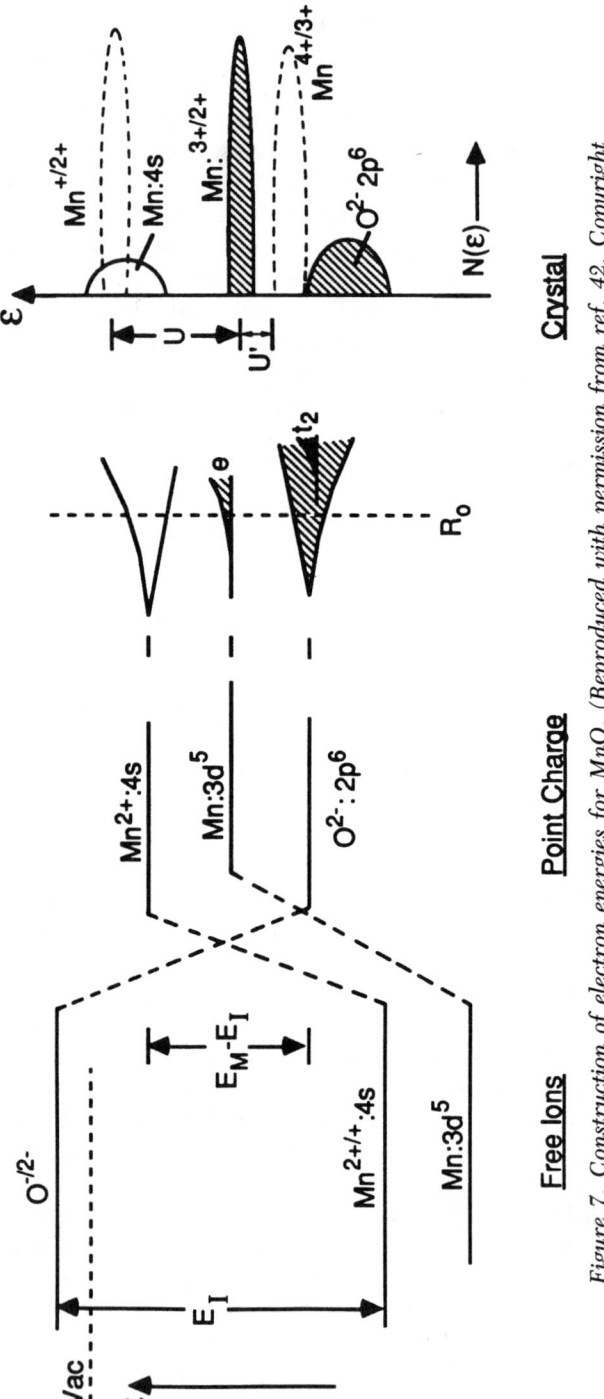

Figure 7. Construction of electron energies for MnO. (Reproduced with permission from ref. 42. Copyright 1987 Pennsylvania State University.)

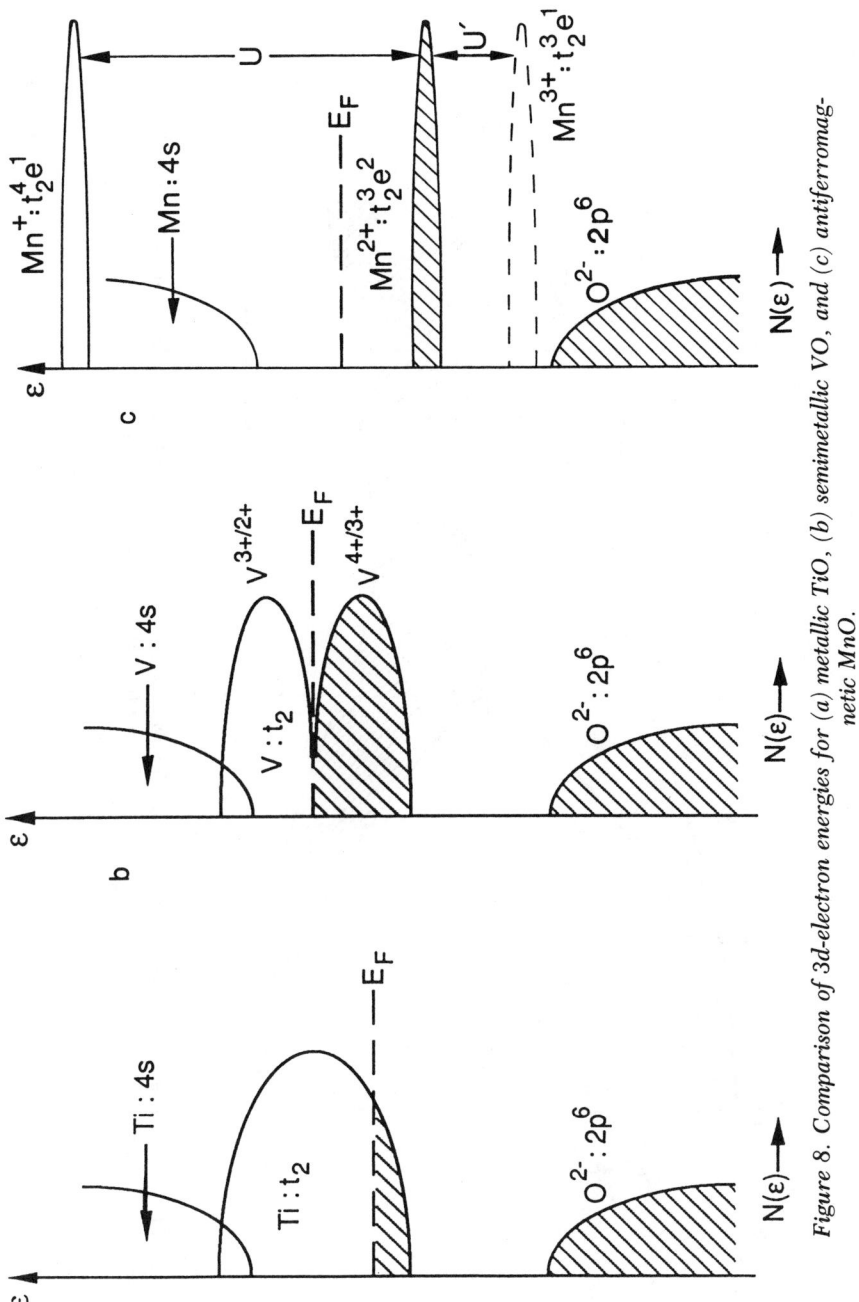

Figure 8. Comparison of 3d-electron energies for (a) metallic VO, (b) semimetallic VO, and (c) antiferromagnetic MnO.

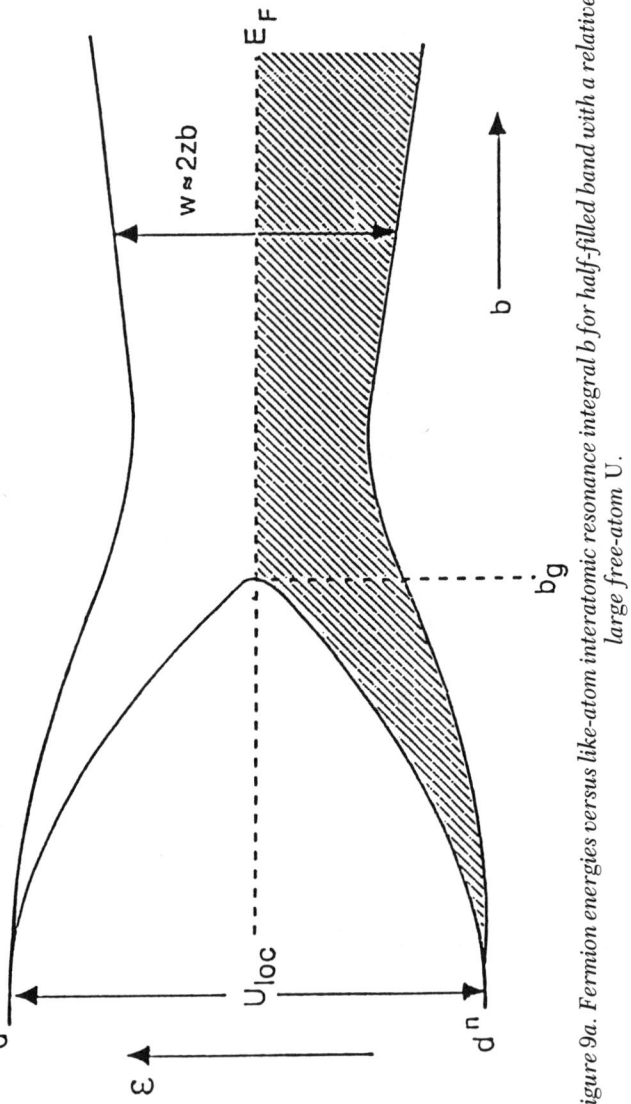

Figure 9a. Fermion energies versus like-atom interatomic resonance integral b for half-filled band with a relatively large free-atom U.

16. GOODENOUGH *Ceramic Superconductors*

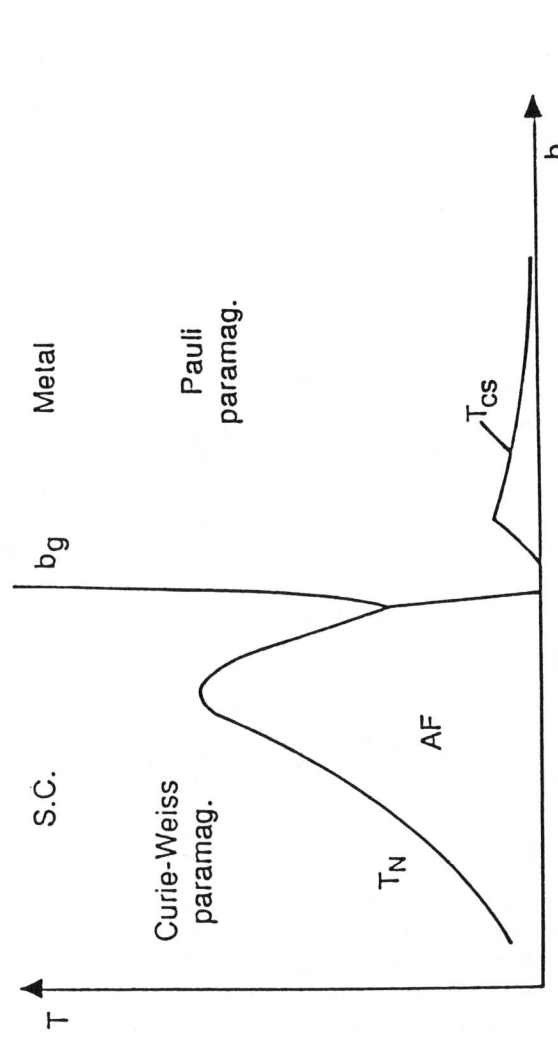

Figure 9b. Transition temperatures versus like-atom interatomic resonance integral b for half-filled band with a relatively large free-atom U. (Reproduced with permission from ref. 42. Copyright 1987 Pennsylvania State University.)

presence of spontaneous magnetic moments associated with a correlation splitting $U > W$ is incompatible with superconductivity.

Mixed-Valent Configuration. Again we must ask what happens when the system is changed from a single-valent to a mixed-valent configuration. Oxidation, for example, would lower E_F into the lower of the correlation-split bands, and reduction would raise it into the upper of the correlation-split bands. In this case, it is necessary to distinguish whether the partially filled band, according to the Heisenberg uncertainty principle, gives rise to an electron-hopping time between neighboring atoms

$$\tau_h \approx \frac{\hbar}{\frac{1}{2} W} \qquad (18)$$

that is long or short compared to the period ω_R^{-1} of the optical-mode vibration that would trap it—by a small-polaron reorganization energy—in a local deformation. As illustrated in Figure 10, the mixed-valent spinel $Li[Mn_2]O_4$ contains small polarons and retains localized magnetic moments on the Mn atoms, whereas isostructural $Li[Ti_2]O_4$ is metallic and a superconductor. Also shown are the $N(E)$ versus E curves for A_xMO_3 bronzes (M = Mo or W) that are metallic superconductors as a result of strong M–O–M interactions; in $Li[Ti_2]O_4$ the strong interatomic interactions are Ti–Ti interactions across shared octahedral-site edges.

Of course, CDW transitions can occur in transition-metal oxides with $W > U$. This situation is richly illustrated in the chemistry of the molybdenum oxides (13). But my purpose now is to turn to the high-T_c copper oxide superconductors.

High-T_c Copper Oxides

All the high-T_c copper oxide superconductors have a common structural feature: they contain intergrowths of copper oxide layers and other oxides. The simplest of these structures is the tetragonal phase of Figure 11(a), which is the room-temperature structure of $La_{1.85}Sr_{0.15}CuO_4$ with a $T_c \sim 40$ K. Between room temperature and T_c, the structure undergoes a displacive transition to orthorhombic symmetry in which the tetragonal (c/a > 1) CuO_6 octahedra rotate cooperatively, as indicated in Figure 11(b) (14, 15). This simplest of the cuprate superconductor systems will serve to illustrate the essential features considered in this chapter.

The phase diagram for the system $La_{2-x}Sr_xCuO_4$ is shown in Figure 12 for $0 \le x \le 0.4$; to retain a full oxygen content with $0.15 < x < 0.27$, it is

16. GOODENOUGH *Ceramic Superconductors* 309

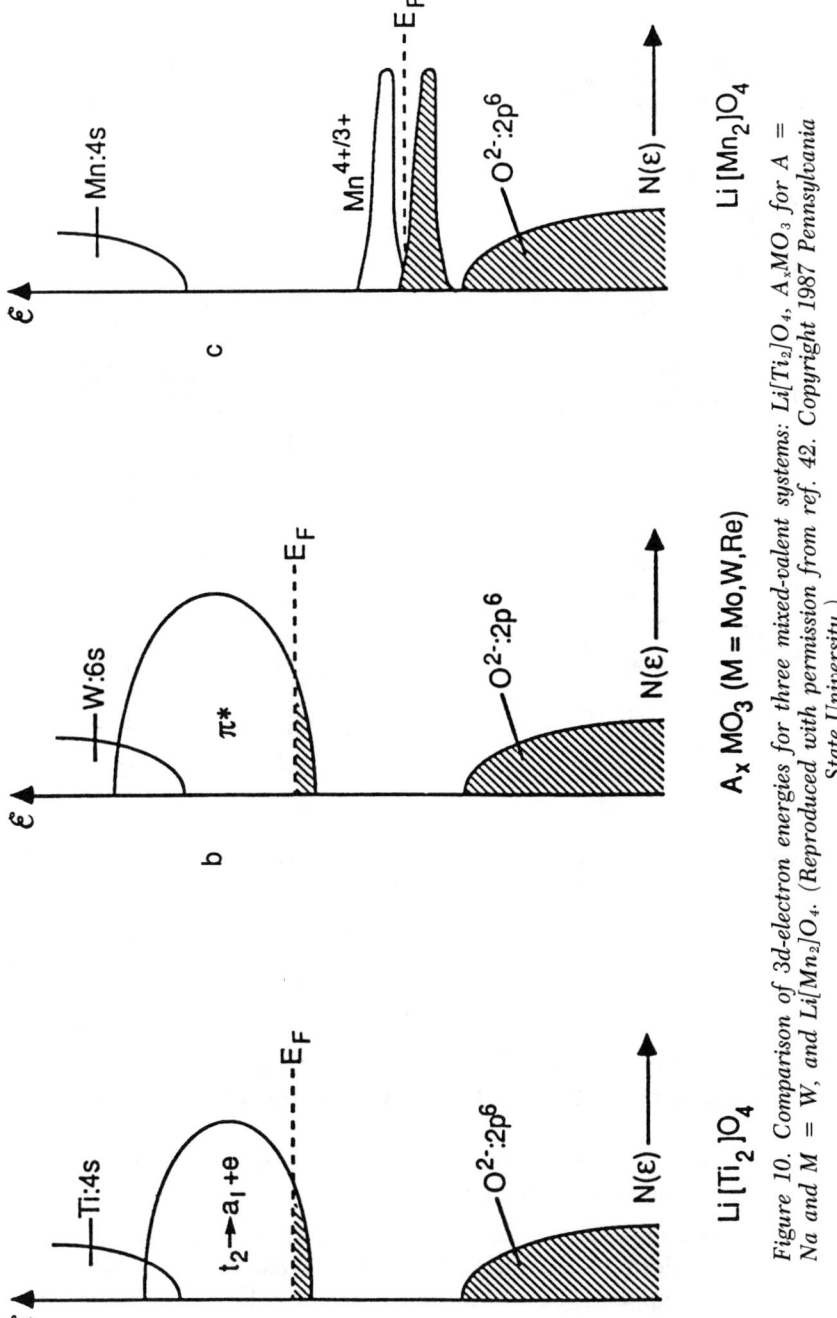

Figure 10. *Comparison of 3d-electron energies for three mixed-valent systems:* $Li[Ti_2]O_4$, A_xMO_3 *for A =
Na and M = W, and* $Li[Mn_2]O_4$. *(Reproduced with permission from ref. 42. Copyright 1987 Pennsylvania
State University.)*

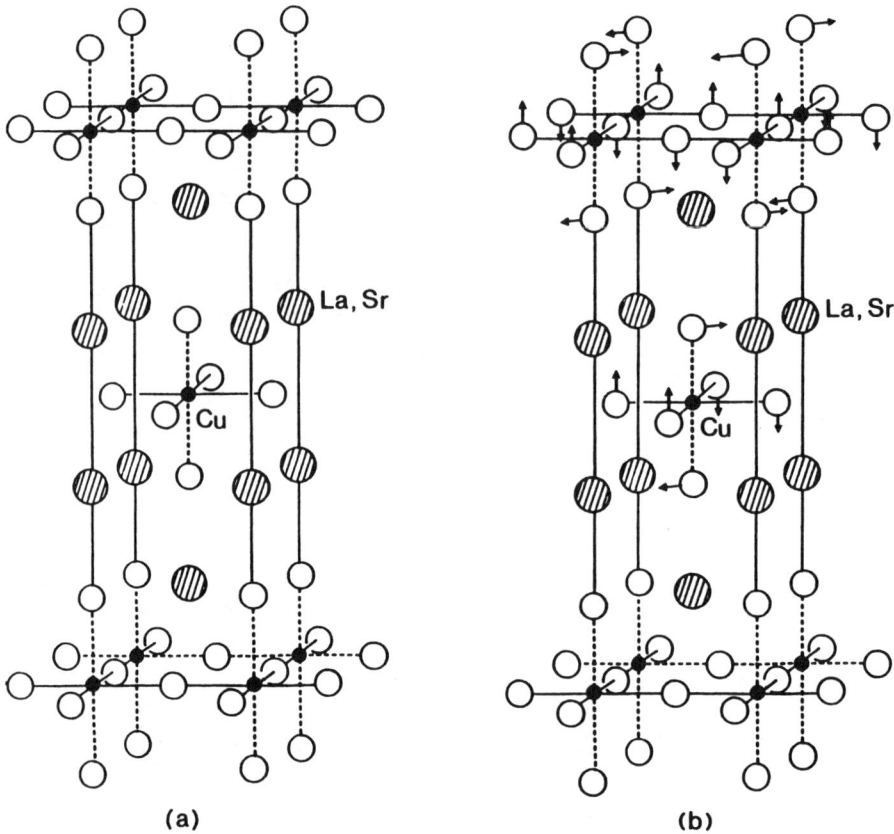

Figure 11. (a) Tetragonal T structure of $La_{1.85}Sr_{0.15}CuO_4$, and (b) oxygen displacements below the orthorhombic–tetragonal transition temperature T_t.

necessary to prepare the samples under pure O_2 and for $x > 0.27$ under high oxygen pressure (~3 kbar) (16–18). Five features should be noted:

1. La_2CuO_4 is orthorhombic at room temperature and an antiferromagnetic semiconductor.

2. In the compositional range $0 < x < 0.08$, the antiferromagnetic-ordering temperature T_N decreases sharply with increasing x, disappearing at a semiconductor–metal boundary, which is probably a narrow two-phase region.

3. Superconductivity is confined to a narrow compositional range

$0.1 \lesssim x < 0.3$, within which the system becomes orthorhombic below a $T_t > T_c$.

4. Dynamic, short-range antiferromagnetic fluctuations have been observed (19) in the superconductor compositions, the coherence length decreasing with increasing x and the spin fluctuations exhibiting no change as the temperature was reduced through T_c.

5. The normal metal state changes from p-type to n-type on doping beyond the superconductive compositional range.

Figure 13(a) shows a schematic representation of the one-electron density of states $N(E)$ versus energy E, as calculated from band theory for La_2CuO_4 (20–22). However, La_2CuO_4 is antiferromagnetic, with a copper magnetic moment $\mu_{Cu} = 0.5\ \mu_B$ (23, 24). Consequently, it is necessary to amend the diagram by the introduction of a correlation splitting of the σ^* band, as shown in Figure 13(b). From the tetragonal ($c/a > 1$) symmetry of the CuO_6 octahedra, the top of the σ^* band has $3d_{x^2-y^2}$ parentage (z axis parallel to c axis), but with strong covalent hybridization with the O-$2p_x$ and O-$2p_y$ σ-bonding orbitals in the CuO_2 planes. The observation of a copper magnetic moment shows that the empty σ^* band states are primarily Cu-3d in character, which means an $E_M > E_I$ applies for the σ-bonding orbitals, where the energies E_M and E_I have the same meaning as in Figure 1, but with the second electron removed from copper originating from the Cu-$3d_{x^2-y^2}$ orbital. However, with a

$$U_{eff} = U' + \Delta_t > W_\sigma \qquad (19)$$

where Δ_t is the tetragonal component of Δ_{cf}, it is necessary to inquire whether an $E_M > E_I$ can still apply on deeper oxidation of the system.

Mobile Holes. The spectroscopic evidence suggests that, in all the p-type copper oxide superconductors, oxidation creates mobile holes in band states that are primarily O-2p in character, the particular 2p orbitals being the p_x and p_y σ- or π-bonding orbitals of the oxygen atoms in the CuO_2 planes (25, 26). How this can be so is illustrated schematically in Figure 13(c). This peculiar situation arises because of the cross-over from $E_M > E_I$ for Cu^{2+} to $E_M < E_I$ for Cu^{3+}. The correlation splitting of the σ^* band ensures that, on oxidation, E_F drops into the π^* band rather than into the lower σ^* band. Moreover, the observation of short-range spin fluctuations on the copper in the superconductive phase shows the persistence of this correlation splitting to higher values of x in $La_{2-x}Sr_xCuO_4$. Preliminary data show that for $x > 0.3$, where the system remains tetragonal and a normal metal to lowest temperatures, E_F has moved out of the π^* band into only the σ^* band as the correlation splitting becomes $U < W_\sigma$, where Figure

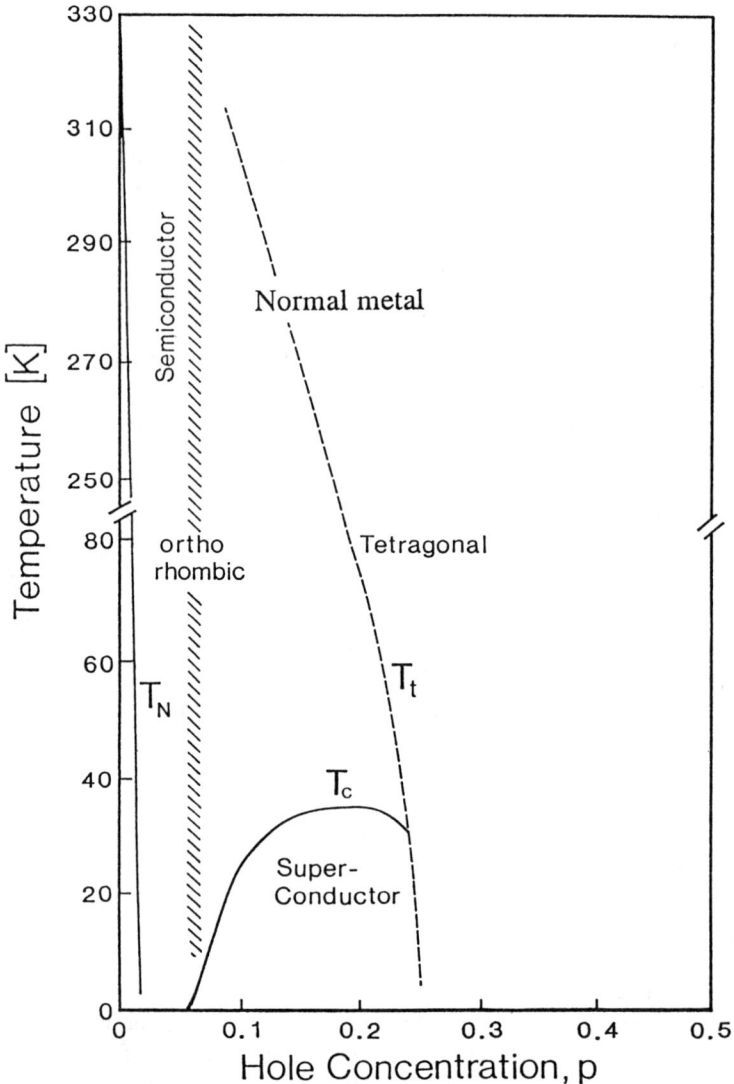

Figure 12. Tentative phase diagram for the system $La_{2-x}Sr_xCuO_4$. (Reproduced with permission from ref. 41. Copyright 1989 Materials Resource Society.)

13(a) applies. Without the correlation splitting, the holes should be stabilized in the σ^* band states, as these are the most strongly antibonding. The unsplit $\sigma^*_{x^2-y^2}$ band is less than half-filled, so the charge carriers become n-type.

Thus, a common feature of the electronic structure of the high-T_c $Ba_{1-x}K_xBiO_3$ and $La_{2-x}Sr_xCuO_4$ (and other copper oxide systems as well) is a Fermi energy that intersects (or nearly intersects) two bands, a σ^* band of primarily cationic character (Bi-6s or Cu-3d) and a π or π^* band of primarily

O-$2p_\pi$ character. Mobile holes in either the π^* or lower $\sigma^*_{x^2-y^2}$ bands are responsible for the superconductivity in the copper oxides (27); they make possible an abrupt transition from an antiferromagnetic semiconductor to a superconductor with increasing oxidation, and from a superconductor to a normal metal where a collapse of the correlation splitting of the $\sigma^*_{x^2-y^2}$ band occurs.

In this connection, attention should be drawn to the evidence for a sharp transition in the Cu–O bond length on going from the antiferromagnetic state to the superconductive state (28, 29). In the case of the superconductor phase $La_2CuO_{4.05}$—prepared at 800 °C under 23 kbar of pressure in a double cell consisting of CrO_3 on one side of a $Zr_{1-x}Ca_xO_{2-x}$ separator and the sample on the other—heating in air yields a first-order phase change to antiferromagnetic La_2CuO_4 with loss of oxygen (28). The excess oxygen is incorporated as O^{2-} ions in the tetrahedral sites of the double LaO layers of Figure 11(a) (30, 31), where it oxidizes the CuO_2 layers (28). The change in lattice parameters on traversing the transition is shown in Figure 14.

The a-axis increases on passing from the superconductor to the antiferromagnetic semiconductor. The increase in Cu–O distance is even more marked, as the tilting of the CuO_6 octahedra is greater in the antiferromagnetic phase. The shorter Cu–O distance is reflected in a stronger Cu–O–Cu interaction, which moves the system to the right from a $b_\sigma < b_g$ in the energy-band diagram of Figure 9 for the half-filled $\sigma^*_{x^2-y^2}$ band; in fact, it appears to move b_σ to a $b_\sigma \approx b_g$. Such a change would not normally introduce superconductivity; any band exhibiting short-range antiferromagnetic correlations would have a deep minimum in the $N(E)$ versus E curve, which would suppress T_c according to conventional BCS theory.

Suppression of Cu Magnetic Moment. The discussion thus far has concentrated on the electronic properties that may be peculiar to the high-T_c copper oxides. We have found evidence for an important increase in the strength of the Cu–O–Cu interactions in the CuO_2 sheets, and hence for a partial suppression of the copper magnetic moment, on passing from an antiferromagnetic phase to a superconductor phase on oxidation from $(CuO_2)^{2-}$. Where the Cu–O–Cu bond angles are bent from 180°, the mobile holes responsible for the high-T_c superconductivity occupy band states of primarily O-2p character; where the bond angles at low T are 180°, the charge carriers occupy $\sigma^*_{x^2-y^2}$ band states and the oxides exhibit normal metallic behavior to lowest temperatures. In copper oxides with the $YBa_2Cu_3O_{6+x}$ structure, the CuO_2–Y–CuO_2 layers containing double CuO_2 sheets retain bond angles that are bent from 180° at all oxidation states, and T_c is found to increase with the concentration of mobile holes (32). We have also argued that where the Fermi energy cuts both a π^* and a σ^* band, there the orbital hybridizations that optimize Cu–O bonding and bond angles can readily adjust themselves—via electron-lattice coupling—into bond

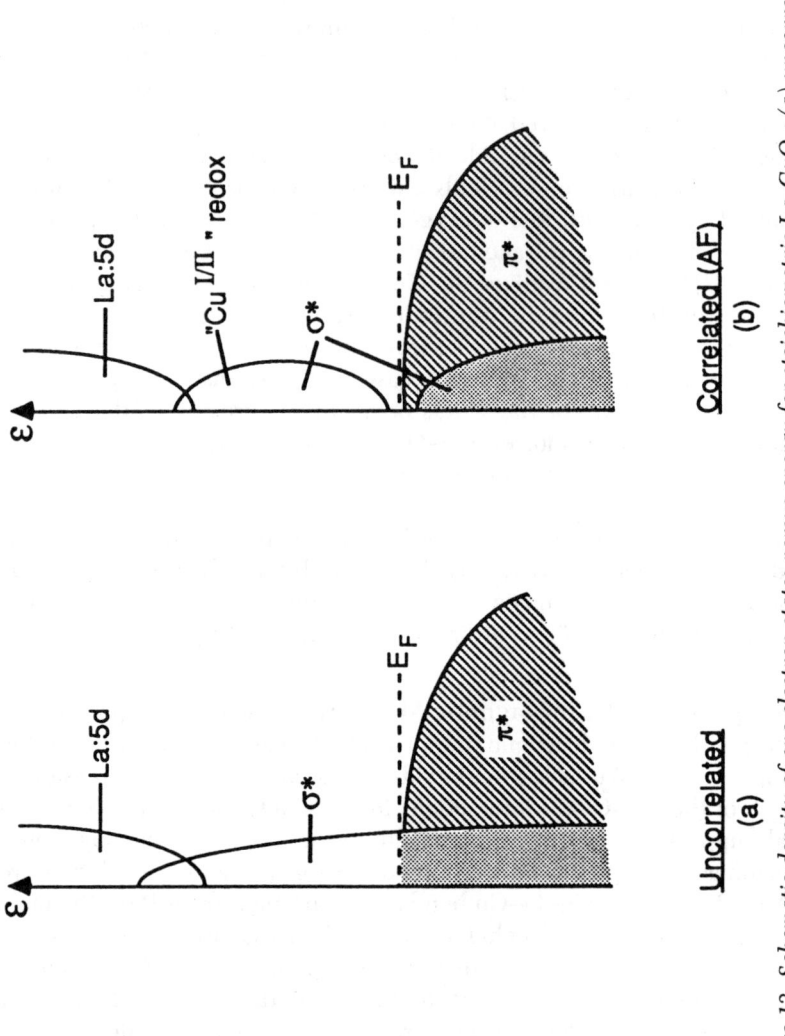

Figure 13. Schematic density of one-electron states versus energy for stoichiometric La_2CuO_4: (a) uncorrelated band calculations; (b) with correlation splitting of the Cu(II/I) and Cu(III/II) couples.

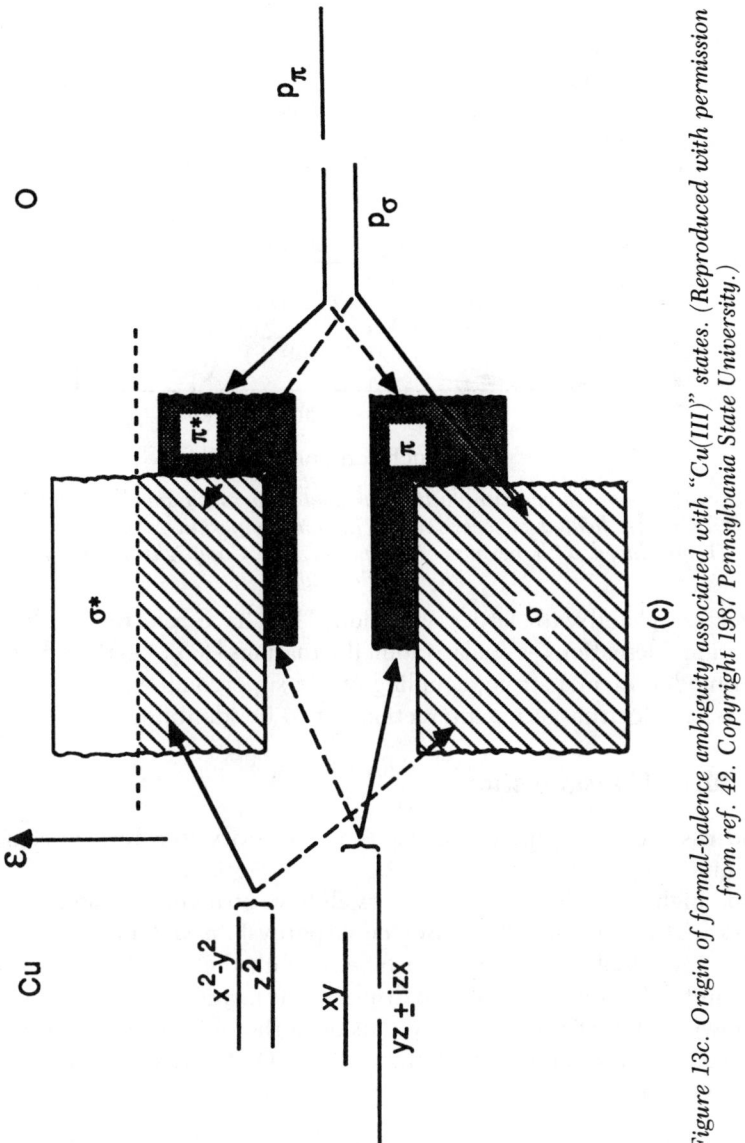

Figure 13c. Origin of formal-valence ambiguity associated with "Cu(III)" states. (Reproduced with permission from ref. 42. Copyright 1987 Pennsylvania State University.)

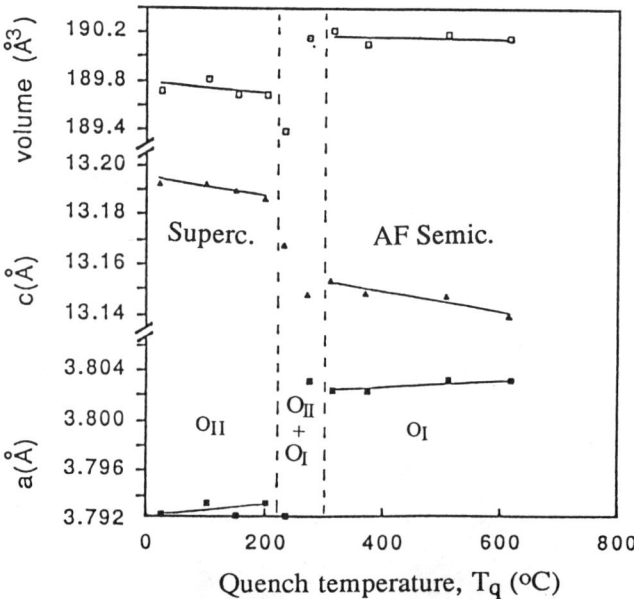

Figure 14. Room-temperature lattice parameters versus quench temperatures for $La_2CuO_{4.05}$ heated successively to higher temperatures before quenching. (Reproduced with permission from ref. 28. Copyright 1989 American Physical Society.)

changes associated with lattice vibrations. We have referred to these hybridization fluctuations as "polarization fluctuations" (33). On the other hand, such a model infers a strong coupling of the superconductive Cooper pairs via a strong electron-lattice interaction yet to be identified.

Structural Considerations

To address the remaining questions, it is necessary to return to structural considerations.

The high-T_c oxide superconductors all have structures related to that of perovskite, $CaTiO_3$. The ideal ABO_3 cubic perovskite contains a simple cubic BO_3 framework of corner-shared octahedra with the A cation at the center of the unit-cell cube. A variety of displacive transitions to lower symmetry are known (34); $CaTiO_3$, for example, is orthorhombic at room temperature as a result of a cooperative buckling of the TiO_3 framework that optimizes the Ca–O bonding.

Perovskite Structure. The perovskite structure is remarkable in two fundamental respects. It may sustain large concentrations of vacancies on any of the three atomic arrays; many of the high-T_c copper oxide phases not discussed contain ordered oxygen vacancies, for example. On traversing a [001] axis, BO_2 planes alternate with rock-salt-type AO planes; this structural feature requires a matching of bond lengths that, once met, allows ready stabilization of intergrowth structures. La_2CuO_4 is the simplest example of such an intergrowth structure.

The criterion for bond-length matching has traditionally been expressed by the Goldschmidt tolerance factor

$$t = \frac{r_A + r_O}{\sqrt{2}(r_B + r_O)} \tag{20}$$

where r_A, r_B, and r_O are the respective ionic radii. If t is 1, there is a perfect epitaxial match for intergrowth of (AO) rock-salt layers with BO_2 planes. However, the thermal expansions of the softer A–O bonds are greater than those of the B–O bonds, so the factor t decreases with decreasing temperature. The existence of a $t \approx 1$ at the firing temperature in a high-temperature synthesis may allow the structure to form, but at lower temperatures the B–O bonds are subjected to a compressive stress and the A–O bonds to a tensile stress. This bond mismatch can be relieved by a displacive transition such as the octahedral-site buckling found in $CaTiO_3$.

Orthorhombic Distortion. In La_2CuO_4, the distortion from tetragonal to orthorhombic symmetry represents an improvement of the bond-length matching; the La–O bond lengths are decreased and the Cu–O bond lengths are increased by the tilting of the CuO_6 octahedra. If the smaller Nd^{3+} ion is substituted for La^{3+}, the bond-length mismatch is accommodated by a removal of the c-axis oxygen to the plane of tetrahedral sites of Nd^{3+} ions, thus creating a fluorite-type intergrowth layer. In this structure (Figure 15) the Cu–O–Cu bond angle is 180°, and the oxide does not become a superconductor if these CuO_2 layers are oxidized (35). However, it does become a superconductor if these layers are reduced (36–39).

What is remarkable is that all of the p-type superconductors (those that become superconductors on oxidation) have Cu–O–Cu bond angles <180°. Moreover, the disappearance of superconductivity with increasing x in the system $La_{2-x}Sr_xCuO_4$ is associated with the disappearance of the orthorhombic distortion at a $T_t > T_c$. Even the superconductor $La_2CuO_{4.05}$, which is nearly tetragonal at room temperature, becomes orthorhombic at $T_t > T_c$.

In addition to this structural correlation, the copper oxide superconductors are characterized by a small coherence length ($\xi \sim 10$ Å) and a weak isotope effect, which are characteristic of the strong-coupling limit where it should be possible to think about the coupling of Cooper pairs in potential wells created by real-space lattice deformations. The structural correlations appear to provide a clue as to what those real-space lattice deformations may be.

Consider first a p-type orthorhombic $La_{1.85}Sr_{0.15}CuO_4$. The orthorhombic distortion is created to relieve the compressive stress on the Cu–O bond

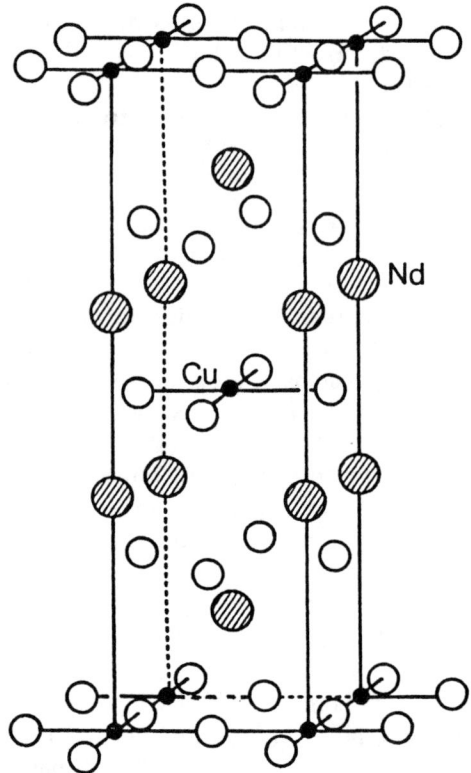

Figure 15. Tetragonal T structure of Nd_2CuO_4. (Reproduced with permission from ref. 41. Copyright 1989 Materials Resource Society.)

length. In this situation, Sr^{2+} substitution for a smaller La^{3+} ion relieves this stress—and hence reduces the magnitude of the orthorhombic distortion—both by increasing the mean A–O bond length and by introducing holes (O-$2p_\pi$ holes) into the Cu–O bonds. In this environment, a hole in the antibonding π^* bands of O-p_x,p_y and Cu-d_{xy} parentage would shorten the Cu–O bond length in the basal plane and thus would tend to increase the Cu–O–Cu bond angle toward 180°. Therefore, a cooperative change in the bond angle within a restricted area of a CuO_2 sheet would tend to localize the hole to that area of the sheet.

Polarons. The extended hole and associated lattice deformation would represent a "large polaron" because the hole would be trapped in a region including multiple Cu and O centers. The mobility of a large polaron is not activated; it moves with a momentum vector **k** that remains a good quantum number. Collision with a second large polaron of momentum –**k** and opposite spin creates a reinforced potential well in which the hole pair becomes trapped as a "large bipolaron". In an external electric field, the potential well with its two trapped holes moves in the superconducting state with a drift velocity without scattering by the phonons of the crystal. What makes strong the coupling between the paired holes is an enhancement of the electron lattice coupling by the internal stresses built into the system through a bond-length mismatch and by the relief of this stress with a bending of the Cu–O–Cu bond and easy bond rehybridization. Only if the Cu–O–Cu angle is bent from 180° can the structure respond strongly to the stress relief associated with π^* band holes.

In Nd_2CuO_4, the 180° Cu–O–Cu bonds tend to be stretched, so there can be no potential well associated with a hole. Lengthening of the Cu–O–Cu bond length reduces the Cu–O–Cu interactions, which keeps a $b_\sigma < b_g$ and hence a copper magnetic moment. However, introduction of electrons into the upper σ^* band relieves any tensile stress in the Cu–O–Cu bonds and creates a mixed valence on the copper that lowers the critical bandwidth for a spontaneous copper magnetic moment (1). The long-range antiferromagnetic order and μ_{Cu}—as well as any tensile stress on the Cu–O bonds—disappears with an increasing electron concentration in the σ^* bands. Therefore, these materials can be doped n-type. Moreover, square–coplanar coordination is not optimal for Cu^+ ions; the electrons may become trapped as large polarons by a dynamic bending of the Cu–O–Cu bond angles from 180°. Thus, n-type superconductivity can be stabilized and have a high T_c because of enhanced electron-lattice coupling where the Cu–O–Cu bond angle is 180°.

Summary

A survey of electronic-conduction processes in solid oxides was outlined and applied to the problem of high-T_c superconductivity in ceramic materials. The peculiar structural features of the perovskites were highlighted in an attempt to identify the origin of the enhanced pairing potential of the superconductive pairs in perovskite-related structures. The competitive nature vis á vis superconductivity of CDWs or a correlation splitting associated with spontaneous atomic moments was also stressed. The following features appear to be common to all the high-T_c copper oxides:

1. intergrowth structures, which introduce internal stresses, internal electric fields, and a strong anisotropy of the electronic properties;
2. a $W \approx U$, which is manifest by the appearance of superconductivity in a narrow compositional range between an antiferromagnetic semiconductor and a normal metal;
3. a $W \approx \hbar\omega_R$, which results in strong electron-lattice interactions; and

4. an $E_M \approx E_I$—particularly in the p-type superconductors—that lowers the energy of polarization fluctuations that can contribute to screening out the coulombic repulsion between electrons of a Cooper pair.

Probably several (and perhaps all) of these features are playing critical roles in the stabilization of high-T_c superconductivity.

References

1. Goodenough, J. B. *Prog. Solid State Chem.* **1972**, *5*, 145.
2. Chaillout, C.; Santoro, A.; Remeika, J. P.; Cooper, A. S.; Espinosa, G. P.; Marezio, M. *Solid State Commun.* **1988**, *65*, 1363.
3. Cava, R. J.; Batlogg, B.; Krajewski, J.; Ferrel, R. C.; Rupp, L. W.; White, A. E.; Peck, W. F.; Komentani, T. W. *Nature* (London) **1988**, *332*, 814.
4. Mattheiss, L. F.; Gyorgy, E. M.; Johnson, D. W. *Phys. Rev. B* **1988**, *37*, 3745.
5. Hinks, D. G.; Dabrowski, B.; Jorgensen, J. D.; Mitchell, A. W.; Richards, D. R.; Pei, S.; Shi, D. *Nature* (London) **1988**, *333*, 836.
6. Wignacourt, J. P.; Swinnea, J. S.; Steinfink, H.; Goodenough, J. B. *Appl. Phys. Lett.* **1988**, *53*, 1753.
7. Hinks, D. G.; Richards, D. R.; Dabrowski, B.; Marx, D. T.; Mitchell, A. W. *Nature* **1988**, *335*, 419; Kondoh, S.; Sara, M.; Ando, Y.; Sato, M. *Physica C* **1989**, *157*, 469.
8. Hegde, M. S.; Barboux, P.; Chang, C. C.; Tarascon, J. M.; Venkatesan, T.; Wu, X. D.; Inman, A. *Phys. Rev. B* **1989**, *39*, 4752.
9. Oliver, M. R.; Dimmock, J. O.; McWhorter, A. L.; Reed, T. B. *Phys. Rev. B* **1972**, *5*, 1078.
10. Goodenough, J. B. In *Defects and Transport in Oxides;* Seltzer, M. S.; Jaffee, R. J., Eds; Plenum: New York, 1974; p 55.
11. Goodenough, J. B. In *Physics and Chemistry of Electrons and Ions in Condensed Matter;* Acrivos, J. V. et al., Eds; D. Reidel: Amsterdam, 1984; p 1.
12. Goodenough, J. B.; Thackeray, M. M.; David, W. I. F.; Bruce, P. G. *Rev. Chim. Miner.* **1984**, *21*, 435.
13. Goodenough, J. B. In *Proceedings of Climax Fourth International Conference on the Chemistry and Uses of Molybdenum;* Barry, H. F.; Mitchell, P. C. H., Eds; Climax Molybdenum: Ann Arbor, 1982; p 1.
14. Takagi, H.; Uchida, S.; Kitazawa, K.; Tanaka, S. *Jpn. J. Appl. Phys.* **1987**, *26*, L123.
15. Jorgensen, J. D.; Schüttler, H.-B.; Hinks, D. G.; Capone, D. W., II; Zhang, K.; Brodsky, M. D.; Scalapino, D. J. *Phys. Rev. Lett.* **1987**, *58*, 1024.
16. Torrance, J. B.; Tokura, Y.; Nazzal, A. J.; Bezinge, A.; Huang, T. C.; Parkin, S. S. P. *Phys. Rev. Lett.* **1988**, *61*, 1127.
17. Yoshizaki, R.; Ishikawa, N.; Akamatsu, M.; Fujikami, J.; Kurahashi, H.; Saito, Y.; Abe, Y.; Ikeda, H. *Physica C* **1988**, *156*, 297.
18. Kitaoka, Y.; Ishida, K.; Hiiramatsu, S.; Asayama, K. *J. Phys. Soc. Jpn.* **1988**, *57*, 734.
19. Birgeneau, R. J.; Gabbe, D. R.; Jenssen, H. P.; Kastner, M. A.; Picone, P. J.; Thurston, T. R.; Shirane, G.; Endoh, Y.; Sato, M.; Yamada, K.; Hidaka, Y.; Oda, M.; Enomoto, Y.; Suzuki, M.; Murakami, T. *Phys. Rev. B* **1988**, *38*, 6614.
20. Mattheiss, L. F. *Phys. Rev. Lett.* **1987**, *58*, 1028.
21. Yu, J.; Freeman, A. J.; Xu, J.-H. *Phys. Rev. Lett.* **1987**, *58*, 1035.

22. Guo, Y.; Langlois, J.-M.; Goddard, W. A. *Science* **1988**, *239*, 896.
23. Vaknin, D.; Sinha, S. K.; Moncton, D. E.; Johnston, D. C.; Newsam, J. M.; Safinya, C. R.; King, M. E., Jr. *Phys. Rev. Lett.* **1987**, *58*, 2802.
24. Freltoft, T.; Fischer, J. E.; Shirane, G.; Moncton, D. E.; Suiha, S. K.; Vaknin, D.; Remeika, J. P.; Cooper, A. S.; Harshman, D. *Phys. Rev. B* **1987**, *36*, 826.
25. Nücker, N.; Fink, J.; Fuggle, J. C.; Durham, P. J.; Temmerman, W. M. *Phys. Rev. B* **1988**, *37*, 5158.
26. Himpsel, F. J.; Chandrashekar, G. V.; McLean, A. B.; Shaefer, M. W. *Phys. Rev. B* **1988**, *38*, 11946.
27. Dai, Y.; Manthiram, A.; Campion, A.; Goodenough, J. B. *Phys. Rev. B* **1988**, *38*, 5091.
28. Zhou, J.; Sinha, S.; Goodenough, J. B. *Phys. Rev. B* **1989**, *39*, 12331.
29. Goodenough, J. B.; Manthiram, A. *Physica C* **1989**, *157*, 439.
30. Jorgensen, J. D.; Dabrowski, B.; Pei, S.; Richards, D. R.; Hinks, D. G. *Phys. Rev. B* **1989**, *40*, 2187.
31. Chaillout, C.; Cheng, S. W.; Fisk, Z.; Lehmann, M. S.; Marezio, M.; Morosin, B.; Schirber, S. E. *Physica C* **1989**, *158*, 183.
32. Shaefer, M. W.; Penney, T.; Olson, B. L.; Greene, R. L.; Koch, R. H. *Phys. Rev. B* **1989**, *39*, 2914 and references therein; Uemura, Y. J. et al. *Phys. Rev. Lett.* **1989**, *62*, 2317.
33. Goodenough, J. B. *Mater. Res. Bull.* **1988**, *23*, 401.
34. Goodenough, J. B.; Longo, J. M. *Landolt-Börnstein Tabellen III* **1970**, *4a*, 126.
35. Gopalakrishnan, J.; Subramanian, M. A.; Torardi, C. C.; Attfield, J. P.; Sleight, A. W. *Mater. Res. Bull.* **1989**, *24*, 321.
36. Tokura, T.; Takagi, H.; Uchida, S. *Nature* **1989**, *337*, 345.
37. Takagi, H.; Uchida, S.; Tokura, Y. *Phys. Rev. Lett.* **1989**, *62*, 1197.
38. James, A. C. W. P.; Zahurak, S. M.; Murphy, D. W. *Nature* **1989**, *338*, 240.
39. Markert, J. T.; Maple, M. B. *Solid State Commun.* **1989**, *70*, 145.
40. Goodenough, J. B. In *Encyclopedia of Physical Science and Technology 1989 Yearbook*; Academic: New York, 1989; pp 3–33.
41. Goodenough, J. B. In *Identifying the Pairing Mechanism in High-T_c Superconductors*; Torrance, J. B. et al., Eds.; Materials Resource Society Symposium 156; Materials Resource Society: Pittsburgh, PA, 1989; in press.
42. Goodenough, J. B. *J. Mater. Educ.* **1987**, *9*(6), 619–673.

RECEIVED for review April 14, 1989. ACCEPTED revised manuscript September 27, 1989.

17

Geometrical Control of Superconductivity in Copper Oxide Based Superconductors

Jeremy K. Burdett and Gururaj V. Kulkarni

Department of Chemistry and The James Franck Institute, University of Chicago, Chicago, IL 60637

The geometrical control of the electronic structure of some of the recently synthesized copper oxide based high-critical-temperature (T_c) superconductors is presented. Although current thinking concerning these fascinating materials is in a state of flux and unconventional theories abound, it is shown how many of the properties of these systems may be understood by using rather conventional orbital ideas that have been used for a long time by the chemical community.

THE SERIES OF HIGH-TEMPERATURE COPPER OXIDE CONTAINING superconductors that has been synthesized over the past 3 years (1–7) has attracted considerable speculation concerning the nature of the mechanism behind their novel electrical properties. A question of parallel importance is how the electronic structure of these materials leads to a situation that is favorable for the operation of a superconducting mechanism. We have long known that geometrical structure intimately controls the electronic structure of both molecules and solids. In this chapter we want to show that some of the fundamental observations concerning the structure of these electronically novel systems are readily interpreted in terms of conventional orbital ideas (although at the present time some would regard them as a little speculative). We shall comment too on some of the salient features of the electronic state of affairs in these systems that we feel are important.

Electronic Structure of Copper Oxides

The structures of the known copper oxide containing superconductors have been reviewed elsewhere (8). They all have a common feature, a CuO_2 plane that is not always flat, of four-coordinate, approximately square-planar, copper atoms and two-coordinate oxygen atoms, linked in the way shown in Figure 1a. Sheets of this type, if they were linked via two apical oxygen atoms as in Figure 1b, would lead to the simple ReO_3 structure type. Insertion of a large cation (A) in the cavity of this structure leads to the structure of perovskite, of ABO_3 stoichiometry. Thus, the retention of the CuO_2 plane leads to a commonly used description of these systems as one derived from the perovskite arrangement.

Figures 2a–2c show different environments found in known structures.

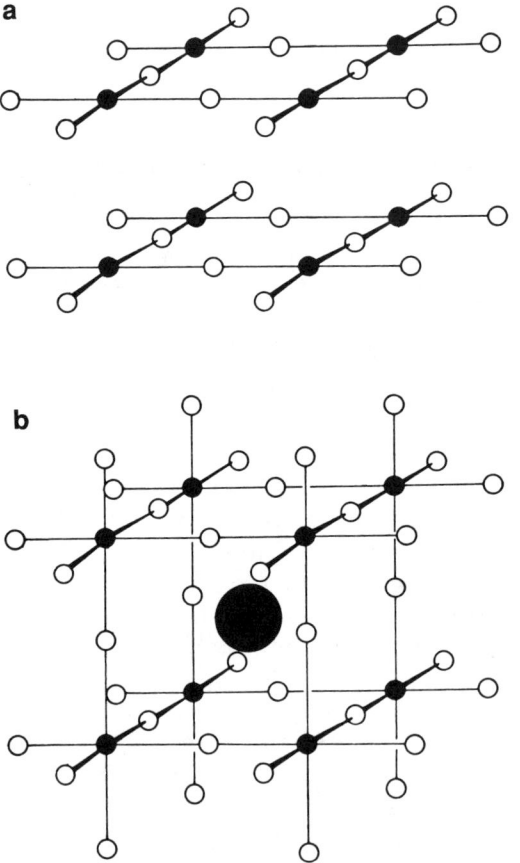

Figure 1. *a*, The CuO_2 sheet, a fundamental constituent of all high-temperature copper oxide superconductors; *b*, the relationship of the CuO_2 sheet to the perovskite structure.

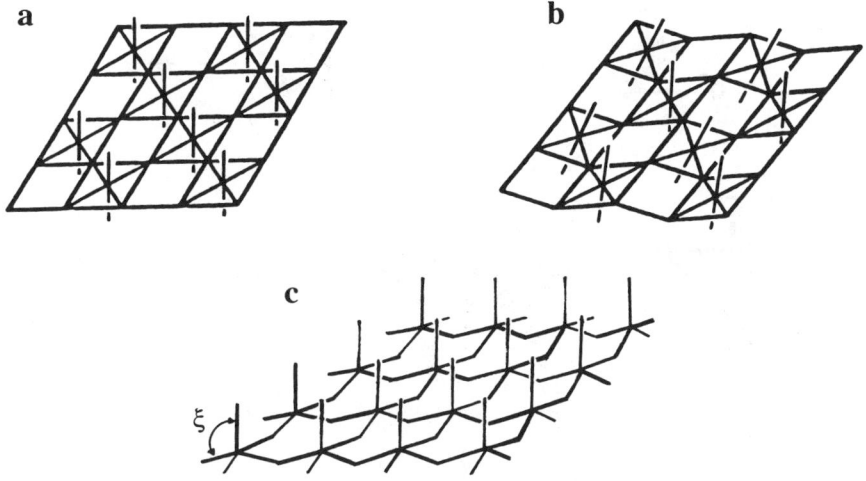

Figure 2. Geometrical arrangements found in copper oxide superconductors.

Figures 2a and 2b show the environment found in the tetragonal and orthorhombic forms, respectively, of the 2–1–4 compound ($La_{2-x}Sr_xCuO_4$). Here the copper atom is six-coordinate. The CuO_2 planes contain an extra two oxygen atoms attached to copper at somewhat longer distances than those in the plane. The sheets are flat in the tetragonal form and rumpled in the orthorhombic form. In this chapter we shall present the first electronic explanation of this interesting transformation. Figure 2c shows the structure of the puckered CuO_2 sheet ($\zeta > 90°$) with a five-coordinate copper atom found in the 1–2–3 compound ($YBa_2Cu_3O_{7-\delta}$). As in the 2–1–4 compound, the apical Cu–O distance is longer than that in the plane. This apical oxygen is connected to another part of the structure in a way we will describe later. Figure 1a shows the bare CuO_2 plane itself, found in $(TlO)_2Ba_2Ca_2Cu_3O_8$, for example.

Coordination Shell Plasticity. The structural chemistry of Cu^{II} is one of the most complex and fascinating of the d-block elements. Particularly interesting is the plasticity of the coordination shell. In Scheme I, a diagram adapted from a well-known review article (9), we show some pathways connecting the variety of geometries found for species with oxygen-containing ligands. Traditionally such distortions have been described as resulting from the Jahn–Teller instability of octahedral d^9 Cu^{II}, but there are clearly other factors at work (10) involving the 4s and 4p higher-energy orbitals on copper. In these systems it is difficult to understand, and thus predict, the size of the distortion away from octahedral for a given system. There is a spectrum of Cu–O distances in systems that are closely related chemically, and quite different distances are often found in polymorphs of the same material. For

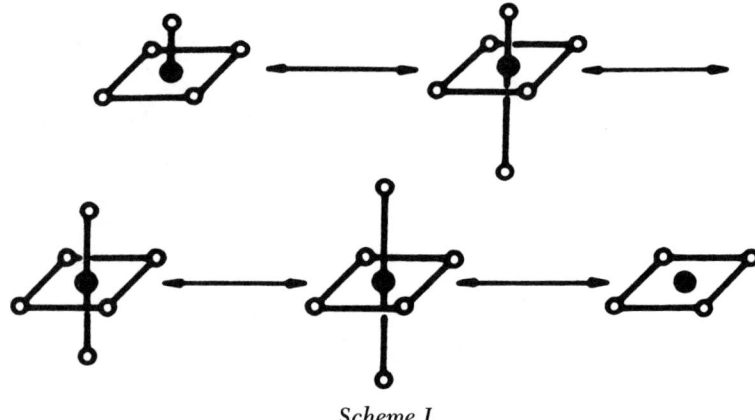

Scheme I

example (9), in the α form of $Cu_2(NH_3)Br_2$ the Cu–N distances are (a pair at) 1.93 Å, and the Cu–Br distances (a pair at) 2.45 Å and (a pair at) 3.08 Å. In the β form they are (a pair at) 2.03 Å and (a set of four at) 2.87 Å, respectively.

What is clear is that the distortion coordinates are invariably soft, probably quite anharmonic, and apparently influenced by the ligand environment via the ubiquitous "crystal-packing forces". For the extended arrays under discussion, this means that the geometrical picture is controlled not only by direct Cu–O interactions but by the nuances of O–O interactions and by the effect of the cations of different sizes and charges. Equilibrium Cu–ligand distances are, as a result, extremely difficult to predict. These interactions have been modeled (*11*) for the present series of compounds with some success. This chapter concentrates on orbital interactions. The structural chemistry of d^8 systems, (for example, Cu^{III}) is very different. Low-spin d^8 complexes (those of Pt^{II} and Pd^{II} are the best known) are never octahedral, but always square-planar. This may be interpreted (*12*) in terms of a larger Jahn–Teller driving force (twice as large by using the angular-overlap model) away from octahedral for the d^8 configuration, compared to that for d^9.

Local Coordination Geometries. From a chemical point of view, the electronic structure of these materials may be built up by consideration of the energy levels associated with local coordination geometries. We will see that this provides a very useful aid in understanding the level shifts of the bands themselves as the structure is changed. Through-bond coupling in oxides is generally considerably smaller than in sulfides, and because the angles of the structure are close to either 90° or 180°, σ/π separability is a reasonable approximation for many purposes.

Figure 3 shows the energies of the d orbitals by using the angular-overlap model (*12*) for a set of relevant geometries. Recall that the energy

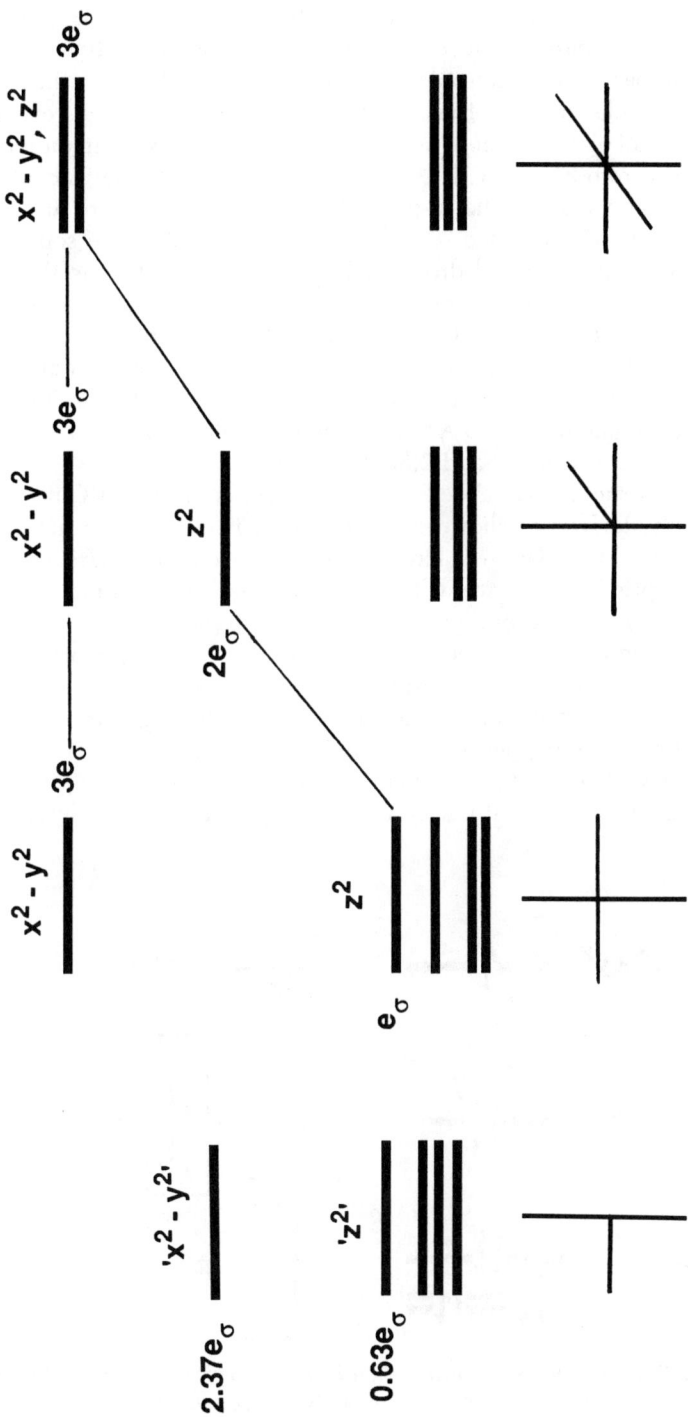

Figure 3. Angular-overlap energy-level diagrams for some coordination environments important in these materials.

parameter e_σ depends upon the bond length. It increases in magnitude as the internuclear separation decreases. Because of the geometry at copper, the important part of the electronic structure (that associated with the highest occupied molecular orbital of an isolated fragment, the $x^2 - y^2$ orbital) is very similar in all of the geometrical environments found. We can understand this by reference to Figure 3. The energy of the $x^2 - y^2$ orbital is controlled solely by the e_σ values of the in-plane ligands, of which there are always four. The energy of the filled, deeper-lying z^2 is much more sensitive to geometry. The energy of z^2 drops to lower energy, either as the apical distances are increased, or as one or both apical ligands are removed. In all of the structure types found (Figures 1 and 2), either there are no apical ligands attached to the CuO_2 planes or the ones that are present are bound at large internuclear separations. Typical values for the 2–1–4 compound are 1.90 Å (in-plane) and 2.45 Å (apical); typical values for the 1–2–3 compound are 1.93 Å (in-plane) and 2.30 Å (apical).

The band structure associated with the copper atoms in the CuO_2 planes is thus simply derived as shown in Figure 4. It is clear that a many-body picture of these materials will have to be employed to understand these systems in depth, but here we will use a traditional chemical model, where many-body terms in the energy are added to a basically one-electron approach to highlight some aspects of their structure. The simple orbital ideas we have described lead to the result that the highest-energy electrons lie in an orbital of $x^2 - y^2$ symmetry, heavily mixed with oxygen σ orbitals and antibonding between copper and oxygen. This view is not universally held. Arguments have been put forward (13) for holes in an oxygen $2p\pi$ derived band. In $La_{2-x}Sr_xCuO_4$ (with $x = 0$) the copper atoms are clearly well-

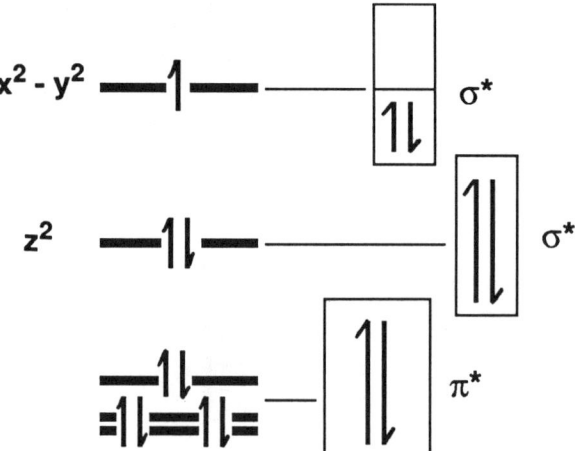

Figure 4. Diagram showing construction of the energy bands of the infinite CuO_2 sheet from the orbitals of a fragment.

described as Cu^{II}, and this $x^2 - y^2$ band is exactly half full of electrons. Such electronic situations, invariably unstable, are usually alleviated (*14, 15*) either by a Peierls-type distortion or by the generation of an antiferromagnetic insulating state.

An example of a Peierls-type distortion is found (*16*) in the mixed-valence platinum chain compounds, $Pt^{II}L_4 \cdot Pt^{IV}L_4X_2$, where X is a halide and L is an amine (Scheme II). Here Pt^{III} atoms disproportionate to alternating oc-

$$....X-Pt-X-Pt-X-Pt-X-Pt.... \longrightarrow ....X-Pt-X\cdots Pt-X-Pt-X\cdots Pt....$$

Scheme II

tahedral Pt^{IV} and square-planar Pt^{II} centers. The energetics driving the distortion are a combination of one-electron terms (leading to a stabilization of the full z^2 band on Pt^{II}) and many-body terms (leading to a coulombic repulsion between the two electrons in the same orbital). Peierls-type distortions may sometimes proceed without any obvious coulombic penalty. Consider the (Jahn–Teller) instability associated with singlet cyclobutadiene, or the analogous (Peierls) structural instability associated with an infinite chain of equidistant hydrogen atoms (*15*). The opening of a Hubbard gap, a feature of the antiferromagnetic insulating state, involves no geometrical change (in first order, at least) and is controlled by many-body effects. An example is found in the undoped material La_2CuO_4. In fact, in all of the superconductors made to date there is some mechanism, either by doping (making $x \neq 0$ in $La_{2-x}Sr_xCuO_4$) or by overlap with orbitals associated with other structural features (as in $YBa_2Cu_3O_{7-\delta}$), which removes some of the $x^2 - y^2$ density. This feature is important because it reduces the possibility of these half-filled band effects. As the charge transfer increases, the tendency for the opening of a Hubbard gap, and the tendency for a Peierls distortion, or other instability is reduced. Thus, a diamagnetic metallic state may result. The properties of this state, from which (on this model) superconductivity occurs, are important here. The electron occupancy of this $x^2 - y^2$ band is crucial to the electronic properties of the material and, as we shall see, is important in the geometrical control of the structure.

The 2–1–4 Compound

In the 2–1–4 compound, $La_{2-x}Sr_xCuO_4$, the occurrence of superconductivity is associated with the value of x. As noted earlier, for $x = 0$, the material is an antiferromagnetic insulator. It becomes a superconductor only for $x > 0.05$, presumably when enough electron density is removed from the $x^2 - y^2$ band to allow the existence of a stable metallic state. This compound exists in (at least) two forms, one of tetragonal and the other of orthorhombic symmetry. The relative stability of the two is strongly controlled by x (17). As the amount of strontium doping (x) increases (i.e., as the average copper d count decreases), the tetragonal structure gradually becomes more stable than the orthorhombic structure. This manifests itself in a drop in the transition temperature for the transformation as x increases (17, 18). For example, with $x = 0$ the orthorhombic-to-tetragonal transition occurs at 533 °C; with $x = 0.15$ it appears at 190 °C; and with $x = 0.2$ it is the only arrangement known. Increasing x leads to a depletion of the highest energy band of the system, $x^2 - y^2$. Figure 5 shows an energy difference curve between the two forms, from tight-binding calculations on the two solids (19) as a function of electron count, designed to explore this transformation.

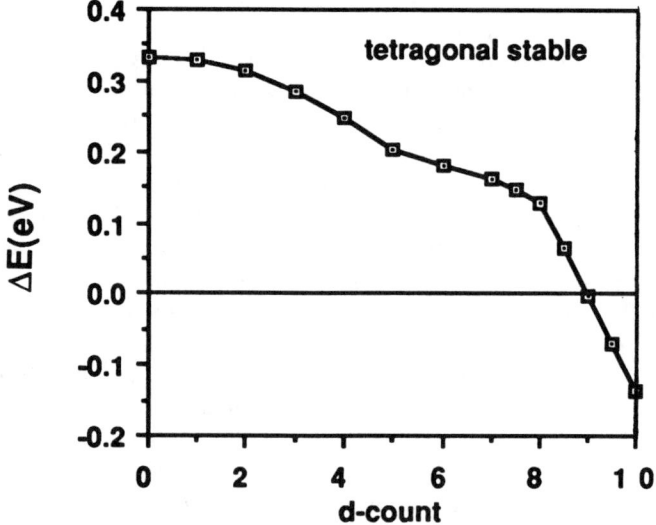

Figure 5. *Calculated energy difference between the tetragonal and orthorhombic forms of the 1–2–3 compound from a tight-binding calculation.*

In good agreement with experiments (17), the tetragonal structure is predicted to be increasingly favored with decreasing d count. The energetically important part of the plot (around d^9), where the slope is largest, occurs in that region where the $x^2 - y^2$ band is being populated. One of the puzzling things about this distortion is that no gap is opened up at the Fermi level, as in the Peierls-type distortion noted in the previous section. (Parenthetically, we note that the shape of this d-count-dependent curve from moments considerations (20) also tells us that the distortion is not of the Peierls type.) What is the driving force for the distortion, and why does it appear to be so dependent on d count?

Puckering Distortion. An important clue comes from studies of the puckering distortion of the five-coordinate arrangement shown in Figure 2c.

A similar curve is found for this distortion. For d^6 systems $\zeta = 90°$, but for d^9 systems $\zeta > 90°$. There is not a good collection of experimental data to compare with theory for these copper-based systems, but the calculated plot is exactly what would be expected from the well-established variation in molecular geometry with d electron count for five-coordinate transition metal systems. A molecular orbital diagram for the puckering distortion (shown in Figure 6b) comes from the work of Rossi and Hoffmann (21). The stabilization associated with the $x^2 - y^2$ orbital during the distortion, driven by the relief of strong metal–ligand antibonding interactions (Figure 6), shifts the energy minimum of the structure away from those geometries with 90° and 180° O–Cu–O angles. Destabilization of the occupied levels favors the $\zeta = 90°$ structure. In Figure 6a the mixing of σ interactions into these yz and xz orbitals as the angle ζ increases leads to their destabilization. As the population of the $x^2 - y^2$ orbital (band) decreases, the driving force for increasing ζ is reduced.

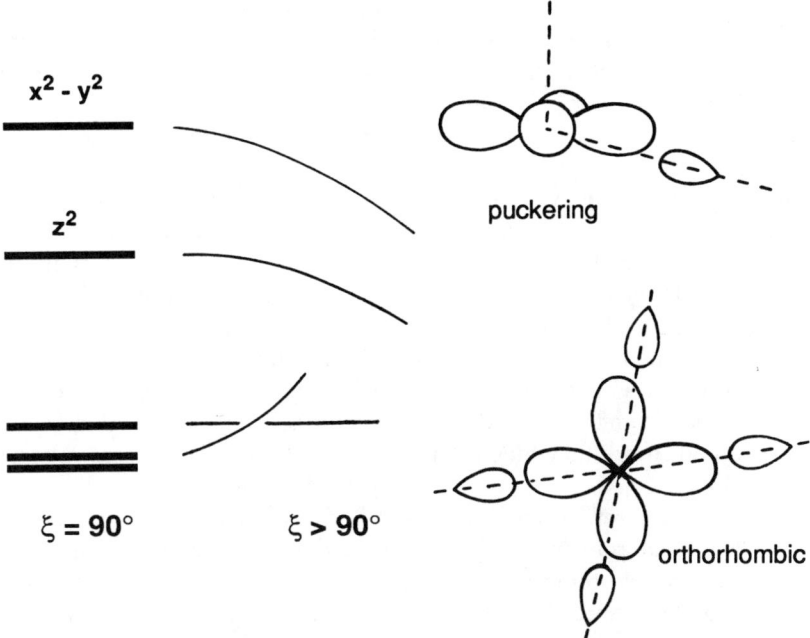

Figure 6. *Energy-level diagrams during puckering distortion.*

There are other solid-state examples of the same type. One that involves a π-type interaction (rather than the σ-type interaction described here) is the question of planar or pyramidal geometry at oxygen in the MO_2 structures of rutile (TiO_2) (planar) and $CaCl_2$ (pyramidal). The planar structure is found for low electron counts where antibonding orbitals are not occupied. However, by the time the π^* levels are full (at d^6), the pyramidal structure (the analog of the puckered structure in the copper oxides) is favored (22). Such orbital ideas are useful in many areas of chemistry; see ref. 23, for example.

Orthorhombic Distortion. A very similar mechanism applies to the distortion in the 2–1–4 compound. Here the geometrical change is more complex, but the principle is the same. The distortion changes the O–Cu–O angles (Figure 6c) such that metal–ligand overlap is reduced and the destabilization associated with the antibonding $x^2 - y^2$ band is lowered. As x increases from zero in the 2–1–4 compound, the electron occupation of the $x^2 - y^2$ band is reduced, the stabilization energy associated with the orthorhombic structure decreases, and eventually the tetragonal structure is found as the lowest energy alternative. Depletion of the $x^2 - y^2$ band on increasing x is associated with a decrease in metal–ligand antibonding interactions on this model. Correlation with experimentally determined Cu–O distances provides supporting evidence for this viewpoint. They are (17, 24) 1.9035(1) Å for $x = 0$ and 1.8896(1) Å for $x = 0.15$.

Thus, the puckering distortion of the 1–2–3 compound and the orthorhombic distortion of the 2–1–4 compound are manifestations of the same electronic phenomenon. The difference between the two lies clearly in the coordination number at copper. A puckering distortion is less likely for six-coordinate systems than for five-coordinate ones, simply because in the former there are oxygen atoms on both sides of the CuO_2 plane. A more complex distortion, at first sight difficult to understand, appears to be the energetically favored alternative here. It is interesting to see that these structural variations are well described by a simple conventional chemical mechanism. Central to the argument is that these highest-energy electrons are located in an $x^2 - y^2$ orbital, certainly heavily mixed with oxygen, but with a strong Cu–O antibonding interaction that is angle-dependent. There is no indication from this analysis of the structural problem, that the electrons are located on an orbital which is largely oxygen π in character, as suggested by other models.

Sleight (e.g., ref. 8) has long identified the local copper geometry as playing a crucial role in these systems and has made some correlations between T_c (critical temperature) and Cu–O distances and angles. He has emphasized, however, the importance of the π-type metal–oxygen interactions in controlling geometry. Sleight's model calls for relief of π^* interactions associated with yz and xz orbitals on bending, in a way exactly analogous to the relief of σ interactions (via the $x^2 - y^2$ orbital) here. It is not possible on such a model to directly build in the effect of electron count changes. For the model described here, the Cu–O distance is sensitive to the population of the $x^2 - y^2$ band. This sensitivity leads us to ask the question whether the Sleight correlation of Cu–O distance with T_c is really one which is measuring $x^2 - y^2$ density.

Once we have explored the results of an orbital model, it is important to put the results in perspective. The computed result of Figure 5 is actually quite sensitive to the geometries chosen for the calculation. Experience tells us that if this is the case, there are usually energetic contributions not properly modeled by using such a one-electron theory. In the present case, nonbonded interactions between the oxygen atoms are an obvious source of the problem. Our study (25) of the details of the geometry of rutile (TiO_2) found an interplay between the direct Ti–O interactions and the matrix O–O interactions of the approximately close-packed solid. Short Cu–O distances in the present series of systems signal an important contribution from this source. These results and the "molecular mechanics" results of ref. 11 therefore present a complementary picture of this structural problem.

The 1–2–3 Compound $YBa_2Cu_3O_{7-\delta}$

The basic geometrical arrangement of $YBa_2Cu_3O_{7-\delta}$, the first above-liquid-nitrogen-temperature superconductor, is simply derived from the perovskite structure (it would have the formula $YBa_2Cu_3O_9$) by the omission of some of the oxygen atoms. The structure is shown for the $\delta = 0$ system in Figure 7 (δ varies from 0 to 1.) Defect perovskites are well known. $Ca_3Mn_3O_{7.5}$

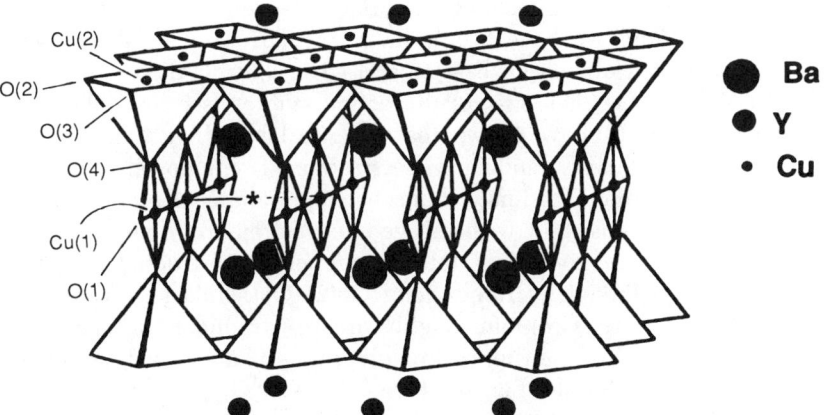

Figure 7. The 1–2–3 structure, $YBa_2Cu_3O_{7-\delta}$, with $\delta = 0$.

(i.e., $CaMnO_{2.5}$), which may be generated by heating the parent perovskite $CaMnO_3$, contains square-pyramidal metal atoms. In the 1–2–3 compound there are square planes and square pyramids for $\delta = 0$. The vacancy ordering in $CaMnO_{2.5}$ may be readily understood (26) using chemical ideas. We shall see that they are also useful for the vacancy problem in the 1–2–3 compound. T_c varies (27) with oxygen stoichiometry in a most interesting way (Figure 8). For low δ it appears to be almost flat, but at around $\delta = 0.3$ there is a rapid drop. This drop is followed by another plateau region before T_c sharply drops to 0 at around $\delta = 0.6$.

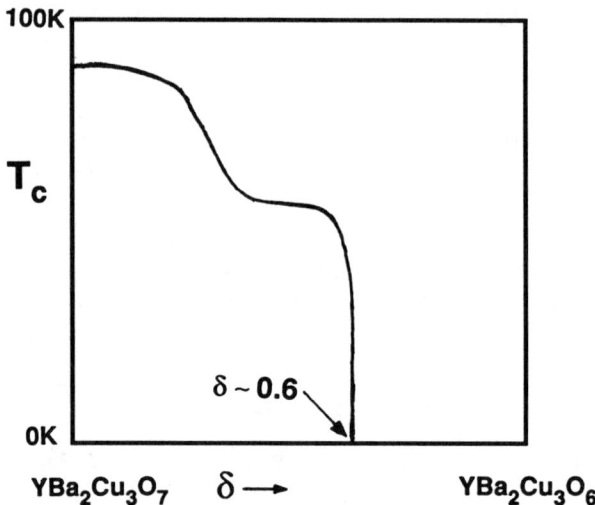

Figure 8. Variation of T_c with oxygen stoichiometry in the 1–2–3 compound.

This material clearly has a different structural chemistry from the 2–1–4 compound, and one complicated by a variable oxygen stoichiometry. There are many questions we have to ask. In particular, why are the vacancies ordered in the way shown for δ = 0, and what controls the ordering for δ ≠ 0? In YBa$_2$Cu$_3$O$_{7-\delta}$ for δ = 0, diffraction studies (*see* the collection of references in ref. 28) have shown that the site labeled with an asterisk in Figure 7 [O(5)] is empty, and thus the structure is composed of five coordinate planes and four coordinate chains of copper atoms. A picture emerges for the structure consistent with the geometrical environments we expect to see for the various oxidation states of copper. The structure consists of Cu^{II}(2)O$_2$ sheets linked by rather long Cu(2)–O(4) linkages to Cu^{III}(1)O$_3$ chains. The sheets thus contain roughly square-pyramidal Cu^{II} atoms. The chains contain Cu^{III} atoms in approximately square-planar coordination, leading overall to an orthorhombic structure (YBa$_2$(Cu^{II}O$_2$)$_2$(Cu^{III}O$_3$)). For δ = 1, all of the sites labeled as O(1,5) are unoccupied, the square planes have been replaced by linear O–Cu^I–O dumbbells, and now the structure is tetragonal (YBa$_2$(Cu^{II}O$_2$)$_2$(Cu^{I}O$_2$)).

Computed Band Structure. This viewpoint also comes from the computed band structure. Figure 9 shows how the orbital diagrams associated with the local (square and square-pyramidal) units get broadened into bands in the extended solid. Because the electronic area of interest lies between d^8 and d^{10}, we only show the $x^2 - y^2$ and $z^2 - y^2$ bands. The unconventional labeling of the latter comes about simply from the axis choice for the solid. It is really only an $x^2 - y^2$ orbital turned on its side. There are twice as many $x^2 - y^2$ levels as there are $z^2 - y^2$ levels, reflecting the

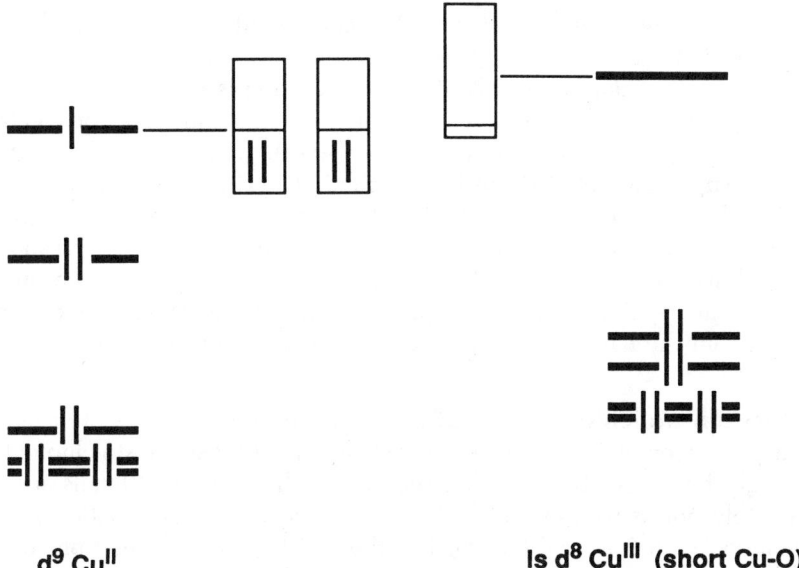

d⁹ Cu^{II}   **ls d⁸ Cu^{III} (short Cu-O)**

Figure 9. Broadening of the levels of local fragments into energy bands in the extended solid for the 1–2–3 structure. Only the two highest levels are shown.

stoichiometry of the material. The picture that results shows two roughly half-full bands corresponding to $x^2 - y^2$ orbitals on the two Cu^{II} atoms, and an almost empty $z^2 - y^2$ band on the Cu^{III} atom. We should ask why the $z^2 - y^2$ band associated with the chain atoms lies to higher energy than the $x^2 - y^2$ band associated with the plane atoms. This is a natural consequence of the shorter average Cu–O distance in the chains (1.86 and 1.94 Å) compared to the planes (1.93 and 1.96 Å), which gives rise to a larger value of e_σ for the former. If the Cu(1)–O distances were the same as the Cu(2)–O distances, the two bands would occupy very similar energetic positions and each copper would have approximately the same oxidation state.

This small amount of overlap that we show is vitally important because this is the mechanism by which the $x^2 - y^2$ bands of the planes avoid the half-filled electron configuration, with all of the drawbacks described earlier. The connectivity of the atoms in this structure allows an interesting flexibility for the chain atom coordination, which then feeds back into the structural problem. Because the Cu(1)–O(4) distance is not fixed by the coordination demands of other atoms, it is able to adjust to such a length that the $z^2 - y^2$ band lies to higher energy than the $x^2 - y^2$ bands, and, in so doing, preserves the approximate Cu^{II} and Cu^{III} labels. The exact details of the plane-to-chain charge transfer are difficult to compute, as they will be sensitive to small variations in the structure. However, the important point is that while in the 2–1–4 compound ($La_{2-x}Sr_xCuO_4$), the electron density in the $x^2 - y^2$ bands is broadly controlled by the amount of strontium doping;

the corresponding feature in the 1–2–3 compound is determined by the geometrical effects we have just described.

The result of doping of this material with zinc is interesting. Zn^{II} is a d^{10} system, and substitution of Cu(2) by zinc will tend to increase the electron density in the $x^2 - y^2$ band. Eventually, when the band is close enough to half-full, superconductivity will be switched off. (This switching off occurs (29) for close to 14% zinc.) With the assumption of approximately equal densities of states for the three bands at the Fermi level, this corresponds to the addition of about 0.043 electron per Cu(2). We should recall a similar figure (around 0.05 electron per copper) for the critical removal of electron density from the half-filled copper band in the 2–1–4 compound.

Oxygen Stoichiometry. Of particular interest in this system is the variation in geometrical and electronic structure with oxygen stoichiometry. Although there have been many structural studies, the fine details are unfortunately not as well established as we would like. The variation in cell parameters with δ is well known, but the identity of the sites from which oxygen is lost is still not completely determined, because diffraction experiments are not sensitive to local ordering features. For $\delta = 0$ there are ordered square planes of copper atoms in the chains, and for $\delta = 1$ there are ordered dumbbells, but the details of the structure for arbitrary δ are somewhat elusive.

Most studies show loss of oxygen from the planes perpendicular to z that contain the Cu(1) atoms, but both O(1) and O(5) sites seem to be occupied for $\delta > 0$. For some values of δ the crystal symmetry appears to be tetragonal rather than orthorhombic. There are probably locally ordered regions of the structure so that such a result may be understood in terms of chains of square-planar copper atoms of varying lengths running in both the a and b directions. One way of regarding the tetragonal structure is thus as an equal mixture of the two. Experimental evidence supports this view (30, 31). A schematic picture of this situation is shown in Figure 10, in which the two coordinate Cu^I atoms are represented by empty space and the Cu^{III} atoms by lines to indicate chains.

One important question concerning the changes in properties with oxygen stoichiometry is how this should be treated theoretically (32). Is it, for example, realistic to add 2δ electrons to the Fermi level of a calculation, with $\delta = 0$ reflecting the loss of δ oxygen atoms? (This is the so-called rigid-band model.) Consider first a model where a single oxygen atom is lost from one of the O(1) sites of the material. Scheme III shows how two originally square-planar copper atoms now lie in a T-shaped environment. Figure 3 shows that the energy levels associated with such a geometry lie deeper than the $x^2 - y^2$ band of the square plane, a result that leads to a loss of electrons from the Fermi level rather than the gain of the rigid band approach. This comes about simply because there are two T-shaped copper atoms, whose

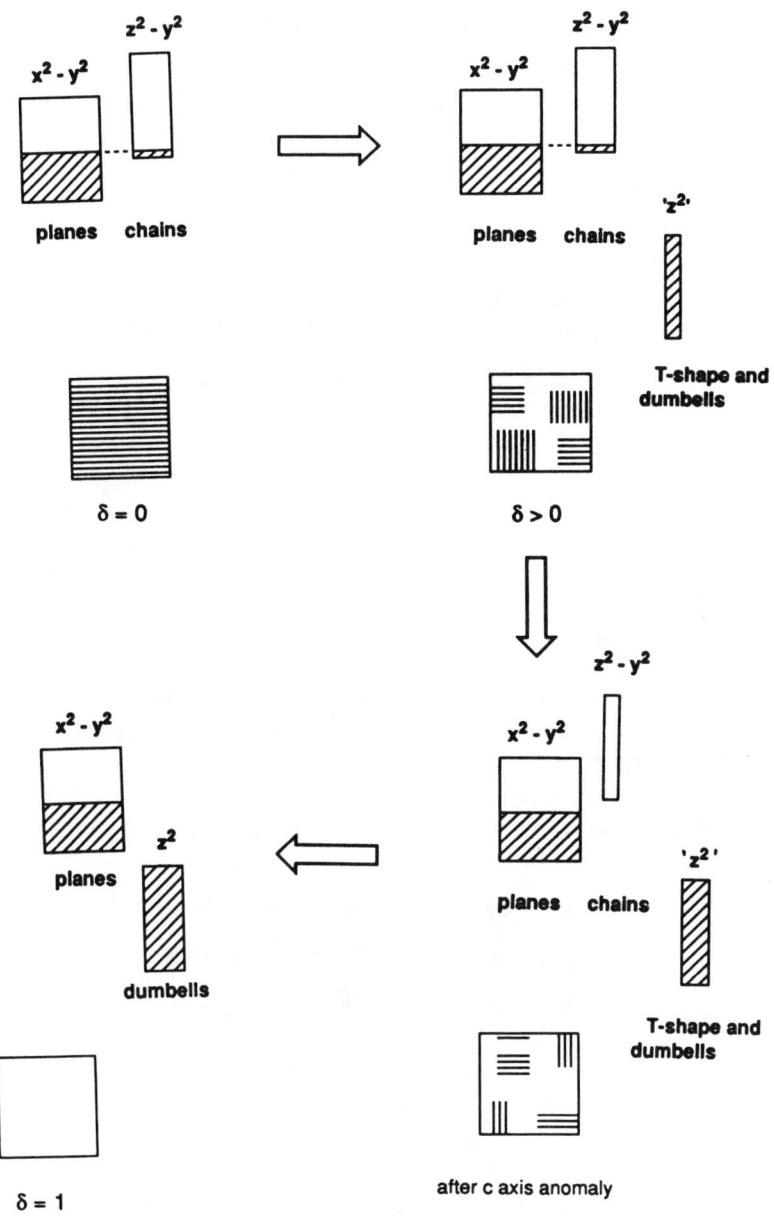

Figure 10. Schematic variation of the band structure of the 1–2–3 compound with oxygen stoichiometry. At the bottom of each figure is a schematic picture showing the arrangement of regions of chain (Cu^{III}) atoms (lines) and dumbbell (Cu^{I}) atoms (space).

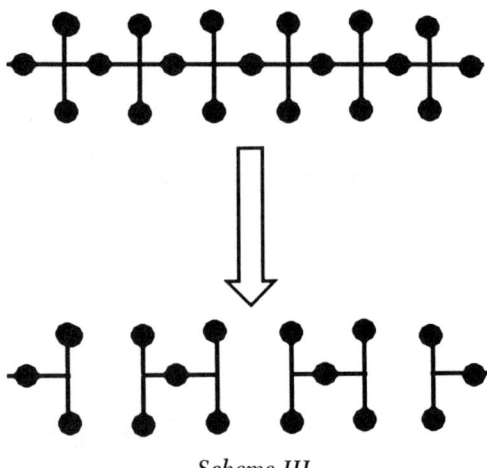

Scheme III

levels require four extra electrons to be filled, but there are only two electrons made available by the loss of a single oxygen atom. If, however, the oxygen vacancies order so that chains of copper dumbbells are formed, as in Scheme IV, only one copper level per oxygen atom drops to low energy and needs to be filled. Therefore, there is no change in the number of electrons at the Fermi level. The way the vacancies order in the structure thus crucially controls the electronic description at the Fermi level. (Elsewhere (19) we suggest that an ordering process is responsible for the first dip at $\delta = 0.3$ in T_c, as shown in Figure 8).

Figure 10 shows in a schematic way how the band structure of the material changes with increasing δ. The number of levels (represented by

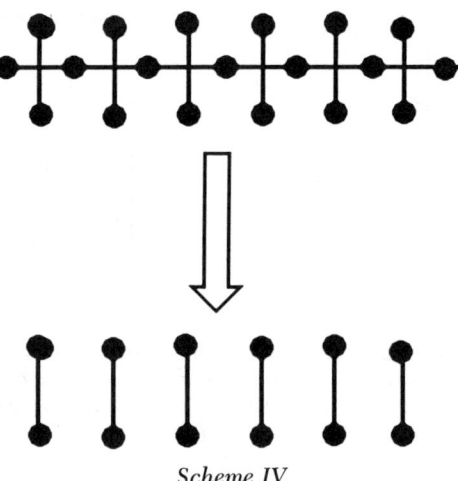

Scheme IV

the breadth of the box in the picture) associated with the $x^2 - y^2$ bands of the planar copper (Cu^{II}) atoms remains constant, but loss of oxygen atoms from the chains generates Cu^I atoms at the expense of Cu^{III} atoms. At $\delta = 1$ there are of course no Cu^{III} $z^2 - y^2$ levels, and the $x^2 - y^2$ bands of the planar copper atoms are exactly half full. Raveau et al. (33) called the T-shape geometry around copper an abnormal one. (In molecular chemistry it is found for a single low-spin d^8 molecule (34).) One way of envisaging the ordering process is thus one which minimizes the number of such local structures.

Some other evidence suggests ordering in the structure as the oxygen is removed. Figure 11 shows the experimentally determined variation (28) in atomic c-axis parameters with δ. The absence of linear variation with δ in these values suggests that a cooperative effect is controlling the structure. On this model, then, for low δ the dimensions are controlled by Cu^{III} and for high δ by Cu^I. There is another feature of these plots which is also interesting, and leads to further correlation with experiment. In Figure 11 a smooth curve has been drawn through the experimental points. Equally well at around $\delta = 0.6$ the data could be interpreted as suffering a sharp change. This has been called (35) "the c-axis anomaly". Interestingly, this is the point where T_c rapidly drops to zero (Figure 8). Of particular importance to us and our model is that at this point, the Cu(1)–O(2) distance abruptly shortens from 1.85 to 1.80 Å. As a result, we calculate that the $z^2 - y^2$ band is pushed to higher energy, and electron transfer to the chains is suddenly cut off. Now the planar copper atoms have exactly half-filled $x^2 - y^2$ bands, and we have a simple explanation for the disappearance of superconductivity. Figure 10 puts these ideas together.

Charge Movement on Distortion

The distortions described in the section on the 2–1–4 compound left all of the copper atoms equivalent in the structure. The change in electron density associated with each atom during the distortion is probably small. Such a comment does not apply to the distortion of the platinum halide chain shown in Scheme II. In this disproportionation process, considerable formal charge transfer has taken place. The distortion of the CuO_2 planes, shown in Scheme V and described as a "breathing" motion, has similar features. A dynamic motion of this type is favored by many in the chemical community as a mechanism for superconductivity, which is associated with very strong electron–phonon coupling. (See, for example, ref. 36 and the series of articles in ref. 6.) Figure 12 shows how the densities of states associated with the $x^2 - y^2$ bands change during such a motion from a calculation on the 1–2–3 compound.

As the distances around one copper atom contract, the band (A) is pushed to higher energy (recall our earlier discussion of the variation of e_σ with bond

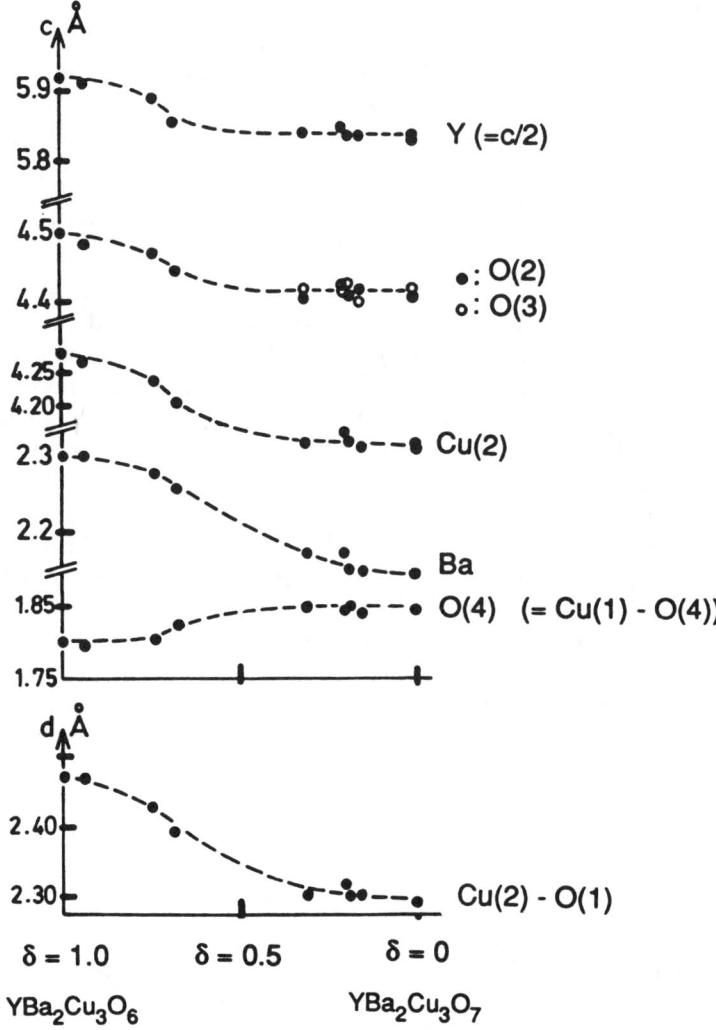

Figure 11. Changes in geometry as a function of oxygen stoichiometry for the 1-2-3 compound. Shown are the changes in z coordinate for the relevant atoms.

length) and becomes largely associated with this copper atom. As the distances around the other copper atom expand, the band (B) moves to lower energy and becomes largely associated with this second type of copper atom. Overall, the effect is to transfer electrons from A to B. As a result, the copper atom with the shorter Cu–O distances becomes more Cu^{III}-like and the copper atom with the longer Cu–O distances becomes more Cu^{I}-like. Increasing and decreasing the Cu–O distances by an amount equal to the

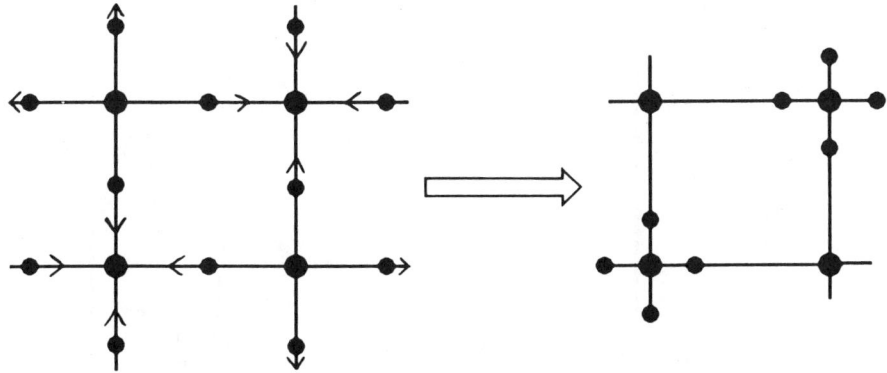

Scheme V

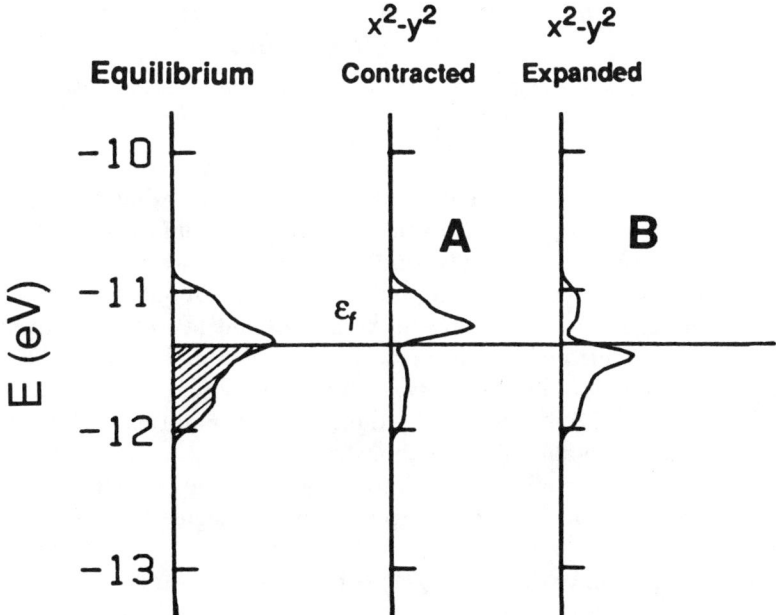

Figure 12. Changes in the $x^2 - y^2$ densities of states on breathing (Scheme V).

magnitude of the thermal vibration parameters associated with the Cu–O distances determined experimentally from neutron diffraction studies (37) of the 1–2–3 compound leads to a calculated charge transfer of 0.22 electron. Our one-electron calculations (Figure 13) show too a stabilization of the distortion for a metal d-count appropriate to the half-filled band on the CuO_2 planes. Such global pictures are very useful in understanding the origin of such distortions. In the language of the moments method (20), the shape of

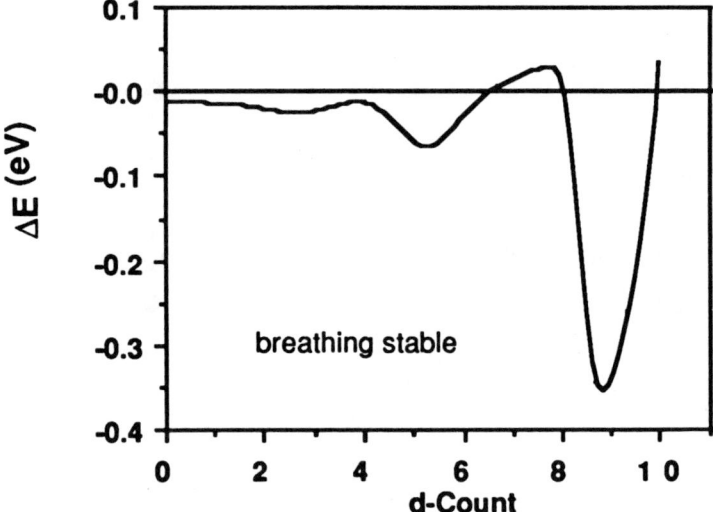

Figure 13. Computed stabilization energy of the breathing mode (Scheme V) in the 1–2–3 compound as a function of the average number of d electrons per copper.

the curve associated with d orbital configurations corresponding to occupancy of $x^2 - y^2$ and z^2 orbitals is a fourth-moment one, typical of Jahn–Teller and Peierls distortions. No static distortion of this type is actually observed, which implies that an important ingredient has been left out of the problem.

We mentioned the importance of both one- and two-electron terms in the energy, in controlling distortions of molecules and solids. Thus, for example, the energy difference between high- and low-spin octahedral Ni^{II} complexes is (crudely) determined by the two-electron terms in the energy, whereas the distortion of the octahedral structure to the square comes about via one-electron terms. Whether a given Ni^{II} complex is octahedral and high-spin or square-planar and low-spin is thus a balance between the two types of forces.

In the distortion of the CuO_2 planes there is a similar balance. Here it involves the cost of placing two electrons on the same copper atom. The effect is largest when the $x^2 - y^2$ band is close to half-full. In this case, on distortion, electrons will have to be paired in some of the orbitals on atom B as a consequence of the Pauli principle. These on-site repulsions associated with the two electrons in the $x^2 - y^2$ band of the Cu^I-like atoms will destabilize such a distortion. Figure 13 indicates that as the electron count moves away from the half-filled band, the driving force associated with the one-electron part of the energy also decreases. Thus, the d-count dependence of these two terms works in opposite directions. Similar comments apply to analogous distortions in the 2–1–4 and other compounds containing sheets of square-planar copper atoms.

The size of these many-body terms is difficult to evaluate. From values of atomic ionization energies (38), the disproportionation reaction $2Cu^{II} \rightarrow Cu^{I} + Cu^{III}$ costs 16.54 eV. Clearly, the actual charge transfer in the solid is considerably less, and the metal orbitals are considerably delocalized via extensive metal–oxygen interactions. (A similar explanation can be offered for the nephelauxetic effect in transition metal complexes.) Disproportionation is observed both in solid $BaBiO_3$ and in $CsAuCl_3$. The atomic values for the energetics of these two cases caution against general use of such numbers in solids. For bismuth, the disproportionation $2Bi^{IV} \rightarrow Bi^{III} + Bi^{V}$ costs much less than a similar disproportionation in copper (4.7 eV). However, for gold, $2Au^{II} \rightarrow Au^{I} + Au^{III}$, the figure is not very different (13.5 eV).

One variant of the breathing motion is shown in structure 1 and Scheme VI for the CuO_2 sheet of the 1-2-3 compound. Here the oxygen atoms

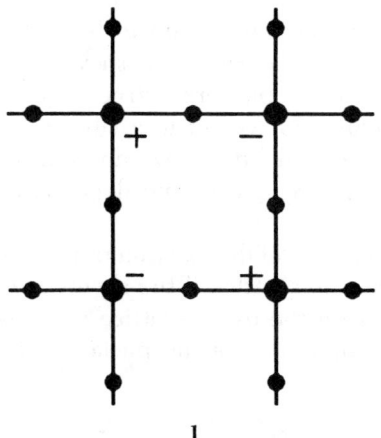

Scheme VI

remain fixed and the Cu–O stretching is really associated with an increase in the puckering of the sheet. (The asymmetric motion we show occurs only at the edge of the Brillouin zone, as shown in structures 2 and 3. At the center of the zone, the two copper atoms move in-phase and chemically remain equivalent.) For this distortion, both an increase in puckering and an increase in Cu–O distance lead to a drop in the energy of the $x^2 - y^2$ band associated with that copper atom, and so the picture is similar to that shown in Figure 10. As before, the unfavorable electron–electron coulombic

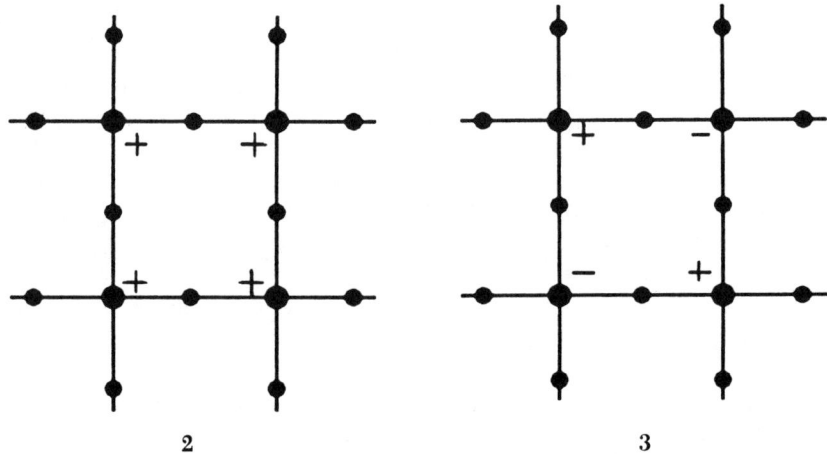

repulsion between two electrons forced to lie on the same copper atom resists the distortion. The particular motion shown in structure 1 and Scheme VI is interesting in that the puckered structures show no motion of the oxygen atoms. Although we have yet to establish the connection between such a motion and superconductivity, we do point out the observation (6, 7) of a zero or near-zero oxygen isotope dependence of T_c in the 1–2–3 compound.

How does the magnitude of the electron transfer on puckering vary with the geometry? It is easy to see that if the sheets are flat then the dynamic electron transfer is zero if the oxygen lattice remains fixed. This is shown in structure 4. If the amplitude of the puckering is the same at the two

centers, then no asymmetry in band position can occur on vibration. Thus, a mechanism of this type suggests that electron transfer will increase with puckering.

The 2–2–1–3 System, $Pb_2Sr_2(Ln_{1-y} M_y^{II})Cu_3O_{8+x}$

$Pb_2Sr_2(Ln_{1-y} M_y^{II})Cu_3O_{8+x}$ (where Ln is a lanthanide or early transition element such as yttrium and M^{II} is a two-valent ion such as Sr or Ca) has recently been characterized (39). It has several features of the 1–2–3 and 2–1–4 compounds. In the stoichiometric compound $Pb_2^{II}Sr_2^{II}Ln^{III}(Cu^{II}O_3)_2Cu^{I}O_2$ ($x = y = 0$), we can envisage (Figure 14)

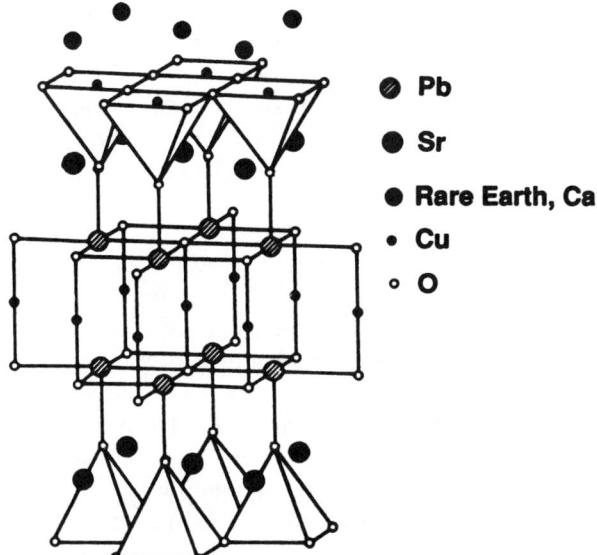

Figure 14. The observed structure of the 2–2–1–3 compound $Pb_2Sr_2(Ln_{1-y} M_y^{II})Cu_3O_{8+x}$.

Cu^IO_2 dumbbells and $Cu^{II}O_3$ square pyramids, just as in the 1–2–3 compound $YBa_2(Cu^{II}O_2)_2(Cu^IO_3)$ with $\delta = 1$. Figure 15 shows a schematic band picture highlighting this. The role of the M^{II} species here is similar to that in the 2–1–4 compound. Electrons are removed from the $x^2 - y^2$ bands of Cu^{II} when $y \neq 0$ but $x = 0$ and the electron occupancy moves away from half-full. The analogy with the 1–2–3 compound may be pushed further to ask whether, when $x > 0$, behavior similar to that described earlier will be

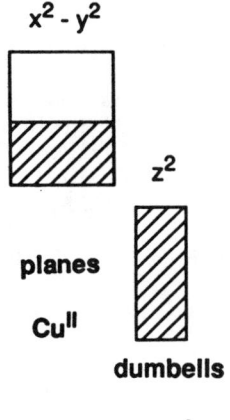

Figure 15. Schematic band picture showing the arrangement of Cu^IO_2 dumbbells and $Cu^{II}O_3$ square pyramids.

found. Addition of oxygen to the dumbbell region could result in chains of square-planar copper atoms if $x = 1$. For x close to 0, T-shaped CuO_3 units will be formed. Thus, geometrically (and perhaps electronically, too), this process is simply the reverse of that for the 1–2–3 compound shown in Schemes III and IV. The situation in $Pb_2Sr_2(Ln_{1-y} M_y^{II})Cu_3O_{8+x}$ is very interesting. The material merits future detailed study of its properties as x and y are varied.

Conclusions

We have shown how many of the electronic aspects of these fascinating systems are controlled by geometrical features of the structure. The focus of attention has been on orbital ideas, but it is clear that other types of interactions are important here too. (*See* ref. 11, for example.) Our computations of the apical–basal angle at the square pyramid in the 1–2–3 compound show a sensitivity of 2° or so, depending on whether the Y atoms are included in the calculation (19). This sensitivity suggests a complex trade-off between different types of interactions. We look forward to the synthesis of more compounds of this type to test the predictions of the model described here.

Acknowledgments

This research was supported by The National Science Foundation and by the University of Chicago. We thank Kathryn Levin for many useful discussions. Figures 2, 5–8, 10–12, and 14 have been adapted from ref. 19. This chapter has been edited for style by members of the ACS Books Department.

References

1. Bednorz, J. G.; Muller, K. A. *Z. Phys.* **1986**, *B64*, 189.
2. Wu, M. K.; Ashburn, J. R.; Torng, C. J.; Horn, P. H.; Meng, R. L.; Gao, L.; Huang, Z. J.; Wang, Y. Q.; Chu, C. W. *Phys. Rev. Lett.* **1987**, *58*, 908.
3. Cava, R. J.; Batlogg, B.; van Dover, R. B.; Murphy, D. W.; Sunshine, S.; Siegrist, T.; Remeika, J. P.; Reitman, E. A.; Zahurak, S.; Espinoza, G. P. *Phys. Rev. Lett.* **1987**, *58*, 1676.
4. Cava, R. J.; van Dover, R. B.; Batlogg, B.; Reitman, E. A. *Phys. Rev. Lett.* **1987**, *58*, 408.
5. Rao, C. N. R.; Ganguly, P.; Raychaudhuri, A. K.; Mahanram, R. A.; Sreedhar, K. *Nature* **1987**, *326*, 856.
6. *Chemistry of High-Temperature Superconductors;* Nelson, D. L.; Whittingham, M. S.; George, T. F., Eds.; ACS Symposium Series 351; American Chemical Society: Washington, DC, 1987.
7. *Chemistry of High-Temperature Superconductors II;* Nelson, D. L.; George, T. F., Eds.; ACS Symposium Series 377; American Chemical Society: Washington, DC, 1988.

8. Sleight, A. W. *Science* **1988**, *242*, 1519.
9. Gazo, J.; Bersuker, I. B.; Garaj, J.; Kabesova, M.; Kohout, J.; Langfelderova, H.; Melnik, M.; Serato, M.; Valach, V. *Coord. Chem. Rev.* **1976**, *19*, 253.
10. Burdett, J. K. *Inorg. Chem.* **1981**, *20*, 1959.
11. Whangbo, M.-H.; Evian, M.; Beno, M. A.; Geiser, U.; Williams, J. M. *Inorg. Chem.* **1988**, *27*, 467.
12. Burdett, J. K. *Molecular Shapes;* Wiley: New York, 1980.
13. Chen, G.; Goddard, W. A., III *Science* **1988**, *239*, 899.
14. Whangbo, M.-H. *Acc.. Chem. Res.* **1983**, *16*, 95.
15. Burdett, J. K. *Prog. Solid State Chem.* **1984**, *15*, 173.
16. Whangbo, M.-H.; Foshee, M. J. *Inorg. Chem.* **1981**, *20*, 113.
17. Sinha, S. K. *Mater. Res. Bull.* **1988**, *13*, 24.
18. Cava, R. J.; Santoro, A.; Johnson, D. W.; Rhodes, W. W. *Phys. Rev.* **1987**, *B35*, 6716.
19. Burdett, J. K.; Kulkarni, G. V. *Phys. Rev. B* **1989**, *80*, 8908.
20. Burdett, J. K. *Struct. Bonding (Berlin)* **1987**, *65*, 29.
21. Rossi, A. R.; Hoffmann, R. *Inorg. Chem.* **1975**, *14*, 365.
22. Burdett, J. K. *Inorg. Chem.* **1985**, *24*, 2244.
23. Albright, T. A.; Burdett, J. K.; Whangbo, M.-H. *Orbital Interactions in Chemistry;* Wiley: New York, 1985.
24. Jorgensen, J. D.; Schuttler, H.-B.; Hinks, D. G.; Capone, D. W.; Zhang, K.; Brodsky, M. B. *Phys. Rev. Lett.* **1987**, *58*, 1024.
25. Burdett, J. K.; Hughbanks, T.; Miller, G. J.; Richardson, J. W.; Smith, J. V. *J. Am. Chem. Soc.* **1987**, *109*, 3639.
26. Burdett, J. K.; Kulkarni, G. V. *J. Am. Chem. Soc.* **1988**, *110*, 5361.
27. Cava, R. J.; Batlogg, B.; Chen, C. H.; Reitman, E. A.; Zahurak, S. M.; Werfer, D. *Nature* **1987**, *329*, 423
28. Renault, A.; Burdett, J. K.; Pouget, J.-P. *J. Solid State Chem.* **1987**, *71*, 587.
29. Xiao, G.; Cieplak, M. Z.; Gavrin, A.; Streitz, F. H.; Bakhshai, A.; Chien, C. L. *Revs. Solid State Science* **1987**, *1*, 323.
30. Gardner, J. A.; Su, H. T.; McKale, A. G.; Kao, S. S.; Peng, L. L.; Warnes, W. H.; Sommers, J. A.; Athreya, K.; Franzen, H.; Kim, S.-J. *Phys. Rev.* **1988**, *B38*, 317.
31. Ourmazd, A.; Spence, J. C. H. *Nature* **1987**, *329*, 426.
32. Burdett, J. K.; Kulkarni, G. V.; Levin, K. *Inorg. Chem.* **1987**, *26*, 3650.
33. Raveau, B.; Michel, C.; Hervieu, M.; Provost, J. *Revs. Solid State Science* **1988**, *2*, 115.
34. Yared, Y. W.; Miles, S. L.; Bau, R.; Reed, C. A. *J. Am. Chem. Soc.* **1977**, *99*, 7076.
35. Cava, R. J.; Batlogg, B.; Rabe, K. M.; Reitman, E. A.; Gallagher, P. K.; Rupp, L. W., Jr. *Physica* **1988**, *C156*, 523.
36. Simon, A. *Angew. Chem. Int. Ed. Engl.* **1987**, *26*, 579.
37. Beno, M. A.; Soderholn, L.; Capone, D. W.; Hinks, D. G.; Jorgensen, J. D.; Schuller, I. K.; Segre, C. V.; Zhang, K.; Grace, J. D. *Appl. Phys. Lett.* **1987**, *51*, 57.
38. Moore, C. E. *Ionization Potentials and Ionization Limits Derived from Analyses of Optical Spectra, NSRDS-NBS 24;* NBS: Washington, DC, 1970.
39. Cava, R. J.; Batlogg, B.; Krajewski, J. J.; Rupp, L. W.; Schneemeyer, L. F.; Siegrist, T.; van Dover, R. B.; Marsh, P.; Peck, W. F., Jr; Gallagher, P. K.; Glarum, S. H.; Marshall, J. H.; Farrow, R. C.; Waszczak, J. V.; Hull, R.; Trevor, P. *Nature* **1988**, *336*, 211.

RECEIVED for review May 1, 1989. ACCEPTED revised manuscript August 14, 1989.

EXPERIMENTAL ASPECTS
OF SOLID-STATE SYSTEMS

18

Organometallic Chemical Vapor Deposition

Strategies and Progress in the Preparation of Thin Films of Superconductors Having High Critical Temperatures

Lauren M. Tonge, Darrin S. Richeson, Tobin J. Marks, Jing Zhao, Jiming Zhang, Bruce W. Wessels, Henry O. Marcy, and Carl R. Kannewurf

Materials Research Center, Northwestern University, Evanston, IL 60208–3113

Highly oriented films of the high-critical-temperature (T_c) superconductors $YBa_2Cu_3O_{7-\delta}$, $Bi_2(Sr,Ca)_3Cu_2O_x$, and $TlBa_2Ca_2Cu_3O_x$ can be prepared by organometallic chemical vapor deposition (OMCVD) by using volatile molecular β-diketonate (acetylacetonate, dipivaloylmethanate, heptafluorodimethyloctanedionate) and organometallic (triphenylbismuth, cyclopentadienylthallium) precursors. After annealing, zero-resistance temperatures for these three films are 86, 75, and 102 K, respectively. Keys to high-quality OMCVD-derived films include the use of fluorocarbon-containing precursors, the use of low deposition pressures, the use of water as a reactant gas, and the use of rapid thermal annealing techniques.

THE RECENT DISCOVERY OF SEVERAL CLASSES of superconducting mixed-metal oxides with critical temperature (T_c) values greater than 77 K has stimulated intense worldwide scientific interest (1–3). One area of great activity has been the development of processes to produce high-quality thin films of these materials (4–6), because it is likely that high-T_c superconductors will first have technological impact in this form. Efforts to date have primarily centered on physical vapor deposition (PVD) techniques such as sputtering, evaporation, molecular beam epitaxy, and laser ablation (4–6).

An attractive alternative chemical vapor deposition (CVD) approach is organometallic chemical vapor deposition (OMCVD) (7, 8).

Such a process produces films by using volatile metal–organic molecular precursors and suitably designed gas-phase reaction–deposition chemistry. In principle, OMCVD offers the advantages over PVD of relatively simple apparatus, amenability to large-scale deposition, ability to coat variously shaped objects, adaptability to a wide range of materials, the possibility of deposition at low temperatures, and the possibility of creating metastable structures. Indeed, OMCVD is the technique of choice for the large-scale fabrication of films of III–V and II–VI semiconductors (7), and it has been used to produce films of a variety of metal oxides (8–12).

With regard to producing high-T_c superconducting thin films by OMCVD, crucial questions concern whether suitable precursors and deposition chemistry can be developed. This chapter reviews recent efforts at Northwestern University to develop OMCVD processes for high-T_c superconducting films and describes some of the properties of the resulting films. Films of the $YBa_2Cu_3O_{7-\delta}$, $(BiO)_2Sr_2Ca_{n-1}Cu_nO_x$, and $(TlO)_mBa_2Ca_{n-1}Cu_nO_x$ classes of superconductors can be prepared on a variety of substrates. These films have excellent phase purity, high degrees of preferential orientation, and good electrical properties.

Strategies for Precursor Design and Deposition Methodology

A requisite property for all OMCVD precursors must be suitable volatility. An attractive strategy to achieve this volatility is to minimize lattice cohesive energies by encapsulating the metal ion in a sterically saturating nonpolar ligand environment (13, 14). Ligand fluorination is known to further promote volatility (13, 14). For large divalent and trivalent ions (especially Ca^{2+}, Sr^{2+}, Ba^{2+}, and Y^{3+}), such requirements can be satisfied only with bulky, multidentate ligands (15). Other important considerations in precursor selection include ease of synthesis, purification, and handling. All factors being equal, air-stable precursors are most desirable because they are more conveniently manipulated and require less specialized apparatus. Finally, OMCVD precursors must display appropriate gas-phase reactivity for the formation of films (i.e., a precursor that is chemically inert will be useless).

For our initial high-T_c OMCVD experiments (16–21), β-diketonate complexes were chosen for Cu, Y, Ca, Sr, and Ba sources (structure 1). $Cu(acac)_2$ (acac is acetylacetonate), $Y(dpm)_3$, $Ca(dpm)_2$, and $Sr(dpm)_2$ (dpm is dipivaloylmethanate) are sufficiently volatile to transport the respective metals. In our hands, however, $Ba(dpm)_2$ (16, 17) has insufficient volatility. Attempts to transport it result in irreproducible vapor pressure characteristics, thermal decomposition, and low deposition yields. In contrast to $Ba(dpm)_2$, we find that $Ba(fod)_2$ (fod is heptafluorodimethyloctanedionate) (16) is more volatile, exhibits steady vapor pressure, and transports Ba at lower temperatures and

$$M\left[\begin{array}{c} O-C{\overset{R_1}{\diagdown}} \\ {}}CH \\ O-C{\diagdown}_{R_2} \end{array}\right]_n$$

$R_1 = R_2 = CH_3$ $M(acac)_n$

$R_1 = R_2 = C(CH_3)_3$ $M(dpm)_n$

$R_1 = C(CH_3)_3$; $R_2 = CF_2CF_2CF_3$ $M(fod)_n$

1

with far less decomposition. It also serves as a beneficial source of fluoride (vide infra). For Bi and Tl sources, the organometallic compounds triphenylbismuth (structure **2**) and cyclopentadienylthallium (structure **3**) offer high volatility, air stability, and reactivity with respect to deposition chemistry (vide infra). All precursors used in this work were rigorously purified by multiple sublimation or recrystallization.

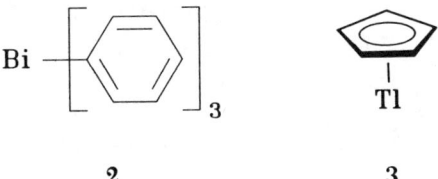

      2   **3**

With regard to designing OMCVD chemistry, the goal has been to employ reagents that cleanly strip ligands from the precursor molecules under as mild conditions as possible, to afford less-volatile products. In principle, simple oxidation (eq 1) offers one such approach.

$$ML_n + O_2 \xrightarrow{\Delta} MO_x + L \text{ oxidation products} \quad (1)$$

where M is metal and L is ligand. However, it appears to require very high deposition temperatures for high-T_c superconductor OMCVD (*19*) and produces significant yields of BaF_2 when applied to $Ba(fod)_2$ transport (eq 2) (*15, 22*).

$$Ba(fod)_2 + O_2 \xrightarrow{\Delta} BaF_2 + \text{ other products} \quad (2)$$

Protonolysis of the precursor metal–ligand bonds with water vapor (eqs 3 and 4) appears to be a cleaner, milder approach, and the hydroxide products are expected to ultimately condense and yield nonvolatile oxides.

$$L_nM-OR + H_2O \xrightarrow{\Delta} L_nM-OH + ROH \quad (3)$$

$$L_nM-R' + H_2O \xrightarrow{\Delta} L_nM-OH + R'H \quad (4)$$

Furthermore, we find that water vapor can be employed to control the amount of BaF_2 produced in $Ba(fod)_2$ OMCVD (eq 5) (16, 22).

$$BaF_2 + H_2O \xrightarrow{\Delta} BaO + 2HF \quad (5)$$

In practice, we find that oxygen saturated with water vapor and water vapor alone are excellent reactant gases for the deposition process.

OMCVD was carried out at low pressure (266–666 Pa [2–5 torr]) in a cold-wall horizontal quartz reactor with separate, parallel, heated quartz inlet tubes for introducing the precursors in an argon stream. Low-pressure operation is particularly useful in maximizing deposition yields at low precursor temperatures. Substrates were located on a carbon susceptor that was heated with an IR lamp. Reactant gases were introduced immediately upstream from the susceptor. Substrates employed include [100] single-crystal MgO, [100] single-crystal $SrTiO_3$, and 9.5 mol % yttria-stabilized zirconia (YSZ) with random orientation. Following deposition, films were annealed under oxygen in a manner appropriate to each material (vide infra).

Microstructural characterization of OMCVD-derived films was carried out by X-ray diffraction (Cu K_α), scanning electron microscopy (SEM), scanning Auger spectroscopy, and energy-dispersive X-ray analysis (EDX). Charge-transport characterization employed four-probe techniques and the automated instrumentation described previously (23, 24).

OMCVD of $YBa_2Cu_3O_{7-\delta}$ Films

Deposition of $YBa_2Cu_3O_{7-\delta}$ films can be carried out by using the precursor combinations $Cu(acac)_2$ + $Y(dpm)_3$ + $Ba(fod)_2$ (16) or $Cu(acac)_2$ + $Y(dpm)_3$ + $Ba(dpm)_2$ (17), with the former precursors preferred (vide supra). In such experiments, source temperatures were 150, 100, and 170 °C for the Cu, Y, and Ba precursors, respectively. The system pressure was 666 Pa (5 torr), the substrate temperature was 700 °C, and water vapor was employed as the reactant gas. The initially deposited films are largely amorphous (as shown by X-ray diffraction, Figure 1A) and insulating. As can also be seen in Figure 1A, the as-deposited films contain traces of BaF_2. The amounts of BaF_2 can be increased by increasing the O_2/H_2O ratios in the reactant gas during deposition (16, 22).

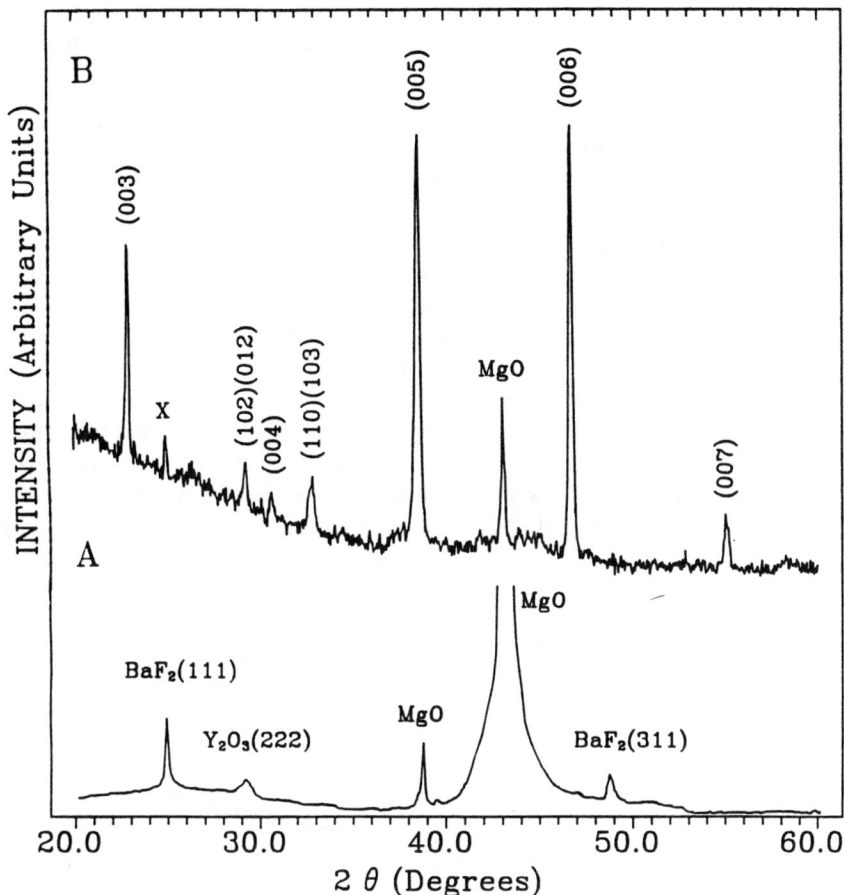

Figure 1. X-ray diffraction patterns of (A) an as-deposited, OMCVD-derived Y–Ba–Cu–O film on [100] MgO and (B) Y–Ba–Cu–O film on [100] MgO after annealing.

Annealing of the film in Figure 1A under flowing oxygen for 10 h at 600 °C, 1.5 h at 900 °C, and 10 min at 960 °C produces crystalline, 0.5–2.0-μm-thick films of the $YBa_2Cu_3O_{7-\delta}$ superconductor (Figure 1B). The enhanced relative intensities of the $(00l)$ reflections versus those in a randomly oriented powder indicate significant preferential orientation of the crystallite c axes perpendicular to the plane of the substrate. Hence, the charge-transporting CuO sheets are oriented parallel to the plane of the substrate. Measurement of the c axis length indicates an oxygen stoichiometry of approximately $YBa_2Cu_3O_{6.6}$. There is no diffractometric evidence for contaminating CuO, $BaCuO_2$, or $Ba_3Y_4O_9$ phases (18). Preliminary studies (22) indicate that increasing the quantity of BaF_2 increases the degree of preferential c axis orientation perpendicular to the substrate plane. A similar effect has been noted in films deposited by pyrolysis of metal carboxylate coatings (25).

Figure 2 shows variable-temperature four-probe resistivity data for an annealed film of the type shown in Figure 1B. The metallike behavior ($d\rho/dT > 0$) at higher temperatures is characteristic of relatively high-quality $YBa_2Cu_3O_{7-\delta}$ films (4–6), as is the onset of the superconducting state at ~90 K. In this particular specimen, $\rho = 0$ occurs at 66.2 K (16). A scanning electron micrograph of a typical annealed $YBa_2Cu_3O_{7-\delta}$ film of the type shown in Figure 1B reveals 1–5-μm grains evenly distributed over the film (Figure 3).

In an effort to improve film charge-transport characteristics and to minimize film–substrate reactions (26), rapid thermal annealing techniques have been applied (22). Annealing as-deposited films in flowing oxygen for 1.5 h at 870 °C and for several seconds at 980 °C considerably enhances grain size and grain overlap (evident in SEM photographs), as well as increasing the $\rho = 0$ temperatures. The resistivity data shown in Figure 4 indicate $\rho = 0$ temperatures for the $YBa_2Cu_3O_{7-\delta}$ films subjected to rapid thermal annealing on $SrTiO_3$ and YSZ of 78 and 87 K, respectively. The latter temperature is

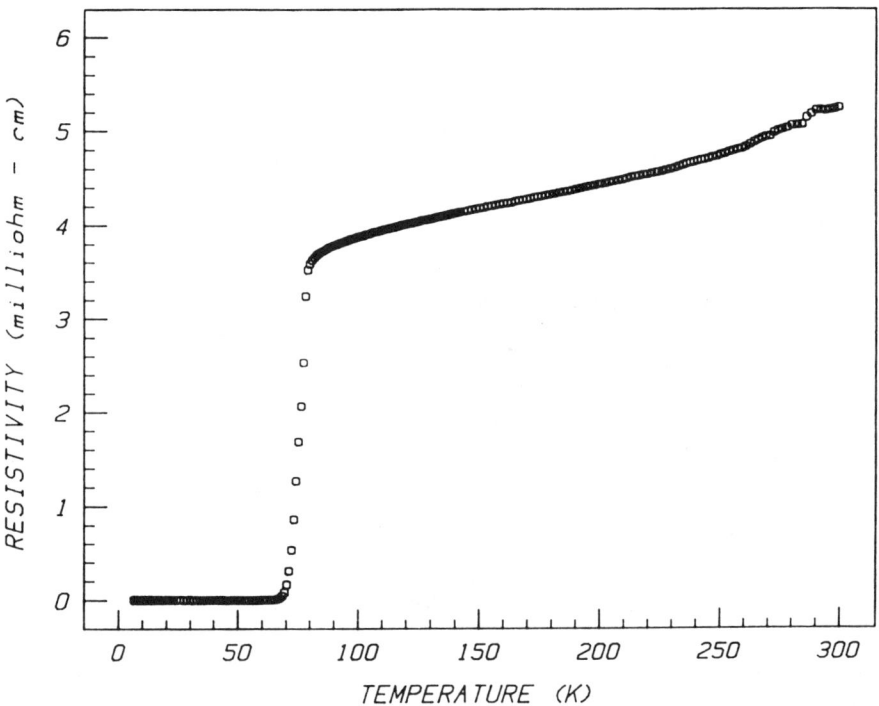

Figure 2. Variable-temperature four-probe resistivity data for an annealed OMCVD-derived $YBa_2Cu_3O_{7-\delta}$ film on [100] MgO.

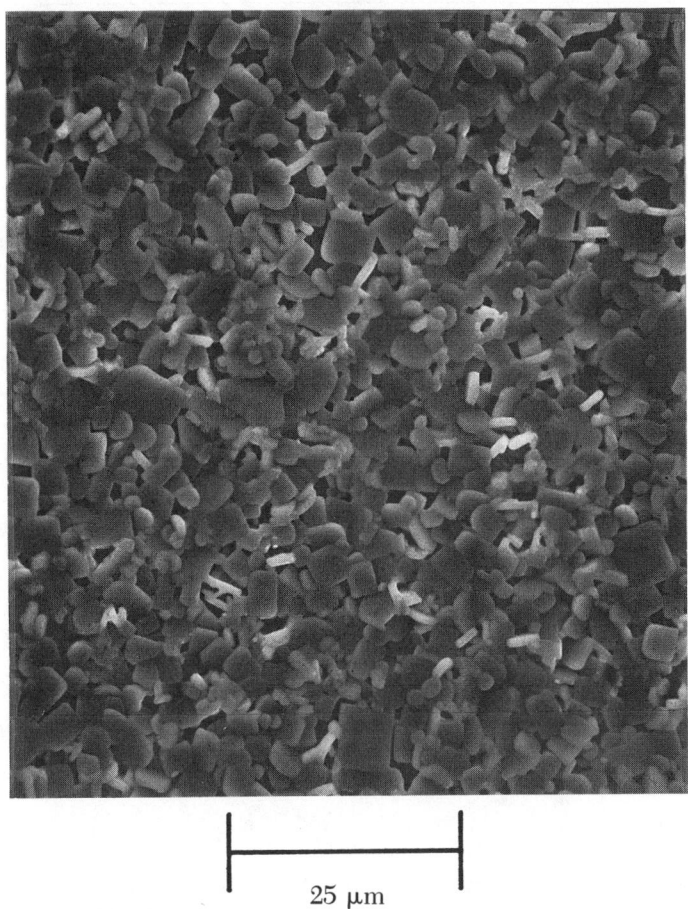

Figure 3. Scanning electron micrograph of an annealed OMCVD-derived $YBa_2Cu_3O_{7-\delta}$ film on [100] MgO.

within a few degrees of the highest values yet reported for $YBa_2Cu_3O_{7-\delta}$ films prepared by PVD techniques (4–6).

OMCVD of $(BiO)_2Sr_2Ca_{n-1}Cu_nO_x$ Films

In contrast to the $YBa_2Cu_3O_{7-\delta}$ system, the Bi–Sr–Ca–Cu–O superconductors exhibit a complex crystal chemistry with T_c values as high as 110 K (27–30). Needless to say, OMCVD, in using four sources, presents a great challenge. Deposition of Bi–Sr–Ca–Cu–O films was carried out at 266 Pa (2 torr) by using the precursors $Bi(C_6H_5)_3$, $Ca(dpm)_2$, $Sr(dpm)_2$, and $Cu(acac)_2$ with source temperatures of 145–230 °C and a substrate temperature of 550 °C (31–34). The reactant gas was water-vapor-saturated oxygen.

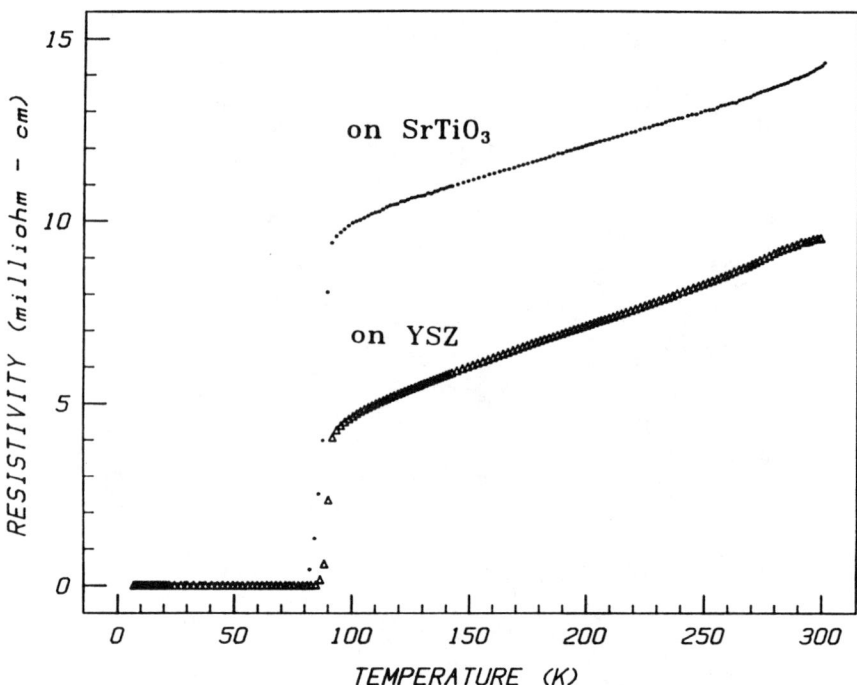

Figure 4. Variable-temperature four-probe resistivity data for OMCVD-derived $YBa_2Cu_3O_{7-\delta}$ films processed by rapid thermal annealing. Substrates are indicated in the figure.

Although water vapor alone gives acceptable deposition yields of Ca, Sr, and Cu oxides, oxygen is necessary to achieve efficient, simultaneous deposition of Bi. Presumably the hydrolytic stability of the Bi–phenyl bond is too great to allow efficient protonolytic cleavage.

Figure 5A shows that the as-deposited Bi–Sr–Ca–Cu–O films are crystalline. However, the lattice parameters of $c = 24.4$ Å and $a = b = 5.4$ Å indicate a previously observed (35, 36) semiconducting phase. SEM indicates an irregular, probably heterogeneous, film microstructure (Figure 6A) for the as-deposited films. Annealing of the as-deposited films in flowing oxygen for 0.5 h at 600 °C and 0.5 h at 865 °C yields 3–6-μm-thick films that (by X-ray diffraction, Figure 5B) consist largely of the $T_c = 85$ K $Bi_2(Sr,Ca)_3Cu_2O_x$ phase (27–30, 37, 38). The enhanced intensities of the (00*l*) reflections indicate significant preferential orientation of the crystallite *c* axes perpendicular to the surface of the substrate. Thus, as in the case of the $YBa_2Cu_3O_{7-\delta}$ films, the charge-carrying CuO planes are formed parallel to the substrate plane. The diffraction data also indicate traces of the $c = 37$ Å, $T_c = 110$ K $Bi_2(Sr,Ca)_4Cu_3O_x$ phase (27–30, 37, 38), as well as of the

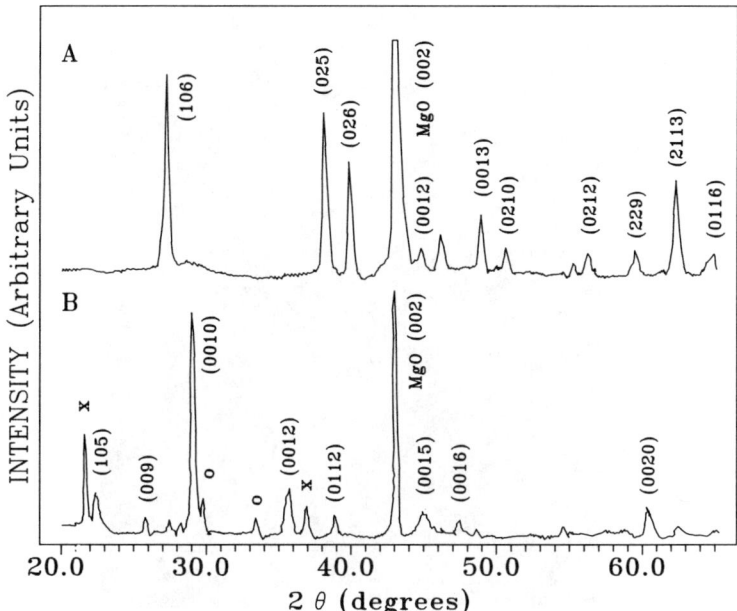

Figure 5. A. X-ray diffraction pattern of an unannealed Bi–Sr–Ca–Cu–O film on [100] MgO. The tentative indexing assumes a tetragonal cell with c = 24.4 Å and a = b = 5.4 Å. B. X-ray diffraction pattern of an annealed Bi–Sr–Ca–Cu–O film on [100] MgO. The reflections assigned to the T_c = 110 K (x) and semiconducting (o) phases are so indicated.

aforementioned c = 24.4 Å semiconducting phase. SEM data for a typical annealed film reveal a rough surface with grain sizes of 5–15 μm (Figure 6B). Results on YSZ substrates are similar.

Figure 7 shows charge-transport data for a typical annealed OMCVD-derived Bi–Sr–Ca–Cu–O film on [100] MgO. As is typical of relatively high-quality films, transport is metallic ($d\rho/dT > 0$) at higher temperatures. The onset of superconductivity begins at ~110 K, with $\rho = 0$ at 75 K. PVD-derived Bi–Sr–Ca–Cu–O films have been reported to exhibit $\rho = 0$ temperatures as high as 100 K (39), although most reports are in the 70–80 K range.

OMCVD of $(TlO)_m Ba_2 Ca_{n-1} Cu_n O_x$ Films

The $(TlO)_m Ba_2 Ca_{n-1} Cu_n O_x$ system (m = 1, 2; n = 1, 2, 3) exhibits the greatest multiplicity of high-T_c superconducting phases (27–30), impressive environmental stability (40–45), a paucity of intergranular "weak links" that appears to limit the critical current densities of other high-T_c materials (44–46), and reported T_c values as high as 122 K (26–29). Nevertheless, the volatility of Tl_2O and Tl_2O_3 complicates the annealing of PVD-derived

Figure 6. A. Scanning electron micrograph of the unannealed Bi–Sr–Ca–Cu–O film in Figure 5A. B. Scanning electron micrograph of the annealed Bi–Sr–Ca–Cu–O film in Figure 5B.

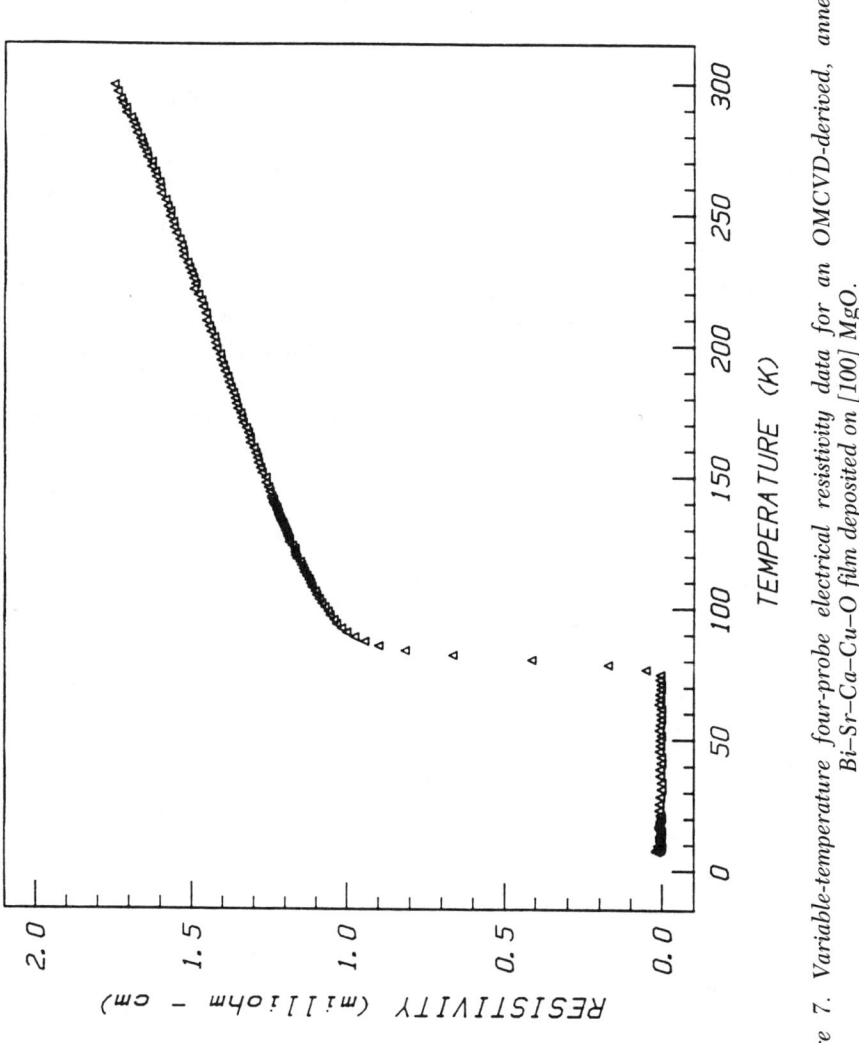

Figure 7. *Variable-temperature four-probe electrical resistivity data for an OMCVD-derived, annealed Bi–Sr–Ca–Cu–O film deposited on [100] MgO.*

Tl–Ba–Ca–Cu–O films, and OMCVD with four sources is expected to be difficult. We find that excellent high-T_c Tl–Ba–Ca–Cu–O films can be prepared by two complementary, OMCVD-based approaches (47). In one approach, a Ba–Ca–Cu–O film is prepared by OMCVD and Tl is then introduced by vapor diffusion of the oxides. In the second approach, Tl is introduced into the Ba–Ca–Cu–O film via an OMCVD process that uses volatile cyclopentadienylthallium.

OMCVD of Ba–Ca–Cu–O films was carried out using Ba(fod)$_2$, Ca(dpm)$_2$, and Cu(acac)$_2$ as precursors at 666 Pa (5 torr) of pressure in the reactor described. The source temperatures were 160–210 °C, the substrate temperature was 600 °C, and water vapor was employed as the reactant gas. The as-deposited films were annealed for 3 h at 800 °C in flowing oxygen saturated with water vapor. The role of the water is to remove excess fluoride from the films (vide supra).

To introduce Tl into the Ba–Ca–Cu–O films by vapor diffusion, the films were heated under air for 3 min at 870 °C (followed by slow cooling) in a closed alumina crucible containing a pellet of bulk TlBa$_2$Ca$_2$Cu$_3$O$_x$. This material serves as a source of Tl$_2$O and Tl$_2$O$_3$ (40–45). X-ray diffraction measurements (Figure 8A) reveal that the films are composed primarily of the TlBa$_2$Ca$_2$Cu$_3$O$_x$ $T_c = 110$ K phase (40–46). The enhanced (00l) reflection intensities indicate, as for the aforementioned Y–Ba–Cu–O and

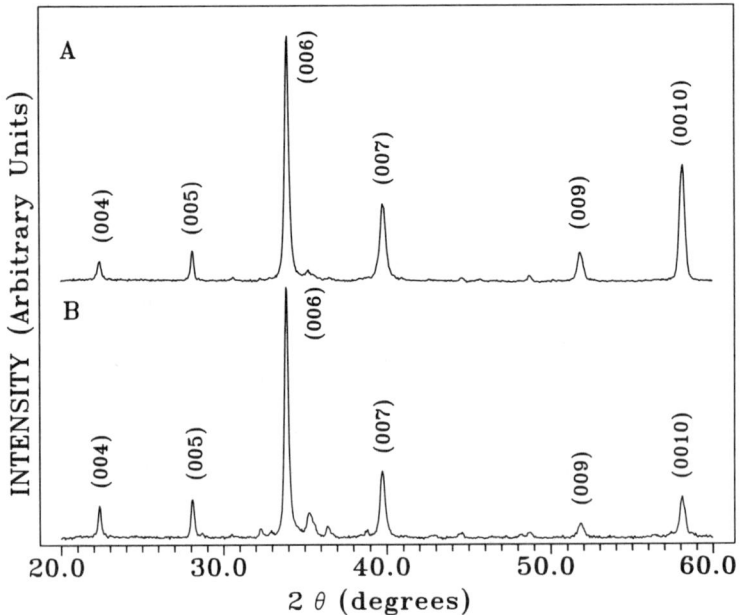

Figure 8. X-ray diffraction patterns of (A) a Tl–Ba–Ca–Cu–O film on YSZ prepared by OMCVD of a Ba–Ca–Cu–O film followed by vapor diffusion of thallium and (B) a Tl–Ba–Ca–Cu–O film on YSZ prepared by OMCVD.

Bi–Sr–Ca–Cu–O films, high preferential orientation of the crystallite c axes perpendicular to the substrate plane. SEM data (Figure 9A) reveal a highly irregular surface with 1–5-μm grains. Charge-transport data (Figure 10A) for these films reveal metallike character at higher temperatures and a transition to the superconducting state beginning at ~120 K. The $\rho = 0$ temperature is 102 K. With one recent exception (46), this temperature is, to our knowledge, comparable to the best achieved with PVD-derived films (40–45).

Tl introduction into the Ba–Ca–Cu–O films by OMCVD was carried out by using $Tl(C_5H_5)$ in an atmospheric-pressure reactor. The source temperature was 50–80 °C, the substrate temperature was 300 °C, the carrier gas was argon, and the reactant gas was water-saturated oxygen. The as-deposited films are amorphous by X-ray diffraction and are semiconducting. Annealing of these films can be carried out by rapid heating (~1 min at 800 °C) under air in a closed crucible or, more satisfactorily, in the presence of bulk $TlBa_2Ca_2Cu_3O_x$ to avoid thallium loss. The X-ray diffraction patterns of these films (Figure 8B) are similar to those described. This similarity indicates the predominance of the $TlBa_2Ca_2Cu_3O_x$ $T_c = 110$ K phase, with the possible presence of traces of other phases. Again, preferential orientation of the crystallite CuO planes parallel to the substrate surface is evident. The charge-transport characteristics of the OMCVD-derived Tl–Ba–Ca–Cu–O films (Figure 10B) are metallike at higher temperatures, with a superconducting onset at ~120 K and $\rho = 0$ at 100 K. Electron micrographs (Figure 9B) reveal a smoother, more regular surface than in the case of films prepared by OMCVD + Tl_2O–Tl_2O_3 vapor diffusion.

Conclusions

The initial results of this research effort show that OMCVD is a viable approach to the synthesis of high-T_c superconducting films. Indeed, highly oriented films of the $YBa_2Cu_3O_{7-\delta}$, $Bi_2(Sr,Ca)_3Cu_2O_x$, and $TlBa_2Ca_2Cu_3O_x$ superconductors have been prepared with zero-resistance temperatures of 86, 75, and 102 K, respectively. The following points are particularly noteworthy:

- the use of $Ba(fod)_2$ as a volatile Ba precursor and source of film-orienting BaF_2;
- the use of low-pressure deposition procedures;
- the use of water as a protonolytic reactant gas and BaF_2 moderator;
- the observation of amorphous Y–Ba–Cu–O and Tl–Ba–Ca–Cu–O films prior to annealing;

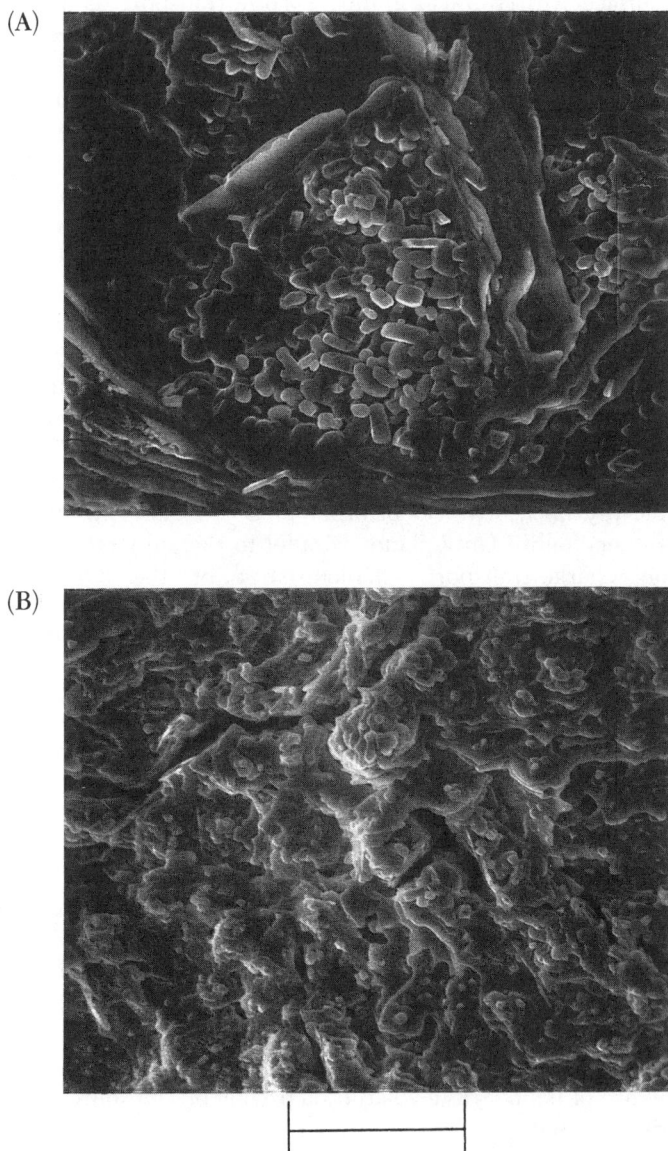

Figure 9. Scanning electron micrograph of (A) the Tl–Ba–Ca–Cu–O film shown in Figure 8A and (B) the Tl–Ba–Ca–Cu–O film shown in Figure 8B.

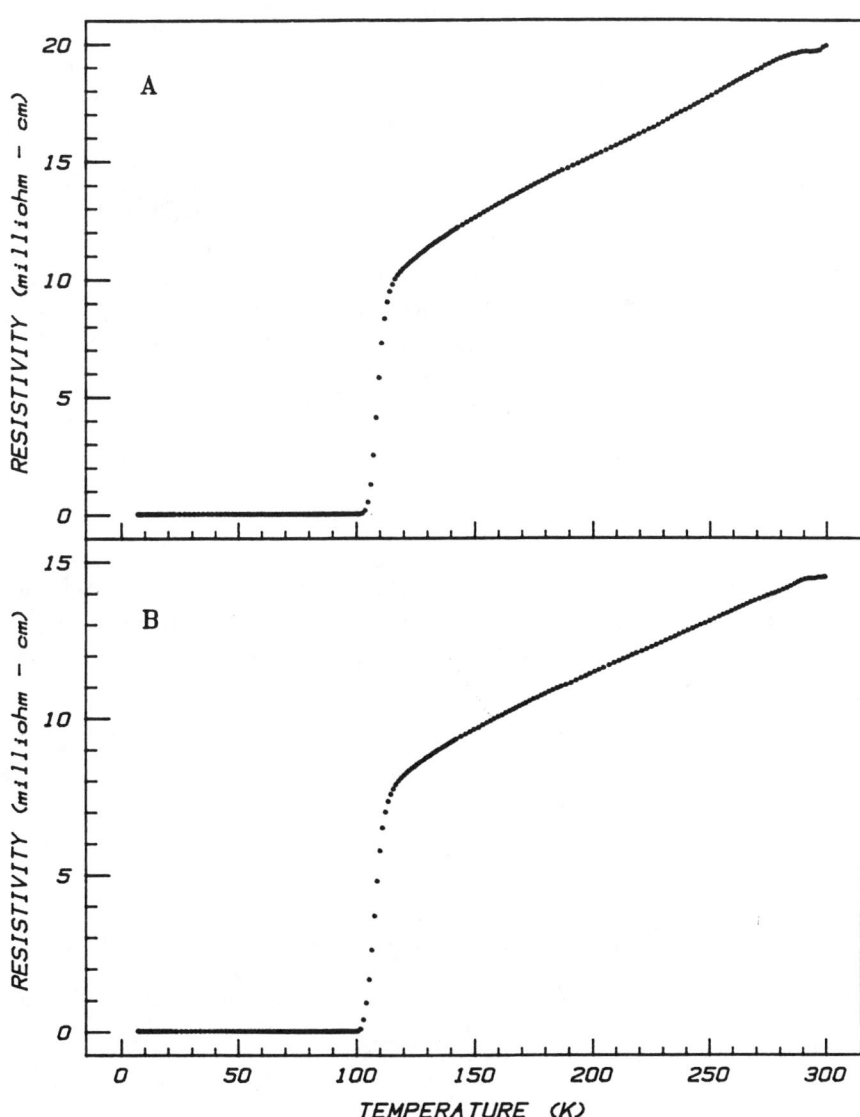

Figure 10. Four-probe variable-temperature electrical resistivity data for (A) the Tl–Ba–Ca–Cu–O film shown in Figure 9A and (B) the Tl–Ba–Ca–Cu–O film shown in Figure 9B.

- the use of Bi(C$_6$H$_5$)$_3$ and Tl(C$_5$H$_5$) as volatile Bi and Tl sources, respectively;
- the use of rapid thermal annealing techniques to improve film grain structure and electrical properties; and
- the use of combined OMCVD and vapor diffusion techniques for film deposition.

These results indicate that many factors in the OMCVD process influence high-T_c superconductor film stoichiometry and microstructure (hence, charge-transport characteristics) in a way that is not yet well understood. Important issues to be addressed concern the development of more volatile and reactive precursors, understanding the deposition chemistry, understanding the role in film growth of substrate and added reagents (e.g., BaF$_2$), understanding the annealing–crystallization chemistry, and understanding film microstructure–charge-transport (especially critical current density) relationships. Further research will focus upon these issues and on developing OMCVD routes to other novel electronic materials.

Acknowledgments

This research was supported by the National Science Foundation through the Northwestern Materials Research Center (Grants DMR8520280 and DMR8821571) and in part by the Office of Naval Research and Argonne National Laboratory.

References

1. *The Chemistry of High-Temperature Superconductors;* Nelson, D. L.; Whittingham, S. M.; George, T. F., Eds.; ACS Symposium Series 351; American Chemical Society: Washington, DC, 1987.
2. *The Chemistry of High-Temperature Superconductors II;* Nelson, D. L.; George, T. F., Eds.; ACS Symposium Series 377; American Chemical Society: Washington, DC, 1988.
3. Maple, M. B., Ed. *MRS Bull.* **1989,** *XIV,* No. 1, 20–71, and references therein.
4. *Thin Film Processing and Characterization of High-Temperature Superconductors;* Harper, J. M. E.; Colton, R. J.; Feldman, L. C., Eds.; American Vacuum Society Series No. 3; American Institute of Physics Conference Proceedings No. 165; American Institute of Physics: New York, 1988.
5. *High-Temperature Superconductors II;* Capone, D. W., II; Butler, W. H.; Batlogg, B.; Chu, C. W., Eds.; Extended Abstracts; Materials Research Society: Pittsburgh, PA, 1988.
6. Laibowitz, R. B. In ref. 3; Maple, M. B., Ed.; pp 58–62, and references therein.
7. Dapkus, P. D. *Ann. Rev. Mater. Sci.* **1982,** *12,* 243–269.
8. Prakash, H. *Prog. Cryst. Growth Charact.* **1983,** *6,* 371–391.
9. Ryabova, L. A. *Curr. Topics Mater. Sci.* **1981,** *7,* 587–642.
10. Curtis, B. J.; Brunner, H. R. *Mater. Res. Bull.* **1975,** *10,* 515–520.

11. Itoh, H.; Takeda, T.; Naka, S. *J. Mater. Sci.* **1986**, *21*, 3677–3680.
12. Souletie, P.; Bethke, S.; Wessels, B. W.; Pan, H. *J. Cryst. Growth, (Proc. Third Int. Conf. II–VI Compounds)* **1988**, *86*, 248–251.
13. Cuellar, E. A.; Miller, S. S.; Marks, T. J.; Weitz, E. *J. Am. Chem. Soc.* **1983**, *105*, 4580–4589.
14. Cuellar, E. A.; Marks, T. J. *Inorg. Chem.* **1981**, *20*, 2129–2137.
15. Sievers, R. E.; Sadlowski, J. E. *Science* **1978**, *201*, 217–223, and references therein.
16. Zhao, J.; Dahmen, K.-H.; Marcy, H. O.; Tonge, L. M.; Marks, T. J.; Wessels, B. W.; Kannewurf, C. R. *Appl. Phys. Lett.* **1988**, *53*, 1750–1752.
17. Zhao, J.; Dahmen, K.-H.; Marcy, H. O.; Tonge, L. M.; Wessels, B. W.; Marks, T. J.; Kannewurf, C. R. *Solid State Commun.* **1989**, *69*, 187–189.
18. Berry, A. D.; Gaskill, D. K.; Holm, R. T.; Cukauskas, E. J.; Kaplan, R.; Henry, R. L. *Appl. Phys. Lett.* **1988**, *52*, 1743–1745.
19. Yamane, H.; Masumoto, H.; Hirai, T.; Iwasaki, H.; Watanabe, K.; Kobayashi, N.; Muto, Y.; Kurosawa, H. *Appl. Phys. Lett.* **1988**, *53*, 1548–1550.
20. Panson, A. J.; Charles, R. G.; Schmidt, D. N.; Szedon, J. R.; Machiko, G. J.; Braginski, A. I. *Appl. Phys. Lett.* **1988**, *53*, 1756–1758.
21. Zhang, K.; Kwak, B. S.; Boyd, E. P.; Wright, A. C.; Erbil, A. *Appl. Phys. Lett.* **1989**, *54*, 380–382.
22. Zhao, J.; Marcy, H. O.; Tonge, L. M.; Wessels, B. W.; Marks, T. J.; Kannewurf, C. R. *Physica C,* **1989**, *159*, 710–714.
23. Lyding, J. W.; Marcy, H. O.; Marks, T. J.; Kannewurf, C. R. *IEEE Trans. Instrum. Meas.* **1988**, *37*, 76–80.
24. Kanatzidis, M. G.; Marks, T. J.; Marcy, H. O.; McCarthy, W. J.; Kannewurf, C. R. *Solid State Commun.* **1988**, *65*, 1333–1337.
25. Gupta, A.; Jagannathan, R.; Cooper, E. I.; Geiss, E. A.; Landman, J. I.; Hussey, B. W. *Appl. Phys. Lett.* **1988**, *52*, 2077–2079.
26. Cheung, C. T.; Ruckenstein, E. *J. Mater. Res.* **1989**, *4*, 1–15, and references therein.
27. Sleight, A. W.; Subramanian, M. A.; Torardi, C. C., ref. 3, pp 45–48, and references therein.
28. Schuller, I. K.; Jorgensen, J. D. In ref. 3; Maple, M. B., Ed.; pp 27–30, and references therein.
29. Maeda, H.; Tanaka, Y.; Fukutomi, M.; Asano, T. *Jpn. J. Appl. Phys.* **1988**, *27*, L209–L210.
30. Tarascon, J. M.; LePage, Y.; Barboux, P.; Bagley, B. G.; Greene, L. H.; McKinnon, W. R.; Hull, G. W.; Giroud, M.; Hwang, D. M. *Phys. Rev. B* **1988**, *37*, 9382–9389.
31. Zhang, J.; Zhao, J.; Marcy, H. O.; Tonge, L. M.; Wessels, B. W.; Marks, T. J.; and Kannewurf, C. R. *Appl. Phys. Lett.* **1989**, *54*, 1166–1168.
32. Yamane, H.; Kurosawa, H.; Hirai, T.; Iwasaki, H.; Kobayashi, N.; Muto, Y. *Jpn. J. Appl. Phys.* **1988**, *27*, L1495–L1497.
33. Gaskill, D. K.; Berry, A. D.; Cukauskas, E. J.; Fatemi, M.; Fox, W. B.; Holm, R. T.; Kaplan, R., presented at the Materials Research Society Fall Meeting, Boston, MA, Nov. 28–Dec. 2, 1988; Abstract F9.8.
34. Kimura, T.; Ihara, M.; Yamawaki, H.; Ikeda, K.; Ozeki, M., presented at the Materials Research Society Fall Meeting, Boston, MA, Nov. 28–Dec. 2, 1988; Abstract F9.7.
35. Mikalsen, D. J.; Roy, R. A.; Yee, D. S.; Shivashanker, S. A.; Cuomo, J. J. *J. Mater. Res.* **1988**, *3*, 613–618.
36. Rice, C. E.; Levi, A. F. J.; Fleming, R. M.; Marsh, P.; Baldwin, K. W.; Anzlowar, M.; White, A. E.; Short, K. T.; Nakahara, S.; Stormer, H. L. *Appl. Phys. Lett.* **1988**, *52*, 1828–1830.

37. Hazen, R. M.; Prewitt, C. T.; Angel, R. J.; Ross, N. L.; Finger, L. W.; Hadidiacos, C. G.; Veblen, D. R.; Heaney, P. J.; Hor, P. H.; Meng, R. L.; Sun, Y. Y.; Wang, Y. Q.; Xue, Y. Y.; Huang, Z. J.; Gao, L.; Bechtold, J.; Chu, C. W. *Phys. Rev. Lett.* **1988**, *60*, 1174–1177.
38. Marshall, A. F.; Oh, B.; Spielman, S.; Lee, M.; Eom, C. B.; Barton, R. W.; Hammond, R. H.; Kapitulnik, A.; Beasley, M. R.; Geballe, T. H. *Appl. Phys. Lett.* **1988**, *53*, 426–428.
39. Fukutomi, M.; Machida, J.; Tanaka, Y.; Asano, T.; Yamamoto, T.; Maeda, H. *Jpn. J. Appl. Phys.* **1988**, *27*, L1484–L1486.
40. Ichikawa, Y.; Adachi, H.; Setsune, K.; Hatta, S.; Hirochi, K.; Wasa, K. *Appl. Phys. Lett.* **1988**, *53*, 919–921.
41. Nakao, M.; Yuasa, R.; Nemoto, M.; Kuwahara, H.; Mukaida, H.; Mizukami, A. *Jpn. J. Appl. Phys.* **1988**, *27*, L849–L851.
42. Lee, W. Y.; Lee, V. Y.; Salem, J.; Huang, T. C.; Savoy, R.; Bullock, D. C.; Parkin, S. S. P. *Appl. Phys. Lett.* **1988**, *53*, 329–831.
43. Shih, I.; Qiu, C. X. *Appl. Phys. Lett.* **1988**, *53*, 523–525.
44. Ginley, D. S.; Kwak, J. F.; Hellmer, R. P.; Baughman, R. J.; Venturini, E. L.; Morosin, B. *Appl. Phys. Lett.* **1988**, *53*, 406–408.
45. Ginley, D. S.; Kwak, J. F.; Hellmer, R. P.; Baughman, R. J.; Venturini, E. L.; Mitchell, M. A.; Morosin, B. *Physica C* **1988**, *156*, 592–598.
46. Hong, M.; Liou, S. H.; Bacon, D. D.; Grader, G. S.; Kwo, J.; Kortan, A. R.; Davidson, B. A. *Appl. Phys. Lett.* **1988**, *53*, 2102–2104.
47. Richeson, D. S.; Tonge, L. M.; Zhao, J.; Zhang, J.; Marcy, H. O.; Marks, T. J.; Wessels, B. W.; Kannewurf, C. R. *Appl. Phys. Lett.* **1989**, *54*, 2154–2156.

RECEIVED for review May 1, 1989. ACCEPTED revised manuscript August 2, 1989.

19

Centered Cluster Halides for Group-Three and Group-Four Transition Metals

A Versatile Solid-State and Solution Chemistry

Friedhelm Rogel, Jie Zhang, Martin W. Payne, and John D. Corbett

Department of Chemistry, Iowa State University, Ames, IA 50011

A large number and variety of M_6X_{12}-type cluster halides constructed from a rare-earth metal or zirconium (X is Cl, Br, or I) are stable only when centered by an interstitial element Z. At least 22 elements are known to function as Z, and over a dozen structure types are represented. New results are presented in three different areas: the encapsulation of 3d transition elements in zirconium chlorides to give $Li_x(Zr_6Cl_{12}Z)Cl_{6/2}$ phases in the Nb_6F_{15} structure with two independent, interpenetrating cluster networks or analogues of the $K_2Zr_7Cl_{18}H$ structure; bonding of heavy transition metals (Ru, Rh, Pd, Re, Os, Ir, or Pt) within rare-earth-metal (R) clusters with $R_7I_{12}Z$ or $R_6I_{10}Z$ compositions (R is Pr, Gd, or Y); the preparation of new cluster phases from solid-state cluster compounds via room-temperature solution chemistry in CH_3CN. The products include the optimal 14-electron cluster $((C_2H_5)_4N^+)_4Zr_6Cl_{18}C^{4-}$, the 18-electron analogue $((C_2H_5)_4N^+)_4[(Zr_6Cl_{12}Fe)Cl_6^{4-}]$, and three examples containing the new 12-electron cluster $Zr_6Cl_{12}Be^{4-}$, in which cluster-bonding electrons have clearly been removed.

CHEMICAL CLUSTERS OF THE $M_6X_{12}{}^{n+}$ AND $M_6'X_8{}^{m+}$ TYPES have been known for many years for the more-central transition elements (M is Nb or Ta; M' is Mo; X is Cl, Br, I, or, in one case, F). For the more-relevant M_6X_{12} type, varied binary and ternary niobium and tantalum compounds

0065–2393/90/0226–0369$06.25/0
© 1990 American Chemical Society

have been identified that contain $(Nb,Ta)_6X_{12}^{2+,3+}$ clusters along with additional halide anions (1). The electronic requirements of these clusters seem to be relatively restrictive, with only 16 or 15 electrons left in metal–metal bonding orbitals in the two charge types, respectively, after the lower-lying halogen valence orbitals have been filled. The $(Nb,Ta)_6X_{12}^{2+,3+}$ cluster phases are generally synthesized at high temperature and therefore represent equilibrium species. On the other hand, the reversible oxidation of these clusters to $(Nb,Ta)_6X_{12}^{4+}$ species can be accomplished in aqueous solution near room temperature (2, 3). Presumably, these products are only kinetically stable with respect to disproportionation.

Given the generality that something close to 15 or 16 electrons is required for cluster stability, it is clear that analogous clusters of either zirconium (Ti, Hf) or yttrium (Sc, La, etc.) will be seriously electron-deficient. This shortage might be alleviated by forming $Zr_6X_{12}^{3-}$ cluster anions instead, but these have not yet been found. Rather, a very wide collection of cluster phases constructed from most of the earlier transition metals can be achieved if particular interstitial elements are also bound in the center of each cluster. One reason for this requirement is presumably the additional electrons contributed to the cluster-bonding manifold, although strong bonding between the host metal and the added interstitial element must also be an important component of this kind of "solution" to cluster stability.

This new chemistry was discovered by accident, as is often the case. "Adventitious impurities" led to new scandium, yttrium, and zirconium phases that, as luck would have it, were exceedingly well crystallized and therefore easy to recover for crystal-structure studies, even with quite small yields. A residual electron density was observed in the centers of some, but not all, of the clusters that were found within the new phases. These observations, coupled with both the dimensional implications of interstitial elements that had been found in some extended cluster systems for which the host structure was already known (4, 5) and parallel work by others on gadolinium halide clusters containing dicarbon (6, 7), led to the recognition of both the phenomenon and its extensive implications (8–10). Subsequent synthetic explorations with the intentional introduction of potential interstitial elements have revealed the remarkable breadth and versatility of this new chemistry.

Some of the first mysteries to be clarified were a few novel and unexpected zirconium and scandium phases that naturally had been obtained in low and erratic yields. These turned out to be $Zr_6I_{12}C$, $CsZr_6I_{14}C$, and $Sc(Sc_6Cl_{12}N)$ cluster compounds, as well as infinite condensed cluster chain examples like $Sc_7Cl_{10}C_2$ and Sc_4Cl_6B (10, 11). A much larger group of centered clusters was uncovered when the studies were extended to the zirconium chlorides (9, 12–17). In this case, the range of possible interstitial elements was not much greater (comprising just H, Be, B, C, or N), but the structural variety was impressively more diverse.

The variety arose from both a greater versatility in intercluster bridging modes by chlorine and the opportunity to accommodate alkali-metal cations within that bridged cluster network. The basis for the bridging options is shown on the left side of Figure 1, a centered zirconium cluster to which are bonded not only the customary 12 inner chlorine atoms (Cl^i) that bridge edges of the (nominal) Zr_6 octahedron, but also six more outer chlorines (Cl^a) at the vertices. The latter appear in some way in all cluster halide structures, a fact indicating the presence of a strongly bonding orbital that is directed outward (exo) from each cluster vertex. Thus, the increase in the amount of additional chlorine bonded to an M_6Cl_{12} cluster compound from 0 to 6 (n) is accompanied by a regular change in the character of the chlorines occupying these terminal positions, starting at $n = 0$ with inner, edge-bridging types "borrowed" from adjacent clusters (Cl^{i-a}), then those that only bridge between clusters (Cl^{a-a}), and ending with chlorine atoms that are terminal at only one vertex (Cl^a).

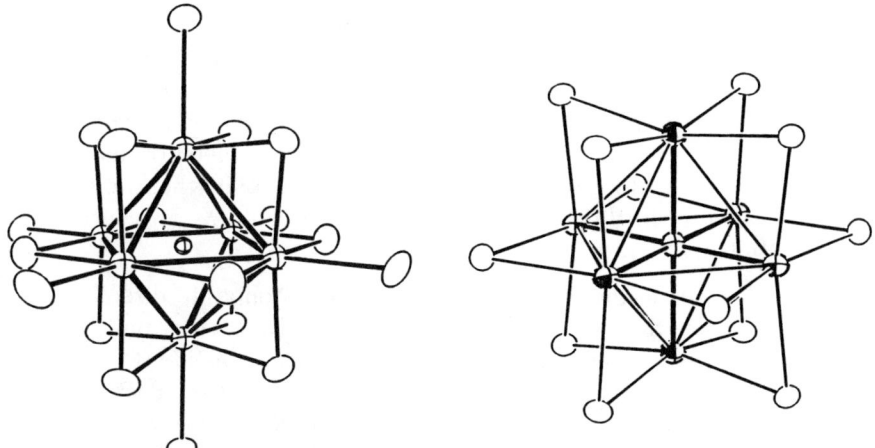

Figure 1. Zr_6X_{12} clusters containing interstitial atoms Z. Left, the $(Zr_6X_{12}Z)X_6$ unit with the essential terminal halide atoms bound at the six metal vertices; right, the cluster core with a better representation of the strong Zr–Z bonding with the interstitial atom. Halogen is represented by open ellipsoids in both.

Zirconium chloride phases are known that contain all possible amounts of additional halide, $0 \leq n \leq 6$. In addition, the intercluster bridging that is present for all cases but $n = 6$ generates cluster matrices that may be somewhat flexible and even afford several interconnection motifs for a given n. One can also accommodate cations within these networks, alkali-metal ions (M^I) in nearly all instances. Thus, all three variables (x, Z, and n) may be altered to generate a large family of compounds: $M_x^I[Zr_6Cl_{12}(Z)]Cl_n$.

Syntheses of these compounds at 750–850 °C generally give single-phase products under thermodynamic control, and so the particular products obtained are largely controlled by stoichiometric proportions that fix n and x and by $n(Z)$, the number of electrons provided by a workable Z. Electronic control of these syntheses appears to focus quite well on 14 bonding electrons in clusters centered by main-group element (viz., $14 = 6 \cdot 4 - (12 + n) + n(Z) + x$). Molecular orbital (MO) calculations are consonant with experimental observation and correspond to an occupancy of Zr–Z bonding levels a_{1g}^2 and t_{1u}^6 plus the Zr–Zr bonding highest occupied molecular orbital (HOMO) t_{2g}^6; the lowest unoccupied molecular orbital (LUMO) is a_{2u}. The view on the right in Figure 1 emphasizes the somewhat greater importance of the Zr–Z bonding developed.

Several additional factors also appear important in the stability of not only ~50 such zirconium chlorides in about 12 different structures, but also in a somewhat dissimilar group of zirconium iodide clusters. These factors have recently been generalized and summarized (17). A particular subset of the following factors will be important in our subsequent discussions:

1. Larger interstitial atoms naturally expand the cluster, and thus bring each metal vertex closer to the plane of the four neighboring X^i atoms (Figure 1). This condition does two important things, particularly with iodides: It allows the necessary exo halogen at each vertex to come closer to the metal (reduced X^a–X^i repulsion), and it enhances the probability that the a_{2u} will remain the LUMO orbital because of Zr–X^i π^* contributions thereto.

2. Changes in the cluster proportions in the opposite direction increase the likelihood of forming 15- or 16-electron units in the presence of a smaller metal (Nb), a larger halide (I), or a smaller interstitial element. This so-called matrix effect (18) does well in correlating zirconium chloride results with the different stabilities of centered zirconium iodides and also the empty niobium and tantalum cluster halides and oxides.

3. The sizes and numbers of intercluster cavities suitable for countercations are determined both by the size of the clusters (via Z, X) and by the angle at intercluster bridging halide. Conversely, the need for cavities of a given number and size, and therefore the structure adopted, may be forced by the reactants provided and the need for 14 electrons in most clusters.

4. Arguments regarding phase stability are made difficult and tenuous because the most important factor in these syntheses, or their failure, is the stability of alternate phases. These phases may change with the choice of cluster metal, halide, or Z.

The striking operation of these factors, sometimes in inexplicable ways, can be seen in the contrasting behaviors of the centered zirconium iodides. First, these have to date been found with only two structures and compositions, namely $Zr_6I_{12}Z$ and $(M^+)Zr_6I_{14}Z$. Second, the clusters are now remarkably more adaptable than the chlorides regarding the size of Z, which may include K, Al, Si, Ge, and P. The 11% range in extreme Zr–Z distances in the iodides corresponds to 0.38 Å in the average separation $\bar{d}$(Zr–Zr) (8, 19, 20). The larger members of this group reduce a matrix effect that is more severe than in cluster chlorides and therefore improve both $Zr-I^i$ and $Zr-I^a$ overlap. As might be expected from the preceding discussion of electronic stability in the chlorides, a good number of 15- and 16-electron clusters are now found with iodides.

Finally, 3d transition metals may also be encapsulated in both Zr_6I_{12}-type (Cr, Mn) and $(M)Zr_6I_{14}$-type structures (Mn, Fe, Co) (21). With 3d valence orbitals present on Z, the optimum stability simply increases from 14 to 18 electrons as a result of the addition of a nonbonding e_g^4 orbital set on the transition metal Z. (The t_{1u} and t_{2g} cluster orbitals exchange roles in Zr–Z and Zr–Zr bonding.) This type of cluster provides new and interesting versions of intermetallic compounds.

This chapter relates exceptional advances in three different extensions of cluster chemistry.

1. The novel structures achieved when the smaller zirconium chloride clusters are combined with 3d metals, principally in $Zr_6Cl_{15}Z$ and $M_2ZrCl_6 \cdot Zr_6Cl_{12}Z$ (= $M_2Zr_7Cl_{18}Z$) structure types.

2. The remarkable differences found for rare-earth-metal cluster iodides, in which one can encapsulate 4d and 5d, as well as 3d, metal atoms.

3. The beginning of a nonaqueous solution chemistry of centered zirconium chloride clusters that can be achieved at room temperature, including the isolation of more-oxidized 12-electron clusters.

New Zirconium Cluster Chlorides

A sizable difference has long been evident in the stabilities of zirconium cluster chlorides versus iodides regarding enclosure of main-group elements beyond the second period (e.g., Al, Si, P, Ge). In spite of many attempts, no chloride cluster has been found that bonds any of these elements. One possible explanation is that the matrix effect and poor bonding caused by the disparity in sizes of iodine and zirconium is relieved and the Zr–I bonding is improved to a significant extent when larger interstitial atoms are enclosed.

Because iodides readily encapsulate 3d metals (21), the initial expectation was that a similar exclusion would apply to chloride clusters as well, because atoms such as Fe and Co are almost as large as Si, Ge, and similar atoms in the iodides.

It became clear by mid-1986 that our predictions of phase stability on such a simple basis were once again inadequate and that many 3d elements may be so bonded in $Zr_6Cl_{12}Z$ units. Moreover, the combination of the smaller chloride with a larger interstitial element has a significant effect on the shape of the cluster. This difference provides a variety of $Zr_6Cl_{15}Z$ phases in what had hitherto been a singularity, the novel Nb_6F_{15} structure (22). These chloride clusters also yield several variations of the $K_2ZrCl_6 \cdot Zr_6Cl_{12}(H)$ structure (23), wherein a lanthanide or alkaline-earth element substitutes for zirconium in the $ZrCl_6^{2-}$ group. The opportunity to vary the electron count on the interstitial atom once again gives an appreciable versatility to the structural family.

A powerful structure-determining element in this cluster chemistry is the necessity of bonding halide exo at the cluster vertices, a process that generates network structures for compositions with 0–4 halides beyond the basic 12 (17). A large interstitial atom together with a small halogen places the metal vertex closer to the plane of the four adjoining X^i atoms (Figure 1). This configuration provides more access to the vertex both for bonding of X^a and for reaction with nucleophiles. These factors are all important in determining the size and number of cavities between clusters, where countercations can be accommodated. Thus, for $Zr_6Cl_{12}(Z)Cl_{6/2}$ compositions, the Ta_6Cl_{15}-type structure provides the most compact arrangement for small Z (B, C, N), and apparently, only one small cation can be accommodated (14).

Progressively more open arrangements occur for $KZr_6Cl_{15}C$ (12) and $K_2Zr_6Cl_{15}B$ (13) structure types, as an increasing fraction of the Zr_6–Cl^a–Zr_6 bridges become linear. The intercluster cavities become even larger when iron is incorporated. Thus, $KZr_6Cl_{15}Fe$ (but not $CsZr_6Cl_{15}Fe$) exhibits a superstructure of the $KZr_6Cl_{15}C$ type, in which the linear chains have puckered appreciably in order to diminish the oversized cavities available to potassium. In contrast, the structure of $CsZr_6Cl_{15}Co$ is normal ($CsZr_6Cl_{15}C$ type (12)), and $CsZr_6Br_{15}Fe$ and $CsZr_6Br_{15}Co$ exhibit what appears to be cation disorder, presumably because the potential sites are now too large.

Nb_6F_{15} Types. The most novel results along these lines occur with $Zr_6Cl_{15}Z$ compositions when Z is Mn, Fe, Co, or Ni and with only small Li^+ cations, or none at all. The zirconium vertices in these clusters lie nearly in the plane of the four adjacent Cl^i atoms, the trans Cl^i–Zr–Cl^i angles average 175°, and only linear intercluster bridges Zr_6–Cl^a–Zr_6 allow the maximum possible separation of the clusters. As a result, these compounds now contain two interpenetrating but not interconnected

$(Zr_6Cl_{12}Z)Cl_{6/2}$ networks, a configuration that had previously been seen only for $Nb_6F_{12}F_{6/2}$ (22).

New compounds of this type include the dark purple $Li_2Zr_6Cl_{15}Mn$, $LiZr_6Cl_{15}Fe$, $Zr_6Cl_{15}Co$, and $Zr_6Cl_{15}Ni$, with single-crystal results available for all. The first three contain 18 cluster-bonding electrons, and the nickel member has 19. The lithium positions in the manganese-containing phase have been confirmed by NMR spectroscopy. The $Th_6Br_{15}Fe$ phase has been reported to have the same structure (24).

This novel cluster network is shown in Figure 2 for $LiZr_6Cl_{15}Fe$.

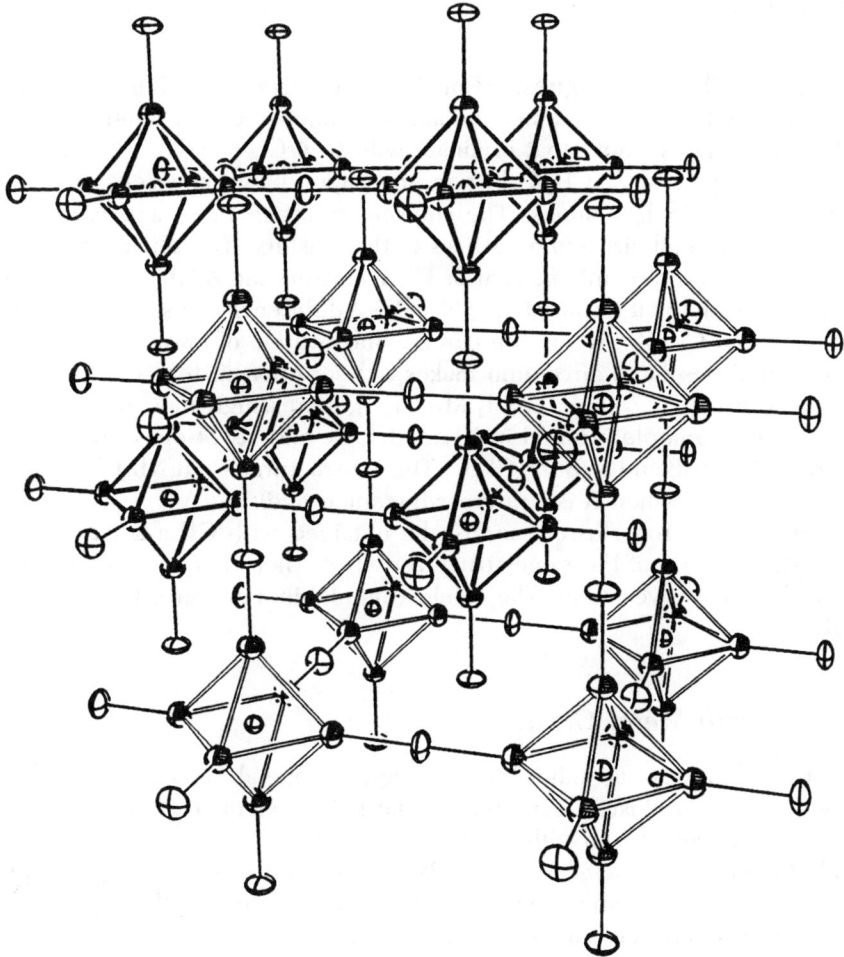

Figure 2. The two interpenetrating ReO_3-like networks in $Li(Zr_6Cl_{12}Fe)Cl_{6/2}$, distinguished by solid and open outlines of the zirconium octahedra. Chlorine atoms (crossed thermal ellipsoids) form linear bridges between the $Zr_6Cl_{12}Fe$ clusters. Inner chlorine atoms in the clusters and the lithium atoms are not shown (90% probability ellipsoids).

The arrangement amounts to a body-centered-cubic array of clusters. Corner clusters are connected only to other corners, with a similar relationship for the body-centered units. Each network is equivalent to the ReO_3 ($= ReO_{6/2}$) structure, with rhenium replaced by the much-larger $Zr_6Cl_{12}Z$ cluster, which allows two such networks to interpenetrate.

Remarkably enough, a single ReO_3-like connectivity has also been found with a $(Zr_6Cl_{12}Mn)Cl_{6/2}$ cluster network. In this case, the arrangement has simply twisted into a rhombohedral array of bridged clusters, completely equivalent to the $ReO_3 \rightarrow PdF_3$ structural relationship (25), while three Cs^+ ions and the new $ZrCl_5^-$ (D_{3h}) prop it open. A [110] section of this result is shown in Figure 3.

$K_2Zr_7Cl_{18}H$ Analogues. Finally, these larger $Zr_6Cl_{12}Z$ units may also be combined with other compound anions to allow access to another variety of cluster phases, those that originate with $K_2ZrCl_6 \cdot Zr_6Cl_{12}(H)$ (23). The cubic-close-packed $Zr_6Cl_{12}Z$ clusters (*see* Figure 4) are now separated by a like array of $ZrCl_6^{2-}$ anions. The chlorine atoms in these anions are also bonded exo to all zirconium vertices of the clusters. The separation of the structure description into neutral $Zr_6Cl_{12}Z$ clusters and $ZrCl_6^{2-}$ ions is somewhat artificial, but it is consistent with a 0.3-Å difference in distances between Cl^a–Zr_6 and Cl^a–$ZrCl_5^{2-}$ in the parent. Substitution of lower-field La^{3+} or Ba^{2+} for the isolated zirconium makes a description in terms of 6–18-like clusters, as in $(Cs^+)_2La^{3+}[(Zr_6Cl_{12}Mn)Cl_6^{5-}]$, increasingly more apt. Another 18-electron example is found in $KLaZr_6Cl_{18}Fe$ and, as 14-electron clusters, in $K_2LaZr_6Cl_{18}B$ and $K_2BaZr_6Cl_{18}C$. The isostructural $K_{0.1}R(Nb_6Cl_{18})$ phases are also known when R is a rare-earth element (26).

An 18-electron cluster can also be obtained with $CsLaZr_6Cl_{18}Fe$ in a distorted version of the same structure, where there is now room for only a single unipositive cation. The 16-electron niobium parent is $CsLuNb_6Cl_{18}$ (27).

Rare-Earth-Metal Clusters

Neither chlorides nor iodides of the rare-earth metals appear to afford the great variety of clusters centered by the lighter elements that have been seen with zirconium. In addition to $Sc(Sc_6Cl_{12}(B,N))$ noted earlier, there are only the analogues $Sc(Sc_6I_{12})(B,C)$ (28) and a dicarbide-containing $Sc_6I_{11}C_2$ (29). The reduced iodine content of the latter is achieved through sharing of I^{i-i} atoms between pairs of clusters, such as

$$\begin{array}{ccc} Sc & & Sc \\ & \diagdown \quad \diagup & \\ & I & \\ & \diagup \quad \diagdown & \\ Sc & & Sc \end{array}$$

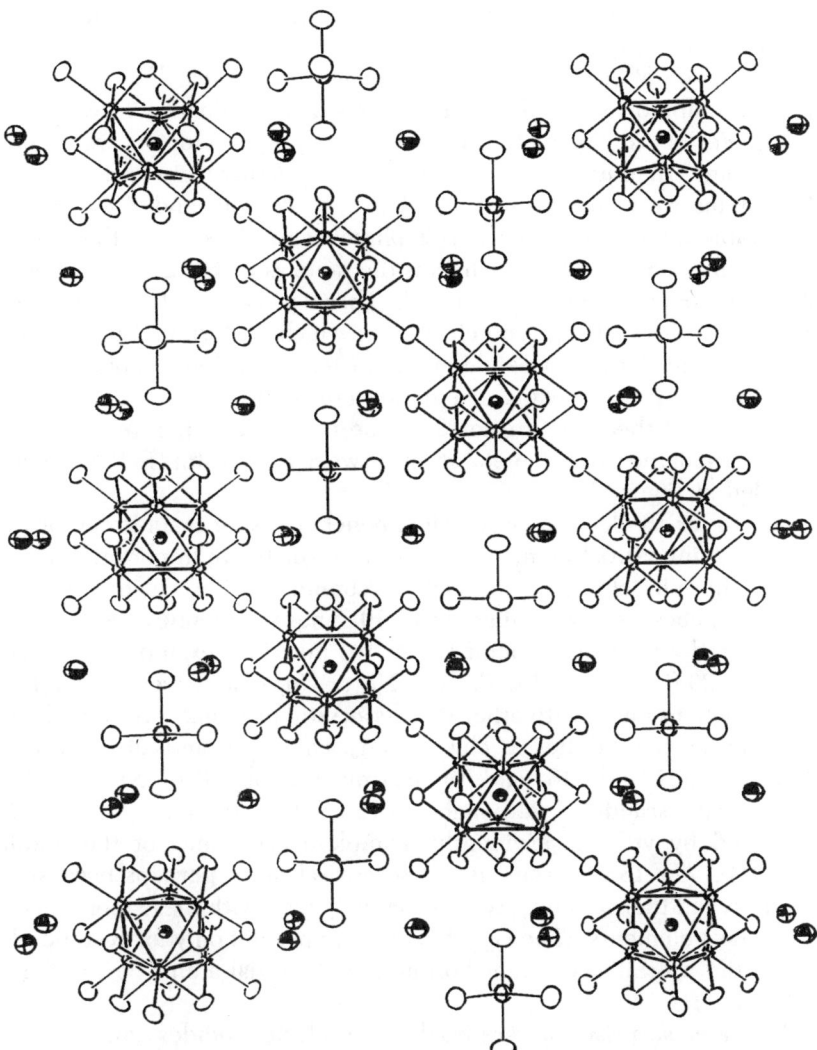

Figure 3. A [110] section of the structure of $(Cs^+)_3[(Zr_6Cl_{12}Mn)Cl_{6/2}]^{2-}ZrCl_5^-$ ($R\bar{3}c$). The cluster network defines a twisted ReO_3-like unit with bridging angles of 133° (vs. 180° in Figure 2). The clusters and the $ZrCl_5^-$ ions lie on vertical threefold axes (90% ellipsoids).

The C_2 interstitial element is already well-known from work elsewhere on a variety of gadolinium halides, for example, in condensed clusters that share metal edges either in dimers, as in $Gd_{10}Cl_{18}(C_2)_2$ (6), or in an infinite chain version $Gd_{12}I_{17}(C_2)_3$ (7, 24). On the other hand, the yttrium chlorides seem to be quite free of centered octahedral cluster examples of all kinds. An exploration of reduced lanthanum chlorides suggests a similar result is probable there, too (H.-J. Meyer, unpublished).

Exploration of 3d metals in cluster iodides of rare-earth metals (R) was initially prompted by the discovery of $Zr_6I_{12}Z$-type clusters in which Z is Cr, Mn, Fe, or Co. As is often the case, many more clusters were subsequently found than just zirconium analogues. Initially, rare-earth-metal iodide systems were investigated for R (Sc, Y, Pr, Gd) and Z (Mn–Ni). A considerable number of black $R_7I_{12}Z$ phases were discovered therein with the structure of $Sc(Sc_6Cl_{12}N)$, which is the same as that of $Zr_6I_{12}C$ when the first (isolated) metal atom is omitted. These phases contain cubic-close-packed (ABC . . .) R_6I_{12}-type clusters in which all terminal positions are bonded to inner iodine in neighboring clusters. Octahedral holes between clusters of the same stacking orientation are occupied by the R^{3+} ion. A [110] section of the structure for $Sc_7I_{12}Co$ is shown in Figure 4 (30). The environment of the seventh metal atom between two clusters in this structure is detailed in Figure 5.

The 14- or 18-electron closed-shell benchmarks for main-group or transition-metal interstitial elements seem to be relatively important with zirconium chloride clusters, although zirconium iodide clusters with 3d transition metals usually contain 18 or 19. These guidelines seem less important for the rare-earth elements, as we now find examples that range from 16 to 20 (i.e., with Mn, Fe, Co, Ni, or Cu, respectively, in $Pr_7I_{12}Z$).

Distinctions found with other R hosts are still puzzling. Presumably they depend on such subtleties as cluster and cavity sizes and alternate phase stabilities. Thus, only 18- and 19-electron examples (Co, Ni) have been obtained with scandium, although 16–18-electron clusters (Mn, Fe, Co) are formed by yttrium. Evidently, complete occupancy of the nominal t_{1u}^6 HOMO (for 18 electrons) is not so critical here, perhaps because the a.o.'s on R and thence the t_{1u} set lie relatively high with these more-electropositive host elements. (The t_{1u} set is really important only for R–R bonding because the p orbitals on transition-metal interstitial atoms lie too high to be effective.)

A greater surprise for rare-earth-metal cluster iodides came with the discovery that these are also stable when encapsulating 4d and 5d metal atoms (31, 32). For example, the following Z are found in $Pr_7I_{12}Z$ phases,

Mn	Fe	Co	Ni	Cu
	Ru	Rh	Pd	
Re	Os	Ir	Pt	

thereby spanning 16- to 20-electron clusters. The same list is found for gadolinium; however, Ag gives clusters with neither host (or the right conditions have not been found). On the other hand, yttrium does not appear to yield $Y_7I_{12}Z$ products with Ni, Rh, Pd, Os, Ir, Pt, or any of the group-

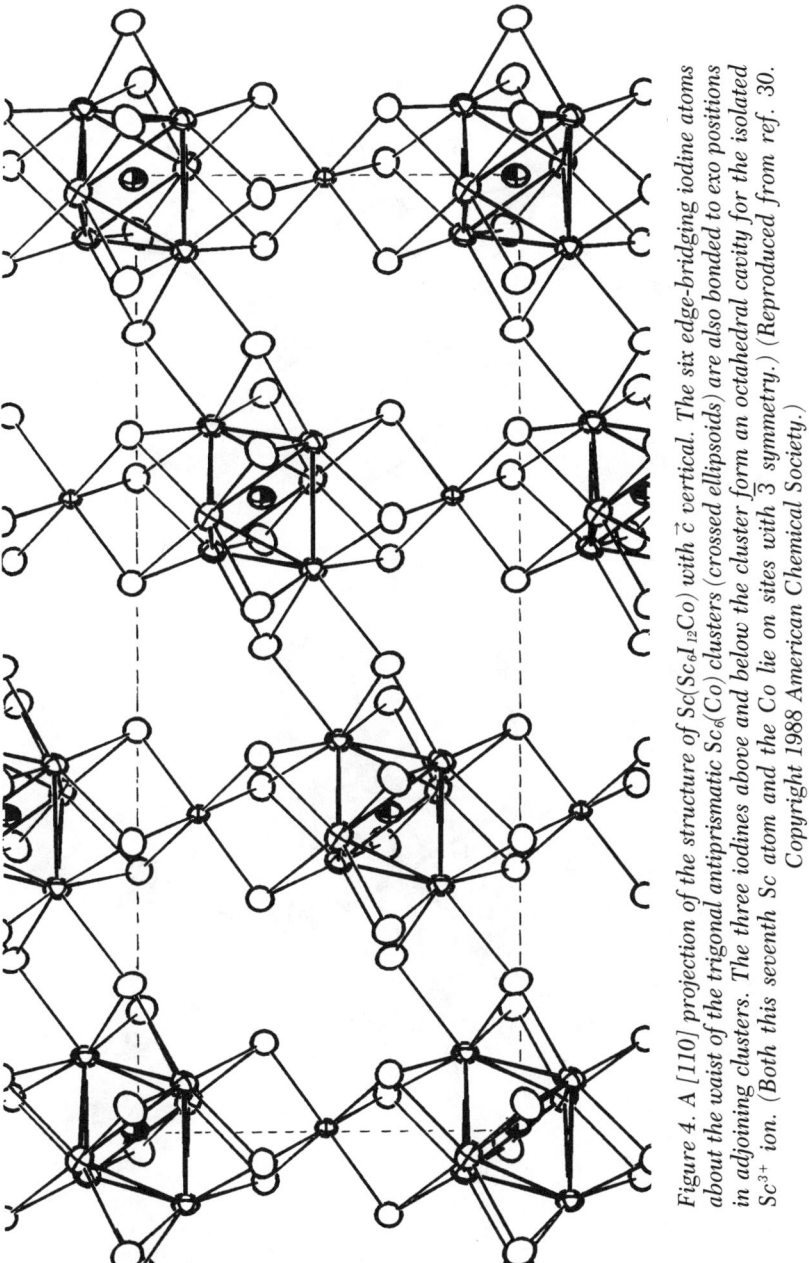

Figure 4. A [110] projection of the structure of Sc(Sc$_6$I$_{12}$Co) with $\vec{c}$ vertical. The six edge-bridging iodine atoms about the waist of the trigonal antiprismatic Sc$_6$(Co) clusters (crossed ellipsoids) are also bonded to exo positions in adjoining clusters. The three iodines above and below the cluster form an octahedral cavity for the isolated Sc^{3+} ion. (Both this seventh Sc atom and the Co lie on sites with $\bar{3}$ symmetry.) (Reproduced from ref. 30. Copyright 1988 American Chemical Society.)

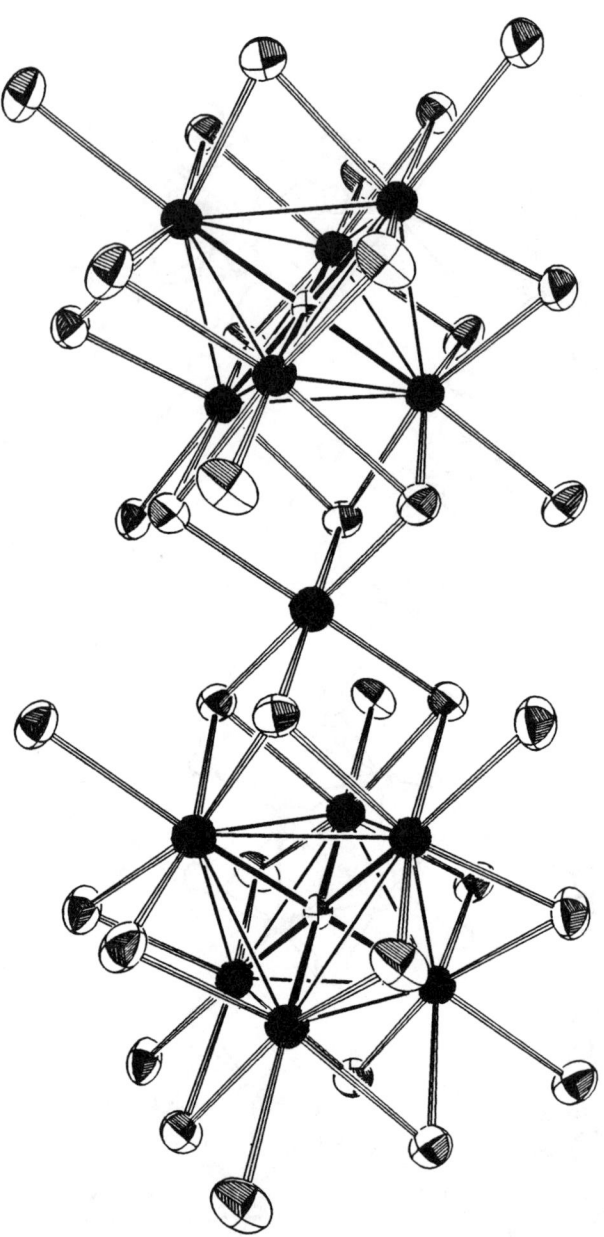

Figure 5. The arrangement of two $R_6X_{12}Z^{3-}$ clusters about the isolated R^{3+} atom in $R_7I_{12}Z$ phases. R atoms are solid and the R–Z bonding is emphasized. New compounds in this structure include those where R is Pr or Gd and Z is Ru, Rh, Pd, Re, Os, Ir, or Pt.

six elements. However, in the cases of Os, Rh, Ir, and Pt, this result is apparently because an alternate cluster phase, $Y_6I_{10}Z$, intrudes. Both structure types occur only with yttrium clusters containing cobalt or ruthenium. Zirconium seems to show very little of an analogous chemistry with most of the 4d and 5d elements.

The versatility of Nature is nicely demonstrated by this new structure type, $Y_6I_{10}Ru$ (31), which occurs for yttrium when Z is Co, Ni, Ru, Rh, Os, Ir, or Pt, but so far only for $Pr_6I_{10}Os$ and not (yet) for gadolinium. The reduced iodine content is achieved, as with $Sc_6I_{11}C_2$, through sharing of inner iodine atoms. In this case, the shared atoms form infinite chains of bridged clusters rather than pairs. A portion of this construction in $Y_6I_{10}Os$ is shown in Figure 6. Pairs of I4 atoms bridge octahedral edges in adjoining clusters, whereas parallel bridges involving I5 are of the more common I^{i-a} type.

The ideal octahedral cluster would exhibit a t_{1u}^4 HOMO, which may be the reason for a significant compression of the isolated octahedra in $Y_6I_{10}Ru$ along what is close to a fourfold axis roughly normal to the chain direction. This distortion is in the direction to produce an e_u^4 ground state and possibly occurs because the p orbitals on Z (which normally would yield the opposite distortion) lie so high as to be unimportant. Instead, the distortion is determined more by Y–Y interactions (30). Although the Y–Ru bonds in $Y_6I_{10}Ru$ differ by 0.21 Å, the difference is only 0.09 Å in $Y_6I_{10}Os$. The smaller difference may mean the behavior is more complex and less obvious than anticipated. There seems to be no significant compression in $Y_6I_{10}Ir$ (17 electrons) (32). The average Y–Os distance in $Y_6I_{10}Os$ is 0.17 Å less than that in the intermetallic Y_3Os. We have observed this general effect with many novel M_6Z clusters (19, 21, 30, 31) in which it might be imagined that we have simply captured the smallest element of an intermetallic interaction.

Zirconium Cluster Chlorides in Nonaqueous Solutions

A large family of centered zirconium cluster chlorides $M_x^I(Zr_6Cl_{12}Z)Cl_n$ can be obtained in about 12 structural arrays in which the clusters are held together by chlorine bridges with Cl^{a-a} or Cl^{a-i} functionalities. The stabilities of these clusters peak sharply at 14 cluster electrons. The question naturally arises as to whether a greater range of electronic stabilities might exist in metastable clusters prepared near room temperature, paralleling the behavior with $Nb_6X_{12}^{n+}$ units where the 15- and 16-electron units obtained in high-temperature phases can be oxidized to 14-electron examples in aqueous solution.

The solution process envisioned requires not just solvation of the product ions, but a suitable nucleophile to open up the Zr_6–Cl–Zr_6 linkages. Extensive explorations of this chemistry have revealed a sizable list of reactions

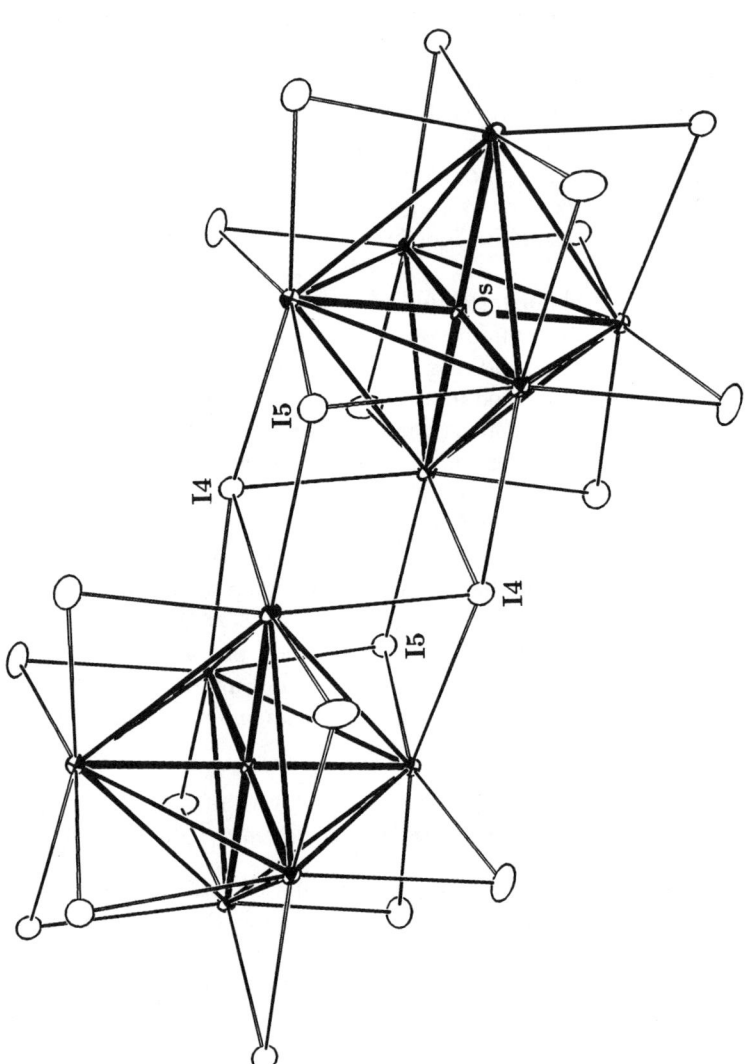

Figure 6. A portion of the infinite cluster chain in $Y_6I_{10}Os$. Pairs of I4 atoms bridge yttrium edges in both clusters. I5 atoms have a more normal I^{i-a} function, bridging one edge and bonding to metal vertex in an adjoining cluster. Inversion centers at the osmium atoms generate infinite chains of such bridged clusters.

that do not work, but some that will. The best quantified systems are those from which single crystals have been isolated and characterized structurally. A substantial number of other solutions exist from which suitable crystals have not yet been obtained.

The clusters are naturally good reducing agents and rapidly react with water, alcohol, and acetone, with both oxidation and solvolysis. Acetonitrile seems quite suitable as a solvent under many circumstances, whereas dimethylformamide and dimethyl sulfoxide oxidize most substrates too readily. Solution of cluster species appears to require not only ligation but also complexation of the alkali-metal cation better than that afforded by the solvent alone. Both crown ethers (18-crown-6, 15-crown-5) and 2,2,2-crypt have served this purpose well, although crystalline derivatives have been isolated more frequently as R_4N^+ or $(C_6H_5)_4P^+$ salts. The reactions are carried out with conventional vacuum techniques and multichambered glass apparatus equipped with poly(tetrafluoroethylene) (Teflon) needle valves.

A common mode of reaction produces a $Zr_6Cl_{18}Z$ anion from a chloride-poorer (bridged) phase. Retention of a 14-electron unit is illustrated by reaction 1.

$$KZr_6Cl_{15}C(s) \xrightarrow[(C_2H_5)_4NBr]{CH_3CN, 18-crown-6} ((C_2H_5)_4N^+)_4(Zr_6Cl_{18}C^{4-}) \cdot 2CH_3CN \quad (1)$$

The conversion takes place over a period of a few days at room temperature and is accompanied by precipitation of a second, poorly crystalline material. The additional chloride necessary to form the "6–18" product must come from the "6–15" reactant, a fact suggesting that the precipitate left may be $Zr_6Cl_{14}C$ (12). The same dark red product has actually been found in both triclinic and monoclinic structures, the choice apparently depending on how rapidly the crystals are grown. Reactions of other clusters containing interstitial Be, B, or C all seem to follow this general route to $Zr_6Cl_{18}Z^{n-}$ solutes plus a chloride-poorer precipitate, in spite of the presence of diverse neutral ligands or, as in reaction 1, excess bromide or chloride.

On the other hand, the incorporation of iron within the cluster is known to expand the metal octahedron relative to the surrounding chlorides. This expansion evidently makes the vertex more accessible to attack. Thus, reaction 2 produces a very dark blue-violet solution of the indicated product rapidly and quantitatively.

$$KZr_6Cl_{15}Fe \xrightarrow[2,2,2-crypt]{CH_3CN_9(C_2H_5)_4NBr} ((C_2H_5)_4N^+)_4[(Zr_6Cl_{12}Fe)Cl_3Br_3^{4-}] \cdot 2CH_3CN \quad (2)$$

Both clusters have the optimal 18-electron count. The solutions do not appear as stable as before.

Questions regarding the accessibility of more oxidized clusters were first answered by crystalline products obtained through accidents or inappropriate chemistry. Twelve-electron $Zr_6Cl_{18}Be^{4-}$ clusters have been obtained in three different black salts originating with two different reactants, $Na_4Zr_6Cl_{16}Be$ and $K_3Zr_6Cl_{15}Be$. Both an air leak and the addition of acetone were originally responsible for the oxidations. The compounds are $[(C_6H_5)_4P^+]_4(Zr_6Cl_{18}Be^{4-}) \cdot 4pyridine$, $PPN^+(Na\text{-}crypt^+)_3(Zr_6Cl_{18}Be^{4-})$ and $((C_2H_5)_4N^+)_4(Zr_6Cl_{18}Be^{4-}) \cdot 2CH_3CN$ (PPN is bis(triphenylphosphoranylidene)ammonium). The last compound is isostructural with both the monoclinic carbon- and the iron-centered cluster products described. Bromide did not substitute for Cl^a in the product obtained from $Na_4Zr_6Cl_{16}Be$. The 12-electron cluster is stable in both tetrahydrofuran and pyridine. Oxidation of $Rb_5Zr_6Cl_{18}B$ by the CH_3CN solvent at ~50 °C was also found to produce the more-complex $[(C_6H_5)_4P^+]_6(Zr_6Cl_{18}B^{4-})ZrCl_6^{2-}$ with a 13-electron cluster. The 12-electron cluster present in the $(C_6H_5)_4P^+$ salt is shown in Figure 7.

The dimensions of the new clusters are nicely consistent with those found in networks with the same cluster-based electron counts, and they

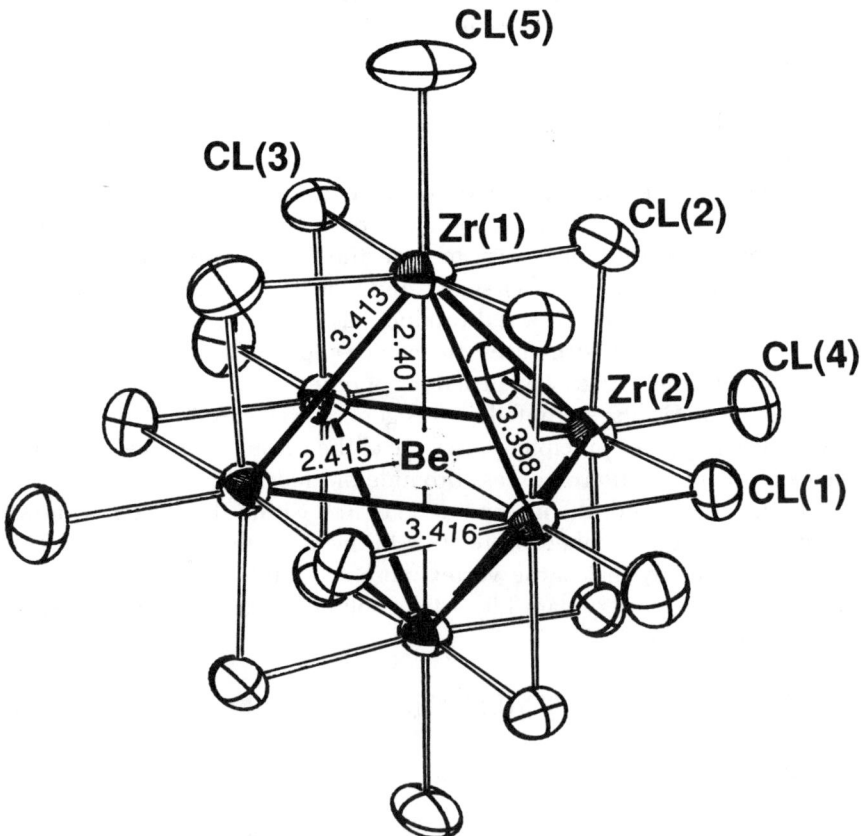

Figure 7. The 12-electron cluster in the compound $[(C_6H_5)_4P^+]_4$ $(Zr_6Cl_{12}\text{-}BeCl_6^{4-}) \cdot 4py$. The cluster has D_{2d} symmetry.

show the appropriate changes on oxidation. A few more-subtle factors that apparently affect cluster distances need to be noted, however. The radially directed (exo) bonds Zr–Cl^a act in opposition to Zr–Z bonds. Long Zr–Cl^a bonds correlate with shorter opposed Zr–Z interactions, and vice versa. This factor has been discussed most for Zr$_6$X$_{14}$Z and Zr$_6$Cl$_{16}$Z structures (8, 15, 16), where there are two different kinds of exo chlorines in each and the clusters are distorted in the expected way. In general, the more-basic chlorine atoms of lower coordination number (Cl^a > Cl^{a-a} > Cl^{a-i}) are bound to the cluster more strongly, so that the new Zr$_6$Cl$_{18}$Z^{n-} clusters described in this chapter should represent the end point even in the presence of large and less-polarizing countercations. Although such effects appear real, they are small (only a few hundredths of an Ångstrom in most cases), whereas the cluster electron count appears to be more important in determining distances.

Table I compares the Zr–C, Zr–Zr, and Zr–Cl^a distances in the two ((C$_2$H$_5$)$_4$N)$_4$(Zr$_6$Cl$_{18}$C) · 2CH$_3$CN structures with those in three solid-state cluster carbides, the last of which contains 15 rather than 14 electrons. The first two extended structures have only bridging chlorines in exo positions. The Zr–Cl^a separations are seen to shorten by 0.04 Å when these become

Table I. Comparison of Zr$_6$Cl$_{18}$C^{4-} Cluster Dimensions with Those in Cluster Networks

Parameter	[(C$_2$H$_5$)$_4$N$^+$]$_4$Zr$_6$Cl$_{18}$C^{4-}		Zr$_6$Cl$_{14}$C^a	KZr$_6$Cl$_{15}$C^b	Cs$_3$Zr$_6$Cl$_{16}$C^c
	Tricl.	Monocl.			
Cluster electrons	14	14	14	14	15
$\bar{d}$(Zr–C)	2.296	2.295	2.286	2.279	2.261
$\bar{d}$(Zr–Zr)	3.248	3.245	3.232	3.223	3.197
$\bar{d}$(Zr–Cl)	2.597	2.586	2.630^d	2.640^d	2.596, 2.689^d

NOTE: All $\bar{d}$ values are given in angstrom units. Standard deviations for all structures listed are ≤0.002 Å in Zr–C and Zr–Zr, and ≤0.006 Å in Zr–Cl.
^{a}Data from ref. 16.
^{b}Data from ref. 12; averages for two independent clusters.
^{c}Data from ref. 15.
^{d}For bridging Zr–Cl^{a-a}.

terminal atoms in the new derivatives. The Zr–C distances in the new Zr$_6$Cl$_{18}$C^{4-} clusters then average slightly greater (0.013 Å) than those in bridged structures, and the Zr–Zr separations closely follow a $\sqrt{2}\ \bar{d}$(Zr–Z) relationship appropriate to an octahedron. The extra cluster electron present in Cs$_3$Zr$_6$Cl$_{16}$C is clearly bonding for Zr–C interactions, decreasing that distance by about 0.020 Å.

The dimensions of the 13-electron cluster Zr$_6$Cl$_{18}$B^{4-} are similarly compared with those for several 14-electron solid-state boron analogues in Table II. The evidence seems clear that a cluster-bonding electron is being removed, with the Zr–B distance increasing by nearly 0.05 Å. MO calculations

Table II. Comparison of Dimensions of a 13-Electron $Zr_6Cl_{18}B^{4-}$ Cluster with 14-Electron Solid-State Examples

Separation	$[(C_6H_5)_4P^+]_6[Zr_6Cl_{18}B^{4-}]/ZrCl_6^{2-}$	$K_2Zr_6Cl_{15}Ba$	$CsKZr_6Cl_{15}B^b$	$Rb_5Zr_6Cl_{18}B^c$	$Ba_2Zr_6Cl_{17}B^d$
$\bar{d}$(Zr–B)	2.352[e]	2.304	2.304	2.317	2.300
$\bar{d}$(Zr–Zr)	3.327	3.253	3.260	3.277	3.250
$\bar{d}$(Zr–Cl)	2.661	2.635[f]	2.66[f]	2.679	2.655

NOTE: All values are given in angstrom units.
[a]Data from ref. 13.
[b]Data from ref. 12; averages for two independent clusters.
[c]Data from ref. 17. The larger cluster may result from a larger number of cations.
[d]Data from unpublished research, Zhang, J.; Corbett, J. D.
[e]Cluster is trigonal antiprismatic, with a 0.045-Å difference of $\bar{d}$(Zr–B).
[f]Zr–Cl^{a-a} distance.

Table III. Dimensions of 12-Electron $Zr_6Cl_{18}Be^{4-}$ Clusters Relative to Those in 14-Electron Analogs in Extended Structures

Parameter	Cations with $Zr_6Cl_{18}Be^{4-}$				$K_3Zr_6Cl_{15}Be$[a]	$Na_4Zr_6Cl_{16}Be$[b]
	$(C_6H_5)_4P^+$	$PPN^+(Na\text{-}Cp^+)_3$	$(C_2H_5)_4N^+$			
Cluster electrons	12	12	12		14	14
$\bar{d}(Zr-Be)$	2.410	2.397	2.404		2.333	2.333
$\bar{d}(Zr-Zr)$	3.409	3.390	3.399		3.300	3.299
$\bar{d}(Zr-Cl)$[a]	2.572	2.556	2.568		2.683[c]	2.667, 2.773[c]

NOTE: All $\bar{d}$ values are given in angstrom units.
[a] Data from ref. 13.
[b] Data from ref. 15.
[c] Distance for $Zr-Cl^{a-a}$.

(*17*) show that the Zr–Zr bonding HOMO in the parent is clearly t_{2g}^6, with the Zr–Z (and Zr–Zr) bonding sets t_{1u}^6 and a_{1g}^2 lying lower.

Twelve-electron clusters have been characterized in three different compounds, all beryllides. Table III shows that this trend continues; removal of two electrons from the nominal t_{2u}^6 causes $\overline{d}$(Zr–Be) to increase by 0.070 Å, with $\overline{d}$(Zr–Zr) changing by 0.10 Å. The parallel decrease in $\overline{d}$(Zr–Cl^a) is 0.11 Å, but this is confused by the intrusion of bridging Zr–Cl^{a-a} functions in the reference structures. Another factor may pertain in the Zr–Cl^a interactions as well: the so-called matrix effect caused by close Cl–Cl contacts all around the cluster. To some extent, this could mean that the chlorine framework remains relatively fixed as the Zr$_6$ unit moves within it in response to bonding and size effects. In this case, the zirconium would in part just move out toward the Cl^a atom in response to the loss of two cluster-bonding electrons. Such a behavior is particularly clear with iodides (*19, 21, 30*).

Finally, dimensions of the new 18-electron iron-centered clusters (Zr$_6$Cl$_{12}$Fe)Cl$_3$Br$_3^{4-}$ and (Zr$_6$Cl$_{12}$Fe)Cl$_6^{4-}$, both isolated as the (C$_2$H$_5$)$_4$N$^+$ salts, are comparable to those in the bridged LiZr$_6$Cl$_{15}$Fe, Zr$_6$Cl$_{14}$Fe, and the starting material KZr$_6$Cl$_{15}$Fe. The Zr–Fe distances in the new derivatives average 2.440 Å, 0.016 Å longer than those in which the opposed terminal chlorine is shared. The Zr–Cl^a separations are 0.14 Å less than in the first reference phase described, about as expected.

Closure

The foregoing sections represent snapshots of some major results in three ongoing research investigations of halide cluster chemistry that were taken in February 1989. The completion dates of all of these are some time away, and so the presentations are necessarily fragmentary and incomplete. The continued discovery of new and unexpected chemistry in all three areas is especially noteworthy and encouraging.

Acknowledgments

We are indebted to C. Perrin for crystallographic data furnished in advance of publication. This research was supported by the National Science Foundation, Solid-State Chemistry, via grant DMR-8318616, and was carried out in facilities of the Ames Laboratory, U.S. Department of Energy.

References

1. Wells, A. F. *Structural Inorganic Chemistry*; 5th ed.; Clarendon Press: Oxford, 1984; pp 432–437.
2. McCarley, R. E.; Hughes, B. G.; Cotton, F. A.; Zimmerman, R. *Inorg. Chem.* **1965**, *4*, 1491.

3. Fleming, P. D.; Dougherty, T. A.; McCarley, R. E. *J. Am. Chem. Soc.* **1967**, *89*, 159.
4. Ford, J. E.; Corbett, J. D.; Hwu, S.-J. *Inorg. Chem.* **1983**, *22*, 2789.
5. Seaverson, L. M.; Corbett, J. D. *Inorg. Chem.* **1983**, *22*, 3202.
6. Warkentin, F.; Masse, R.; Simon, A. *Z. Anorg. Allg. Chem.* **1982**, *491*, 323.
7. Simon, A.; Warkentin, E. *Z. Anorg. Allg. Chem.* **1983**, *497*, 79.
8. Smith, J. D.; Corbett, J. D. *J. Am. Chem. Soc.* **1985**, *107*, 5704.
9. Ziebarth, R. P.; Corbett, J. D. *J. Am. Chem. Soc.* **1985**, *107*, 4572.
10. Hwu, S.-J.; Corbett, J. D.; Poeppelmeier, K. R. *J. Solid State Chem.* **1985**, *57*, 43.
11. Hwu, S.-J.; Corbett, J. D. *J. Solid State Chem.* **1986**, *64*, 331.
12. Ziebarth, R. P.; Corbett, J. D. *J. Am. Chem. Soc.* **1987**, *109*, 4844.
13. Ziebarth, R. P.; Corbett, J. D. *J. Am. Chem. Soc.* **1988**, *110*, 1132.
14. Ziebarth, R. P.; Corbett, J. D. *J. Less-Common Met.* **1988**, *137*, 21.
15. Ziebarth, R. P.; Corbett, J. D. *Inorg. Chem.* **1989**, *28*, 626.
16. Ziebarth, R. P.; Corbett, J. D. *J. Solid State Chem.* **1989**, *80*, 56.
17. Ziebarth, R. P.; Corbett, J. D. *J. Am. Chem. Soc.* **1989**, *111*, 3272.
18. Corbett, J. D. *J. Solid State Chem.* **1981**, *37*, 335.
19. Smith, J. D.; Corbett, J. D. *J. Am. Chem. Soc.* **1986**, *108*, 1927.
20. Rosenthal, G.; Corbett, J. D. *Inorg. Chem.* **1988**, *27*, 53.
21. Hughbanks, T.; Rosenthal, G.; Corbett, J. D. *J. Am. Chem. Soc.* **1988**, *110*, 1511.
22. Schäfer, H.; Schnering, H.-G.; Niehues, K.-J.; Nieder-Vahrenholz, H. G. *J. Less-Common Met.* **1965**, *9*, 95.
23. Imoto, H.; Corbett, J. D.; Cisar, A. *Inorg. Chem.* **1981**, *20*, 145.
24. Simon, A. *Angew. Chem. Int. Ed. Engl.* **1988**, *27*, 160.
25. Babel, D. *Struct. Bonding (Berlin)* **1967**, *3*, 40.
26. Ihmaine, S.; Perrin, C.; Peña, O.; Sergent, M. *J. Less-Common Met.* **1988**, *137*, 323.
27. Ihmaine, S.; Perrin, C.; Sergent, M. *Acta Cryst.* **1989**, *C45*, 705.
28. Dudis, D. S., Corbett, J. D.; Hwu, S.-J. *Inorg. Chem.* **1986**, *25*, 3434.
29. Dudis, D. S., Corbett, J. D.; Hwu, S.-J. *Inorg. Chem.* **1987**, *26*, 1933.
30. Hughbanks, T.; Corbett, J. D. *Inorg. Chem.* **1988**, *27*, 2022.
31. Hughbanks, T.; Corbett, J. D. *Inorg. Chem.* **1989**, *28*, 631.
32. Payne, M. W.; Corbett, J. D. *Inorg. Chem.*, in press.

RECEIVED for review May 1, 1989. ACCEPTED revised manuscript August 28, 1989.

Oxidative Intercalation of Graphite by Fluoroanionic Species

Evidence for Thermodynamic Barrier

Neil Bartlett, Fujio Okino, Thomas E. Mallouk, Rika Hagiwara, Michael Lerner, Guy L. Rosenthal, and Kostantinos Kourtakis

Department of Chemistry, University of California, Berkeley, CA 94720
Materials and Chemical Sciences Division, Lawrence Berkeley Laboratory, Berkeley, CA 94720

Whether oxidative intercalation of graphite by fluoroanions can occur may be estimated from the electron affinity, E ($-\Delta H_{298}$ in kcal mol^{-1}), of the oxidizing half reaction. For first-stage salts with MF_6^- guests, E must exceed 120 kcal mol^{-1}. The values of $E(e^- + \frac{3}{2} MF_5 \rightarrow MF_6^- + \frac{1}{2} MF_3)$ for AsF_5 and PF_5 (125 and 87 kcal mol^{-1}, respectively) account for the spontaneous intercalation of graphite by AsF_5 and the failure of PF_5 to do so. Spontaneous intercalation by $PF_5 + F_2$ occurs because $E(e^- + PF_5 + \frac{1}{2} F_2 \rightarrow PF_6^-) = 165$ kcal mol^{-1}. The thermodynamic nature of the barrier to intercalation is demonstrated by the reduction of C_xPF_6 salts by PF_3: $C_xPF_6 + \frac{1}{2}PF_3 \rightarrow xC + \frac{3}{2} PF_5$. Chlorine with certain fluoroacids can also bring about intercalation of fluoroanions. Polarizable neutral molecules, by improving the lattice energetics of graphite salts, may also be spontaneously intercalated (e.g., $C_{14}AsF_6 + \frac{1}{2}AsF_5 \rightarrow C_{14}AsF_6 \cdot \frac{1}{2}AsF_5$). When high positive charge at carbon occurs, F^- is transferred from the fluoroanion to the carbon and novel graphite fluorides (e.g., $C_{1.3}F$) are formed.

MOST OF THE REACTANTS THAT BRING ABOUT INTERCALATION of graphite are either strong oxidants or reductants such as nitric acid or alkali metal (1–3). The oxidation or reduction of the graphite that accompanies the intercalation was clearly shown by Ubbelohde and co-workers (4–6) to be accompanied by an increase in the in-plane resistivity of the graphite. This resistivity was commonly an order of magnitude smaller in fully intercalated material than in the pristine graphite from which the intercalation compounds had been obtained. The graphite band structures of Coulson and co-workers (7, 8) nicely account for the conductivity in terms of increase in the number of electron carriers in the conduction band of the reduced graphite or an increased number of electron holes in the valence band of oxidized graphite.

The preparation of $C_6F_6^+$ salts (9) and the derivation from them of polycyclic cation salts (10, 11) suggested that very high positive charges might be sustainable in graphite by AsF_6^- and other stable perfluoroanions and that excellent electrical conductors made of high-oxidation-state carbon could thereby be obtained. This development led to synthesis of graphite fluoroarsenates via interaction of graphite with $O_2^+AsF_6^-$ and to the discovery (12) that the guests in the product of the interaction of graphite with AsF_5 suggested that AsF_5 acts as an electron oxidizer according to equation 1.

$$e^- + \tfrac{3}{2} AsF_5 \rightarrow AsF_6^- + \tfrac{1}{2} AsF_3 \qquad (1)$$

The graphite–AsF_5 intercalation compound was prepared first by Chun-Hsu et al. (13). The compound aroused much interest because of the finding (14) that the in-plane specific conductivity of some of the graphite–AsF_5 materials exceeded that of copper. Although the (graphite)$^+AsF_6^-$ salts appeared to involve higher electron withdrawal from the graphite, higher conductivities than those of their graphite–AsF_5 relatives were not observed (15). Indeed, when the (graphite)$^+AsF_6^-$ salts were prepared from the interaction of graphite with AsF_5-F_2 gaseous mixtures (16), the product salts were often less conductive than their partially oxidized relative, graphite–AsF_5.

It appeared that the poor electrical conductivity must be associated with bonding of fluorine to carbon throughout the graphite structure, the fluorine migration throughout the galleries being facilitated by the intercalated species. Direct evidence (15) for a decrease in conductivity was obtained in the case of graphite intercalation by IrF_6 when the first stage was reached. This and the finding from Mössbauer spectroscopy (17) that the first-stage C_xIrF_6 salt contained some Ir(V) in lower than octahedral site symmetry were consistent with transfer of F^- from the anion to the carbon.

Aside from this matter of the transfer of fluorine to the highly oxidized carbon, it appeared to Bartlett and McQuillan (18) that the intercalation of graphite by oxidizing fluorospecies was thermodynamically determined. Although Born–Haber cycles have long been applied to intercalation compounds of graphite [e.g., see G. R. Hennig review (1)], a salt formulation for graphite compounds formed with guests capable of yielding fluoroanions (e.g., AsF_5) was not generally accepted at that time. Bartlett and McQuillan (18) noted that, at least for MF_6^- salts, where the nearly isodimensional nature of MF_6^- would result in similar lattice energetics for a given level of charge, the free energy of formation with change in metal atom (M) would depend most upon the electron affinity of the oxidizing half reaction. They estimated a threshold electron affinity of approximately 125 kcal mol^{-1}, but they did not establish the thermodynamic nature of that threshold.

This chapter provides further evidence for the importance of the electron affinity of the oxidizing half reaction in the intercalation of graphite by fluorospecies. Moreover, the simple thermodynamic model that accounts for the oxidative intercalation of graphite by fluoroanions is confirmed by reversal of the intercalation in reactions of known enthalpy change. Threshold values of electron affinity (E) for the onset of intercalation and for first-stage intercalation by MF_6^- have also been assessed.

Certain salts (e.g., $C_{28}^+AsF_6^-$) provide for the access of fluorine throughout the galleries of the graphite and its extensive fluorination. This fluorine access has resulted in the separation of a C_xF phase that has proved to be fluorinated graphite, retaining sp^2 carbon. That phase, which is an insulator, can have a fluorine content as high as $C_{1.3}F$.

Discussion

Electron Affinity Values. The hexafluorides of the third transition series elements (M^1) have interatomic distances that change only slightly with atomic number (19), and their effective packing volumes (20) are similar. Therefore, we assume that the work done in separating the graphite-atom sheets (W) will be the same for all M^1 in making a salt of given composition $C_xM^1F_6^-$. In addition, the lattice energies (U) for all M^1 would be alike. Because the work function (I) would be the same for a given level of charging (C_x^+), we can expect the enthalpy change for reaction 2 to change in step with the electron affinity, E, of M^1F_6, in accord with the cycle illustrated in Figure 1.

$$\Delta H°_{298}[xC(\text{graphite}) + M^1F_6 \rightarrow C_x^+M^1F_6^-] \quad (2)$$

George and Beauchamp (21) measured $E(WF_6)$ as 81 kcal mol^{-1} and Nikitin et al. (22) evaluated $E(PtF_6)$ to be 184 kcal mol^{-1}. Because the observed chemistry (23) requires a smooth increase in E with atomic number for these hexafluorides, the $E(MF_6)$ for the other fluorides can be obtained by interpolation. They provide an excellent basis for assessment of the threshold for intercalation to form $C_x^+M^1F_6^-$. Magnetic data (15) for the Os and Ir salts $C_xM^1F_6$ have established the quinquevalence of the metals. Moreover, WF_6 will not by itself bring about intercalation, whereas both OsF_6 and IrF_6 form first-stage salts and ReF_6 forms a higher-stage salt (24). Therefore, the $E(M^1F_6)$ threshold for intercalation of graphite by MF_6^- has to be between 81 and 106 kcal mol^{-1} [the $E(WF_6)$ and $E(ReF_6)$ values]. The threshold

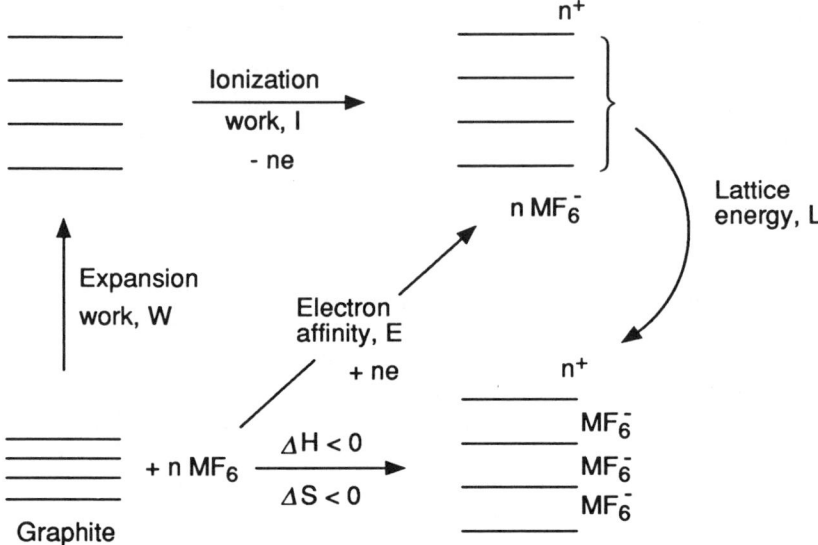

Figure 1. A thermodynamic cycle for oxidative intercalation of graphite.

$E(MF_6)$ value for first-stage $C_x{}^+M^1F_6{}^-$ formation has to be between $E(ReF_6)$ and $E(OsF_6)$ (i.e., between 106 and 131 kcal mol^{-1}).

All $MF_6{}^-$ ions have M–F distances (25) between 1.7 and 1.9 Å, so it is probably not a gross approximation to extend the assumptions of the constancy of W and U, used for the third transition series $M^1F_6{}^-$, to all $MF_6{}^-$. This approximation appears to be valid, as the following examples and the data given in Figure 2 illustrate.

Although PF_5 (in contrast to AsF_5) does not itself bring about intercalation of graphites, first-stage salts are easily formed in admixture with fluorine (26–28) at approximately 20 °C. This salt formation happens because (29) $E(PF_5 + \frac{1}{2}F_2 + e^- \to PF_6{}^-) = 163$ kcal mol^{-1}, which is slightly greater than $E(IrF_6)$. Because the electron affinity of the analogous AsF_5–F_2 reaction is even greater (30), the easy formation (12) of first-stage $C_x{}^+AsF_6{}^-$ salts is explained. But the $C_x{}^+PF_6{}^-$ salts also provide for the demonstration of the thermodynamic basis of the electron-affinity threshold.

Treatment of second- and higher-stage $C_x{}^+PF_6{}^-$ salts with PF_3 at approximately 20 °C reduces the salts with elimination (29, 31) of PF_5 according to equation 3,

$$\tfrac{1}{2}PF_3 + PF_6{}^- \to \tfrac{3}{2} PF_5 + e^- \qquad (3)$$

for which the enthalpy change (29) of 87 kcal mol^{-1} represents the electron affinity for the reverse reaction. This value is not much greater than $E(WF_6)$. Because WF_6 by itself is not intercalated by graphite, such electron affinities

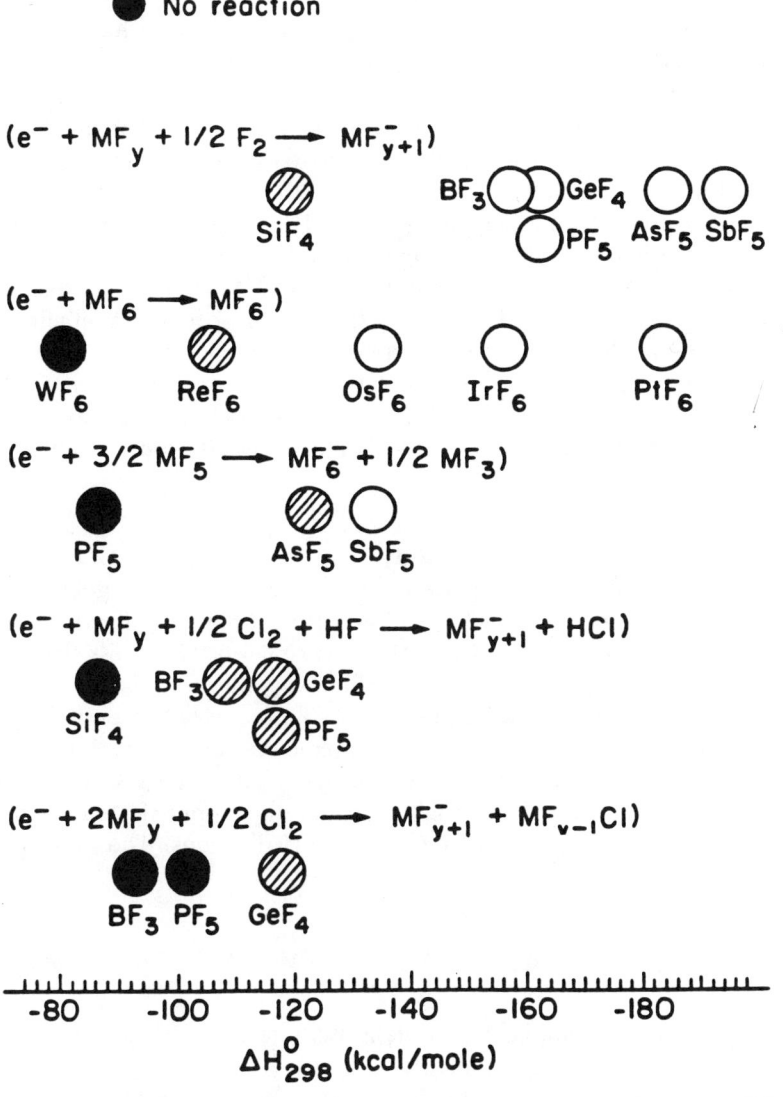

Figure 2. Intercalation of graphite by perfluoroanions as a function of the electron affinity of the oxidizing half reaction.

are inadequate to meet the requirements for graphite intercalation by MF_6^-. On the other hand, the spontaneous intercalation of graphite by AsF_5 (13) is evidently a consequence of the high electron affinity of the oxidizing half reaction (29) $[E(e^{-1} + \frac{3}{2}AsF_5 \rightarrow AsF_6^- + \frac{1}{2}AsF_3) \geq 123$ kcal mol^{-1}.

The vacuum-stable C_xAsF_6 product obtained by removing AsF_5 and AsF_3 from the graphite–AsF_5 material (32) is a mixture of first- and second-stage C_xAsF_6 salts. Even repeated treatment of graphite with AsF_5 does not yield a pure first-stage salt, and treatment of first-stage $C_x^+AsF_6^-$ (made via $AsF_5 + F_2$) with a half-molar quantity of AsF_3 leads to a mixture of second- and first-stage C_xAsF_6. This result gives clear evidence that the E for the AsF_5 electron-oxidation half reaction is close to that for the first-stage C_xAsF_6 thermodynamic threshold.

Apparently, this simple model for $C_x^+MF_6^-$ salts also fits nonoctahedral MF_y^-, so long as the effective height of MF_y^- in the graphite galleries is close to that of MF_6^-. Recent examples (33) from these laboratories include the intercalation of GeF_5^- from GeF_4–Cl_2 mixture [the electron affinity for the appropriate reaction, $E(2 \text{ GeF}_4 + \frac{1}{2}Cl_2 + e^- \rightarrow GeF_5^- + GeF_3Cl)$, is 118 kcal mol^{-1}] and PF_6^- from PF_5–Cl_2–HF mixture [for which $E(PF_5 + \frac{1}{2}Cl_2 + HF + e^- \rightarrow PF_6^- + HCl)$ is 117 kcal mol^{-1}]. In conformity with these electron affinities, each intercalation is spontaneous, but first-stage salts are not produced.

Volume and Lattice Enthalpy. In the example of GeF_5^- intercalation just described, we departed from the case of MF_6^- guest species. GeF_5^- has one fewer F ligand; therefore, it must have an effective volume, based on Zachariasen's criterion (34), approximately 17 Å^3 less than MF_6^-. For closely packed guest species, this effective volume would be important. First let us examine the consequences of volume on lattice enthalpy.

Figure 3 illustrates a remarkably simple correlation between lattice enthalpies and the formula unit volume for A^+X^- salts recognized by Mallouk et al. (30) as being within the usual errors of rather sophisticated evaluations from known structures and partial charge distributions. The empirical relationship

$$U(A^+X^-)(\text{in kcal mol}^{-1}) = 556.3[V(\text{in Å}^3)]^{-1/3} + 26.3 \quad (4)$$

should be limited in application to light-atom hard-ligand cases because the equation does not contain explicitly a closed-shell repulsion term. The advantage of the relationship is that it immediately provides for an assessment of the lattice energy of an A^+X^- salt if the close-packing volume (V) is known.

If the cation is large (e.g., [(n-butyl)$_4$N]$^+$), changing the anion from PF_6^- to BF_4^- changes the lattice enthalpy less than in the case of K^+ salts. Indeed, for KPF_6 and KBF_4 salts, the difference in lattice enthalpy is ap-

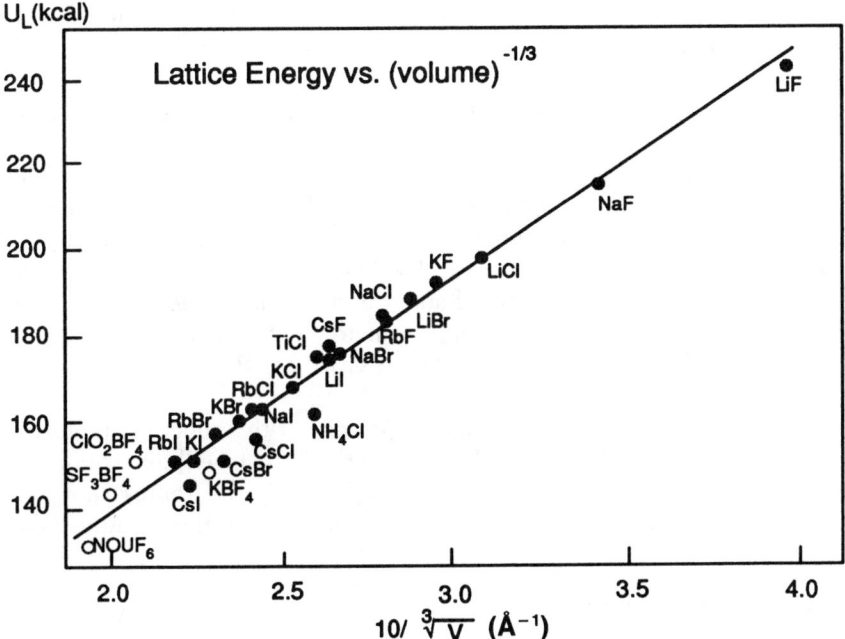

Figure 3. Correlation of lattice energies for A^+B^- salts with formula unit volume. Open circles represent lattice energies calculated as described in ref. 28. The closed circles are lattice energies from ref. 35.

proximately 10 kcal to the advantage of KBF_4. But the fluoride ion affinities (30) ($-\Delta H°_{298}$, $MF_y + F^{-1} \rightarrow F_{y+1}^-$) are BF_3, 92 and PF_5, 101 kcal mol^{-1}). Therefore,

$$\Delta H°_{298} (KPF_6 + BF_3 \rightarrow KBF_4 + PF_5) \approx -1 \text{ kcal mol}^{-1} \quad (5)$$

Although the entropy change is slightly unfavorable, it is not sufficient to render $\Delta G°_{298}$ positive (36). For fluorides, $S°_{298}$ is proportional to formula unit volume ($V_{F.U.}$), with $S°_{298}$ (in entropy units) equal to 0.42 $V_{F.U.}$ (in Å^3). On the other hand, in the case of the $[(n\text{-butyl})_4N]^+$ salts, the lattice energy favors the BF_4^- much less than in the K^+ case. The entropy change is nearly the same as in the K^+ salt situation. The consequences are dramatic to the chemistry. KPF_6 is quantitatively converted (37) to KBF_4 by BF_3 at 20 °C, whereas the mixed butylammonium salts are in equilibrium with BF_3 and PF_5 at ordinary pressures.

Graphite has a much smaller effective radius than $[(n\text{-butyl})_4N]^+$, at least normal to the sheet ($r_{eff} \sim 1.7$ Å). Therefore, the impact of volume of the anion upon the lattice energy is important. Thus the graphite hexafluorophosphate salt, like KPF_6 and unlike $[(n\text{-butyl})_4N]^+PF_6^-$, is quantitatively converted (37) by BF_3 to the fluoroborate.

$$C_xPF_6 + (y + 1)BF_3 \rightarrow C_xBF_4 \cdot yBF_3 + PF_5 \quad (6)$$

As eq 6 demonstrates, more BF_3 is taken up than PF_5 is liberated. This result is an oddity of the graphite salt situation. Here, in contrast to the K^+ and butylammonium salts, the anions are exposed to one another in the galleries between the carbon sheets. Evidently the additional BF_3 uptake provides for increased screening of anion from anion. Indeed, y can be as great as 3 with ordinary pressures (approximately 1 atm) of BF_3 and a second-stage C_xPF_6 salt can yield a first-stage salt $C_xBF_4 \cdot 3BF_3$.

Nestled Salts. A more dramatic instance of the effect of "dielectric spacers" in graphite salts is provided (32) by the first-stage fluoroarsenate of composition $C_{14}AsF_6$. This material is prepared by the routes shown in Scheme I and is identical to the first-stage component of the vacuum-stable

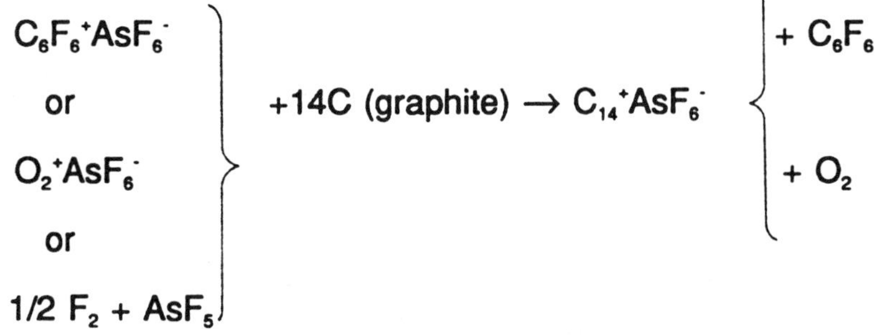

$$\left.\begin{array}{l} C_6F_6^+AsF_6^- \\ \text{or} \\ O_2^+AsF_6^- \\ \text{or} \\ 1/2\,F_2 + AsF_5 \end{array}\right\} +14C\,(\text{graphite}) \rightarrow C_{14}^+AsF_6^- \left\{\begin{array}{l} + C_6F_6 \\ \\ + O_2 \\ \\ \end{array}\right.$$

Scheme I

product of the interaction of graphite and AsF_5 gas. The AsF_6^- ions are nestled in ordered domains within the graphite galleries, as shown in Figure 4. This arrangement is an accidental consequence of the F–F nearest-neighbor distance in AsF_6^- being similar to the distance between the centers of contiguous hexagons in the graphite-sheet structure. Figure 4 shows that closer packing of the nestled AsF_6^- is not possible. This arrangement is in harmony with the observed composition of this phase, $C_{14}AsF_6$. When the solid is exposed to a sustained pressure of AsF_5 (approximately 1 atm at approximately 20 °C), the X-ray diffraction pattern changes dramatically and the composition changes in accord with eq 7.

$$C_{14}^+AsF_6^- + \tfrac{1}{2}AsF_5 \rightarrow C_{14}^+AsF_6^- \cdot \tfrac{1}{2}AsF_5 \quad (7)$$

The X-ray data show that this uptake of AsF_5 by the $C_{14}AsF_6$ expands the

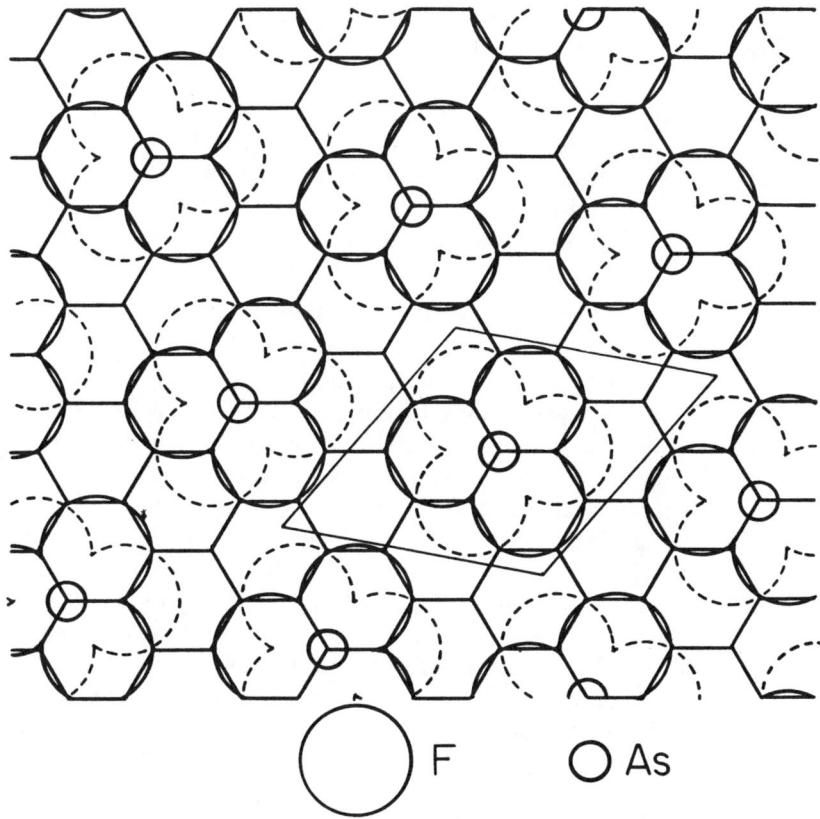

Figure 4. The idealized nestled structural model for a single layer of AsF_6^- guests in $C_{14}AsF_6$. The unit cell is outlined.

carbon-sheet separation from 7.6 Å in the nestled salt to 8.1 Å in the $C_{14}AsF_6 \cdot \frac{1}{2}AsF_5$. Evidently the dielectric screening effect [or bonding of the AsF_5 to yield a μ-fluoro-bridged species (38) such as $(F_5As-F-AsF_5)^-$] provides sufficient favorable energy to more than compensate for the diminished attraction energy, which must accompany the expansion, and the unfavorable entropy change associated with the uptake of gaseous AsF_5. The observed diffraction pattern of the $C_{14}^+AsF_6^-$ is satisfactorily accounted for (32) with the structure illustrated in Figure 4. This structure requires that adjacent carbon-atom sheets be in staggered relationship to each other. As atoms, of course, always reside midway between eclipsed carbon atoms of these adjacent staggered sheets.

With the uptake of AsF_5 (to composition $C_{14}AsF_6 \cdot \frac{1}{2}AsF_5$), the carbon-atom sheets move into registry with one another, as well as farther apart. The observed diffracted X-ray intensities require that the As atoms be placed midway between the enclosing carbon sheets. In the ab plane, all of the guest atoms are fully disordered in this phase. In $C_{14}AsF_6 \cdot \frac{1}{2}AsF_5$ the guests are therefore like a two-dimensional liquid. Removing the AsF_5 restores the nestled, relatively ordered, $C_{14}AsF_6$ phase.

Much the same sort of structural change as observed for $C_{14}AsF_6$ to $C_{14}AsF_6 \cdot \frac{1}{2}AsF_5$ occurs when AsF_3 is added to the nestled salt. On removal of volatiles, AsF_5 is observed as a consequence of the reduction of the carbon already alluded to, and the resulting solid is a mixture of first- and second-stage C_xAsF_6 nestled AsF_6 salts.

$$\frac{1}{2}AsF_3 + AsF_6^- \rightarrow \frac{3}{2} AsF_5 + e^- \tag{8}$$

Fluorinated Graphite. Perhaps the most unexpected aspect of graphite fluorosalt chemistry was the finding that salts such as C_xAsF_6 rapidly consume additional gaseous fluorine (16, 32) at approximately 20 °C. This uptake is accompanied by a marked loss of electrical conductivity. Okino (32) found that this material of composition $C_{14}AsF_6 \cdot \sim 2F$ was less conductive than the graphite from which it was made. Because the carbon sheets in $C_{14}AsF_6$, which itself shows a specific conductivity comparable with that of aluminum metal, bear a charge C_{14}^+, they are unlikely to be attacked by elemental fluorine, which is an electrophile. The attack to generate C–F bonds probably involves transfer of F^- to a positive carbon atom. The fluorination of $C_{14}AsF_6$ and other C_xAsF_6 salts may therefore occur via a species AsF_7^{2-}. This, like most seven-coordinate species (39) (e.g., IF_7 and ReF_7) would undergo intramolecular and intermolecular fluorine–ligand exchange. The latter, via transient fluorine-bridged species to AsF_6^- (i.e., $[F_6As-F-AsF_6]^{3-}$), could provide for the observed facile distribution of fluorine throughout the graphite galleries. Such mechanisms could also in part provide a plausible basis for understanding the formation of a fluorinated graphite of composition $C_{1.3}F$, which is formed (40) as a black first-stage solid concurrently with first-stage fluorinated graphite salt $C_{14}AsF_6 \cdot xF$, when second- or third-stage C_xAsF_6 salts are fluorinated at ~ 20 °C in the presence of anhydrous hydrogen fluoride (AHF). However, the separation of the C_xF phase in the presence of AHF must also involve fluoride ion transport as $(HF)_nF^-$ species.

The $C_{1.3}F$ is a more highly fluorinated relative of the sp^2 carbon fluorides first described by Rudorff and Rudorff (41). It possesses remarkable kinetic stability, and its resistance to oxidation by perchloric acid at 160 °C provides for its separation from $C_{14}AsF_6 \cdot xF$, which the acid destructively oxidizes at that temperature. Because the X-ray diffraction data (40) for $C_{1.3}F$ show the hexagonal carbon-atom array to be only slightly expanded relative to graphite itself, with a graphite-cell a_o equal to 2.46 Å, most of the fluorine ligands in this material must have neighbors at 2.46 Å. This distance is remarkably short. In harmony with the use of so many of the π-system electrons in binding F ligands (each C to which F is bound must be an insulating point in the network), $C_{1.3}F$ is an insulator. Its specific conductivity, $\sigma \sim 10^{-7} \, \Omega^{-1} \, cm^{-1}$, is attributable to F^- transport in the gallery.

Conclusions

Evidently the sp^2 carbon network is remarkably stable to electron oxidation to high charge levels (at least $C_{14}{}^+$). Good conductors are thereby generated. For high positive charge development, however, it is essential to have anions of very low fluorobasicity; otherwise, F^- transfers to the carbon. The sp^2 bonding of graphite persists even when 70% of the π bonding electrons are consumed in bonding F ligands, as in the insulator $C_{1.3}F$.

Acknowledgments

This work was supported by the Director, Office of Energy Research, Office of Basic Energy Sciences, Chemical Science Division of the U.S. Department of Energy, under Contract No. DE–AC03–76SF00098.

References

1. Hennig, G. R. *Prog. Inorg. Chem.* **1959**, *1*, 125.
2. Rudorff, W. *Adv. Inorg. Chem. Radiochem.* **1959**, *1*, 224.
3. Herold, A. In *Mater. Sci. Eng.* **1977**, *31*, 1; and In *Physics and Chemistry of Materials with Layered Structures, Intercalated Layered Materials*; Levy, F. A., Ed; Reidel: Dordrecht, Netherlands, 1979; Vol. 6, pp 323–421.
4. Bottomley, M. J.; Parry, G. S.; Ubbelohde, A. R. *Proc. Roy. Soc. (London)* **1964**, *A279*, 291.
5. McDonnell, F. R. M.; Pink, R. C.; Ubbelohde, A. R. *J. Chem. Soc.* **1951**, 191.
6. Ubbelohde, A. R. *Nature* **1971**, *232*, 43.
7. Coulson, C. A.; Taylor, R. *Proc. Phys. Soc. (London)* **1952**, *65*, 815.
8. Coulson, C. A.; Duncanson, W. E. *Proc. Phys. Soc. (London)* **1952**, *65*, 825.
9. Richardson, T. J.; Bartlett, N. *J. Chem. Soc., Chem. Commun.* **1974**, 427.
10. Richardson, T. J.; Tanzella, F. L.; Bartlett, N. *J. Am. Chem. Soc.* **1986** *108*, 4937.
11. Richardson, T. J.; Tanzella, F. L.; Bartlett, N. In *Polynuclear Aromatic Compounds*; Ebert, L. B., Ed; Advances in Chemistry 217; American Chemical Society: Washington, DC, 1988; p 169.
12. Bartlett, N.; Biagioni, R. N.; McQuillan, B. W.; Robertson, A. S.; Thompson, A. C. *J. Chem. Soc., Chem. Commun.* **1978**, 200.
13. Chun-Hsu, L.; Selig, H.; Rabinovitz, M.; Agranat, I.; Sarig, S. *Inorg. Nucl. Chem. Lett.* **1975**, *11*, 601.
14. Falardeau, E. R.; Foley, G. M. T.; Zeller, C.; Vogel, F. L. *J. Chem. Soc., Chem. Commun.* **1977**, 389.
15. Bartlett, N.; McCarron, E. M.; McQuillan, B. W.; Thompson, T. E. *Synth. Met.* **1979–80**, *1*, 221.
16. Thompson, T. E.; McCarron, E. M.; Bartlett, N. *Synth. Met.* **1981**, *3*, 255.
17. Kaindl, G.; Mallouk, T. E.; Bartlett, N., unpublished.

18. Bartlett, N.; McQuillan, B. W. In *Intercalation Chemistry;* Whittingham, S.; Jacobson, A., Eds.; Academic: New York, 1982; pp 19–53.
19. Kimura, M.; Schomaker, V; Smith, D. W.; Weinstock, B. *J. Chem. Phys.* **1968**, *48*, 4001.
20. Siegel, S.; Northrop, D. A. *Inorg. Chem.* **1966**, *5*, 2187.
21. George, P. M.; Beauchamp, J. L. *Chem. Phys.* **1979**, *36*, 345.
22. Nikitin, M. I.; Sidorov, L. N.; Korobov, M. V. *Int. J. Mass Spectrom. Ion. Phys.* **1981**, *37*, 13.
23. Bartlett, N. *Angew. Chem. Int. Ed. Engl.* **1968**, *7*, 433.
24. Selig, H.; Vaknin, D.; Davidov, D.; Yeshurun, Y. *Synth. Met.* **1985**, *12*, 479.
25. *Interatomic Distances* and *Supplement* (Special Publications Nos. 11 and 18); The Chemical Society: London, 1958, 1965.
26. Ebert, C. B.; Selig, H. *Mater. Sci. Eng.* **1977**, *31*, 177.
27. Ebert, C. B.; Selig, H. *Synth. Met.* **1981**, *3*, 53.
28. McCarron, E. M.; Grannec, Y. J.; Bartlett, N. *J. Chem. Soc., Chem. Commun.* **1980**, 890.
29. Rosenthal, G. L.; Mallouk, T. E.; Bartlett, N. *Synth. Met.* **1984**, *9*, 433.
30. Mallouk, T. E.; Rosenthal, G. L.; Muller, G.; Brusasco, R.; Bartlett, N. *Inorg. Chem.* **1984**, *23*, 3167.
31. Rosenthal, G. L. Ph.D. Thesis, University of California—Berkeley, 1984.
32. Okino, F. Ph.D. Thesis, University of California—Berkeley, 1983; Okino, F.; Bartlett, N., to be submitted to *Chemistry of Materials.*
33. Hagiwara, R.; Lerner, M.; Bartlett, N., to be published.
34. Zachariasen, W. H. *J. Am. Chem. Soc.* **1948**, *70*, 2147.
35. Kapustinskii, A. F. *Zh. Fiz. Khim.* **1934**, *5*, 59.
36. Mallouk, T. E. Ph.D. Thesis, University of California—Berkeley, 1984.
37. Kourtakis, K. Ph.D. Thesis, University of California—Berkeley, 1987.
38. Dean, P. A. W.; Gillespie, R. J.; Hulme, R. *J. Chem. Soc., Chem. Commun.* **1969**, 990.
39. Bartlett, N.; Beaton, S.; Reeves, L. W.; Wells, E. J. *Can. J. Chem.* **1964**, *42*, 2531.
40. Hagiwara, R.; Lerner, M.; Bartlett, N. *J. Chem. Soc., Chem. Commun.* **1989**, 573.
41. Rudorff, W.; Rudorff, G. *Chem. Ber.* **1947**, *80*, 413.

RECEIVED for review June 7, 1989. ACCEPTED revised manuscript September 27, 1989.

21

Intramolecular Electron Transfer and Electron Delocalization in Molybdophosphate Heteropoly Anions

Julie N. Barrows[1] and Michael T. Pope[2]

Department of Chemistry, Georgetown University, Washington, DC 20057

The electronic structures of mixed-valence heteropoly 12-molybdophosphate anions (heteropoly blues) were investigated by electronic absorption spectroscopy and by ^{17}O and ^{31}P NMR spectroscopy in aqueous and acetonitrile solutions. The one-, two-, and four-electron reduced forms of $\alpha\text{-}PMo_{12}O_{40}{}^{3-}$ (αI, αII, and αIV), are shown to retain, on the NMR time scale, the tetrahedral symmetry of the oxidized Keggin structure. These anions are class II mixed-valence species, with the additional electrons weakly trapped on molybdenum centers. The particularly stable four-electron reduced species $\beta\text{-}H_xPMo_{12}O_{40}{}^{(7-x)-}$ (βIV) is derived from the C_{3v} isomer of the Keggin structure. This anion is shown, on the basis of its intense narrow optical absorption band, ^{17}O NMR, and anomalous ^{31}P chemical shift, to contain delocalized valences (class III). Upon deprotonation, βIV reverts to a class II species with "normal" spectroscopic properties.

USE OF THE MOLYBDENUM BLUE METHOD for spectrophotometric determination of phosphate, silicate, and related species is widespread and has a

---

[1]Current address: Division of Colors and Cosmetics, U.S. Food and Drug Administration, Washington, DC 20204
[2]Address correspondence to this author.

0065–2393/90/0226–0403$06.00/0
© 1990 American Chemical Society

long history (1, 2). Although experimental procedures vary in detail, the main feature of such determinations involves the addition of molybdate and a reducing agent (such as ascorbate or tin(II)) to an acidified solution of the analyte. An intense blue color develops in the solution and its absorbance is measured. The species responsible for the blue color is one of a family of mixed-valent heteropolymolybdate anions known as heteropoly blues (3, 4).

In the absence of a reducing agent, reaction of molybdate(VI) and phosphate ions in acidic media leads to the formation of the 12-molybdophosphate anion, $\alpha\text{-PMo}_{12}O_{40}^{3-}$ ($\alpha 0$), with the well-known Keggin structure (5–7) illustrated in Figure 1.

As with many other Keggin-structure polymolybdates and polytungstates, the reduction behavior of $\alpha 0$ has been studied by electrochemical methods (8–11). Fruchart and Souchay (9) reported that $\alpha 0$, on a rotating platinum electrode in 1 M $HClO_4$ in 50% aqueous dioxane to prevent hydrolysis, showed four reversible two-electron reductions leading to species identified as αII, αIV, αVI, etc. (The roman numerals signify the number of added electrons.) When the reduced species were generated by controlled potential electrolysis in aqueous acidic solution, the polarograms underwent modification and ultimately showed a different set of two-electron waves that were attributed to a β isomer. Corresponding reduced derivatives of the β series were labeled βII, βIV, etc. The oxidized $\beta 0$ species may be partially stabilized in organic solvents (8). Polarograms of $\alpha 0$ and βIV are illustrated in Figure 2. The βIV species is generated by electrolytic reduction of $\alpha 0$ in aqueous media or by direct reaction of molybdate(VI), molybdenum(V), and phosphate (12). The βIV complex is formed in the molybdenum blue method for phosphate analysis.

We reported (13) the structure of the βIV anion from X-ray analysis of a crystal of a hemicalcium acid–salt, $Ca_{0.5}H_6PMo_{12}O_{40} \cdot 17.5H_2O$. The structure, illustrated in Figure 3, confirmed the assumption that the α–β isomerism was based upon rearrangement of an edge-shared triad of MoO_6 octahedra. In the β anion, one such triad has been rotated by $\pi/3$ with respect to its position in the α (Keggin) structure. Closer inspection of the βIV structure revealed that the molybdenum atoms in the ring of six edge- and corner-shared MoO_6 octahedra adjacent to the rotated group had been displaced from their positions observed in the oxidized α anion, and that the oxygens linking the edge-shared octahedra were protonated.

The Mo•••Mo separations were 3.56 and 3.67 Å for octahedral edge- and corner-shared connections, respectively, in contrast to values of 3.41 and 3.71 Å found in $\alpha 0$ (5). Because there was no crystallographic evidence of either static or dynamic disorder, it appeared that the four additional electrons had become delocalized over part or all of the structure. The present investigation was undertaken to define the electronic structure of βIV and to contrast this with what is known about the electronic structures of other heteropolymolybdate blues.

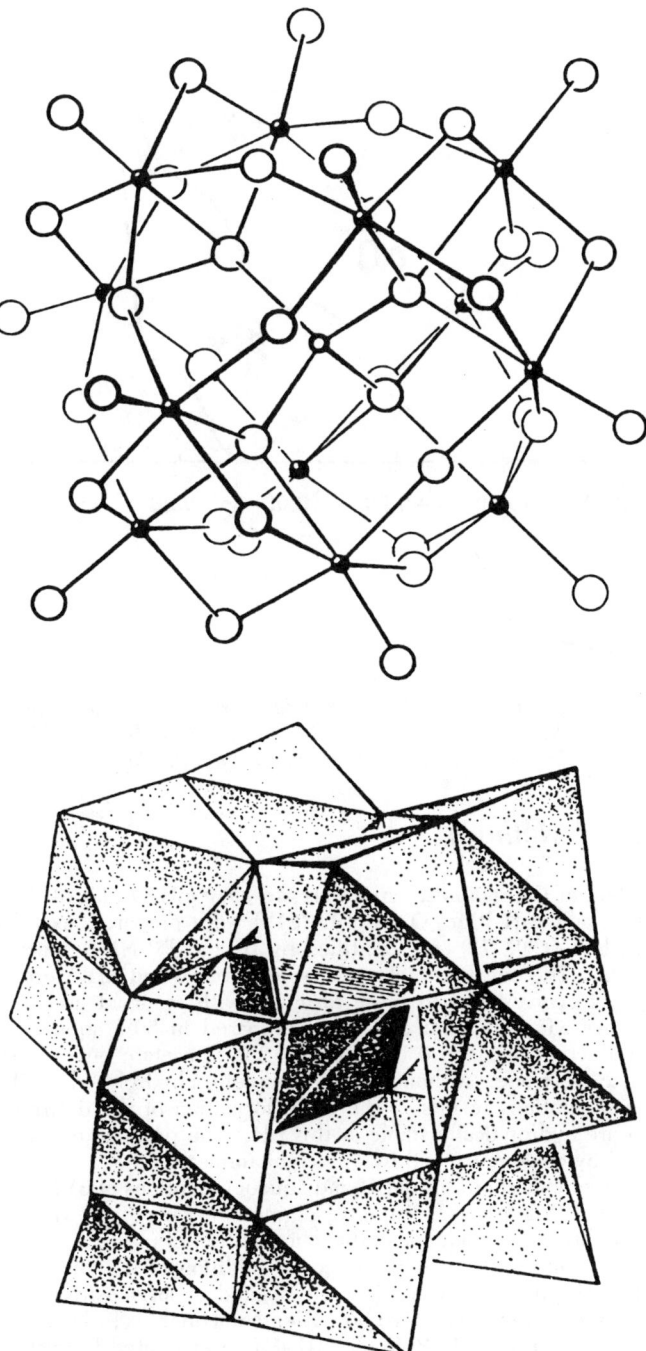

Figure 1. Bond and polyhedral representations of the Keggin structure observed for α-PMo$_{12}$O$_{40}^{3-}$.

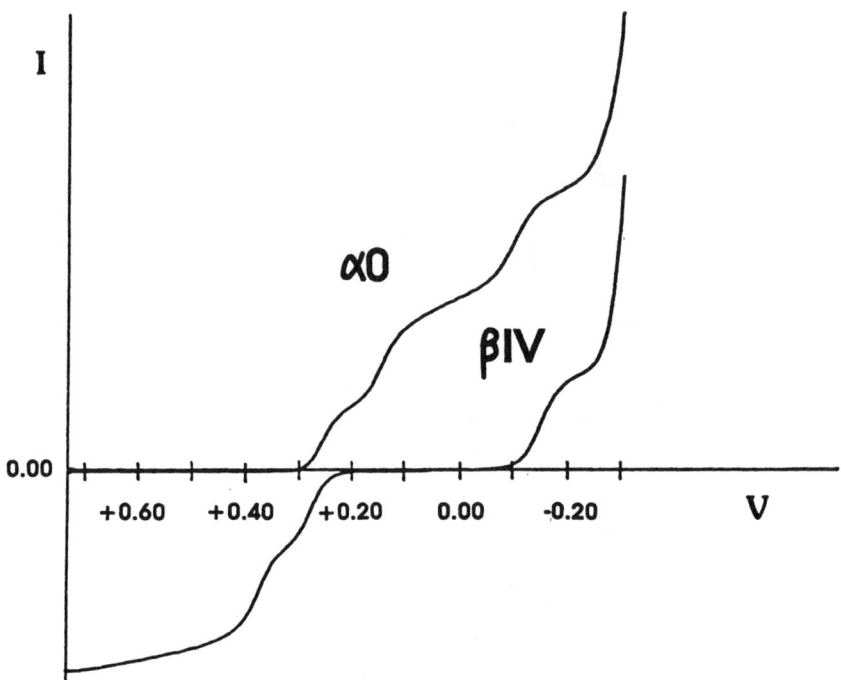

Figure 2. Rotating platinum electrode polarograms, vs. a saturated calomel electrode, of α-$Na_2HPMo_{12}O_{40}$ (α0) and 0.02 M β-$H_7PMo_{12}O_{40}$ (βIV) in 50% ethanol–H_2O. Supporting electrolyte: 0.5 M HCl.

Experimental Methods

Preparation of Compounds. α-$Na_2HPMo_{12}O_{40} \cdot mH_2O$ (α0), β-$[(n$-$C_4H_9)_4N]_3$-$PMo_{12}O_{40}$ (β0), and β-$H_7PMo_{12}O_{40} \cdot nH_2O$ (βIV) were prepared according to Rocchiccioli-Deltcheff et al. (12) and identified by IR and ^{31}P NMR spectroscopy and electrochemistry. The effective molecular weights of α0 and βIV were determined by coulometry to correspond to these formulas with $m = 31$ and $n = 21$. The heteropoly blues αII, αIV, and αVI were prepared in 50% aqueous ethanol by electrolytic reduction of α0 in 0.1 M HCl or 2.0 M acetate buffer, pH 4.7, on a platinum gauze electrode. (Although α0 is completely hydrolyzed at pH 4.7, electrolysis of such solutions results in the clean formation of the desired heteropoly blues, which are hydrolytically stable at this pH.) The progress of the electrolyses was monitored by coulometry and by cyclic voltammetry.

Aqueous ethanolic solutions of βII and aqueous solutions of βVI were prepared analogously from βIV at pH 4.7. Tetrabutylammonium salts of α0, αII, and βII were prepared by addition of an aqueous ethanolic solution of $(n$-$C_4H_9)_4NBr$. The tetrabutylammonium salt of βIV was prepared by addition of a 1 M HCl solution of $(n$-$C_4H_9)_4NBr$. The heteropoly blues were isolated as air-sensitive acid salts and were handled under argon in a glove box. Elemental analysis [calculated (found), in percent] for α-$[(n$-$C_4H_9)_4N]_3H_2PMo_{12}O_{40}$ (H_2αII, precipitated from 0.5 M HCl in 50% ethanol): C, 22.59 (23.17); H, 4.34 (4.34); N, 1.65 (2.06); P, 1.21 (1.21).

Potentiometric titration with $(n$-$C_4H_9)_4NOH$ confirmed both acidities, and the deprotonated anion gave clean voltammograms in acetonitrile. Elemental analysis [calculated (found), in percent] for β-[(n-C$_4$H$_9$)$_4$N]$_4$HPMo$_{12}$O$_{40}$ (HβII, precipitated from pH 4.7 buffer in 50% ethanol): C, 27.52 (26.81); H, 5.23 (5.06); N, 2.01 (2.33); P, 1.11 (1.12). The acidic proton was confirmed by potentiometric titration with (n-C$_4$H$_9$)$_4$NOH. Elemental analysis [calculated (found), in percent] for β-[(n-C$_4$H$_9$)$_4$N]$_3$H$_4$PMo$_{12}$O$_{40}$ (H$_4$βIV, precipitated from 1 M HCl, product is very air-sensitive): C, 22.58 (21.36); H, 4.42 (4.17); N, 1.65 (1.90); P, 1.21 (1.24).

Potentiometric titration with (n-C$_4$H$_9$)$_4$NOH revealed only one clear endpoint, corresponding to one replaceable proton in acetonitrile solution. The tetrabutylammonium salt of αI was prepared by electrolytic reduction of the tetrabutylammonium salt of α0 in acetonitrile (0.02 M, 0.1 M (n-C$_4$H$_9$)$_4$ClO$_4$). The dark green air-stable product spontaneously precipitated and was filtered off, washed with dichloromethane, and recrystallized from acetonitrile–dichloromethane.

α-[(n-C$_4$H$_9$)$_4$N]$_4$SiW$_{12}$O$_{40}$ was prepared as previously described (12). α-[(n-C$_4$H$_9$)$_4$N]$_4$PMo^{V}W$_{11}$O$_{40}$ and α-[(n-C$_4$H$_9$)$_4$N]$_3$PMo^{VI}W$_{11}$O$_{40}$ were prepared by modification of previously reported procedures (14–16): α-[(n-C$_4$H$_9$)$_4$N]$_3$PW$_{12}$O$_{40}$ (A) was prepared as previously described (12). [(n-C$_4$H$_9$)$_4$N]MoOCl$_4$ (B) was prepared by adding (n-C$_4$H$_9$)$_4$NBr to molybdenum(V) in aqueous HCl ("Mo(V) solution" (12)). The bright green solid was filtered off, washed with 6 M HCl, and recrystallized from methanol containing 10% 12 M HCl. To 4 g of A in 150 mL of hot acetonitrile were added 5 mL of 1 M (n-C$_4$H$_9$)$_4$NOH in methanol, then 1.2 g of B. The purple product, α-[(n-C$_4$H$_9$)$_4$N]$_4$PMo^{V}W$_{11}$O$_{40}$ (C), was precipitated with ethyl acetate, filtered off, and recrystallized from acetonitrile. α-[(n-C$_4$H$_9$)$_4$N]$_3$PMo^{VI}W$_{11}$O$_{40}$ was prepared by bromine oxidation of an acetonitrile solution of C. The yellow product was precipitated with ethyl acetate, filtered off, and recrystallized from acetonitrile.

^{17}O-enriched samples of tetrabutylammonium salts of α0, αI, and αII were obtained by first preparing an aqueous solution of enriched α0 (17). A mixture of 1.65 mL of 20% ^{17}O-enriched H$_2$O and 1.65 mL of water was carefully added to 0.150 g of PCl$_5$. To the resulting solution was added 1.0 g of Na$_2$MoO$_4$·2H$_2$O, and the solution was allowed to stand for 4 h. Concentrated (12 M) HCl (0.5 mL) was then added and after 1 h the solution was transferred to the electrolysis cell with 7.7 mL of H$_2$O, 0.5 mL of 12 M HCl, and 12.0 mL of ethanol.

The solution was electrolytically reduced to αII and the reduced solution treated with (n-C$_4$H$_9$)$_4$NBr to yield α-[(n-C$_4$H$_9$)$_4$N]$_3$H$_2$PMo$_{12}$O$_{40}$ (enriched H$_2$αII). Air oxidation of an acetonitrile solution of H$_2$αII yields α0, which was recovered by solvent stripping. Careful neutralization of the protons of H$_2$αII by stoichiometric addition of (n-C$_4$H$_9$)$_4$NOH in acetonitrile gives a solution that is air-oxidized to αI. ^{17}O-enriched β0 was obtained by adding 0.6 mL of 20% ^{17}O-enriched H$_2$O to 0.3 g of β0 in 20 mL of CH$_3$CN. The solution was allowed to stand overnight. The solid was recovered by solvent stripping. (The oxygens bonded to the central phosphorus atom do not undergo noticeable oxygen exchange and do not give a detectable ^{17}O NMR line under these conditions.)

Elemental analyses were carried out by E+R Analytical Laboratory, Corona, NY.

Electrochemistry. Polarography and cyclic voltammetry were performed with apparatus assembled from McKee–Pederson modules, or on a voltammetric analyzer (IBM model EC–225). Rotating and stationary platinum working electrodes, a platinum wire counter electrode, and a saturated calomel reference electrode were used. Electrolyses were carried out on a platinum gauze electrode with a potentiostat (Brinkmann–Wenking model 70 TST) and a coulometer (Koslow Scientific model 541).

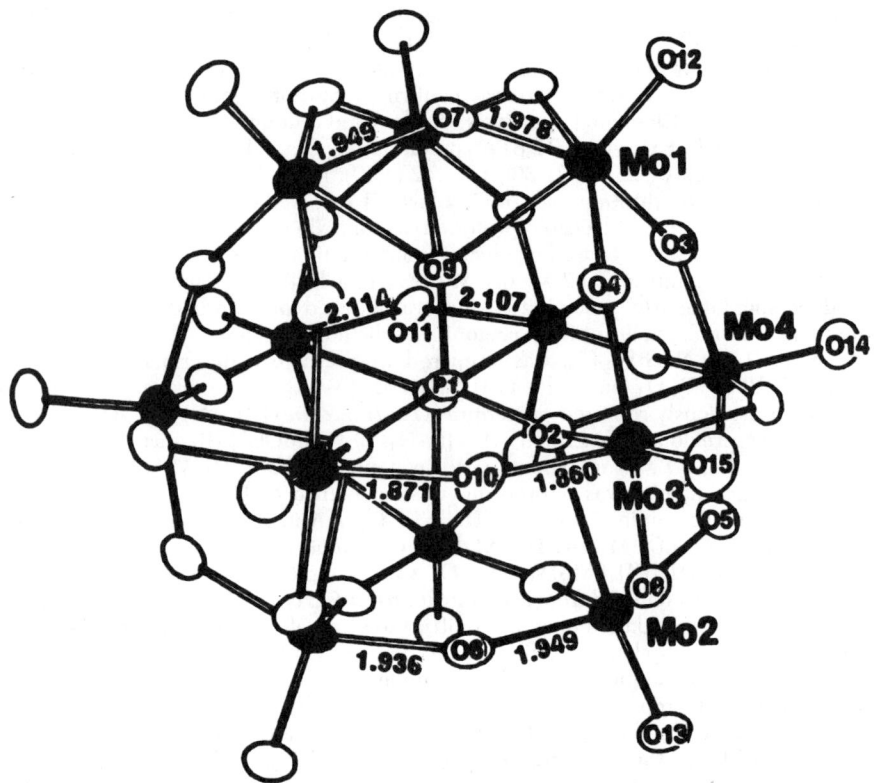

Figure 3a. Bond representations of the structure of the anion in β-$Ca_{0.5}H_6PMo_{12}O_{40} \cdot 17.5H_2O$, viewed perpendicular to the three-fold axis. The three oxygen atoms, O11, at the edge-shared junctions of the equatorial ring, are protonated. (Reproduced from ref. 13. Copyright 1985 American Chemical Society.)

NMR Spectroscopy. A spectrometer (Bruker AM–300WB) operating at 7.05 T (300.13 MHz for protons) was used. Resonance frequencies were 40.688 MHz for ^{17}O and 121.496 MHz for ^{31}P. Pulse widths (90°) were 36 μs for ^{17}O and 11 μs for ^{31}P. Chemical shifts are reported with respect to external H_2O and 85% H_3PO_4; this positive sign indicates a resonance at higher frequency than the standard. Longitudinal relaxation times (T_1) were determined by the inversion–recovery method. Solid-state magic angle sample spinning (MASS) and nonspinning ^{31}P spectra of βIV were recorded on the same instrument. The ^{17}O NMR spectrum of β0 was recorded on a spectrometer (400-MHz, Bruker) at the University of South Carolina NSF NMR Facility.

Results and Discussion

One- and Two-Electron Blues. α0 undergoes a series of two-electron reductions in protic solvents leading to protonated αII, αIV, etc.,

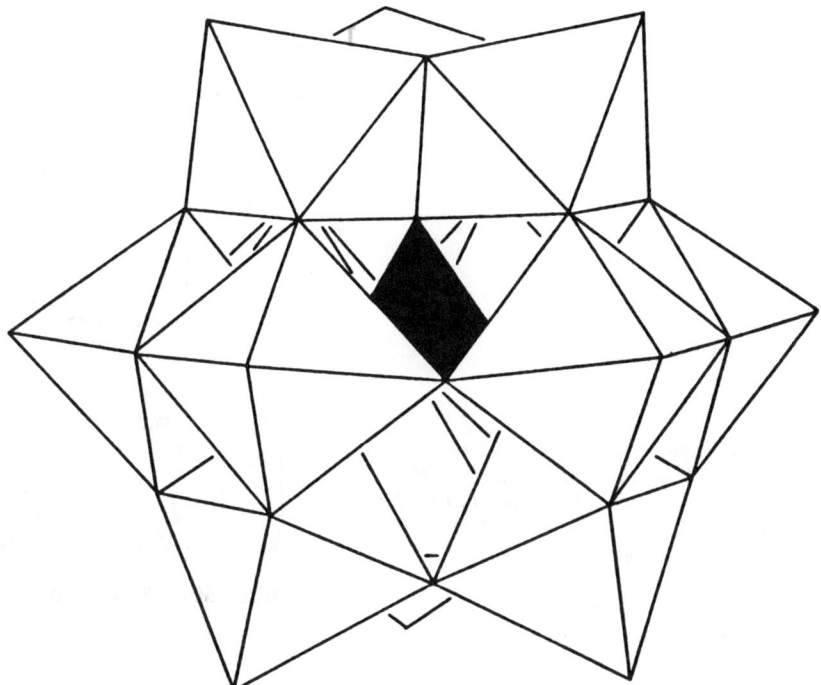

Figure 3b. Polyhedral representation of the anion in Figure 3a.

but as the pH is raised, the two-electron polarographic waves split into pH-independent one-electron waves (9). The one-electron reduced species, $\alpha\text{-PMo}_{12}\text{O}_{40}^{4-}$ (αI), is most conveniently studied by reduction of α0 in an aprotic solvent. The electron spin resonance (ESR) spectrum of αI, reported several years ago (18), consists (at T < ~60 K) of an axial line with hyperfine structure resulting from interaction of the electron with one molybdenum atom (^{95,97}Mo, ~25%, $I = 5/2$, where I is the nuclear spin quantum number).

The hyperfine structure disappeared as the temperature was raised. The complete spectrum broadened rapidly above 77 K and was not observable at room temperature. This behavior, and analogous behavior for corresponding reduced tungstates, has been interpreted according to a class II (19) mixed-valence model (i.e., one in which the unpaired electron is trapped on a single metal atom at low temperatures, but undergoes rapid intramolecular hopping among the other metal atoms as the temperature is raised). The electronic spectrum of αI (and analogues) has been interpreted according to the same model (20).

Both αI and αII anions in acetonitrile solution give well-resolved ^{17}O NMR spectra that are shown in Figure 4 and summarized in Table I. These spectra verify that the reduced anions retain the tetrahedral symmetry of the Keggin structure on the NMR time scale. A similar spectrum for H$_2\alpha$II

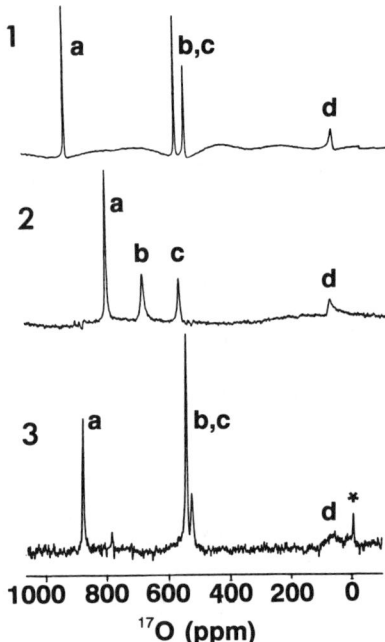

Figure 4. Oxygen-17 NMR spectra of ^{17}O-enriched tetrabutylammonium salts of the oxidized and one-, and two-electron reduced forms of α-$PMo_{12}O_{40}^{3-}$: 1, $\alpha 0$; 2, αI; 3, αII; in CD_3CN, 338 K; a, terminal oxygen; b, oxygen linking corner-shared octahedra; c, oxygen linking edge-shared octahedra; d, interior phosphate oxygen; *, H_2O.

in water–dioxane was reported earlier by Kazansky et al. (21) and interpreted in terms of a delocalized electron pair.

Electron Exchange. The well-resolved spectra of unprotonated αI and αII (Figure 4) verify that intramolecular electron transfer is rapid. Furthermore, the spectrum of αI is consistent with the conclusion drawn from ESR spectroscopy that the unpaired electron undergoes rapid spin relaxation at

Table I. Oxygen-17 NMR Data for Oxidized and Reduced Forms of $PMo_{12}O_{40}^{3-}$ Isomers

Anion	T, K	Chemical Shifts, ppm (Line Widths, Hz)
$\alpha 0$	297	938 (70), 583 (80), 551 (165), 80 (660)
	338	937 (40), 584 (50), 552 (100), 80 (360)
$\beta 0^a$	333	999 (ND), 585 (ND), 555 (ND)
αI	297	784 (210), 706 (570), 567 (580), 80 (660)
	338	801 (160), 686 (410), 570 (290), 80 (500)
αII	297	878 (190), 539 (760), 523 (375), 70 (1000)
$H\alpha II$	297	902 (190), 560 (760), 533 (665), 70 (800)
$H_2\alpha II$	297	917 (270), 571 (620), 537 (700), 70 (1000)

NOTE: Measured on enriched tetrabutylammonium salts in CD_3CN, 0.0025 M ($\alpha 0$, αII, $H\alpha II$, $H_2\alpha II$), ~0.005 M ($\beta 0$), 0.009 M (αI).
[a]Measured at the University of South Carolina NSF NMR Facility; line widths were not determined (ND).

room temperature ($T_e \sim 10^{-9}$ s). This is confirmed by observation of a relatively narrow ^{31}P NMR line for αI at 297 K ($\Delta\nu_{1/2}$, 2.6 Hz; T_1, 0.75 s; T_2, 0.12 s; δ, +3.5 ppm). Chemical shifts and line widths for α0 and αII under the same conditions were −0.4 ppm, 0.1 Hz, and −3.3 ppm, 0.1 Hz, respectively.

Kozik et al. (22) observed narrow ^{31}P NMR lines for paramagnetic tungstophosphate heteropoly blues in aqueous solution and determined intermolecular electron-exchange rates. In the present case, measurements were made of the line width of α0 in the presence of varying concentrations of αI. The ionic strength (μ) of the solution was maintained with α-[(n-C$_4$H$_9$)$_4$N]$_4$SiW$_{12}$O$_{40}$. Evaluated rate constants at 297 K are 5.3±0.3 × 10^3 M^{-1} s^{-1} in acetonitrile (μ = 0.04 M) and 3.8±0.2 × 10^3 M^{-1} s^{-1} in dimethyl sulfoxide (μ = 0.32 M). Diffusion-controlled rates for exchange between −3 and −4 anions of diameter 11.2 Å in solvents of dielectric constant 36.8 and 47.6 are 1 × 10^4 and 7 × 10^4 M^{-1} s^{-1}, respectively.

We have also examined electron exchange between α-PMo^{V}W$_{11}$O$_{40}$$^{4-}$ and α-PMo^{VI}W$_{11}$O$_{40}$$^{3-}$ by the same method. In view of the strongly localized nature of the unpaired electron in α-PMo^{V}W$_{11}$O$_{40}$$^{4-}$ (well-resolved ESR spectrum at room temperature), the ^{31}P NMR line of this anion is broad (120 Hz, acetonitrile; 600 Hz, dimethyl sulfoxide). As anticipated, the exchange rates are slower than those involving α0 and αI: k = 7±2 × 10^2 and 80±6 M^{-1} s^{-1} in acetonitrile and dimethyl sulfoxide, respectively.

Isotropic Shifts. The isotropic shifts of the ^{17}O NMR lines of αI are discussed elsewhere (23, 24). Such shifts are scalar in origin because, in the fast intramolecular electron-exchange limit that pertains here, the anion retains tetrahedral symmetry. The small shift observed for the interior phosphate oxygens is consistent with the weak bonds that link these oxygens to the molybdenums. The remaining lines in the spectrum of αI at 297 K are assigned, by comparison to Mo$_6$O$_{19}$$^{2-/3-}$, as follows: 784 ppm, terminal oxygen; 706 ppm, oxygen linking corner-shared octahedra; 567 ppm, oxygen linking edge-shared octahedra. On the basis of their temperature dependence (25) (Table I), these lines correlate with those in the spectrum of α0 at 938, 551, and 583 ppm, respectively. The isotropic shifts for the two kinds of bridging oxygens have opposite signs. Our observation of only upfield isotropic shifts for Mo$_6$O$_{19}$$^{2-/3-}$ (23), an anion that has edge-shared bridging and terminal oxygens but no corner-shared oxygen, is the basis for the stated assignments for the Keggin structure.

Four-Electron Blues. The spectroscopic properties of βIV are summarized in Tables II and III and Figures 5 and 6. It is convenient to discuss these in light of the previously reported crystal structure. All three types of spectra (^{31}P NMR, ^{17}O NMR, and electronic) show pronounced changes as the anion is deprotonated. The pKs of the final three acidities of β-

$H_7PMo_{12}O_{40}$, corresponding to the loss of the protons located on the ring of six MoO_6 octahedra previously discussed, are 4.5, 6.9, and 9.3 (9).

In 0.1 M HCl the ^{17}O NMR spectrum is consistent with the C_{3v} symmetry of the β structure. At 338 K it shows three resolved resonances (2:1:1) in the terminal oxygen region, as well as five "bridging" resonances (Figure 5). Kazansky et al. (26) described a similar ^{17}O NMR spectrum, with two lines of approximately equal intensity resolved in the terminal region, and attributed it to the two-electron reduced species. However, the reported ^{31}P chemical shift (discussed later) identifies it as βIV. The lower effective NMR symmetry of βIV is caused by an unequal distribution of the four additional electrons, rather than by the molecular structure. This causality is demonstrated by the simple spectrum of β0, which has a total of only three lines in the terminal and doubly-bridging oxygen regions (Table I). The ^{17}O NMR spectrum of αIV, recorded in 50% ethanol–D_2O at 297 K to retard decomposition, also shows three lines (Table II), as expected for an exchange-averaged Keggin structure. Even at 297 K, the spectrum of βIV is still partially resolved in acidic solution, but at pH 7.6 only two broad resonances are detected.

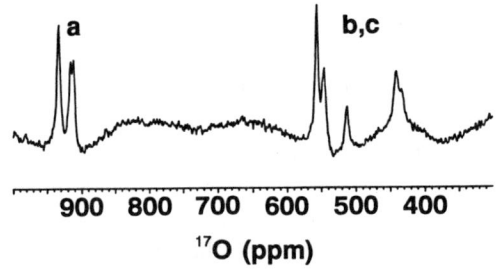

Figure 5. Natural abundance ^{17}O NMR spectrum of βIV anion in 0.1 M HCl, D_2O, 338 K; oxygen labels as in Figure 4.

As also observed by others (21, 26, 27), the ^{31}P resonance of βIV at pH ~0 appears at −12.5 ppm. This represents an upfield shift of 9 ppm from that of α0, whereas other reduced species (αII, αIV, αVI, βII, and βVI) have corresponding shifts of <4 ppm (δ −4.4 to −7.6). The chemical shift of βIV is unchanged between pH 0 and 4.7, but thereafter the line broadens somewhat and the chemical shift decreases (see Table III).

Proton Exchange. Solutions of the tetrabutylammonium salt of βIV, β-$[(n\text{-}C_4H_9)_4N]_3H_4PMo_{12}O_{40}$, in acetonitrile show a broad (60 Hz) ^{31}P NMR line that broadens still further (210 Hz) when a single proton is removed by the addition of the stoichiometric quantity of $(n\text{-}C_4H_9)_4NOH$. Further addition of base led to multiple-line spectra that indicate anion decomposition. We attribute the broad lines to slow proton exchange (inter- or intramolecular) for the following reasons. The values of T_1 for the aqueous and

Table II. Oxygen-17 NMR Data for Four-Electron Reduced Forms of $PMo_{12}O_{40}^{3-}$ Isomers

Anion	T, K	Chemical Shifts, ppm (Line Widths, Hz)
αIV[a]	297	911 (480), 465 (1200), 438 (1400)
βIV[b]	297	934 (550), 909 (500), 530 (1000), 440 (590)
	338	933 (200), 915, 911 (335[c]), 555 (170)
		545 (335), 512 (225), 441, 433 (580[c])
βIV[d]	297	868 (1230), 513 (2000)

NOTE: Measured at natural abundance.
[a]0.2 M αIV, 0.1 M HCl, 50% ethanol–D_2O.
[b]0.25 M βIV, 0.1 M HCl, D_2O.
[c]Total width of two overlapping lines.
[d]0.1 M βIV, 1.0 M Tris buffer, pH 7.6, D_2O.

nonaqueous solution spectra are similar (2.5 and 3.0 s, respectively); this similarity rules out possible paramagnetic effects. A broad line, observed in the ^1H NMR spectrum of the tetraprotonated species in acetonitrile at 4 ppm, is attributed to the anion's protons because the line disappears upon the addition of a drop of D_2O.

Table III. Phosphorus-31 NMR Data for Oxidized and Reduced Forms of $PMo_{12}O_{40}^{3-}$ Isomers

Anion	Aqueous Solvent[a]	pH	Chemical Shifts, ppm (Line Widths, Hz)	
			Aqueous[b]	Acetonitrile[c]
α0	A	4.7	−3.2 (<1)	−0.4 (0.1)
αI				+3.5 (2.6)
αII	A	4.7	−5.9 (0.8)	−3.3 (0.1)
αIV	A	4.7	−5.0 (1.5)	
αVI	A	4.7	−7.6 (1.6)	
β0				−0.1 (0.3)
βII	A	4.7	−5.8 (1.8)	−2.4 (8)
βIV	B	0	−12.5 (1.2)	
	C	2.7	−12.4 (0.8)	
	A	4.7	−12.6 (<1)	
	D	5.1	−11.5 (4.3)	
	E	6.9	−10.6 (3.2)	
	F	7.6	−9.8 (5.4)	
	G	9.5	−9.4 (3.8)	
H_4βIV				−9.5 (60)
H_3βIV				−8.6 (210)
βVI	D	4.7	−6.4 (17)	

NOTE: Data were obtained at 297 K.
[a]Solvents: A, 2 M acetate buffer, 50% ethanol–D_2O; B, 0.5 M HCl–D_2O; C, 1 M $CH_2ClCOOH$ buffer–D_2O; D, 2 M acetate buffer–D_2O; E, 1 M $KHPO_4$ buffer–D_2O; F, 1 M Tris buffer–D_2O; and G, 1 M $NaHCO_3$ buffer–D_2O.
[b]0.02 M (α0, αII, αIV, αVI, βII, βVI), 0.01 M (βIV).
[c]0.002 M (α0, β0), 0.01 M (αI, αII, βII, βIV).

The optical spectrum of βIV is also strongly pH dependent, in contrast to those of other molybdophosphate blues (see Figure 6). The intense (molar absorptivity (ϵ_{max}) of 27,000 M^{-1} cm^{-1}), relatively narrow band at 830 nm, which is the reason for the analytical value of the molybdenum blue procedure, is lost as the anion is deprotonated. Similar spectral changes have been reported previously (28), and also for the corresponding "βIV" derivative of SiMo$_{12}$O$_{40}$$^{4-}$ (29). The spectrum of the deprotonated anion shows broader absorption bands with $\epsilon_{max} \sim$ 8000 M^{-1} cm^{-1}, which are similar to those observed for other heteropoly blue species (3, 29–31), such as the protonated and deprotonated αII illustrated in Figure 6a.

With the exception of βIV in acidic solution, the maximum intensities of heteropoly blue spectra are roughly proportional to the number of added electrons, ~2000 M^{-1} cm^{-1} per electron for polymolybdates. This observation may be explained on the basis of virtually independent intervalence charge-transfer transitions within the polyanion structure, assuming a class II mixed-valence model for these species.

Electronic Structure. The unique βIV spectrum suggests a change of electronic structure, perhaps to class III with complete ground-state delocalization of the four electrons over all or part of the anion structure. The 830-nm absorption band is much narrower than would be calculated from the conventional Hush treatment if it were assumed to be an intervalence charge-transfer transition of a class II complex (Table IV).

The delocalized nature of βIV is further supported by the X-ray structural results that indicate three-fold symmetry for an anion with four additional electrons. On the basis of these data, the observation of three terminal ^{17}O NMR resonances with relative intensities 2:1:1, and the large chemical shift anisotropy ($\sigma_\perp - \sigma_\parallel$ = 10 ppm) observed in the solid-state ^{31}P NMR spectrum of βIV, there is good reason to believe that the electrons are predominantly delocalized over the equatorial ring of six MoO$_6$ octahedra. ($\sigma_\perp$ and $\sigma_\parallel$ are the perpendicular and parallel components of the nuclear shielding tensor.)

Ring-Current Effects. Such electron delocalization should lead to the observation of ring-current effects. Indeed, the situation in βIV would then be analogous to that of benzene, because the four extra electrons can be considered to be involved in π orbitals (derived from Mo "d_{xy}-" and O π orbitals) that are topologically analogous to the π orbitals of benzene. By using the conventional Johnson–Bovey tables (33, 34) and a ring radius of 3.44 Å (= Σ(Mo•••Mo)/2π), we can estimate a ring-current shielding of the central phosphorus nucleus of 4.1 ppm.

Kozik et al. (35) and Casañ-Pastor (36) attributed enhanced diamagnetic susceptibilities of two-electron reduced heteropolytungstate blues to ring currents. Our general observation that chemical shifts of central ^{31}P nuclei are decreased as the anions are reduced is consistent with, although does

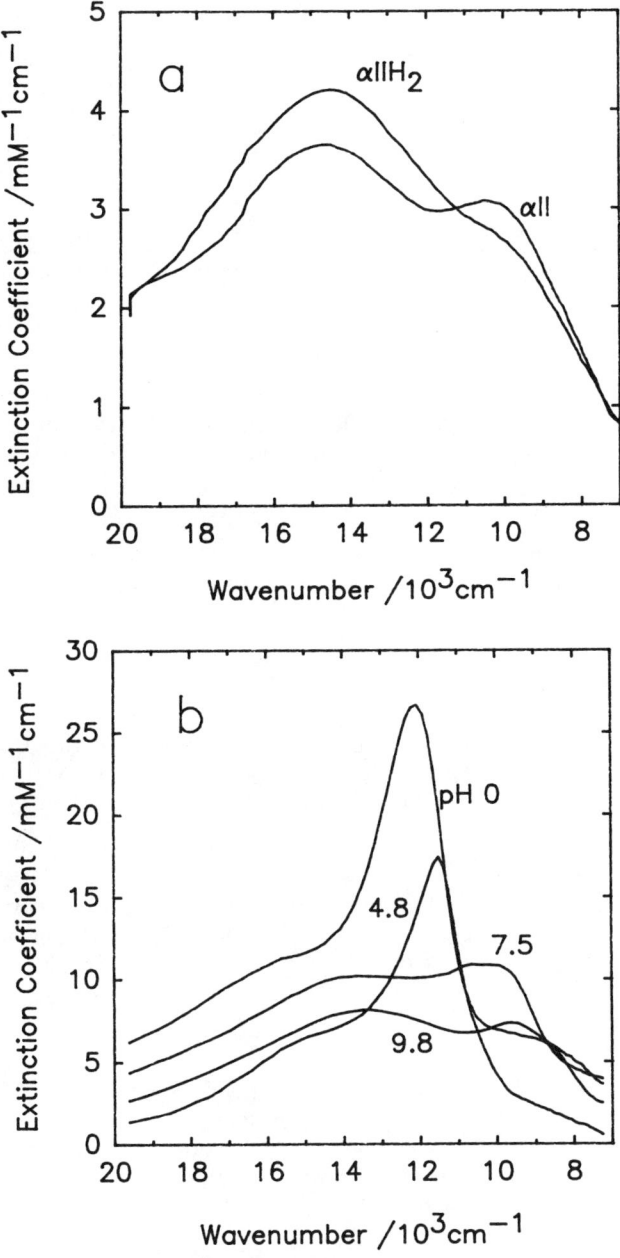

Figure 6. Electronic spectra of a, $\alpha\text{-PMo}_{12}O_{40}^{5-}$ (αII) and $\alpha\text{-}H_2PMo_{12}O_{40}^{3-}$ ($H_2\alpha II$) as tetrabutylammonium salts in CH_3CN; and b, $\beta\text{-}H_7PMo_{12}O_{40}$ (βIV) in aqueous solution at varying pH.

Table IV. Intervalence Charge-Transfer Band Profiles (cm^{-1}) for Reduced Forms of PMo$_{12}$O$_{40}$$^{3-}$ Isomers

Anion	E_{op}^{a}	$\Delta\nu_{1/2}$, obs.	$\Delta\nu_{1/2}$, calc.^{b}
H$_2\alpha$II	10120	4910	4815
αII	10120	4650	4815
βII	9460	4480	4650
βIV (pH 0)	9230	4240	4600
	12050^c	2900	5250
βIV (pH 9.8)	9580	3550	4680

^{a}Lowest energy of intervalence charge-transfer absorption band, unless specified otherwise.
$^b\Delta\nu_{1/2} = (16 \, kT \ln 2 \, E_{op})^{1/2}$ (ref. 32).
^{c}Prominent 830-nm band.

not require, that interpretation. Deprotonation of βIV results in both loss of the intense optical absorption band and decreased shielding of the phosphorus nucleus. We speculate that the deprotonated anion has more conventional Mo•••Mo separations in the equatorial ring (i.e., alternating between 3.4 and 3.7 Å for edge- and corner-shared octahedra) and that this conformation results in less electron delocalization and a class II electronic structure.

Acknowledgments

We thank Paul Ellis, Helga Cohen, and Alan Benesi at the University of South Carolina NSF NMR Facility for recording the ^{17}O NMR spectrum of β0; Michael Geckle of Bruker Instruments for recording the solid-state ^{31}P NMR spectra of βIV; and C. F. Hammer for technical advice. Acknowledgment is made to the donors of The Petroleum Research Fund, administered by the American Chemical Society, for support of this research through Grant No. 18537–AC3. The research is part of the Ph.D. dissertation of JNB (Georgetown University, 1989).

References

1. Boltz, D. F.; Mellon, M. G. *Anal. Chem.* **1947**, *19*, 873.
2. Murphy, J.; Riley, J. P. *Anal. Chim. Acta* **1962**, *27*, 31.
3. Pope, M. T. In *NATO Advanced Study Institute Series, Series C*; Brown, D. B., Ed.; Reidel: Dordrecht, Netherlands, 1980; p 365.
4. Pope, M. T. *Heteropoly and Isopoly Oxometalates*; Springer-Verlag: New York, 1983.
5. Strandberg, R. *Acta Chem. Scand.* **1975**, *A29*, 350.
6. D'Amour, H.; Allmann, R. *Z. Kristallogr.* **1976**, *143*, 1.
7. Clark, C. J.; Hall, D. *Acta Crystallogr. B* **1976**, *B32*, 1545.
8. Hori, T.; Fujinaga, T. *Bull. Chem. Soc. Jpn.* **1985**, *58*, 1380.
9. Fruchart, J. M.; Souchay, P. *C. R. Seances Acad. Sci., Ser. C* **1968**, *266*, 1571.

10. Tanaka, N.; Unoura, K.; Itabashi, E. *Inorg. Chem.* **1982**, *21*, 1662.
11. Unoura, K.; Tanaka, N. *Inorg. Chem.* **1983**, *22*, 2963.
12. Rocchiccioli-Deltcheff, C.; Fournier, M.; Franck, R.; Thouvenot, R. *Inorg. Chem.* **1983**, *22*, 207.
13. Barrows, J. N.; Jameson, G. B.; Pope, M. T. *J. Am. Chem. Soc.* **1985**, *107*, 1771.
14. Altenau, J. J.; Pope, M. T.; Prados, R. A.; So, H. *Inorg. Chem.* **1975**, *14*, 417.
15. Prados, R. A.; Meiklejohn, P. T.; Pope, M. T. *J. Am. Chem. Soc.* **1974**, *96*, 1261.
16. O'Donnell, S. E., Ph.D. Dissertation, Georgetown University, 1975.
17. Filowitz, M.; Ho, R. K. C.; Klemperer, W. G.; Shum, W. *Inorg. Chem.* **1979**, *18*, 93.
18. Launay, J. P.; Fournier, M.; Sanchez, C.; Livage, J.; Pope, M. T. *Inorg. Nucl. Chem. Lett.* **1980**, *16*, 257.
19. Robin, M. B.; Day, P. *Adv. Inorg. Chem. Radiochem.* **1967**, *10*, 247.
20. Sanchez, C.; Livage, J.; Launay, J. P.; Fournier, M.; Jeannin, Y. *J. Am. Chem. Soc.* **1982**, *104*, 3194.
21. Kazansky, L. P.; Potapova, I. V.; Spitsyn, V. I. *Proceedings of the 3rd International Conference on the Chemistry and Uses of Molybdenum;* Climax Molybdenum: Ann Arbor, MI, 1979; p 67.
22. Kozik, M.; Hammer, C. F.; Baker, L. C. W. *J. Am. Chem. Soc.* **1986**, *108*, 7627.
23. Piepgrass, K.; Barrows, J. N.; Pope, M. T. *J. Chem. Soc., Chem. Commun.* **1989**, 10.
24. Barrows, J. N., Ph.D. Dissertation, Georgetown University, 1989.
25. Kazansky, L. P.; Fedotov, M. A. *Koord. Khim.* **1988**, *14*, 939.
26. Kazansky, L. P.; Fedotov, M. A.; Potapova, I. V.; Spitsyn, V. I. *Doklady Chem. Engl. Trans.* **1979**, *244*, 36.
27. Aoshima, A.; Yamaguchi, T. *J. Chem. Soc. Jpn., Chem. Ind. Chem.* **1986**, 5, 641.
28. Fruchart, J. M.; Hervé, G.; Launay, J. P.; Massart, R. *J. Inorg. Nucl. Chem.* **1976**, *38*, 1627.
29. Massart, R. *Ann. Chim. (Paris)* **1969**, *4*, 441.
30. Massart, R. *Ann. Chim. (Paris)* **1968**, *3*, 507.
31. Massart, R. *Ann. Chim. (Paris)* **1969**, *4*, 285.
32. Hush, N. S. *Prog. Inorg. Chem.* **1967**, *8*, 391.
33. Johnson, C. E.; Bovey, F. A. *J. Chem. Phys.* **1958**, *29*, 1012.
34. Emsley, J. W.; Feeney, J.; Sutcliffe, L. H. *High Resolution Nuclear Magnetic Resonance Spectroscopy;* Pergamon: Oxford, 1965; Vol. 1, p 595.
35. Kozik, M.; Casañ-Pastor, N.; Hammer, C. F.; Baker, L. C. W. *J. Am. Chem. Soc.* **1988**, *110*, 7697.
36. Casañ-Pastor, N., Ph.D. Dissertation, Georgetown University, 1988.

RECEIVED for review May 1, 1989. ACCEPTED revised manuscript August 14, 1989.

22

Organometallic Electron-Transfer Salts with Tetracyanoethylene Exhibiting Ferromagnetic Coupling

Joel S. Miller[1] and Arthur J. Epstein[2]

[1]Central Research and Development, E. I. du Pont de Nemours and Company, Experimental Station–E328, Wilmington, DE 19880–0328
[2]Department of Physics and Department of Chemistry, The Ohio State University, Columbus, OH 43210–1106

Some molecular organic solids comprising linear chains of alternating total spin angular momentum quantum number $S = \frac{1}{2}$ metallocenium donors, D, and cyanocarbon acceptors, A (i.e., $\cdots D^{\cdot+}A^{\cdot-}D^{\cdot+}A^{\cdot-}\cdots$) exhibit cooperative magnetic phenomena (i.e., ferro-, antiferro-, ferri-, and metamagnetism). For $[Fe^{III}(C_5Me_5)_2]^{\cdot+}[TCNE]^{\cdot-}$ (Me is methyl; TCNE is tetracyanoethylene), bulk ferromagnetic behavior is observed below the Curie temperature of 4.8 K. Replacement of Fe^{III} with Cr^{III}, Ni^{III}, and Fe^{II} leads to complexes with antiferromagnetic coupling, ferrimagnetic behavior, and almost no magnetic interaction, respectively. These results are consistent with a model of configuration mixing of the lowest charge-transfer excited state with the ground state developed earlier to understand the magnetic coupling of such systems. The model, which predicts the magnetic coupling as a function of electron configuration and direction of charge transfer, is a useful guide in the design of new organic and organometallic complexes with cooperative magnetic coupling. New TCNE-based electron-transfer salts were prepared to test the model and identify new materials with ferromagnetic coupling.

MOLECULAR AND ORGANIC FERROMAGNETIC COMPOUNDS, although postulated in the 1960s, have only recently been synthesized and characterized

(1–5). This development, which parallels the discovery of molecular and organic-based superconductors, extends the study of cooperative phenomena in molecular organic materials. The broad range of phenomena in the molecular organic solid state, combined with the anticipated modification of physical properties via conventional synthetic organic chemistry and the ease of fabrication enjoyed by soluble materials, may ultimately lead to their use in future generations of electronic and photonic devices.

This chapter summarizes the configuration mixing of a virtual triplet excited state with the ground state for an alternating donor–acceptor (D–A) one-dimensional chain model for the stabilization of ferromagnetic coupling of this electron-transfer compound (1–6). A discussion of the common idealized magnetic behaviors expected in materials (2) and a more comprehensive discussion of several models for ferromagnetic coupling in molecular polymeric materials can be found in published reviews (1–5).

Stabilization of Ferromagnetic Coupling by Configuration Mixing

Spin alignment throughout the solid is necessary for bulk ferromagnetism. Several mechanisms (1–5) have been proposed for the pairwise stabilization of ferromagnetic coupling among spins. However, these schemes are insufficient to account for three-dimensional ferromagnetic behavior. A mechanism to account for this three-dimensional interaction is described in this chapter.

The model of configuration mixing of a virtual triplet charge-transfer excited state with the ground state for a $\cdots D^{\cdot +}A^{\cdot -}D^{\cdot +}A^{\cdot -}D^{\cdot +}A^{\cdot -}\cdots$ chain to stabilize ferromagnetic coupling was originally introduced by McConnell (6). For a $D^{\cdot +}A^{\cdot -}$ pair with a half-occupied nondegenerate highest occupied molecular orbital (HOMO), the spins couple antiferromagnetically (Figure 1, Ia). (Presumably, the virtual charge transfer involves only the highest-energy partially occupied molecular orbital (POMO). Circumstances in which virtual excitation from a lower-lying filled (or to a higher-lying filled) orbital dominate the admixing exciting state are conceivable, and the orbital degeneracy and symmetry restrictions are relaxed.)

Admixture of the higher energy charge-transfer states with the ground state lowers the total electronic energy and stabilizes antiferromagnetic coupling. Figure 1 (IIb and IIc) illustrates an electron being delocalized onto an adjacent site. This energy reduction and delocalization does not occur when the two electron spins are parallel (ferromagnetically aligned), in accord with the Pauli exclusion principle. Thus, antiferromagnetic and (for other electron configurations) ferromagnetic coupling can be achieved along and between chains (2, 5, 7, 8).

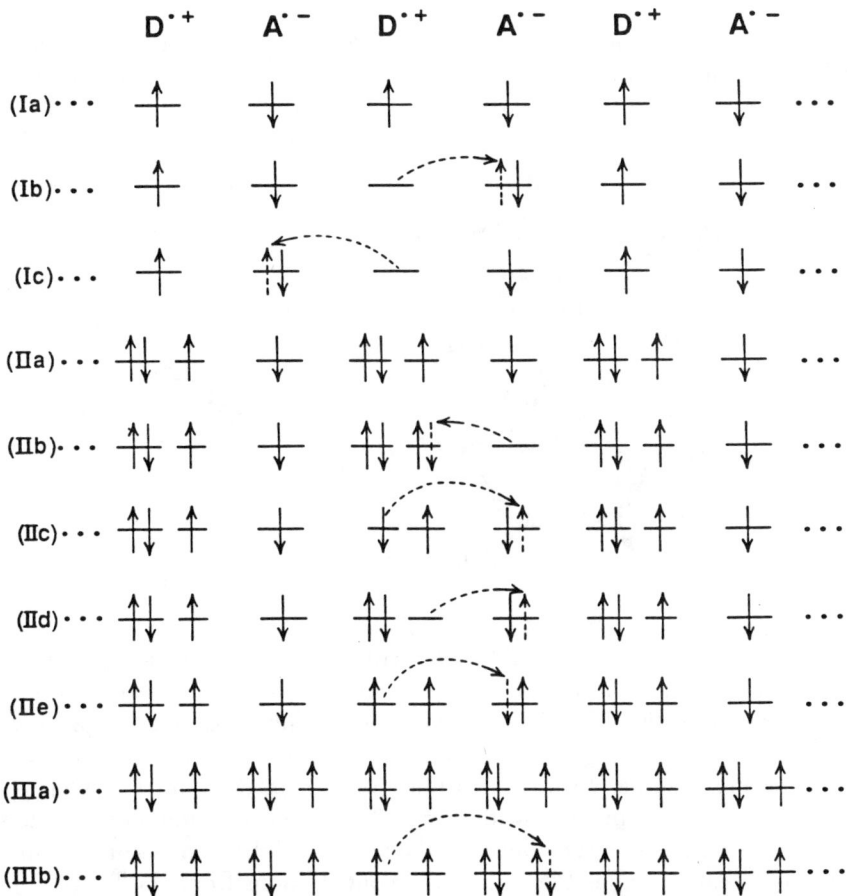

Figure 1. Schematic illustration of stabilization of antiferromagnetic or ferromagnetic coupling. If both the D and A have a half-filled nondegenerate POMO (s^1) (Ia), then the A←D (or D←A) charge-transfer excited state (Ib or equivalently Ic) stabilizes antiferromagnetic coupling. If either D or A has a non-half-filled degenerate POMO (e.g., d^3, assumed here to be the D) (IIa), then the D←A charge-transfer excited state formed (IIb) or A←D charge-transfer excited states formed via excitation of a "spin up" D electron (IIc or IId) will stabilize antiferromagnetic coupling. In contrast, the excited state formed via the D←A charge transfer (IIe) stabilizes ferromagnetic coupling. Hund's rule predicts this to be the dominant excited state that admixes with the ground state. If the D and A have non-half-filled degenerate POMO (e.g., d^3) (IIIa), then the A←D (or D←A) charge-transfer (disproportionation) excited state (IIIb) stabilizes ferromagnetic coupling.

$[Fe^{III}(C_5Me_5)_2]^{\cdot+}[TCNE]^{\cdot-}$ as a Model System Exhibiting Bulk Ferromagnetic Behavior

Experimental evidence for ferromagnetic ground-state behavior in a molecular compound has been limited to the charge-transfer salt of decamethylferrocene, $Fe^{II}(C_5Me_5)_2$, **1**, with tetracyanoethylene, **2** (Me is methyl; TCNE is tetracyanoethylene) (9, 10). This electron-transfer salt possesses both the alternating $\cdots D^{\cdot+}A^{\cdot-}D^{\cdot+}A^{\cdot-}D^{\cdot+}A^{\cdot-}\cdots$ (Figure 2) crystal and electronic structures prescribed by the configuration mixing mechanism already described (9).

The high-temperature susceptibility of $[Fe^{III}(C_5Me_5)_2]^{\cdot+}[TCNE]^{\cdot-}$ fits the Curie–Weiss expression with $\theta = +30$ K (Figure 3) and indicates dominant ferromagnetic interactions (9). The susceptibility and saturation magnetization calculated as the sum of the contributions from $[Fe^{III}(C_5H_5)_2]^{\cdot+}$ parallel to the C_5 molecular axis and $[TCNE]^{\cdot-}$ is 6.46 millielectromagnetic units per mole (memu/mol) at 290 K and 16.7 electromagnetic units·kilogram per mole (emukG/mol), respectively. These values are in excellent agreement with the observed values of 6.67 memu/mol and 16.3 emukG/mol for single crystals aligned parallel to the chain axis (10).

A spontaneous magnetization is observed for polycrystalline samples below 4.8 K in the Earth's magnetic field (9). The magnetization of these crystals is 36% greater than that of iron metal on a per-iron basis, and it agrees with the calculated saturation moment for ferromagnetic alignment of the donor and the acceptor spins. The critical (Curie) temperature, T_c, is 4.8 K, and hysteresis loops characteristic of ferromagnetic materials are observed (1–5). A large coercive field of 1 kG is recorded at 2 K (10). The physical properties are summarized in Table I.

Single-crystal susceptibility can be compared with different physical

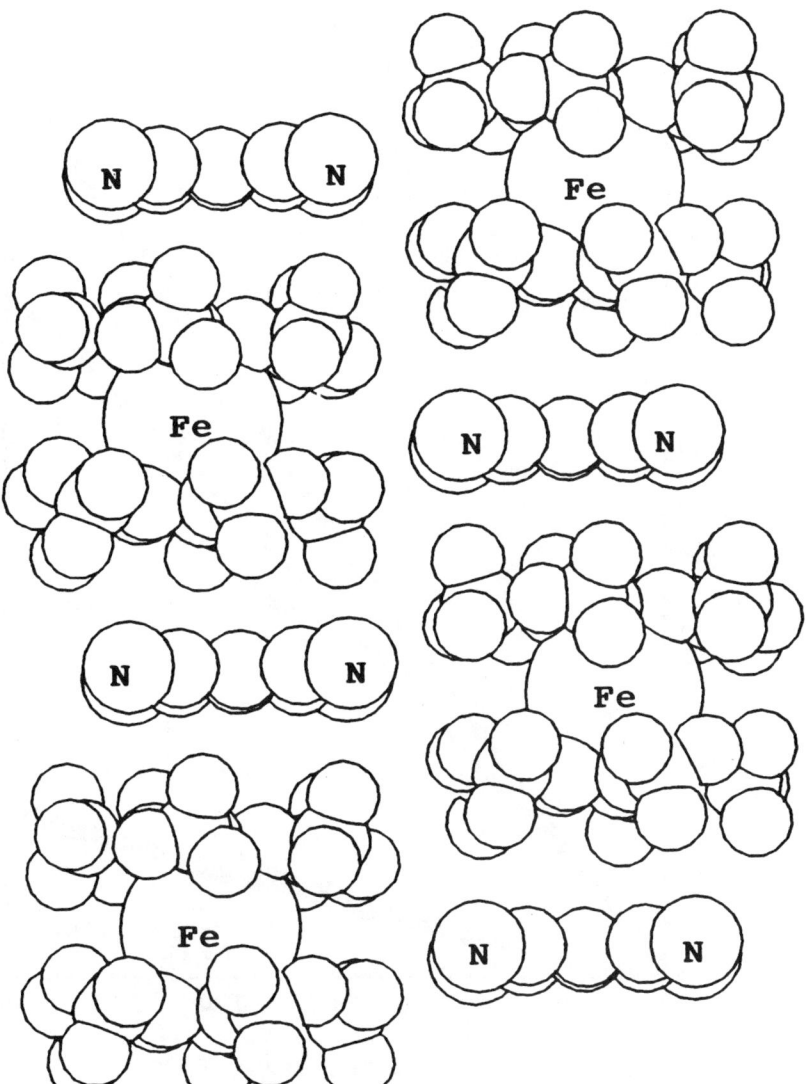

Figure 2. Alternating donor–acceptor, •••D·⁺A·⁻•••), linear chain structure of [Fe^III(C₅Me₅)₂]·⁺[A]·⁻ (A is TCNQ, TCNE, DDQ, or C₄(CN)₆), [Fe^II(C₅H₅)₂][TCNE], and [Fe^III(C₅Me₅)₂]·⁺[C₃(CN)₅]·⁻. The structure shows adjacent out-of-registry chains for TCNE.

models to aid the understanding of microscopic spin interactions. For samples oriented parallel to the field, the susceptibility above 16 K fits a one-dimensional Heisenberg model with a ferromagnetic exchange, J, of 19 cm⁻¹ (10). Variation of the low-field magnetic susceptibility with temperature for an unusually broad temperature range above T_c [$\chi \propto (T - T_c)^{-\gamma}$], magneti-

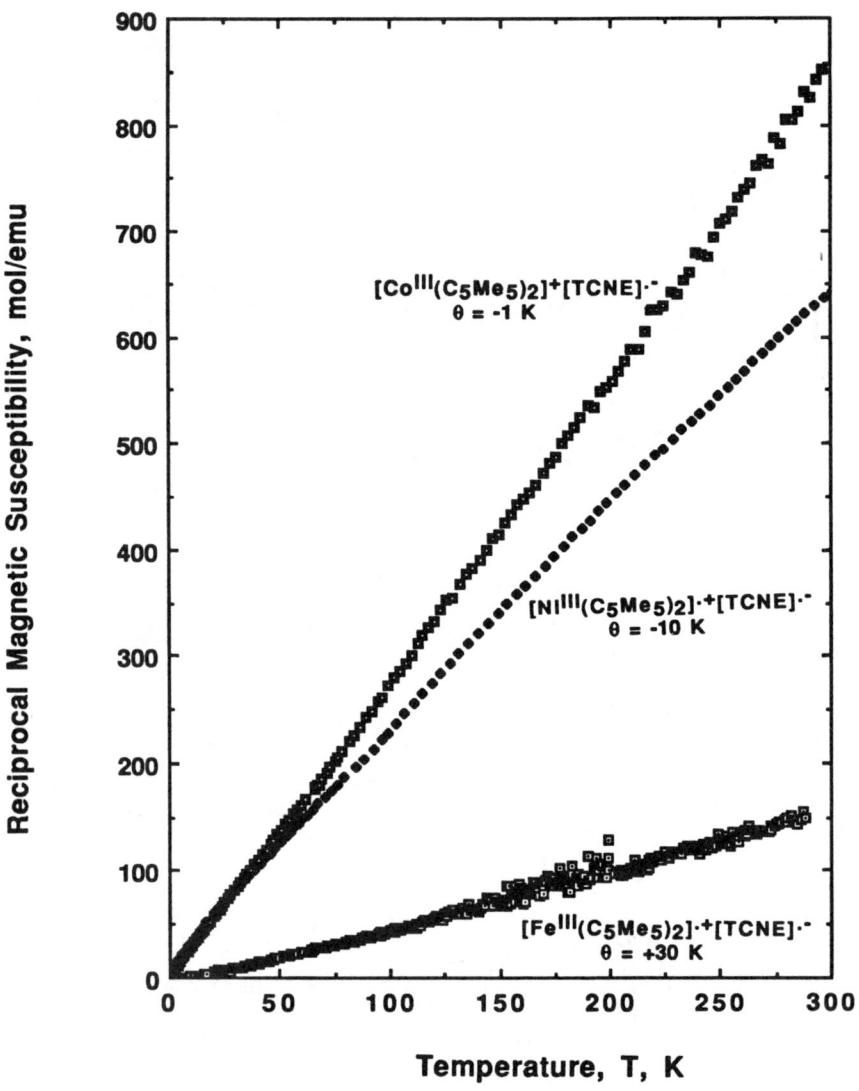

Figure 3. Reciprocal susceptibility, χ^{-1}, extrapolated from the high-temperature data for $[M^{III}(C_5Me_5)_2]^{\cdot+}[TCNE]^{\cdot-}$ [M is Fe (ferromagnetic: $\theta = 30$ K), Ni (antiferromagnetic: $\theta = -10$ K), and Co (paramagnetic: $\theta = -1$ K)].

zation with temperature below T_c [$M \propto (T_c - T)^{-\beta}$], and the magnetization with magnetic field at T_c ($M \propto H^{1/\delta}$) enabled the estimation of the β, γ, and δ critical exponents. The values of 1.2, ~0.5, and 4.4, respectively, were determined for the magnetic field parallel to the chain axis. These values are consistent with a mean-field-like three-dimensional behavior. Thus, above 16 K, one-dimensional nearest-neighbor spin interactions are sufficient

Table I. Physical Properties of $[Fe(C_5Me_5)_2]^{\cdot+}[TCNE]^{\cdot-}$

Formula	$C_{26}H_{30}N_4Fe$
Structure	1-D··· ·D·⁺·A·⁻·D·⁺·A·⁻·D·⁺·A·⁻··· Chains
Solubility	Conventional organic solvents
Critical/Curie temperature	4.8 K
Curie-Weiss θ constant	+30 K
Spontaneous magnetization	Yes, in zero applied field
Magnetic susceptibility (∥ to 1-D chains)	0.00667 emu/mol (obs, 290 K)
Magnetic susceptibility (∥ to 1-D chains)	0.00640 emu/mol (calc, 290 K)
Magnetic susceptibility (⊥ to 1-D chains)	0.00180 emu/mol (obs, 290 K)
Magnetic susceptibility (⊥ to 1-D chains)	0.00177 emu/mol (calc, 290 K)
Saturation magnetization (∥ to 1-D chains)	16,300 emuG/mol (calc 16,700 emuG/mol); 36% greater than iron (iron basis)
Saturation magnetization (⊥ to 1-D chains)	6,000 emuG/mol (calc 8,800 emuG/mol)
Intrachain exchange interaction (∥ to 1-D chains)	27.4 K (19 cm⁻¹)
Intrachain exchange interaction (⊥ to 1-D chains)	8.1 K (5.6 cm⁻¹)
Hysteresis curves	Yes (1000 G coercive field; cf. 1 G for iron metal)
β Critical constant	0.5 (cf. 0.38 for iron metal)
γ Critical constant (∥ and ⊥ to 1-D chains)	1.22, 1.19 (cf. 1.33 for iron metal)
δ Critical constant	4.4 (cf. unknown for iron metal)
Ferromagnetic ordering	Yes; neutron diffraction studies on polycrystalline deuterated samples
⁵⁷Fe Mossbauer Zeeman splitting	Yes; in zero applied field [large internal field: 424,000 G (4.2K)]
Physical model	New mechanism for ferromagnetism appears to be operative; predictive model based upon configurational mixing of the lowest charge transfer excited state developed
Additional magnetic behaviors observed:	Chemical modification leads to meta-, antiferro-, ferri-, para-, and diamagnetic behavior

to understand the magnetic coupling, but near T_c three-dimensional spin interactions are dominant (10).

The ^{57}Fe Mossbauer spectra of the TCNE electron-transfer salt of Fe(C$_5$Me$_5$)$_2$ are informative. Atypical six-line Zeeman split spectra are observed in zero applied magnetic field at low temperature as the radical anions provide an internal dipolar field. For example, a Zeeman split spectrum with an internal field of 424 kG (4.2 K) is observed for the [TCNE]$^{\cdot-}$ (9) salt. The internal fields are substantially greater than the expectation of 110 kG/spin/Fe.

Similarly structured electron-transfer complexes based on TCNE and organometallic donors were investigated in an effort to understand the structural features necessary to stabilize their bulk ferromagnetic behavior. Our study included complexes with substitution of the Me groups on the cyclopentadienide ring with H, increasing the ring size to six by using bis(arene)chromium, and substitution of Fe with Ru and Os.

Stable radicals are needed to form ferromagnetically coupled chains. Thus, electron transfer must occur to enable closed-shell donors and acceptors to be candidates for magnetic materials. The one-electron solution reversible reduction potential, $E°$, provides a means to gauge whether or not electron transfer might occur for a solid. For example, ferrocene is more difficult to oxidize (by 0.5 V) than decamethylferrocene, and it is unable to reduce TCNE (11–15). Nevertheless, the diamagnetic ferrocene analog of [Fe^{III}(C$_5$Me$_5$)$_2$]$^{\cdot+}$[TCNE]$^{\cdot-}$ (i.e., [Fe^{II}(C$_5$Me$_5$)$_2$][TCNE]) forms (12–14) and possesses the identical structural motif (16–18) (Figure 3). Either a temperature- or pressure-induced "neutral–ionic" transition (19–22) might be sufficient to lead to the stabilization of ferromagnetic behavior. However, above 2 K at ambient pressure, only Fe^{II} is observed via Mossbauer spectroscopy, and no discontinuity is observed in the susceptibility data (15).

The Co^{III} analog, [Co^{III}(C$_5$Me$_5$)$_2$]$^{\cdot+}$[TCNE]$^{\cdot-}$, has been prepared and exhibits essentially the Curie susceptibility anticipated for $S = \frac{1}{2}$ [TCNE]$^{\cdot-}$ ($\theta = -1.0$ K) (9). Because the cation is diamagnetic, the electron-transfer complex has only one spin per formula unit. It appears that the $\cdots$D$^{\cdot+}$A$^{\cdot-}$ D$^{\cdot+}$A$^{\cdot-}\cdots$ structure type with both $S \geq \frac{1}{2}$ D and $S \geq \frac{1}{2}$ A is necessary, but insufficient, for stabilizing cooperative highly magnetic behavior. Attempts to prepare [M^{III}(C$_5$Me$_5$)$_2$]$^{\cdot+}$ (M is Ru, Os) salts of [TCNE]$^{\cdot-}$ have yet to lead to suitable compounds for comparison with the highly magnetic Fe^{III} phase (23). Formation of [Ru^{III}(C$_5$Me$_5$)$_2$]$^{\cdot+}$ is complicated by disproportionation to Ru^{II}(C$_5$Me$_5$)$_2$ and [Ru^{IV}(C$_5$Me$_5$)(C$_5$Me$_4$CH$_2$)]$^+$ (24). The Os^{III} analog led to the preparation of a salt with TCNE; however, low susceptibilities and crystals unsuitable for single-crystal X-ray studies (23) have hampered progress in this area. Replacement of Fe^{III} in [Fe^{III}(C$_5$Me$_5$)$_2$]$^{\cdot+}$[TCNE]$^{\cdot-}$ with Ni^{III} ($S = \frac{1}{2}$) or Cr^{III} ($S = \frac{3}{2}$) leads to compounds exhibiting cooperative magnetic properties (25).

The motivation for studying these complexes emanated from the model for the stabilization of ferromagnetic coupling in molecular solids (26). Antiferromagnetic coupling is predicted for d^1–s^1 complexes with $s^1 \leftarrow d^1$ charge transfer and ferromagnetic coupling for $d^1 \leftarrow s^1$ charge transfer [s^1 is one electron in a nondegenerate orbital on, for example, the donor; d^1 is one electron in a doubly degenerate (or accidentally degenerate) orbital on, for example, the acceptor]. The $[Ni^{III}(C_5Me_5)_2]^{\cdot+}[TCNE]^{\cdot-}$ complex possesses this electronic configuration, and its susceptibility obeys the Curie–Weiss expression with θ = –10 K (Figure 3). This situation is consistent with dominant antiferromagnetic interactions and $s^1 \leftarrow d^1$ charge transfer (1–5, 10). The model predicts antiferromagnetic coupling if each site possesses a half-filled POMO.

For heterospin systems ($S_D \neq S_A$) with lower-symmetry s and d POMOs, only two-electron configurations support ferromagnetic coupling. Illustrative systems, however, have yet to be identified for these electron configurations. Because of an accidental degeneracy of the cation's e_{2g} and a_{1g} orbitals (11), $[Cr^{III}(C_5Me_5)_2]^{\cdot+}$ possesses a t^3 POMO [t^3 is three electrons in a triply degenerate (or accidentally degenerate) orbital on, for example, the donor]. Thus, the $[TCNE]^{\cdot-}$ salt of $[Cr^{III}(C_5Me_5)_2]^{\cdot+}$ is predicted to exhibit antiferromagnetic behavior leading to ferrimagnetic coupling (as the spin state cannot cancel) for either $A^{\cdot-} \leftarrow D^{\cdot+}$ or $D^{\cdot+} \leftarrow A^{\cdot-}$ charge transfer (1–5). An investigation of the magnetic properties of the Cr^{III} system is in progress. However, the preliminary magnetization data are consistent with ferrimagnetic behavior (25).

To test the necessity of a ^2E ground state, the TCNE electron-transfer salt with the lower-symmetry $Fe(C_5Me_4H)_2$ donor was prepared (27). The magnetic susceptibility can be fit by the Curie–Weiss expression between 2.2 and 320 K, and the moment is consistent with two independent spins. The susceptibility is not field-dependent and θ ~ 0 K. The absence of three-dimensional ferromagnetic or antiferromagnetic ordering above 2.2 K in $[Fe(C_5Me_4H)_2]^{\cdot+}[TCNE]^{\cdot-}$ contrasts with the behavior of $[Fe(C_5Me_5)_2]^{\cdot+}[TCNE]^{\cdot-}$. This is in accord with the ^{57}Fe Mossbauer data, which only shows nuclear quadrupole splitting for the $[Fe(C_5Me_4H)_2]^{\cdot+}$ salts and not zero-field Zeeman splitting. The lack of magnetic ordering may arise from poorer intra- and intermolecular overlap within and between the chains leading to substantially weaker magnetic coupling for $[Fe(C_5Me_4H)_2]^{\cdot+}[TCNE]^{\cdot-}$. This weak magnetic coupling would suppress any spin-ordering temperature. Alternatively, because of the overall C_{2v} symmetry, the $[Fe(C_5Me_4H)_2]^{2+}$ charge-transfer excited state may be a single, not a triplet as expected for $[Fe(C_5Me_5)_2]^{2+}$. The admixture of a singlet (not a triplet) charge-transfer excited state should lead to antiferromagnetic (not ferromagnetic) coupling (1–5).

The TCNE electron-transfer salt of $Fe(C_5Me_5)(C_5H_5)$ was sought, as the donor had C_5 symmetry and was sufficiently strong to reduce TCNE. However, unlike the TCNE charge and electron-transfer salts of $Fe(C_5H_5)_2$ and

$Fe(C_5Me_5)_2$, respectively, which possess the desired •••DADADA••• motif in the solid state, the simple (1:1) $Fe(C_5Me_5)(C_5H_5)^{\cdot+}$ salt could not be isolated. The isolated complex (2:3) salt does not have the desired •••DADADA••• one-dimensional chain structure. This structural motif is the only one reported to support cooperative magnetic interactions (e.g., ferromagnetic). Therefore, we were unsuccessful in employing $Fe(C_5Me_5)(C_5H_5)$ as a donor to prepare new materials to extend our understanding of cooperative magnetic behavior in molecular materials. The formation of [FeCpCp*]$_2$[TCNE]$_3$ • THF (THF is tetrahydrofuran) emphasizes the current inability to predict solid-state compositions, let alone structure types (28).

The infrared spectra and magnetic susceptibility of the $[Cr^I(C_6Me_x\text{-}H_{6-x})_2]^{\cdot+}[TCNE]^{\cdot-}$ ($x = 3, 6$) salts are consistent with two unpaired electrons per formula unit. These salts exhibit dominant ferromagnetic coupling, as evidenced from a fit of the high-temperature susceptibility to the Curie–Weiss expression with $\theta = \sim +11.4$ K (29). Crystals suitable for single-crystal X-ray analysis, however, have not been prepared; thus, the structures of these salts are unknown. With the observation of ferromagnetic coupling in other charge-transfer salts with •••$D^{\cdot+}A^{\cdot-}D^{\cdot+}A^{\cdot-}D^{\cdot+}A^{\cdot-}$••• linear chains (1–5) and the infrared evidence for isolated $[TCNE]^{\cdot-}$ and not $[TCNE]_2^{2-}$, we propose that these ferromagnetically coupled complexes also have this structural arrangement. Specific details of the intra- and interchain interactions arising from the canting and interchain registry of the chains, as well as the interatomic separations, must await the structural determinations.

This study was designed to probe the effect of the electronic structure on magnetic behavior. The present understanding of the mechanism for stabilization of ferromagnetic coupling in molecular-based donor–acceptor complexes is configurational mixing of a charge-transfer excited state with the ground state. The model predicts that, for excitation from the HOMO of a donor to an acceptor (both with a half-filled nondegenerate HOMO, as is the case for these $[TCNE]^{\cdot-}$ salts), only antiferromagnetic coupling is stabilized. Because the $[Cr(arene)_2]^{\cdot+}$ cation has an $e_g^4 a_{1g}^1$ electronic structure and a $^2A_{1g}$ ground state (30), antiferromagnetic behavior is predicted. Thus, the observed ferromagnetic coupling suggests that the model is inadequate. The model, however, is consistent with the observed data if we consider that the $A^{\cdot-} \leftarrow D^{\cdot+}$ charge-transfer excitation results from the next highest occupied molecular orbital (not the POMO of the cation) to the radical anion POMO. Thus ferromagnetic coupling, which may ultimately lead to bulk ferromagnetic behavior, is achievable for systems where both the donor and acceptor have ^2A_g ground states.

Recently a TCNE electron-transfer salt of the quadruply bonded Mo_2L_2 was described in the literature (31). Because the infrared $\nu(C\equiv N)$ absorption was characteristic of isolated $[TCNE]^{\cdot-}$ radical anions and the cation was also a radical with potentially doubly degenerate HOMO (i.e., d^3), we sought

                    Me   Me
                     \\   //
                      \\   //
                       N   N
                      / \   / \
                 [benzene] [benzene]
                       N   N
                      / \   / \
                      //   \\
                     //   \\
                    Me   Me

L

to study the magnetic properties of this complex. The results of the magnetic susceptibility measurements taken on the Faraday apparatus between 2.2 and 320 K are consistent with two independent spins modestly antiferromagnetically coupled (i.e., θ = −6 K). The observed effective moment is 2.45 μ_B; that expected for two independent $g = 2$, $S = 1/2$ spins is 2.44 μ_B. At low temperature a field-dependent susceptibility is not observed. Crystals suitable for single-crystal X-ray study, however, could not be prepared; thus, the solid-state packing motif is unknown.

Model for the Stabilization of Bulk Ferromagnetic Behavior

The model for magnetic coupling by configuration admixing of a virtual triplet excited state with the ground state is limited to the repeat unit (i.e., $[Fe^{III}(C_5Me_5)_2]^{\cdot +}[TCNE]^{\cdot -}$. Inter- and intrachain spin alignment are required for bulk ferromagnetism. Mixing of a with the E_{gs} ground state lowers the energy to E'_{gs} (Figure 4). Because the cation is essentially equidistant to $[TCNE]^{\cdot -}$ above and below it within a chain, virtual transfer of an e_{2g} electron may occur to form the admixable triplet excited state with either $[TCNE]^{\cdot -}$. Thus, excitations a and b enable two excited states, a and b, to mix with E_{gs} to lower the energy to E''_{gs} (Figure 4) and lead to intrachain spin alignment.

However, even with complete intrachain spin alignment (i.e., ferromagnetic coupled), correlation of spins on adjacent chains in the opposite sense would lead to bulk antiferromagnetic coupling. Macroscopic ferromagnetism will not occur unless interchain spin alignment occurs. If adjacent chains are out of registry by one-half the chain axis length, then $[TCNE]^{\cdot -}$ radicals residing in adjacent chains may be ferromagnetically coupled to the interchain Fe^{III} sites, as are the intrachain $[TCNE]^{\cdot -}$ radicals. Thus, excitations a, b, and c enable three excited states, a, b, and c, to mix with E_{gs} to further lower the energy to E'''_{gs} (Figure 4) and lead to the spin alignment throughout the bulk that is necessary for bulk ferromagnetism.

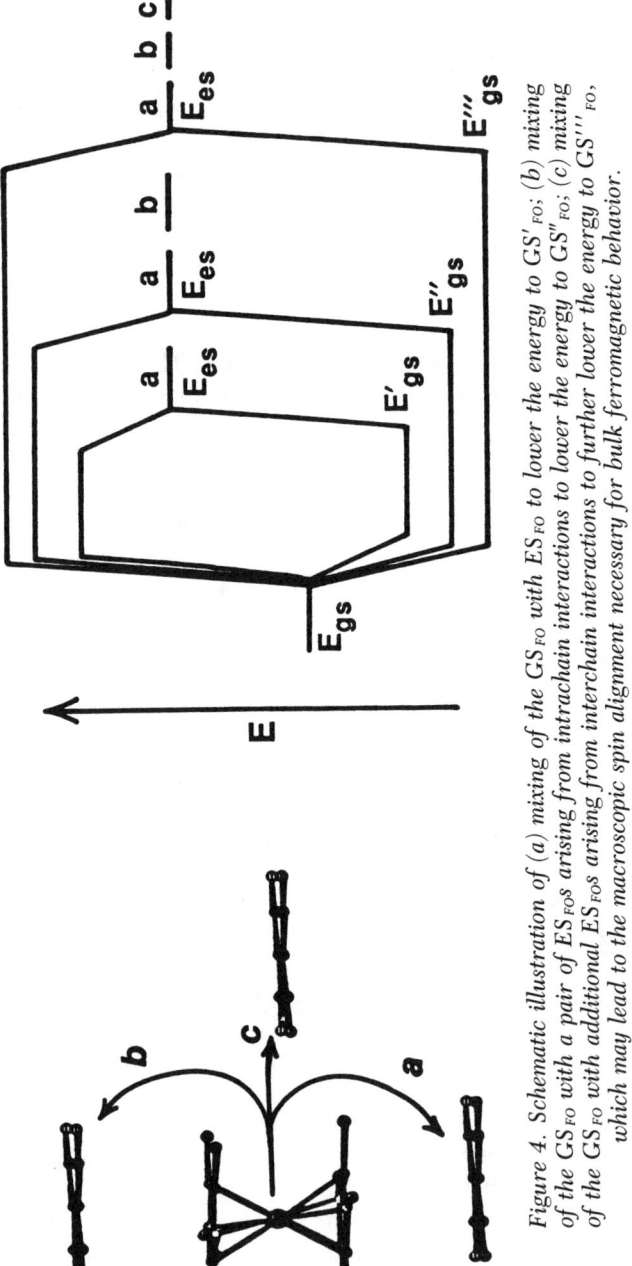

Figure 4. Schematic illustration of (a) mixing of the GS_{FO} with ES_{FO} to lower the energy to GS'_{FO}; (b) mixing of the GS_{FO} with a pair of ES_{FO}s arising from intrachain interactions to lower the energy to GS''_{FO}; (c) mixing of the GS_{FO} with additional ES_{FO}s arising from interchain interactions to further lower the energy to GS'''_{FO}, which may lead to the macroscopic spin alignment necessary for bulk ferromagnetic behavior.

Design Criteria for Ferromagnetic Coupling via the McConnell Mechanism

The McConnell mechanism leads to several important criteria for the design of a molecular organic ferromagnet. Foremost is the requirement that the stable radicals possess a non-half-filled degenerate POMO. (For a charge-transfer salt, this requirement applies to either a donor or acceptor, but not to both.) These radicals must not have structural–electronic distortions that lower the symmetry and significantly break the degeneracies (e.g., the Jahn–Teller effect). Accidentally degenerate systems (e.g., high spin transition, lanthanide, and actinide metal coordination complexes), however, suffice. Furthermore, opposing effects (e.g., retro versus forward virtual charge transfer) or magnitude of the stabilization (e.g., inversely proportional to distance and energy difference between the mixing states) may obscure the effect and lead to other phenomena. Additionally, other mechanisms (*1, 2, 5*) for molecular-based ferromagnetic behavior may be present.

Conclusions

The quest for s/p orbital-based ferromagnets remains the focus of intense worldwide interest. The magnetic data on $[Fe(C_5Me_5)_2]^{\cdot+}[TCNE]^{\cdot-}$ demonstrates that ferromagnetism is achievable in organic-based molecular systems. Replacement of the doublet organic acceptor with a diamagnetic acceptor demonstrates that the organic species is crucial for achieving bulk ferromagnetism. This system contains low-spin Fe^{III}, not high-spin Fe^{II} or Fe^{III} or iron metal. The ferrocenes possess chemical (e.g., reactivity similar to aromatic organic compounds, such as benzene) and physical (e.g., solubility in conventional polar organic solvents) properties akin to organic compounds, not inorganic network solids.

Accidental or intrinsic orbital degeneracies, albeit rare for organic molecules, are needed for stabilization of ferromagnetic coupling by the extended McConnell-like mechanism. Thus, for stable D_{2d} or $C_{\geq 3}$ symmetry, $S \geq 1/2$ radicals with a degenerate POMO are required. It is a challenge to the synthetic chemist to prepare radicals that have nondegenerate POMOs and do not undergo a Jahn–Teller distortion, which would eliminate the desired electronic configuration. In addition to preparation of the desired radicals, their secondary and tertiary solid-state structures must be achieved. Finally, single crystals large enough for the study of their anisotropic magnetic properties must be prepared.

Acknowledgments

The authors gratefully acknowledge partial support by the Department of Energy Division of Materials Science (Grant No. DE–FG02–86ER

45271.A000). We deeply thank our co-workers (R. W. Bigelow, J. C. Calabrese, S. Chittipeddi, A. Chakraborty, K. Ming-Chi, K. R. Cromack, D. A. Dixon, P. J. Krusic, V. L. Goedken, D. M. O'Hare, W. M. Reiff, H. Rommelmann, C. Vazquez, M. D. Ward, D. Wipf, and J. H. Zhang) for the important contributions they have made toward the success of the work reported herein.

References

1. Miller, J. S.; Epstein, A. J. *NATO Adv. Study Ser.* **1987**, *168B*, 159.
2. Miller, J. S.; Epstein, A. J.; Reiff, W. M. *Chem. Rev.* **1988**, *88*, 201.
3. Miller, J. S.; Epstein, A. J.; Reiff, W. M. *Acc. Chem. Res.* **1988**, *23*, 114.
4. Miller, J. S.; Epstein, A. J.; Reiff, W. M. *Science* **1987**, *240*, 40.
5. Miller, J. S.; Epstein, A. J. *Adv. Org. Chem.* in press.
6. McConnell, H. M. *Proc. R. A. Welch Found. Chem. Res.* **1967**, *11*, 144.
7. Radhakrishnan, T. P.; Soos, Z.; Endres, H.; Azevedo, L. J. *J. Chem. Phys.* **1986**, *85*, 1126.
8. E. Dormann, E.; Nowak, M. J.; Williams, K. A.; Angus, R. O., Jr.; Wudl, F. *J. Am. Chem. Soc.* **1987**, *109*, 2594.
9. Miller, J. S.; Calabrese, J. C.; Rommelmann, H.; Chittipeddi, S.; Zhang, J. H.; Reiff, W. M.; Epstein, A. J. *J. Am. Chem. Soc.* **1987**, *109*, 769.
10. Chittipeddi, S.; Cromack, K. R.; Miller, J. S.; Epstein, A. J. *Phys. Rev. Lett.* **1987**, *58*, 2695.
11. Robbins, J. L.; Edelstein, N.; Spencer, B.; Smart, J. C. *J. Am. Chem. Soc.* **1982**, *104*, 1882.
12. Webster, O. W.; Mahler, W.; Benson, R. E. *J. Am. Chem. Soc.* **1962**, *84*, 3678.
13. Rosenblum, M.; Fish, R. W.; Bennett, C. *J. Am. Chem. Soc.* **1964**, *86*, 5166.
14. Brandon, R. L.; Osipcki, J. H.; Ottenberg, A. *J. Org. Chem.* **1966**, *31*, 1214.
15. Miller, J. S.; Zhang, J. H.; Reiff, W. M. in preparation.
16. Adman, E.; Rosenblum, M.; Sullivan, S.; Margulis, T. N. *J. Am. Chem. Soc.* **1967**, *89*, 4540–4542.
17. Foxman, B. private communication.
18. Sullivan, B. W.; Foxman, B. *Organometallics* **1983**, *2*, 187.
19. Batail, P.; LaPlaca, S. J.; Mayerle, J. J.; Torrance, J. B. *J. Am. Chem. Soc.* **1981**, *103*, 951.
20. Mayerle, J. J.; Torrance J. B.; Crowley, J. T. *Acta Crystallograph.* **1979**, *B35*, 2988.
21. Kanai, Y.; Tani, M.; Kagoshima, S.; Tokura, Y.; Koda, Y. *Syn. Met.* **1984–1985**, *10*, 157.
22. Metzger, R. M.; Torrance, J. B. *J. Am. Chem. Soc.* **1985**, *107*, 117.
23. O'Hare, D. M.; Miller, J. S, *Organometallics* **1988**, *7*, 1335.
24. U. Kolle; Grub, J. *J. Organomet. Chem.* **1985**, *289*, 133.
25. Miller, J. S.; Epstein, A. J. in preparation.
26. Miller, J. S.; Epstein, A. J. *J. Am. Chem. Soc.* **1987**, *109*, 3850.
27. Miller, J. S.; Glatzhofer, D. T.; O'Hare, D. M.; Reiff, W. M.; Chakraborty, A.; Epstein, A. J. *Inorg. Chem.* **1989**, *28*, 2930.
28. Miller, J. S.; Glatzhofer, D. T. in preparation.
29. Miller, J. S.; O'Hare, D. M.; Chakraborty, A.; Epstein, A. J. *J. Am. Chem. Soc.* **1989**, *111*, 7853.
30. Anderson, S. E.; Drago, R. S. *J. Am. Chem. Soc.* **1970**, *92*, 4244.
31. Giraudon, J.-M.; Guerchais, J.-E.; Sala-Pala, J.; Toupet, L. *J. Chem. Soc., Chem. Commun.* **1988**, 921.

RECEIVED for review May 1, 1989. ACCEPTED revised manuscript October 10, 1989.

23

Stabilization of Conducting Heteroaromatic Polymers in Large-Pore Zeolite Channels

Thomas Bein[1], Patricia Enzel[1], Francois Beuneu[2], and Libero Zuppiroli[2]

[1]Department of Chemistry, University of New Mexico, Albuquerque, NM 87131
[2]Laboratoire des Solides Irradiés, S. E. S. I., École Polytechnique, F-91128 Palaiseau Cedex, France

Different strategies to encapsulate polymeric chains in ordered hosts are discussed. Pyrrole was polymerized as a model within the crystalline channel system of faujasite (three-dimensional) and mordenite (one-dimensional) zeolite molecular sieves. Polymerization required the presence of an intrazeolite oxidant such as ferric or cupric ions. No reaction was observed with the Na and Fe(II) forms of the zeolites. The yield of the reaction was highest with vapor-phase pyrrole, and it was low in aqueous solvents. The systems were characterized with a combination of electronic, infrared, and Raman spectroscopic data. Electron spin resonance measurements were used to explore the transport properties, and preliminary bulk powder conductivity measurements were performed to evaluate potential conduction paths on the outside of the zeolite crystals. The intrazeolite heteroaromatic polymer chains represent the first example of host-stabilized one-dimensional molecular conductors, or molecular wires.

THE DISCOVERY OF DOPED CONJUGATED POLYMERS has generated substantial research interest, both at the fundamental level and in view of possible applications (1–3). These polymers include doped polyacetylene, polyaniline, polypyrrole, and other polyheterocycles. Potential applications based on the conducting properties of these systems range from light-weight batteries, antistatic equipment, and microelectronics to speculative concepts such as "molecular electronic" devices (4, 5).

0065–2393/90/0226–0433$06.00/0
© 1990 American Chemical Society

Polypyrrole (PPy), for example, has been extensively studied in the form of thin films deposited on electrode surfaces (6). Electrochemical oxidative polymerization of pyrrole with anions (e.g., ClO_4^-, HSO_4^-) present in solution results in relatively air-stable, highly conducting films. Polypyrrole can also be prepared via chemical oxidation of pyrrole with Cu(II) or Fe(III) salts in solution (7–10). The proposed reaction path (6) via cation radicals is shown in Scheme I.

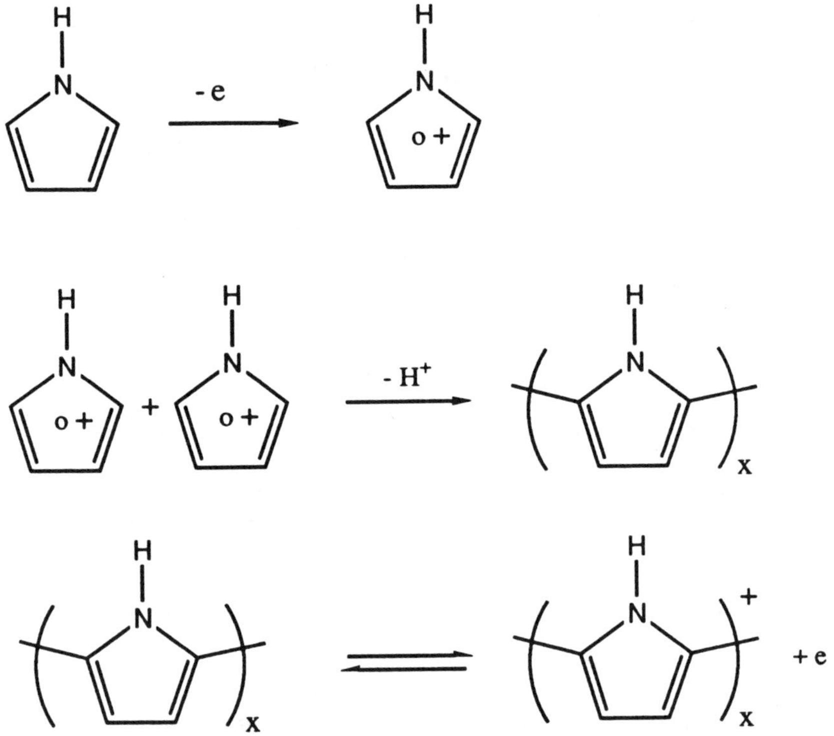

Scheme I. Oxidative polymerization of pyrrole.

In contrast to inorganic semiconductors that are characterized by three-dimensional covalent bonding and high carrier mobilities and are adequately described by rigid-band models, interactions in organic polymers are highly anisotropic (11). Atoms are covalently linked along the chains, whereas interchain interactions are much weaker; these structural features can cause collective instabilities such as Peierls distortions. Doping occurs by charge transfer between the intercalated dopant molecules or atoms and the organic chains, and it can result in substantial local relaxations of the chain geometry. New localized electronic states in the gap are introduced by these charge-transfer-induced local geometric modifications of the polymer.

In heteroaromatic polymers such as PPy, the ground state is nondegenerate; the ground-state geometry with aromatic structure within rings and single bonds between rings is more stable than the corresponding quinoid resonance structure. Because the quinoid structure has a smaller band gap (lower ionization potential and larger electron affinity), the introduction of a charge on the chain can result in relaxation from the aromatic to the quinoid structure. The resulting electronic structure of PPy as a function of doping is depicted schematically in Figure 1.

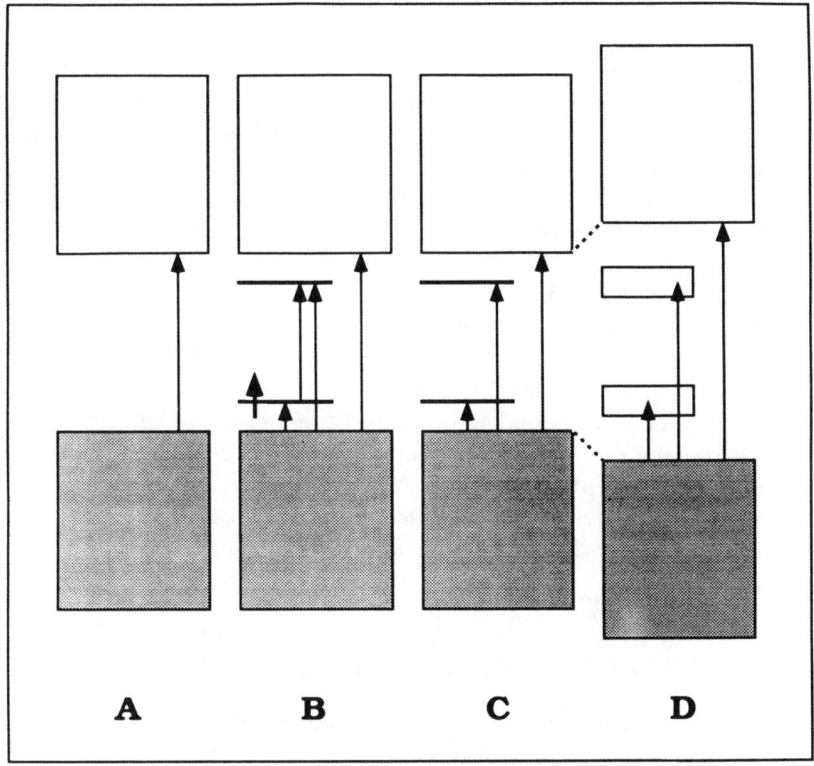

Figure 1. Band structure of polypyrrole as a function of doping level (schematic). Key: A, neutral polymer; B, polaron (+), spin = ½, three new transitions; C, bipolaron (++), spin = 0, two transitions; D, heavily doped, bipolaron bands.

Recent studies (12, 13) conclude that at low doping levels, the chains are ionized to produce a radical cation (polaron) that is "pinned" to the counterion and does not contribute significantly to the conductivity. At higher doping levels, polarons can combine or ionize to form spinless dications (bipolarons) that are associated with a quinoid segment extending over four to five rings. The bipolarons are assumed to transfer charge via interchain hopping that corresponds to the observed spinless conductivity.

This model is consistent with the absence of Pauli susceptibility in the highly conducting form of PPy (13).

The study of the conduction mechanism of these polymers has been impeded by the low level of structural definition of most samples. The amorphous products of electrochemical and chemical oxidative polymerization reactions present a wide range of possible interchain interactions (e.g., islands with microcrystalline order separated by amorphous regions, unknown degrees of cross-linking, or bundles of fibers combined in disordered arrays).

Fundamental studies of the electronic structure and conduction mechanism of conducting polymers would benefit substantially if the low-dimensional structures were available as isolated, structurally well-defined, chemically accessible entities. The goal of our research program in this area is to design corresponding model systems via encapsulation of polymeric chain conductors in low-dimensional, crystalline host lattices, particularly in zeolites. Embedding a single chain in a zeolite matrix is also of interest, because the electronic properties of single chains in the solid state are not easily available. Chain segments, even if short, are interesting because of potential quantum-size effects on the electronic structure that have been observed in colloidal semiconductor systems, both in suspension and stabilized in zeolite host systems (14–17).

This chapter reports on our initial progress (18) in the encapsulation of heteroaromatic conjugated polymers in large-pore zeolites, particularly polypyrrole. We recently succeeded in synthesizing polyaniline (19, 20) and polythiophene (21) in zeolite Y and mordenite. Previously reported strategies for the design of low-dimensional structures are discussed in the following sections.

Stabilization of Low-Dimensional Polymer Structures

Urea and other organic hosts have been explored for the radiation-induced inclusion polymerization of clathrated monomers such as butadiene (22). Well-studied examples include vinyl chloride, acrylonitrile and butadiene in urea, butadiene, pentadiene in deoxycholic acid, and ethylene and propylene in perhydrotriphenylene. Polymerization in the clathrates is usually induced by exposure to high-energy radiation that produces radicals derived from host and guest molecules, and proceeds via a living radical mechanism. Inclusion polymerization can result in a high degree of steric control of the resulting included polymer chains. To our knowledge, no conjugated, conducting clathrate systems have yet been synthesized.

If the dimension of channel-shaped hosts is extended such that unhindered diffusion of monomers can occur, conducting polymers can be formed on an electrode surface. Polypyrrole and poly(3-methylthiophene) fibrils with diameters between 0.03 and 1 μm at 10-μm length have been synthesized in microfilter membranes (Nucleopore) (23, 24). Stretch alignment has often been employed to create directional anisotropy in preformed polymers, such as in polyacetylene (25–27). Liquid crystal polymerization under a magnetic field is an alternative technique to achieve alignment (28, 29).

Inorganic layer compounds have been explored for the polymerization of organic compounds in two dimensions (30). Prominent examples are based on layer perovskite halide salts with the general formula $(RCH_2NH_3)_2MX_4$, where M is a divalent metal such as Mn, Fe, Cu, and Cd, and X is a halide ion (31). The organic cations separate the layers formed by the metal halides. If the organic cations contain 2,4-diene units, the corresponding layered compounds can be polymerized with gamma irradiation to form 1,4-disubstituted *trans*-polybutadienes (32, 33). However, if related diacetylene-containing layer perovskites are polymerized, the original lattice is gradually broken up (30). Recent studies explored the intercalative polymerization of pyrrole, thiophene, and aniline in layered FeOCl (and V_2O_5) (34–36); this polymerization exploits the well-established oxidative intercalation of organic molecules (37, 38) with concomitant reduction of FeOCl. It was found that pyrrole intercalated into FeOCl and polymerized to result in an inorganic-conducting polymer hybrid structure with an interlayer spacing increased by 5.23 Å.

Other two-dimensional systems that will not be discussed here include the electropolymerization of aniline intercalated in montmorillonite clay (39) and polymerizations in Langmuir–Blodgett films or bilayer membranes (40).

Efforts to form composites between polypyrrole and a variety of porous materials such as paper, cloth, or wood have been based on an approach comparable to that used in the present study. Typically, the respective material was impregnated with an oxidant such as $FeCl_3$ (41, 42) and subsequently contacted with pyrrole vapor or solution. Polypyrrole (and polyaniline) have been included in perfluorosulfonated ionomer membranes (Nafion) by stepwise treatment with aqueous ferric chloride and the monomers (43).

This brief survey indicates that no single chains of conjugated systems with long-range order have yet been stabilized in solid matrices. We will show that our approach, based upon intrazeolite polymerization reactions, succeeds in the formation of such systems.

Zeolites as Microporous Host Structures

Zeolites (44–47) are crystalline open framework metal oxide structures (classically aluminosilicates with hydrophilic surfaces) with pore sizes between 3 and 12 Å and exchangeable cations compensating for the negative charge of the framework. The topologies of these systems include one-dimensional channels, intersecting two-dimensional channels, and three-dimensional open frameworks. Recent developments include hydrophobic structures with compositions close to SiO_2 (48), incorporation of transition metals into the framework (49), and the discovery of metal aluminophosphate sieves (50). Alkali metal cations, possibly coordinated to oxygen–metal rings with C_{3v} or C_{2v} symmetry in zeolite Y (Figure 2), can be exchanged for transition metal ions and the system can be heat-treated to induce cation migration and removal of water. Table I presents a brief description of zeolite structure types used in this study.

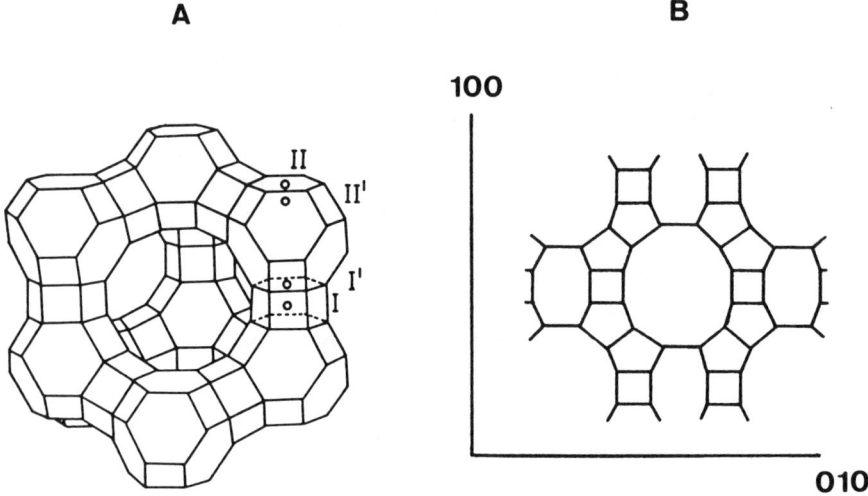

Figure 2. Structure of faujasite (A) and mordenite (B).

These features make zeolites extremely attractive candidates as hosts for polymeric conductors. They offer well-defined, stable crystalline channel structures with dimensions at the molecular level. The nature of the internal surface, as determined by cation type and other factors, will affect the dipolar and redox surface interactions and diffusion rates of polar versus nonpolar monomers. Previous work related to the present study includes the catalytic formation of polyacetylene from acetylene on the external surface of CoY and NiY zeolites (51) and on KX zeolite (52).

This chapter presents recent results on the successful encapsulation of polypyrrole chains in large-pore three-dimensional and one-dimensional zeolite hosts. The host structures examined in this study include zeolite Y (three-dimensional channel system, 7.5-Å pores), mordenite (one-dimensional channels, 7-Å pores), and zeolite A (three-dimensional channels, 4.1-Å pores). Polypyrrole was oxidatively polymerized in Cu(II)- or Fe(III)-containing zeolite pores, such as

$$NaY + FeSO_4(aq) \xrightarrow{N_2} NaFe(II)Y \xrightarrow{O_2/350\ °C}$$

$$NaFe(III)Y \xrightarrow[\text{(vapor/solvent)}]{\text{pyrrole}} PPy/NaFe(II)Y \quad (1)$$

Table I. Representative Zeolite Structure Types

Name	Unit Cell, Composition	Cage Type	Main Channels, Å[a]
LTA, Linde A	$[Na_{12}[(AlO_2)_{12}(SiO_2)_{12}]\ 27\ H_2O]_8$	α, β	4.1 ***
FAU, Faujasite	$Na_{58}[(AlO_2)_{58}(SiO_2)_{134}]\ 240\ H_2O$	β, 26-hedron(II)	7.4 ***
MOR, Mordenite	$Na_8[(AlO_2)_8(SiO_2)_{40}]\ 24\ H_2O$	complex 5-1	6.7 x 7.0 * ↔ 2.9 x 5.7 *

[a] The number of stars (*) at the channel description indicates the dimensionality of channel connections.

The intrazeolite oxidation of Fe(II) with oxygen probably involves the formation of Fe(III)–O–Fe(III) bridges in appropriate coordination sites (53), although generation of some amorphous iron oxide cannot be excluded (54).

The systems were characterized with a combination of electronic, infrared, and Raman spectroscopic data. Electron spin resonance (ESR) and conductivity measurements were used to explore the transport properties.

Experimental Details

Sample Preparation. *Zeolite Precursor Materials.* Fe(II)Y zeolite was prepared from the sodium form of zeolite Y (Linde LZ-Y52, Alfa) by conventional aqueous ion exchange with ferrous sulfate under nitrogen atmosphere and dried at 295 K under nitrogen (54). The resulting ferrous Y zeolite [unit cell compositions, based upon ion exchange, $Fe_6Na_{44}(AlO_2)_{56}(SiO_2)_{136}$ for solution loadings and $Fe_{10}Na_{36}(AlO_2)_{56}(SiO_2)_{136}$ for vapor-phase loadings] was heated under a flow of oxygen (40 mL/min) in a quartz tube reactor at a rate of 1 K/min to 370 K (10 h), then to 620 K (3 h), and subsequently evacuated for 2 h at that temperature to give Fe(III)Y. The color of the zeolite changed from white to light brown during this treatment. Ferrous mordenite (Fe(II)M) was similarly prepared from Na mordenite (LZM-5, Union Carbide) by ion exchange to give white $Fe_3Na_2(AlO_2)_8(SiO_2)_{40}$. This material was oxidized in a scheme similar to that used with ferrous Y and generated light brown Fe(III)M. Cu(II) forms of these zeolites and zeolite A were obtained by ion exchange with $Cu(NO_3)_2$, resulting in samples CuY, CuM, and CuA with 15, 2.5, and 8 Cu ions per unit cell (u.c.), respectively. Dehydrated sodium zeolites (heated at 1 K/min up to 720 K under vacuum, kept at 720 K for 10 h) and the metal-containing dry samples were stored under nitrogen in a glove box prior to use.

Bulk Polypyrrole. Bulk polypyrrole was prepared by chemical oxidation of pyrrole (Aldrich) with $FeCl_3 \cdot 6H_2O$ (Aldrich) according to published procedures (7, 8, 55). Oxidation with an oxidant:pyrrole ratio of 2.4 in aqueous solution at 292 K, carried out under nitrogen, represents the optimal reaction conditions for highest yield and conductivity (7).

Intrazeolite Polymerization of Pyrrole. Pyrrole was diffused into the pore system of the dry zeolites in a variety of solvents and via gas-phase adsorption (Table II). In the solvent reactions, 0.500 g of Fe(III)Y was suspended in a solution of 5.16 mg of pyrrole (0.077 mmol) in 50 mL of the respective solvent and stirred at 295 K for 15 h under nitrogen. Similarly, 0.500 g of Fe(III)M was reacted with 12.5 mg (0.187 mmol) of pyrrole. For the vapor-phase reactions, ~0.5 g of dry zeolite was weighed into a small quartz reactor, evacuated at a vacuum line at 1.33 mPa (10^{-5} torr), and equilibrated with 270 Pa (2 torr) of degassed pyrrole at 295 K for 1 h.

Characterization. Fourier transform infrared (FTIR) spectra were taken at 4-cm^{-1} resolution (Mattson Polaris instrument) and were analyzed with the ICON software. Electronic absorption spectra of the samples dispersed in glycerin were obtained with a spectrophotometer (PE 356) at 2-nm resolution. ESR spectra between 40 and 300 K were obtained with a Varian E-109 instrument operating at X-band frequencies. Conductivity measurements at 295 K were carried out with pressed wafers of the bulk polymers and of the zeolite powders by using the four-point technique (56). Resonance Raman spectra were generated by irradiating the sample with 5 mW at 457.9 nm. Scanning electron micrographs were taken with an Hitachi S-800 microscope.

Table II. Zeolite–Pyrrole Samples

Sample[a]	Loading with Pyrrole[b]	Surface Capacity[c]	Reaction Time, h	Color
NaY–V	41	0.2	1	white
NaM–V	2	0.06	1	white
Fe(II)Y–V	40	0.2	1	white
Fe(II)M–V	2	0.06	1	white
Fe(III)Y–V	39	0.2	1	black/turquoise
Fe(III)Y–W	2	0.2	15	light green
Fe(III)Y–T	2	0.2	15	grey/turquoise
Cu(II)Y–V	50	0.2	1	black/turquoise
Fe(III)M–V	1.0	0.06	1	turquoise
Fe(III)M–W	1.2	0.06	15	light green
Fe(III)M–T	1.2	0.06	15	light green
Cu(II)M–V	0.7	0.06	1	turquoise
Cu(II)A–V	0.3	0.3	1	very light green

[a]Abbreviations: Y, zeolite Y; M, mordenite; A, zeolite A; V, vapor phase; W, water; T, toluene.
[b]Molecules per unit cell.
[c]The surface capacities for 1-μm crystals are based on 0.6-nm diameter for pyrrole; normalized per zeolite unit cell.

Results and Discussion

If pyrrole is allowed to diffuse into large-pore zeolites that contain ferric ions, the color of the resulting adduct changes slowly from light brown to different shades of turquoise green and black. This striking reaction appears to be completed within a few minutes if the pyrrole is admitted as vapor into the dry zeolite, and within a few hours if diffused from solution (Table II). In contrast, pyrrole in zeolites containing only sodium ions or Fe(II) ions does not react to form colored products. Irrespective of pyrrole concentration and medium (different solvents or vapor phase), none of these blank experiments resulted in a color change of the sample (Table II). This behavior strongly suggests that pyrrole polymerizes to polypyrrole in a redox reaction that involves the intrazeolite ferric or cupric ions. Bulk polypyrrole has been synthesized chemically via oxidative coupling of pyrrole in various solutions and suspensions of ferric chloride. An overall reaction scheme in H_2O/C_2H_5OH solution has been proposed (57).

$$4C_4H_5N + 9FeCl_3 \cdot nH_2O \rightarrow (C_4H_3N)_4{}^+Cl^-$$
$$+ 8HCl + 9FeCl_2 \cdot nH_2O \quad (2)$$

The following discussion of the sample characteristics confirms the formation of intrazeolite polypyrrole.

Sample Characteristics. IR Spectra. Bulk polypyrrole is characterized by a broad IR absorption that extends from an electronic absorption band at ~1 eV (8000 cm^{-1}) down to about 1700 cm^{-1} and obscures the C–H and N–H vibrations of the polymer (6). The oxidized and the neutral forms of the polypyrrole have very similar C–C and C–H pyrrole ring vibrations in the 1600–800-cm^{-1} region (58, 59). In contrast to pyrrole, the ring vibrations of the polymer are relatively broad and can be distinguished from remaining unreacted monomer. The intrazeolite polypyrrole shows IR bands quite similar to the typical set of bands characteristic for bulk polypyrrole. However, certain shifts that vary with host and preparation conditions are observed (Figure 3). According to the electrical potential of the zeolite, the polymer chains should not exactly duplicate the bulk spectrum. For instance, bands of sample Fe(III)Y–V prepared from vapor phase occur at 1572 (1540), 1473 (1452), 1309 (1280–1300), and 790 (791) cm^{-1}. Positions of bulk polypyrrole are given in parentheses (58, 59).

The IR spectra of zeolite Y samples show a higher concentration of polypyrrole than the mordenite samples, probably because of the higher pore volume fraction available in the former host. Intrazeolite PPy recovered after dissolution of the host in HF shows features very similar to those of the bulk polymer (Figure 3).

Formation of polypyrrole according to the oxidative reaction mechanism is expected to produce two protons per monomer. These protons must be accommodated in the zeolite, but the IR data of these samples do not show clear hydroxyl features. This result is not surprising in view of the strong electronic absorption extending beyond the high-energy part of the spectrum. Similar observations are noted in the polypyrrole–FeOCl system (60). With its pronounced acid–base and ion-exchange properties, the zeolite framework (or intrazeolite Fe species) is expected to accommodate the additional proton concentration.

Raman Spectra of Intrazeolite Polypyrrole. Bulk polypyrrole is known to exhibit weak Raman spectra, but resolved bands in the 800–1600 cm^{-1} range have recently been obtained with smooth films formed on electrode surfaces (61). Resonance Raman spectra of polypyrrole–zeolite samples (e.g., sample Fe(III)Y–V) show weak but resolved bands at 1598 and 1418 cm^{-1} that correspond to the vibrations at 1591 and 1418 cm^{-1} reported for polypyrrole film grown on a Pt electrode (62).

Location of the Polymer Chains. It is clear from the combined spectroscopic evidence discussed and from the electronic spectra that polypyrrole forms when pyrrole reacts with zeolites containing ferric and cupric ions. In the context of synthetic strategies for the design of molecular "wires", it is important to determine the location of the polypyrrole with respect to the host. A film of polypyrrole at the outside of the zeolite crystals would not constitute a system of isolated molecular chains. Because of the insoluble nature of the polymer, extraction experiments do not provide much information. However, significant indirect evidence confirms the picture of zeolite-encapsulated polypyrrole.

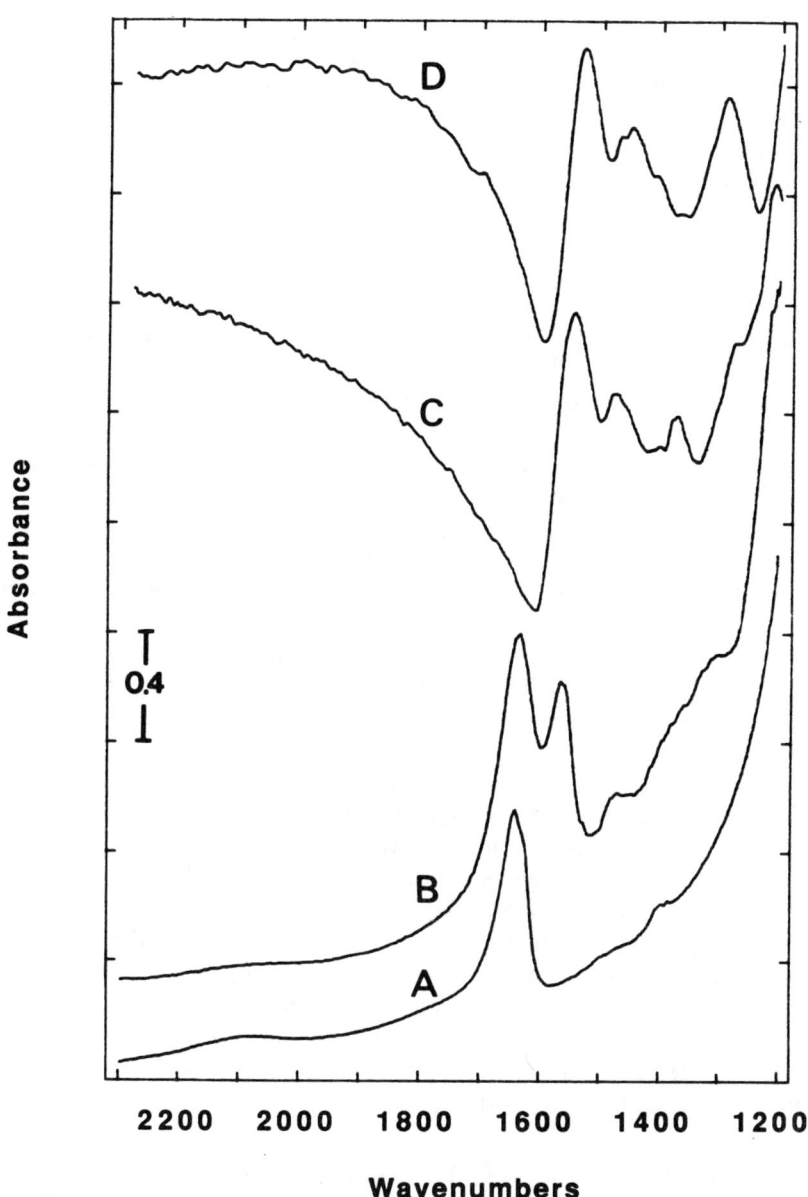

Figure 3. IR spectra (KBr pellets) of samples. Key: A, NaY; B, Fe(III)–Y; C, PPy extracted from Fe(III)–Y; and D, bulk polypyrrole.

The polypyrrole is not formed in either the Na or Fe(II) forms of the zeolites. Therefore the polymerization is dependent upon a redox reaction with the predominantly intrazeolite Fe(III) and Cu(II) ions. The polymerization reaction is one to two orders of magnitude slower in the zeolite host than it is in homogeneous solution. This retardation indicates diffusional limitations for the pyrrole migrating into the channel system. Finally, an analysis of the uptake of monomer into the zeolite confirms the same conclusions. Table II lists the uptake of monomer in different zeolite hosts and compares the values with those expected from monolayer-coating on the surface of crystals with 1.0-μm diameter.

These results show that far more monomer can be accommodated within the zeolite pores than expected for surface adsorption (the values for mordenite are smaller because of a smaller unit cell). Furthermore, we find that pyrrole is not adsorbed into the pore system of zeolite A, as expected from its smaller pore diameter (4.1 Å), but the uptake corresponds very well with the value expected for surface adsorption. Thus, if the monomer is excluded from the interior of the zeolite, not more than about one monolayer is adsorbed at the crystal surface.

On the basis of the stoichiometry of monomer loading versus Fe(III) or Cu(II) present in the zeolite, it can be concluded that in the zeolite Y samples prepared via vapor-phase saturation, only about 20% of the adsorbed pyrrole molecules can be oxidatively coupled. The remainder of the monomer is probably still present in the zeolite pores. The maximum exchange level of ~10 Fe/u.c. was chosen to avoid potential zeolite lattice breakdown (54). In all other samples, the relative excess of Fe(III) is expected to suffice for a complete reaction.

The rate of polymerization as estimated from the rate of sample coloration is much faster in a three-dimensional zeolite (zeolite Y) than in a one-dimensional pore system (mordenite). All these data taken together provide clear evidence that the polymerization of pyrrole proceeds within the zeolite channel system if the monomer has access into the pores.

The reaction medium has an interesting influence on the intrazeolite concentration of the polymer. In contrast to homogeneous solution preparations, water is one of the most unfavorable media for intrazeolite polymerizations. Reactions in hydrocarbon solvents and via vapor-phase loading result in the highest polymer yields. Possibly, water screens the intrazeolite redox-active metal ions from the monomer molecules.

Electronic Structure and Transport Properties. *Electronic Spectra.* The Fe(III)-containing zeolite hosts exhibit a broad absorption between 300 and 450 nm, which accounts for the light brown color of these zeolites. The electronic spectra of the sodium and ferrous forms do not differ significantly. If pyrrole is adsorbed into the sodium forms and ferrous forms of both zeolites, the optical spectra in the visible range remain unchanged.

In contrast, the spectrum of sample Fe(III)Y–V shows two important new features at 400–500 nm (~2.7 eV) and at energies lower than 650 nm (Figure 4D).

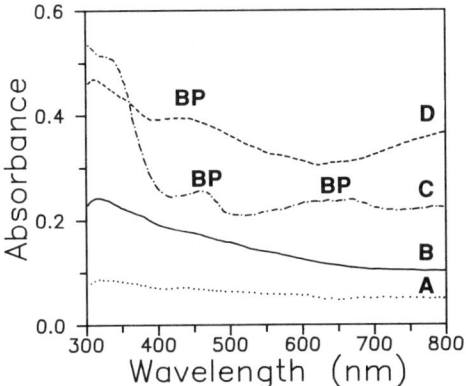

Figure 4. Electronic absorption spectra of samples. Key: A, NaM; B, NaY; C, Fe(III)M–V; and D, Fe(III)Y–V.

The spectral features of the samples prepared in toluene and with vapor-phase pyrrole are more pronounced than those of the other samples. The red absorption of Y zeolite samples is generally stronger than that of mordenite. In addition, the vapor-phase product Fe(III)M–V shows a feature between 520 and 700 nm (~2.0 eV, Figure 4C). Absorption maxima at 2.3 and 0.7 eV have been observed with electrochemically formed, highly doped polypyrrole; the higher energy band shifted to lower values at lower oxidation levels (63). On the basis of the electronic model discussed, we assign the bands at 2.7 and 2.0 eV and that in the near-IR region to bipolaron absorptions that are typical of polypyrrole at different oxidation levels. Thus, we can conclude that the oxidation level of PPy in faujasite is high and that there is probably a bimodal distribution of highly oxidized and medium oxidation levels of polypyrrole in mordenite. The conduction band absorption appears to occur in the near-UV range. Finally, the spectrum of Cu(II)A–V does not exhibit any of the features observed with the large-pore samples (not shown).

ESR Spectra. Polypyrrole in Fe(III)Y–V and Fe(III)M–V shows an ESR signal at proportionality factor $g = 2.0027$, very similar to the bulk value (13). The spin densities calculated from the ESR lines and the monomer loadings are 3×10^{-4} per pyrrole monomer in sample Fe(III)Y–V, and 9×10^{-4} per pyrrole monomer in sample Fe(III)M–V. Because the degree of polymerization is not precisely known, these numbers are lower limits with respect to the polymer content. The corresponding spin susceptibilities show Curie behavior between ~40 and 300 K (Figure 5A). These data are explained

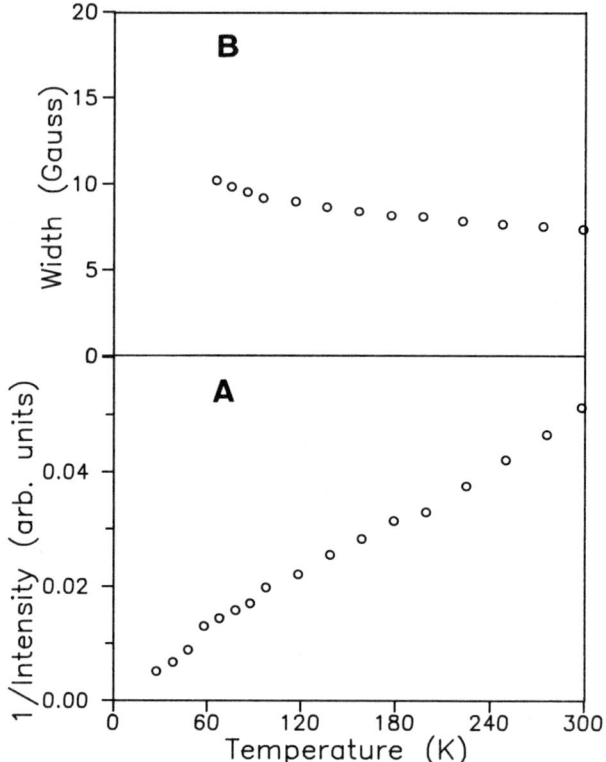

Figure 5. ESR data of sample Fe(III)Y–V. Key: A, Curie behavior; B, line width as function of temperature.

with a small concentration of paramagnetic defects that have also been observed in bulk polypyrrole samples, probably a result of defects in the π system of the polymer chain (13). Surprisingly, the line width of the ESR resonance of intrazeolite polymers is at the order of 10 G (Figure 5B), in contrast to a typical value of 0.2 G for pure bulk polymers. In the bulk, the line width can be reversibly increased to similar values by adding several weight percents of oxygen (13). Because our samples were measured in vacuo, the most likely explanation for the large line width is an intrinsic dipolar interaction between the polymer chains and the charged zeolite channel walls.

Conductivity measurements of polypyrrole–zeolite samples show dramatic differences compared to bulk properties. Although values for bulk polypyrrole are at the order of 10 S/cm (8), the zeolite samples have powder conductivities below 10^{-9} S/cm. These dc measurements are not considered to give the conductivity of the polymer chains because shorter chains could reside in the zeolite cages without providing a continuous conduction path. This striking result is expected, however, because "perfect" samples that have clean, uncoated dielectric crystal surfaces could not provide any continuous conducting path through a pressed wafer. Thus, this observation provides additional evidence for the strictly intrazeolite location of the conjugated polymers. Initial X-ray photoelectron spectroscopy measurements of related intrazeolite polyaniline samples clearly indicate a depletion of nitrogen at the zeolite crystal surfaces.

Conclusions

This study shows that pyrrole can be oxidatively polymerized in the channel system of zeolite molecular sieves. Comprehensive spectroscopic characterization identifies the nature of the polymerization product and provides strong evidence for encapsulation within the channel systems. This approach to isolated "molecular wires" via intrahost polymerization is presently being extended to a variety of other organic and inorganic conducting systems. More detailed studies will address issues such as the intrazeolite polymer chain length, effects of host pore size, and transport properties of the resulting low-dimensional systems.

Acknowledgments

Instrumentation used in this work was acquired with funding from the National Science Foundation. We thank A. Galuska (Sandia National Laboratories) for X-ray photoelectron spectroscopy experiments and B. Crawford (University of New Mexico) for Raman measurements.

References

1. *Proceedings of the International Conference on Science and Technology of Synthetic Metals;* Aldissi, M., Ed.; *Synth. Met.* **1989**, *28*(1–3) and *29*(1).
2. *Handbook of Conducting Polymers;* Skotheim, T. A., Ed.; Marcel Dekker: New York, 1986; Vol. 1.
3. *Conducting Polymers: Special Applications;* Alcacer, L., Ed.; D. Reidel: Dordrecht, Netherlands, 1987.
4. *Molecular Electronic Devices;* Carter, F. L., Ed.; Marcel Dekker: New York, 1982.
5. *Molecular Electronic Devices II;* Carter, F. L., Ed.; Marcel Dekker: New York, 1987.
6. Street, G. B. *Handbook of Conducting Polymers;* Skotheim, T. A., Ed.; Marcel Dekker: New York, 1986; Vol. 1, p 265.
7. Armes, S. P. *Synth. Met.* **1987**, *20*, 365.
8. Rapi, S.; Bocchi, V.; Gardini, G. P. *Synth. Met.* **1988**, *24*, 217.
9. Chao, T. H.; March, J. *J. Polym. Sci., Polym. Chem. Ed.* **1988**, *26*, 743.
10. Pron, A.; Suwalski, J.; Lefrant, S. *Synth. Met.* **1987**, *18*, 25.
11. Bredas, J. L.; Themans, B.; Andre, J. M.; Chance, R. R.; Silbey, R. *Synth. Met.* **1984**, *9*, 265.

12. Genoud, F.; Guglielmi, M.; Nechtschein, M.; Genies, E.; Salmon, M. *Phys. Rev. Lett.* **1985**, *55*, 118.
13. Scott, J. C.; Pfluger, P.; Krounbi, M. T.; Street, G. B. *Phys. Rev. B* **1983**, *28*, 2140.
14. Parise, J. B.; MacDougall, J.; Herron, N.; Farlee, R. D.; Sleight, A. W.; Wang, Y.; Bein, T.; Moller, K.; Moroney, L. *Inorg. Chem.* **1988**, *27*, 221.
15. Wang, Y.; Herron, N. *J. Phys. Chem.* **1987**, *91*, 257.
16. Herron, N.; Wang, Y.; Eddy, M. M.; Stucky, G. D.; Cox, D. E.; Moller, K.; Bein, T. *J. Am. Chem. Soc.* **1989**, *111*, 530.
17. Moller, K.; Eddy, M. M.; Stucky, G. D.; Herron, N.; Bein, T. *J. Am. Chem. Soc.* **1989**, *111*, 2564.
18. Bein, T.; Enzel, P. *Angew. Chem.* **1990**, in press.
19. Bein, T.; Enzel, P. *Synth. Met.* **1989**, *29*, E163.
20. Enzel, P.; Bein, T. *J. Phys. Chem.* **1989**, *93*, 6270.
21. Enzel, P.; Bein, T. *J. Chem. Soc., Chem. Commun.* **1989**, 1326.
22. Farina, M.; Di Silvestro, G.; Sozzani, P. *Crystallographically Ordered Polymers;* Sandman, D. J., Ed.; ACS Symposium Series 337; American Chemical Society: Washington, DC, 1987; 79, and references cited therein.
23. Penner, R. M.; Martin, C. R. *J. Electrochem. Soc.* **1986**, *133*, 2206.
24. Cai, Z.; Martin, C. R. *J. Am. Chem. Soc.* **1989**, *111*, 4138.
25. Mizoguchi, K.; Kume, K.; Masubuchi, S.; Shirakawa, H. *Synth. Met.* **1987**, *17*, 405.
26. Horton, M. E.; Friend, R. H.; Foot, P. J. S.; Billingham, N.; Calvert, P. D. *Synth. Met.* **1987**, *17*, 395.
27. Mulazzi, E.; Brivio, G. P.; Lefrant, S.; Faulques, E.; Perrin, E. *Synth. Met.* **1987**, *17*, 325.
28. Akagi, K.; Katayama, S.; Shirakawa, H.; Araya, K.; Muko, A.; Narahara, T. *Synth. Met.* **1987**, *17*, 241.
29. Park, Y. W.; Lee, Y. S.; Kim, Y. K.; Lee, C. K.; Park, C.; Shirakawa, H.; Akagi, K.; Kitagaki, T.; Katayama, S. *Synth. Met.* **1987**, *17*, 539.
30. Day, P. *Handbook of Conducting Polymers;* Skotheim, T. A., Ed.; Marcel Dekker: New York, 1986, 117.
31. Arend, H.; Huber, W.; Mischgofsky, F. H.; Richter-van Leeuwen, G. K. *J. Cryst. Growth* **1978**, *43*, 213.
32. Tieke, B.; Chapuis, G. *Crystallographically Ordered Polymers;* Sandman, D. J., Ed.; ACS Symposium Series 337; American Chemical Society: Washington, DC, 1987; p 61.
33. Tieke, B.; Wegner, G. *Makromol. Chem. Rapid Commun.* **1981**, *2*, 543.
34. Kanatzidis, M. G.; Tonge, L. M.; Marks, T. J.; Marcy, H. O.; Kannewurf, C. R. *J. Am. Chem. Soc.* **1987**, *109*, 3797.
35. Kanatzidis, M. G.; Hubbard, M.; Tonge, L. M.; Marks, T. J.; Marcy, H. O.; Kannewurf, C. R. *Synth. Met.* **1989**, *28*, C89.
36. Kanatzidis, M. G.; Wu, C.-G.; Marcy, H. O.; Kannewurf, C. R. *J. Am. Chem. Soc.* **1989**, *111*, 4139.
37. Kauzlarich, S. M.; Stanton, J. L.; Faber, J., Jr.; Averill, B. A. *J. Am. Chem. Soc.* **1986**, *108*, 7946.
38. Kauzlarich, S. M.; Teo, B. K.; Averill, B. A. *Inorg. Chem.* **1986**, *25*, 1209.
39. Inoue, H.; Yoneyama, H. *J. Electroanal. Chem. Interfacial Electrochem.* **1987**, *233*, 291.
40. Kuo, T.; O'Brien, D. F. *J. Am. Chem. Soc.* **1988**, *110*, 7571, and references cited therein.
41. Bocchi, V.; Gardini, G. P. *J. Chem. Soc., Chem. Commun.* **1986**, 148.
42. Bjorklund, R. B.; Lundstrom, I. *J. Electron. Mater.* **1984**, *13*, 211.

43. Aldebert, P.; Audebert, P.; Armand, M.; Bidan, G.; Pineri, M. *J. Chem. Soc., Chem. Commun.* **1986**, 1636.
44. Breck, D. W. *Zeolite Molecular Sieves;* R. E. Krieger: Malabar, FL, 1984.
45. *Zeolites: Facts, Figures, Future;* Jacobs, P. A.; van Santen, R. A., Eds.; *Stud. Surf. Sci. Catal.* **1989**, *49*, Elsevier: Amsterdam.
46. *New Developments in Zeolite Science and Technology;* Murakami, Y.; Iijima, A; Ward, J. W., Eds.; Kodansha: Tokyo, 1986.
47. Szostak, R. *Molecular Sieves: Principles of Synthesis and Identification;* Van Nostrand Reinhold: New York, 1989.
48. Kokotailo, G. T.; Lawton, S. L.; Olson, D. H.; Meier, W. M. *Nature* **1978**, *272*, 437.
49. Ball, W. J.; Dwyer, J.; Garforth, A. A.; Smith, W. J. *New Developments in Zeolite Science and Technology;* Murakami, Y.; Iijima, A.; Ward, J. W., Eds.; Kodansha: Tokyo, 1986; p 137.
50. Messina, C. A.; Lok, B. M.; Flanigen, E. M. U.S. Patent 4 544 143, 1985.
51. Dutta, P. K.; Puri, M. *J. Catal.* **1988**, *111*, 453.
52. Heaviside, J.; Hendra, P. J.; Tsai, P.; Cooney, R. P. *J. Chem. Soc., Faraday Trans. 1* **1978**, *74*, 2542.
53. Dalla Betta, R. A.; Garten, R. L.; Boudart, M. *J. Catal.* **1976**, *41*, 40.
54. Pearce, J. R.; Mortier, W. J.; Uytterhoeven, J. B.; Lunsford, J. H. *J. Chem. Soc., Faraday Trans. 1* **1981**, *77*, 937.
55. Myers, R. E. *J. Electron. Mater.* **1986**, *15*, 61.
56. Smits, F. M. *Bell Syst. Tech. J.* **1958**, *37*, 711.
57. Pron, A.; Kucharski, Z.; Budrowski, C.; Zagorska, M.; Krichene, S.; Suwalski, J.; Dehe, G.; Lefrant, S. *J. Chem. Phys.* **1985**, *83*, 5923.
58. Neoh, K. G.; Tan, T. C.; Kang, E. T. *Polymer* **1988**, *29*, 553.
59. Zagorska, M.; Pron, A.; Lefrant, S.; Kucharski, Z.; Suwalski, J.; Bernier, P. *Synth. Met.* **1987**, *18*, 43.
60. Kanatzidis, M. G., personal communication; see also refs. 34, 35.
61. Cheung, K. M.; Smith, B. J. E.; Batchelder, D. N.; Bloor, D. *Synth. Met.* **1987**, *21*, 249.
62. Olk, C. H.; Beetz, C. P., Jr.; Heremans, J. *J. Mater. Res.* **1988**, *3*, 984.
63. Kaufman, J. H.; Colaneri, N.; Scott, J. C.; Kanazawa, K. K.; Street, G. B. *Mol. Cryst. Liq. Cryst.* **1985**, *118*, 171.

RECEIVED for review May 1, 1989. ACCEPTED revised manuscript October 17, 1989.

INDEXES

Author Index

Anderson, D. R., 253
Barrows, Julie N., 403
Bartlett, Neil, 391
Bashkin, J. S., 147
Bein, Thomas, 433
Beuneu, Francois, 433
Brunschwig, Bruce S., 65
Burdett, Jeremy K., 323
Cabana, Leonardo A., 101
Chong, Kul Ryu, 211
Corbett, John D., 369
Dixon, Dabney White, 161
Durham, Bill, 181
Elliott, C. Michael, 211
Enzel, Patricia, 433
Epstein, Arthur J., 419
Ferrere, S., 211
Goodenough, John B., 287
Gozashti, Saeed, 225
Hagiwara, Rika, 391
Hahm, Seung, 181
Headford, C. L. E., 211
Hoffman, Brian M., 125
Hong, Xiaole, 161
Hopkins, J. B., 253
Hush, N. S., 27
Isied, Stephan S., 91
Johnson, Michael K., xv
Kannewurf, Carl R., 351
King, R. Bruce, xv
Kourtakis, Kostantinos, 391
Kulkarni, Gururaj V., 323
Kurtz, Jr., Donald M., xv
Kutal, Charles, xv
Lappin, A. Graham, 237
Lerner, Michael, 391
Lewis, Nita A., 197
Lian, Ping Pan, 181

Long, Joan, 181
Mallouk, Thomas E., 391
Marcy, Henry O., 351
Marks, Tobin J., 351
Marusak, Rosemary A., 237
McLendon, G., 147
Miller, Joel S., 419
Millett, Francis, 181
Norton, Michael L., xv
Okino, Fujio, 391
Ondrechen, Mary Jo, 225
Orman, L. K., 253
Payne, Martin W., 369
Pope, Michael T., 403
Ratner, Mark A., 125
Reimers, J. R., 27
Richeson, Darrin S., 351
Rogel, Friedhelm, 369
Rosenthal, Guy L., 391
Schanze, Kirk S., 101
Schmehl, Russell H., 211
Scott, Robert A., xv
Shields, Thomas P., 237
Sutin, Norman, 65
Taveras, Daniel V., 197
Tonge, Lauren M., 351
Wallin, Sten A., 125
Wessels, Bruce W., 351
Whangbo, Myung-Hwan, 269
Williams, R. J. P., 3
Yabe, T., 253
Zhang, Jie, 369
Zhang, Jiming, 351
Zhang, Li-Tai, 225
Zhao, Jing, 225
Zhou, Feimeng, 225
Zuppiroli, Libero, 433

Affiliation Index

Brookhaven National Laboratory, 65
Colorado State University, 211
E. I. du Pont de Nemours and
  Company, 419
École Polytechnique, 433
Georgetown University, 403
Georgia State University, 161
Iowa State University, 369
Lawrence Berkeley Laboratory, 391
Louisiana State University, 253
North Carolina State University, 269
Northwestern University, 125, 225, 351
Ohio State University, 419
Rutgers, The State University of New
  Jersey, 91
Tulane University, 211
University of Arkansas, 181
University of California—
  Berkeley, 391
University of Chicago, 323
University of Florida, 101
University of Georgia, xv
University of Miami, 197
University of New Mexico, 433
University of Notre Dame, 237
University of Oxford, 3
University of Rochester, 147
University of Sydney, 27
University of Texas at Austin, 287

Subject Index

A

Acceptor–solvent coupling, 29–30
  collision frequency, 30
  rate constant, 30
Activation entropy and enthalpy
  distance dependence, 76
  peptide-linked compounds, 119
Activation parameter, analysis of data, 119
Adventitious impurities, cluster halides, 370
Air-stable precursors, vapor deposition, 352
Alkyl-linked systems, electron transfer, 212
Angular-overlap model
  energies of the d orbitals, 326–328
  energy-level diagrams, 327f
Anion-binding sites, on the protein surface, 249–250
Anisotropy
  motion along the potential surface, 141
  small and large, 142
Annealed film, variable-temperature four-probe resistivity data, 356
Annealing techniques
  film charge-transport characteristics, 356
  grain structure and electrical properties, 366
Anomalous superconductivity, 293
Antiferromagnetic coupling
  predicted, 427
  schematic illustration of stabilization, 421f
  stabilization, 420
Antiferromagnetic fluctuations, superconductor compositions, 311
Antiferromagnetic insulator, 329
Antiferromagnetic ordering
  net spins of the lattice sites, 279
  temperature, 310
Antisymmetric coordinate, 227
Apomyoglobin, interaction of fluorescent probes, 153
Aromatic compounds
  long-range electron tranfer, 16t
  See also Oxidases
Aromatic rings, vibronic-rotational motion, 13
Association constant, proper geometry for electron transfer, 167

B

Back-electron-transfer rate
  estimation, 219
  observed and calculated, 220
  quenching rates, 220
Band calculations, La_2CuO_4, 314f

Band dispersion
  first Brillouin zone, 276
  ratio of resonance integrals, 272–273
Band energies, nature, 272
Band orbital mixing, 269–286
Bath, transfer of energy, 29
Bi–phenyl bond, hydrolytic stability, 358
Bi–Sr–Ca–Cu–O film
  charge-transport data, 359
  electrical resistivity data, 361f
  film microstructure, 358
  resistivity data, 358f
  scanning electron micrograph, 360f
  surface grain sizes, 359
  X-ray diffraction pattern, 359f
Bimolecular electron transfer
  in biological systems, 168
  rate constants, 162
Binding sites, on protein surface, 249–250
Binuclear ruthenium complexes, bridging pyrazine and bipyridine, 97
Biological electron transfer, overview, 3–23
Biological systems, electron transfer, 65–88
Bipolarons, doped polymer chains, 435
Bleaching effect, ground-state bands, 256
Blue copper proteins
  geometry around metal ion, 3–4
  stiffness, 11
  See also Copper proteins
Bond-length mismatch
  fluorite-type intergrowth layer, 317
  internal stresses, 319
Born–Oppenheimer potential energy surfaces, rate processes, 137
Breathing motion
  CuO_2 sheet of 1-2-3 compound, 343
  density changes, 431f
  stabilization energy, 342f
Bridge-length dependence
  contour plots, 56f
  electron-transfer rates, 53–59
  highly conducting bridge, 57
  Joachim's rate constant, 54
  nonexponential, 60
Bridge eigenstates, couplings of the donor and acceptor levels, 58–59
Bridge length, Rabi rate constant and Joachim's effective two-level rate constant, 55f
Bridge states, practical approach to calculations, 51
Bridged chains, 229–230

INDEX

Bridged systems
  acceptor state probability versus reduced time, 43f
  conductance of the bridge, 28
  dimeric complexes
    discrete, 226–229
    efforts to synthesize, 225
    one-dimensional analogues, 226
  electron transfer and energy transfer, 27–63
  formalism, 27–63
  maximum rate constant, 41
  mixed-valence, 225–235
  model system for electron transfer, 28
  not resonant with donor or acceptor states, 41
  pendulation frequency, 42
Bridging
  cluster halides, 371
  infinite cluster chain, 382f
  shared atoms, 381
Bridging ligand
  control of spectroscopic and conductive properties, 233
  derivatized metalloprotein, 74
  donor electron density, 75
  extended Hückel theory, 74–75
  physical properties, 215t
  redox and spectroscopic properties, 214–215
  spectra of the discrete bridged dimer, 226
  tailored to create a discrete dimer, 229
Brillouin zone
  2D rectangular lattice, 272f
  wave vectors, 272–273
Bulk antiferromagnetic coupling, correlation of spins on adjacent chains, 429
Bulk ferromagnetic behavior
  configuration admixing, 430f
  model for stabilization, 429
  model system, 422–429
  structural features, 426

C

c-Axis anomaly, 339
C–F bond, generation, 400
$C_{1.3}F$, stability and characteristics, 400
Carbon-atom sheets
  effect of AsF_5, 399–400
  relationship to each other, 399
Ceramic superconductors, 287–322
Chain atom coordination, flexibility, 335
Chain length
  electron transfer, 212
  photoinduced electron transfer, 215–220
Charge-density wave state
  dynamic and independent, 281
  instability, 280–282

Charge-density wave state—*Continued*
  long-range order, 281–282
  periodic density variation, 278
Charge-transfer salt, *See* Electron-transfer salt
Chiral induction
  inner-sphere reaction, 244
  metal–ion complexes in electron-transfer reactions, 237–252
  outer-sphere reaction, 244
  precursor complex stereoselectivity, 238
Chiral recognition, *See* Chiral induction
Chromatographic separation of products, 240f
Chromophore–quencher compounds, photoinduced intramolecular ET, 102
Circular dichroism spectra, 242f
Clathrated monomers, radiation-induced inclusion polymerization, 436
Cluster(s)
  body-centered-cubic array, 375–376
  bridged network, 371
  condensed, 377
  18-electron, 376
  electronic requirements, 370
  infinite chain, 377
  kinetic stability, 370
  reducing agents, 383
  rhombohedral array, 376
Cluster anions, 370
Cluster halides
  centered, 369–389
  iodides of rare-earth metals, 378
  solution process, 381–383
  stability of alternate phases, 372
  synthesis, 371–372
Cluster network, zirconium chloride, 377f
CO-rebinding problem, 142
1–2–3 Compound
  computed band structure, 334–336
  defect perovskites, 333
  geometrical effects, 336
  structure, 333f, 334f
  vacancy problem, 333
  variation of T_c with oxygen stoichiometry, 334f
2–1–4 Compound
  distortion, 330
  removal of electron density, 336
  strontium doping, 335
  superconductivity and value of x, 329
2–2–1–3 Compound
  analogy with 1–2–3 compound, 345
  observed structure, 345f
  schematic band picture, 345f
  structural features, 344–346
Conducting polymers
  model systems, 436
  stabilization in large-pore zeolite channels, 433–449

Conduction bands, Group B metals, 290–292
Conductivity
  bridged, 229–230
  dependent upon bridging species, 230
  dependent upon electron occupation, 230
  metal center in a protein sphere, 9
  mixed-valent solids, 8
Configuration mixing, model, 420
Conformational changes
  cytochrome c, 17
  diagonal processes, 131
  electron transfer, 77–80
  interconversion, 125–146
  intramolecular electron-transfer phenomena, 143
  linear alkyl polymers, 217
  synchronous reactions, 131
Conformational dynamics, electron-transfer kinetics, 131–136
Conformational energetics, nonadiabatic electron transfer, 130–131
Conformational equilibrium, to modulate electron-transfer rates, 128
Conformational interconversion, See Conformational changes
Conformational isomerization
  electron-transfer properties, 82
  peptide spacers, 107–114
Conformational motion
  characteristic times for the coordinates, 139
  CO rebinding to myoglobin, 139
Conformational reactions, detection methods, 143
Conformational variation, 127–131
Conformers
  detection methods, 132
  gating conformational change, 136
  parallel reactions, 134
  reactive, 136
Conjugated polymers, intrazeolite location, 447
Cooper pairs
  coupling, 316
  electron–phonon interactions, 284
  pairs of electrons, opposite wave vectors, 283
Cooperative magnetic interactions, structural motif, 428
Coordination shell plasticity, 325–326
Coordination sphere, protein fold, 4
$[Co(ox)_3]^{3-}$
  after dialysis, 248f
  obtained in reaction, 248
Copper magnetic moment
  empty σ^* band states, 311
  La_2CuO_4, 311
  suppression, 313–316

Copper oxide superconductors, 323–347
  d electrons, 288
  electronic structure, 324–329
  geometrical arrangements, 325f
  geometrical control of superconductivity, 323
  mobile holes, 311–313
  structural features, 308–313
  structure–property relationship, 284
Copper proteins
  electron transfer, 18
  structure, 18
Correlation splitting
  copper oxide superconductors, 311
  La_2CuO_4, 314f
  mixed-valent configuration, 308
Coupled electron tranfer, 18–20
Coupled nuclear modes of motion, 127
Covalent hybridization
  crystalline field, 301
  effective charge on ions, 288
  repulsion between bonding and antibonding states, 288
Creutz–Taube ion
  changes in intermolecular distances, 200
  electronic coupling between metal centers, 199
Cross reactions
  cytochromes c and b_5, 175–176
  cytochromes c and c_{551}, 173–174
Crystal-packing forces, 326
Crystal structure
  copper atoms of varying lengths, 336
  diffraction patterns, 282
  periodicity, 298
Crystalline field
  splitting, 301
  wave functions, 301–302
Cu–O antibonding interaction, angle-dependent, 332
Cu–O bond
  distances, 325
  in-plane, 284–285
  length
    abrupt transitions, 311–313
    antiferromagnetic phase, 313
    energy changes, 340
    orthorhombic distortion, 318
    thermal vibration parameters, 341
Cubic framework, structural considerations, 316
Cubic perovskites
  conduction bands, 290
  energy densities, 291f
CuO_2 plane
  construction of energy bands, 328f
  perovskite structure, 324f
  stabilization of distortion, 341
  variations, 325

INDEX

Cytochrome–cytochrome reactions,
  electron-transfer rate constants, 174f
Cytochrome b, characterization, 163–166
Cytochrome b_5, structure, 170f
Cytochrome b_{562}, structure, 20f–21f
Cytochrome c
  change in structure, 16f–17f
  characterization, 163–166
  directional electron transfer, 84–86, 91–100
  electron transfer, 74–75
  electrostatic potentials, 15f
  factors that control electron transfer, 161–179
  geometry around metal ion, 3–4
  heme edge and anion-binding sites, 250f
  labeled at specific lysine groups, 181–193
  mobility map, 10f
  properties, 9–12
  reaction with cyanide, 16f–17f
  stiffness, 11
Cytochrome c derivatives
  electron transfer
    (bpy)$_2$(imid)Ru((His)cyt c), 190
    (bpy)$_2$Ru(dcbpy–cyt c), 186–190, 191f, 192f
  synthesis
    (bpy)$_2$(imid)Ru((His)cyt c), 184–186, 187f, 188f
    (bpy)$_2$Ru(dcbpy–cyt c), 182–184, 183f, 185f, 186f
Cytochrome c_{551}
  characterization, 163–166
  reaction with cytochrome c, 173–174
  reorganizational energy, 173
Cytochromes, electron self-exchange reactions, 161–179

D

D and A eigenstates, discontinuities, 44–47
d-Block oxides, 300–308
d-Block transition-metal oxides, 287
Damped solvent motion, nuclear frequency factor, 152
"Dead-end" mechanism, 250
Decamethylferrocene with tetracyanoethylene
  ^{57}Fe Mossbauer spectra, 426
  ferromagnetic ground-state behavior, 422–429
  high-temperature susceptibility, 422
  properties, 425t
  reciprocal susceptibility, 424f
  saturation magnetization, 422
  stabilization of bulk ferromagnetic behavior, 429
  Zeeman split spectra, 426
Decay path, determination, 216

Decay rate constant, partial decoupling of ET and conformational processes, 143
Density distribution
  charge-density wave state, 278
  spin-density wave state, 278
  waves out of phase, 278
Density distributions, site orbitals, 277
Density wave, orbital mixings, 277
Derivatized metalloprotein, cytochrome c, 74
Diamagnetic metallic state, 329
Dielectric-cavity models, permittivity of the space between metals, 203
Dielectric continuum approximations, intermolecular reactions, 198
Dielectric continuum model, solvent reorganization contribution, 198–199
Dielectric spacers, in graphite salts, 398
Diffraction pattern, 1D metal
  temperature dependence, 282
  with CDW, 283f
Diffuse spots, long-range order, 282
Diffusional limitations, pyrrole migrating into zeolite channel system, 444
Diffusive transport, chemical rate processes, 138
Dimerization of cytochromes, 167
Dioxygen molecule, reduction or release, 21
Dipole moments
  calculation from X-ray structures, 169
  function of ionic strength, 169
  self-exchange rate constant and ionic strength, 169
  through the exposed heme edge of cytochromes, 169t
N,N'-Diquaternarized bipyridines, 211–223
Diquaternized species formation, quantum yields, 220t
Directional electron transfer
  complex chemistry associated with protein, 97
  conformational changes, 66
  cytochrome c, 84–86
  low-λ intermediates, 84–86
  mechanisms at solid electrodes, 98, 98f
  molecular events, 98–99
  ruthenium-modified cytochrome c, 91–100
Discontinuity
  identity of eigenstates, 44
  system-dependent effect, 45
Dispersion relation, 273f, 276f
Disproportionation
  Bi^{4+} reaction, 290–292
  energy costs, 343
Distance constraints, electron transfer, 4
Distance dependence
  across oligoproline spacers, 117–118

Distance dependence—*Continued*
  outer-sphere reorganization energy, 120
  peptide-linked compounds, 117f
Distortion
  breathing motion, 339
  charge movement, 339–344
Distortion coordinates, ligand environment, 326
Donor–acceptor complex
  bridged polyproline complexes, 91–93
  transition metal complex at a specific amino acid site, 93–95
Donor–acceptor distance
  predicted reaction rates, 148
  sterically rigid spacer framework, 211
  uncertainty, 148
Doped conjugated polymers, potential applications, 433
Doping
  electronic conductors, 288
  local relaxations of chain geometry, 434
  native defects, 290
  structural stability, 330
  with zinc, 336
Doping levels, production of polarons, 435
Drift mobility, 298–299
Driving force
  electron-transfer reactions, 190
  electron transfer in Os and Ru couple, 92
  photoinduced electron transfer, 105–106
Dynamics
  temporal, 263–264
  thermal, 262–263
  two-color Raman spectroscopy, 262–264

E

Eccentricity of the ellipse, volume of activation, 206f
Effective atomic magnetic moment, 297
Effective resonance integral, electron occupation, 233
Elastic restoring energy, 290
Electrical conductivity, uptake of gaseous fluorine, 400
Electrical conductors, high-oxidation-state carbon, 392
Electron-affinity threshold, thermodynamic basis, 394
Electron-density waves, density distributions, 279
Electron–electron repulsion, 231–232
Electron-exchange reaction, free energy, 67f
Electron-hopping time, correlation-split bands, 308
Electron–phonon interaction
  condensation of Cooper pairs, 290
  coupling, 284
  traditional superconductors, 284

Electron-transfer centers
  blue copper and heme, 20
  Fe and Mn, 20
  Ni–S, 20
  two or more metal ions, 21
  unsolved problems, 22
Electron-transfer distance, recognition zone, 4
Electron-transfer kinetics, conformational dynamics, 131–136
Electron-transfer mechanism, intepretation of stereoselectivity data, 250
Electron-transfer pathways, 85–86
Electron-transfer processes
  general formalism, 30–41
  in terms of system, 59–60
  kinetics, 29–30
Electron-transfer proteins
  analysis, 9–12
  coupled, 5, 7
  examples, 4t
  large-scale motions, 11
  ligands around metal ions, 3
  scavenger, 5–7
  simple, 5–14
  surface dynamics, 13–14
  typical fold, 7f
  uncoupled, 3–5
Electron-transfer rate
  conformation dependence, 122
  cytochrome c, 13t
  dependence on donor–acceptor separation, 211
  dielectric relaxation effects, 221
  direction of electron transfer, 85–86
  distance dependence, 65, 118
  effect of aromatic groups, 12
  electronic factor, 69–75
  enhancement, 60–61
  interior side chains, 12–13
  measurement, 65–66
  nonadiabatic reaction, 69–71
  normal and inverted regimes, 68–69
  number of C–C bonds, 217f
  Phe-82 movement, 12–13
  population-weighted average, 134
  protein conformations, 65
  rate of conformational change and electronic tunneling, 217
  related to conformation, 121
  temperature dependence, 76, 118, 148
Electron-transfer rate constant
  calculation, 107
  acceptor–solvent coupling, 29–30
  chain length, 212
  controlling factors, 161–162
  in weakly coordinating buffers, 246
  measurement, 162

Electron-transfer rate theory
  nonadiabatic limits, 137
  solvent relaxation, 137–138
Electron-transfer reaction
  conformational parameter, 136
  cytochromes c and b_5, 175
  cytochromes c and c_{551}, 174
  gated, 78t
  in solution
    dielectric cavity models, 203
    ellipsoidal cavity model, 204–206
    hard-spheres model, 202
    high-pressure studies, 197–210
    SMH model, 198, 208
    theoretical models, 197–200
  kinetics, 241
  logarithm of the rate constant, 70f
  models for examining distance effects, 217
  nuclear-configuration space, 78
  pathways, 242
  temperature, concentration, and pH dependence, 93
  time dependence, 32
  two-step mechanism, 80
Electron-transfer reaction rate, bridge-length dependence, 53–59
Electron-transfer salt
  decamethylferrocene with tetracyanoethylene, 422–430
  magnetic properties, 428–429
  one-dimensional chain structure, 427–428
  TCNE, 427
Electron-transfer stereoselectivity, determination, 239–240
Electron-transfer theory
  bridged system, 28–30
  two-level model, 28–29
Electron-transfer time, localized 4f electrons, 298
Electron affinity, oxidative intercalation of graphite, 393–396
Electron energies
  construction, 289f
  MnO, 304f
  three mixed-valent systems, 309f
  TiO, VO, and MnO, 305f
Electron occupation
  polymeric systems, 231
  raising and lowering of orbital energies, 232
Electron populations, lower-energy ligand, 264–265
Electron self-exchange
  in cytochromes, 161–179
  rate constants, 162, 164t, 175
Electron superconductors, 285
Electron superexchange mechanism, 71–74
Electron transfer
  acceptor state as a continuum of levels, 38
  barrier-free regime, 182

Electron transfer—*Continued*
  between centers linked by a bridge, 27–63
  between ligands, 253
  between proteins, experimental approaches, 149
  biological systems, 65–88, 161
  chain of the mitochondrial inner membrane, 6f
  directional, 91–100
  distance dependence in proteins and synthetic systems, 116
  gated, 77–78
  kinetic scheme for donor–acceptor couple, 130f
  mediation by the intervening medium or ligands, 71
  metalloproteins, 74–75
  nonadiabatic, conformational variation, 129–130
  one-direction, 97–99
  photoinduced, 101–124
  rate constants, 115–117
  rate constants and activation parameters, 115t
  reaction of intermediate state, 134
  ruthenium-modified cytochrome c, 85f
  schematic potential-energy surface, 128f, 129f
  theoretical model, 66–69
  through protein, distance dependence, 182
Electronic conductivity, electron mobility, 299–300
Electronic coupling
  along the peptide backbone, 117–118
  biological electron-transfer reactions, 147–159
  distance dependence, 118
  ligand-to-metal charge transfer, 73
  matrix element for exchange reaction, 73
  metal centers, 71–74
  nuclear frequency factor, 152
  outer-sphere electron-exchange reactions, 72–73
  pentaammine pyridine complexes, 73–74
  pressure effects on interactions with medium, 200
  redox sites, 66
  through-bond or through-space, 75
Electronic damping factor
  discrepancies in reported values, 147
  estimation, 150
Electronic energies, construction on the basis of an ionic model, 288
Electronic instability
  Fermi surface nesting, 271, 279–280
  low-dimensional metal, 269–286
  metal–insulator or metal–superconductor transition, 270–271

Electronic instability—*Continued*
 superconductivity, 282–285
Electronic stability
 antiferromagnetic insulating state, 329
 Peierls-type distortion, 329
Electronic state, singlet and triplet, 265
Electronic structure
 copper oxides, 324–329
 effect on magnetic behavior, 428
 energy levels, 326–329
 solid, 269–271
 unique βIV spectrum, 414
Electronic transitions, vibrational coordinate on the bridging species, 228
Electronic transmission coefficient, 69
Electronic transmission factor, distance dependence, 122
Electrostatic factors, distribution of conformers in solution, 220
Electrostatic interactions, work term, 168–170
Ellipsoidal cavity model
 calculations, 204–206
 direct integration approach, 206
 pressure-dependent expressions, 205
 volume of activation, 206
Emission yields and lifetimes, peptide-linked compounds, 114t
Energy-transfer processes
 acceptor–solvent coupling, 29–30
 dependence on the bath, 29
 two-level systems, 29
Energy bands
 of a solid, 269–271
 1–2–3 structure, 335f
Energy densities, one-electron states, 291f
Energy difference, tetragonal and orthorhombic forms, 330f
Energy gap
 between ligands, 254
 bridged systems, 47f
 donor-to-bridge, 61
 electron populations on lower-energy ligand, 264
 lowest-lying triplet MLCT states, 265
 reduction potentials, 262
Energy level
 one-electron theory, 333
 x^2-y^2 orbital, 328
Energy transfer, between centers linked by a bridge, 27–63
Enhancement factor, electron transfer at the heme edge, 166
Entropy change, Re–DMAB system, 119
Equilibrium, complex ion with protein, 247
Equilibrium nuclear configuration, electron-transfer reaction, 78–79
Excitation, bpy-localized MLCT state, 262

Excited-state decay, electron-transfer-quenching pathway, 187
Excited-state electron transfer
 coulombic and dynamic conformational changes, 220
 effects of medium, 221t
 flexible chain-linked donor–acceptor complexes, 221
 See also Photoinduced electron transfer
Excited-state manifold, Re–DMAB complexes, 107
Excited states, spectroscopic results, 255–262
Exponential decay, bridge length, 57
Exponential time-decay, donor to acceptor states, 61
Extended-chain bridged complexes, occupation-dependent conductive properties, 233
Extended-chain polymeric system, conductive properties, 225

F

Fast electron transfer, thermal equilibrium between two dissimilar ligands, 262
Faujasite structure, 438f
Fe–S proteins, structural features, 8t
Fe(III)–O–Fe(III) bridges, intrazeolite oxidation, 438–440
Fermi energy, copper oxide superconductors, 312
Fermi surface
 definition, 274
 dimensional representation, 276–277
 distorted circles, 274–275
 filled wave vectors, 274f
 metals and nonmetals, 274
Fermi surface nesting
 description, 271–276
 electronic instability, 279–280
 metal–insulator transition, 285
 superconducting state, 285
 vectors, 275
Fermion energies
 half-filled band, 294f
 like-atom interatomic resonance integral b, 306f
Ferredoxin, 7f
Ferromagnet, EuO, 297
Ferromagnetic coupling, 419–432
 alternating donor–acceptor one-dimensional chain model, 420
 McConnell mechanism, 431
 model system, 422–429
 stabilization
  by configuration mixing, 420
  mechanism, 428
 model, 426–427
 schematic illustration, 421

Ferromagnetic coupling—*Continued*
  two-electron configurations, 427
Ferromagnetically coupled chains, stable radicals, 426
Ferromagnetism, in organic-based molecular systems, 431
Flash photolysis, gated reactions, 127
Flash photolysis experiment, peptide-linked compounds, 113–115
Flash photolytic excitation, intramolecular electron transfer, 126–127
Fluorescence, 2,6-ANSDMA–apomyoglobin complex, 156f
Fluorescence decay curves
  2,6-ANSDMA–apomyoglobin complex, 154f
  experimental method, 157
Fluorescence emission spectra, 2,6-ANSDMA–apomyoglobin complex, 155f
Fluorescent probes
  decay lifetimes, 153
  protein dynamics, 153
Fluorides, electron affinity for intercalation 393–396
Fluorinated graphite, formation mechanism, 400–401
Fluorine–carbon bonding, electrical conductivity, 392
Fluorine–ligand exchange, mechanism, 400
Fluoroanions, spontaneous intercalation of graphite, 391–402
Formal-valence ambiguity, La$_2$CuO$_4$, 315f
Franck–Condon principle, electron-transfer event, 127
Free energy
  electron transfer, 67
  equal barriers, 81f, 82f, 83f
  function of nuclear configurations, 68
  inverted, 69
  normal, 68–69
  rate processes, 137
  role in interligand electron transfer, 253–266
  ruthenium-modified cytochrome c, 86f
  sections through the parabolic basins, 68f
Freedom, in larger molecules, 125
Frictions, motions in proteins, 139–143

G

Gas-phase reactivity, vapor deposition, 352
Gated electron transfer
  gating within intermediate, 135–136
  limiting case of the general solution, 133–134
  mechanisms, 77–78
  preequilibrium, 134

Gated reactions
  degree of conformational control, 131
  energetics and dynamics, 125–146
  general solution to kinetic equations, 129
Geometric modifications of the polymer, charge-transfer-induced, 434
Golden rule regime
  effective energy gap, 49
  effective two-state coupling, 49
  rate constant, 45
  symmetry, 49
Graphite, oxidative intercalation by fluoroanionic species, 391
Graphite–AsF$_5$ intercalation compound, in-plane specific conductivity, 392
Graphite band structures, conductivity, 392
Graphite-sheet structure, nearest-neighbor distance, 398
Groove openings, cyanide Fe(III), 17
Guest species
  disordered, 400
  effective volume, 396
  interaction of graphite with AsF$_5$, 392

H

Hand-in-glove fitting, 14
Hard spheres model, solvent rearrangement, 202
Helical proteins, adjustable, 12
Helix movements
  coupled, 19
  cross-linking, 19
Heme-edge-to-heme-edge complex
  electron transfer, 172
  electrostatic repulsion, 175
  interaction free energy, 171f
  model, 172
Heme–heme distances, crystal structures and solution, 166
Heme exposure, reorganizational energy, 166
Heme proteins
  bimolecular electron transfer, 163
  coupled helix movements, 19f
  exposed heme edge, 163
  factors that control electron transfer, 161–179
  fraction of protein surface, 163–166
  structural changes, 18
Hemoglobin, conformational change, 18
Heteroaromatic polymers
  conduction mechanism, 436
  ground-state geometry, 434–435
  stabilization in large-pore zeolite channels, 433–449
Heteropoly blues, *See* Molybdophosphate heteropoly anions
High-pressure data
  interpretation, 208

High-pressure data—*Continued*
  test of electron-transfer theories, 208
High-pressure studies, electron-transfer reactions in solution, 197–198
High-spin configuration, 302
High-T_c copper oxides
  common features, 319
  superconductors, 316–319
  transition-metal oxide, 308–317
Hole superconductors, 284
Hole superexchange mechanism, 71–74
Hop mechanism, charge movement, 8
Horse cytochrome c(II)
  dependence of rate on pH, 246f
  rate constants, 245t
Hubbard gap, antiferromagnetic insulating state, 329
Hückel Hamiltonian, matrix of eigenvalues, 58
Hydridization fluctuations, 316
Hydrogen bonding through the carboxylate face of the oxidant, 238

I

Inclusion polymerization, steric control, 436
Insulator
  band energy gap at the Fermi level, 280
  See also Semiconductor
Intercalation of graphite, by fluoroanions, 391–402
Inter-complex-ion charge transfer, mixed-valent crystals, 9
Intercluster bridging modes, 371
Intercluster cavities
  size of interstitial atom, 374
  sizes and numbers, 372
Interligand electron transfer
  apparent rate, 265t
  coupling to metal-centered d–d states, 265
  free energy dependence, 254
  rate, 254
  role of free energy, 253–266
Interpenetrating networks, zirconium cluster chlorides, 375f
Interstitial elements,
  C_2, 377
  centered cluster halides, 369
  crystal-structure studies, 370
  effect on shape of the cluster, 374
  14- or 18-electron closed-shell, 376, 378
  zirconium and scandium phases, 370
Intersystem crossing efficiency, relative changes, 216
Intervalence-transfer absorption, 212
Intervalence-transfer band, symmetrical mixed-valence molecule, 198

Intraatomic energy considerations, transition-metal oxides, 302
Intraatomic interactions, 4f electrons, 296
Intramolecular electron transfer
  cyclic kinetic scheme, 126–127, 126f
  distance dependence, 218
  driving force, 93
  environmental effects, 220–221
  medium effects, 207
  methanol solution, 113
  notation, 82–84
  outer-sphere reorganization energy, 93
  photoinduced, 211–223
  rate constant, 94, 162, 216t
  rates and distances, 92t
  ruthenium site and heme site, 93
  temperature dependence, 92, 218
  theoretical model, 66–69
  thermodynamics and kinetics, 105–107
  three-step mechanism, 82–84
Intramolecular redox reactions, 198–199
Intrazeolite polymerization, reaction medium, 444
Intrazeolite polypyrrole
  electronic absorption spectra, 445f
  electronic structure and transport properties, 444–447
  ESR data, 446f
  IR spectra, 443f
  reaction scheme, 441
  sample characteristics, 441–444
  See also Polypyrrole
Inverted region, conformational isomers, 82
Iodide clusters
  infinite chains of bridged clusters, 381
  rare-earth metals, 378
  structure, 379f, 380f
Ion recombination, 221
Iron-centered clusters, dimensions, 388
Iron proteins
  electron transfer, 18
  structure, 16f–17f
Isomerization reactions, control by gating, 143–144
Isotropic shifts, ^{17}O NMR lines of αI, 411

K

Keggin structure, 12-molybdophosphate anion, 404
  bond and polyhedral representations, 405f
  tetrahedral symmetry, 409
Kinetic behavior, Re–DMAB complexes, 106–107

L

Lattice deformation
  coupling of Cooper pairs, 317
  electron–phonon interactions, 279
  moving electron, 284

Lattice energy
  correlation with volume, 397f
  effect of volume, 397–398
  levels, 271–276
Lattice enthalpy, effect of volume, 396–398
Lattice instabilities, semiconductor–metal transitions, 290
Lattice vibration, perturbation causing CDW state, 279
Layer perovskite halide salts, 437
Ligand fluorination, volatility, 352
Ligand parentage, 257
Ligands, in electron-transfer proteins, 3
Linear-chain structure
  alternating donor–acceptor, 423f
  charge-transfer salts, 428
Liquid crystal polymerization, 437
Local coordination geometries, energy levels, 326–329
Local deformation, 308
Localized electron configuration, transition-metal oxides, 303–308
Localized magnetic moment, transition-metal cation, 302
Long-distance electron transfer
  between proteins, 147–159
  rate constant, 147–148
  through-bond electronic coupling, 212
Low-dimensional metals, electronic instability, 269–286
Low-dimensional polymer structures, stabilization, 436–437
Luminescence decay
  in CH_3CN solution, 120–121
  related to conformation, 121
Luminescence-decay-rate constants, temperature dependence, 218f
Luminescence lifetimes, 216t

M

Macromolecule, conformational interconversion, 125
Macroscopic ferromagnetism, interchain spin alignment, 429
Madelung stabilization, MnO, 303
Magnetic coupling, configuration admixing, 429
Magnetic ordering temperature, antiferromagnetic interaction, 302
Magnetic phenomena, cooperative, 419
Magnetic susceptibility
  TCNE electron-transfer salt, 427
  variation with temperature, 423–424
Manifold energies, conduction and valence bands, 298f, 299f
Marcus inverted region, 103
Matrix effect
  cluster halides, 372
  Zr–Cl interactions, 388

McConnell mechanism
  design of a molecular organic ferromagnet, 431
  nonresonant bridge limit, 51
Mechanisms, means for differentiating, 251
Medium effects, volume of activation, 207
Metal, energy bands, 270
Metal–bridge electronic coupling, delocalization of unpaired electron, 229
Metal coordination spheres, electron-transfer sites, 4
Metal–insulator transition
  nested Fermi surface, 276
  one-electron band theory, 271
  orbital mixing destroys Fermi surface, 280
  series of steps, 280–282
Metal ions, sites in proteins, 3
Metal–ligand electronic coupling, electron occupation, 233
Metal orbitals, delocalization, 343
Metal–oxygen interactions, controlling geometry, 332
Metal–superconductor transition, 1D metal, 282
Metal-to-ligand charge transfer
  photoinduced, 253
  pyridine, 73
  to metal-centered d–d states, 266
Metallic antiferromagnet, GdO, 297
Metallic state
  electronic states derived by orbital mixing, 276–279
  nested Fermi surface, 276
Metalloproteins
  chiral induction, 238
  conformational isomers, 82
  electron transfer, 74–75
  electron transfer in biological systems, 3
  electron-transfer properties, 77
  electron-transfer reactions, 181
Microscopic reversibility, reaction rate constant, 50
Microscopic spin interactions, single-crystal susceptibility, 422–423
Mixed-valence configuration, transition-metal oxides, 308
Mixed-valence dinuclear complexes, ellipsoidal cavity model, 204–206
Mobile holes
  abrupt transitions, 313
  band states, 313
  concentration, 300
  copper oxide superconductors, 311–313
Modified protein, characterization, 93
Molecular dynamics, proteins, 11
Molecular electronics
  bridged systems, 27
  electron-transfer compounds, 28

Molecular electronics—*Continued*
 maximum rate constant, 41
 properties desired in device, 52
Molecular modeling program, peptide structures, 113
Molecular orbitals, metal–ligand bonds, 72–73
Molecular system, surrounding bath, 60
Molecular wires
 host-stabilized one-dimensional molecular conductors, 433
 organic and inorganic conducting systems, 447
Molybdenum blue method, phosphate analysis, 403–404
12-Molybdophosphate anion, reduction behavior, 404
Molybdophosphate heteropoly anions
 bond representation, 408f
 electrochemistry, 407
 four-electron blues
  crystal structure, 411–412
  electronic structure, 414
  intervalence charge-transfer band profiles, 416t
  NMR data, 413t
  optical spectrum of βIV, 414
  proton exchange, 412–414
  ring-current effects, 414–416
 intramolecular electron transfer and electron delocalization, 403–417
 one- and two-electron blues, 408–411
  electron exchange, 410–411
  hyperfine structure, 409
  intramolecular hopping, 409
  isotropic shifts, 411
  oxygen-17 NMR data, 410t
 polyhedral representation, 409f
 preparation of compounds, 406–407
 spectroscopy, 408
 tetrabutylammonium salts, 410f
Mo•••Mo separation
 βIV crystal, 404
 ring-current shielding, 414–416
Monoclinic–tetragonal transition, charge-density wave state, 293
Mordenite
 rate of polymerization, 444
 structure, 438f
Multimode reaction dynamics, 137–143

N

Natural-abundance spectrum of βIV anion, 412f
Néel temperature, antiferromagnetic interaction, 302
Nephelauxetic effect, transition metal complexes, 343
Nesting vector
 Fermi surface extended in two wave-vector directions, 275–276
 Fermi surfaces, 275, 279
Nestled salt
 ions in ordered domains, 398–400
 structural model, 399f
Neutral–ionic transition, stabilization of ferromagnetic behavior, 426
Nonadiabatic electron transfer
 across peptide spacers, 121
 distance dependence, 116
Nonexponential behavior, rate constant, 57
Nonmetallic electronic states, orbital mixing, 276
Nonmonotonic decay
 bridge length, 54–57
 discontinuities in Joachim's expression, 57
Nonresonant bridge limit, McConnell's approach, 51
Nuclear configuration, energy of reactants and products, 79f
Nuclear frequency factor
 electron-transfer rate, 151
 solvent dynamics, 152
Nuclear modes of motion, 127
Nuclear prefactor, for protein systems, 155
Nuclear reorganization energy, 75–84

O

Occupation-dependent resonance energy, 232–233
Octahedral cluster, compression of isolated octahedra, 381
Oligoproline peptides, conformational changes, 103
On-site repulsion
 perturbation causing SDW state, 279
 two electrons in the x^2-y^2 band, 342
One-color spectrum, MLCT state, 264f
One-dimensional chains
 bridged dimeric complexes, 226
 metal ions and bridging ligands, 229–230
One-dimensional metal
 CDW instability as function of temperature, 281f
 open Fermi surface, 274
One-electron calculations, band structure of copper atoms in CuO_2 planes, 328
One-electron terms, prediction of properties, 231
One-electron transfer, horse cytochrome c(II) and $[Co(ox)_3]^{3-}$, 245
Optical density
 depletion of solvent bands, 257
 transient MLCT absorption, 256

Index

Orbital degeneracies
  stabilization of ferromagnetic coupling, 431
  t^3 POMO, 427
Orbital mixing
  CDW, SDW, or superconducting state, 285
  ideal 1D band structure, 276
  near the Fermi level of normal metallic state, 269–286
  perturbation introduced to a metallic system, 279
  spin and charge-density changes, 277
  superconducting state or insulating state, 271
  superconductivity, 282
Ordered oxygen vacancies, high-T_c copper oxide phases, 316
Organometallic chemical vapor deposition, 351–368
  advantages, 352
  Bi–Sr–Ca–Cu–O films, 357–359
  description of process, 354
  development, 353
  issues to be addressed, 366
  precursor design and deposition methodology, 352–354
  Tl–Ba–Ca–Cu–O films, 359–363
  Y–Ba–Cu–O films, 354–357
Organometallic electron-transfer salts with tetracyanoethylene, 419–432
Orthorhombic distortion
  bond-length matching, 317–318
  compressive stress on the Cu–O bond length, 317–318
  coordination number at copper, 332
  stabilization, 332–333
  superconductivity, 317
Orthorhombic symmetry
  displacive transition, 308f
  La_2CuO_4, 310
Oscillator strength, pressure, 199–200
Oscillatory behavior, rate constant, 57
Outer-sphere electron-transfer reactions, 197–198
Outer-sphere stereoselectivity, symmetry and conformational flexibility, 244
Oxalate-bridged inner-sphere intermediate, 244f
Oxidases
  electron-transfer reactions, 14–16
  See also Aromatic compounds
Oxidation, deposition precursors, 353
Oxidative intercalation of graphite
  electron affinity of the oxidizing half reaction, 395f
  electron affinity values, 393–396
  fluorine access, 393
  fluoroanions, 391–402

Oxidative intercalation of graphite—
  Continued
  in-plane resistivity, 392
  thermodynamic cycle, 394f
  thermodynamic model, 393
  volume and lattice enthalpy, 396–398
Oxides
  main-group, 288–295
  metallic conductivity, 288
  rare-earth, 296–300
  single-valent versus mixed-valent, 287–322
  transition-metal, 300–308
Oxidized clusters, accessibility, 383–384
Oxygen displacements
  ideal-perovskite positions, 292f
  orthorhombic symmetry, 310f
Oxygen stoichiometry
  1–2–3 compound, 334
  changes in geometry, 340f
  changes in properties, 336
  geometrical and electronic structure, 336–339
  rigid-band model, 336
  variation of band structure, 337f
Oxygen vacancies, structural order, 338

P

Pair functions, interactions, 284
Pauli paramagnetic susceptibility
  transition from an antiferromagnetic semiconductor, 303
  VO, 303
Peierls distortions, in organic polymers, 434
Peierls-type distortion, 329
Pendulation frequency, two-level and bridged systems, 42
Peptide, ability to mediate electronic coupling, 117–118
Peptide-linked electron-donor–acceptor systems, structures, 104
Peptide spacer
  activation entropy, 119
  conformational isomerization, 107–114
  distance dependence of k_{ET}, 103
  electron transfer, 101–124
Perovskite
  arrangement, 324
  layer perovskites, 437
  structural considerations, 316–319, 324f
Perturbation, causing the CDW state, 279
Perturbation expansion
  direct-coupling and through-bridge terms, 49
  fourth-order terms, 48
  nonresonance condition, 45
Perturbation theory
  evaluation of rate constant, 49

Perturbation theory—*Continued*
  second-order, 302
  two-level systems, 33–41
Phase diagram
  $Ba_{1-x}K_xBiO_3$, 296f
  $La_{2-x}Sr_xCuO_4$, 312f
Phase stability, zirconium cluster chlorides, 374
Phonon
  low-frequency, 284
  perturbation causing CDW state, 279
Phosphate ion
  binding on protein surface, 246
  rate dependence, 247f
Photoexcitation, in interligand electron transfer, 253
Photoexcited Ru(II) diimine complexes, 211–223
Photoinduced electron transfer
  across peptide spacers, 101–124
  bimolecular systems, 102
  chain-length dependence, 212–213, 215–220
  chromophore–quencher compounds, 102
  dependence on number of bonds, 212
  driving force, 106t
  electrochemical potentials, 106t
  emission energies, 106t
  energy-level diagram, 106f
  kinetics, 101–124, 132–134
  kinetics across rigid oligoproline spacers, 103–105
  porphyrin–quinone complexes, 211–223
  processes, 131
  rate constants, 213
  See also Excited-state electron transfer
Photolysis, rate of back electron transfer, 218–219
Photosynthesis
  bridged system, 28
  systems that mimic, 103
Physical vapor deposition techniques, high-T_c superconductors, 351
Planar copper atoms, $x_2–y_2$ bands, 339
Plane-to-chain charge transfer, variation in structure, 335
Point-charge model, 288
Polar-state stabilization, 292–293
Polarizable bridging ligands, spectroscopic and conductive properties, 225–235
Polarization fluctuations, 316
Polaron
  doped polymer chains, 435
  lattice deformation, 318–319
  mobility, 318–319
Polymer chains
  conductivity, 446–447

Polymer chains—*Continued*
  encapsulation in ordered hosts, 433
  zeolite matrix, 436
Polymerization in clathrates, high-energy radiation, 436
Polypeptide donor–acceptor complexes, directional electron transfer, 91–93
Polypyridyl complexes of Ru(II), photochemistry, 253
Polypyrrole
  band structure as a function of doping level, 435f
  composites with porous materials, 437
  conductivity measurements, 446–447
  electrochemical oxidative polymerization of pyrrole, 434
  formation of protons, 442
  intrazeolite oxidation, 438–440
  oxidation level in faujasite and mordenite, 445
  paramagnetic defects, 446
  sample characteristics, 441–444
    concentration in zeolites, 442
    location of the polymer chains, 442–444
    Raman spectra, 442
    ring vibrations, 442
  See also Intrazeolite polypyrrole
Porphyrin–quinone complexes, 211–223
Potential surface
  anisotropy, 141
  CO rebinding to myoglobin, 139
  dynamics on, 137
  ligand protein rebinding reaction, 140f, 141f
  nonuniform frictional forces, 139
Potential well, large bipolaron, 318
Precursor complex, model, 238
Precursor concentration, time dependence, 143
Probe laser, low energy density, 255
Products formed, proportion, 241f
Proline ring carbons, ^{13}C NMR spectra, 109f–112f
Protein(s)
  compared with ordered crystal solids, 9–10
  fluctuation, 10
  molecular dynamics, 11
  relaxation times, 11, 153–155
  structure and function, 11
Protein conformational changes
  distance dependence, 66
  rate constant, 94
  reaction rates, 128
Protein dimerization, 167
Protein donor–acceptor complexes, directional electron transfer, 93–95
Protein dynamics
  biological electron-transfer reactions, 147–159

Protein dynamics—*Continued*
  fluorescent probes, 153–156
  influence on electron-transfer rates, 155–156
  picosecond–nanosecond time scales, 147
Protein fold, geometry around the metal ion, 3–4
Protein matrix, advantages, 8
Protein mobility
  anisotropic diffusion, 144
  classical transition state, 139
  effective frictions, 139
  internal motion, 138–139
  investigations, 166–167
Protein–protein electron transfer, dependence, 172
Protein structure, associated with change of oxidation state, 5
Protein surface, anion-binding sites, 250
Proton-transfer reactions, control by gating, 143–144
Protonation–redox switch, conformational change, 19
Protonolysis, precursor metal–ligand bonds, 354
Puckering distortion
  coordination number at copper, 332
  driving force, 330–332
  electron repulsion, 343–344
  energy-level diagrams, 331f
  energy stabilization, 331
  magnitude of electron transfer, 344
Pulse radiolysis, 127
Pyrolysis of metal carboxylate coatings, preferential orientation, 356
Pyrrole
  diffusion into large-pore zeolites that contain ferric ions, 441
  oxidative polymerization, 434f, 447
  uptake by zeolites, 444

Q

Quenching
  histidine derivatives, 190
  kinetics, 115
  luminescence intensity and lifetime, 187
  peptide-linked compounds, 113
  predominant mechanism, 115
Quinoid structure, doped aromatic polymer, 435

R

Rabi maximum regime, nonresonant bridge, 50
Radial extension, 4f electrons, 297
Radiationless decay, mixing of singlet state with all electronic states available, 266
Rare-earth ions, valence, 297
Rare-earth metal clusters, 376–381
  iodide systems, 378
  stability with 4d and 5d metal atoms, 378–381
Rare-earth oxides, 296–300
  characteristics, 297–300
  electronic structure, 296–297
Rate constant
  acceptor-to-bath coupling, 39
  complete resonance, 53
  expression for a bridged system, 45
  H_{eff} method and "Exact" method, 42
  interpolating function, 39
  microscopic, 53
  microscopic reversibility, 50
  nonexponential region, 57
  nonmonotonic behavior, 54
  observed and theory, 189–190
  Rabi maximum, 39–41
  rate law for two-level systems, 49
Reaction driving force, electron-transfer proteins, 5
Reaction rate
  control, 125–146
  presence of resonances, 61
Redox potential
  cytochrome c and cytochrome c_{551}, 174
  electron-transfer proteins, 5
Reorganization barrier, electron transfer within a derivative, 190
Reorganization energy
  consecutive conformation changes and electron transfer, 77–80
  cytochromes c and b_5, 175
  dependence on steric factor, 173
  distance dependence, 75–77, 218
  electron-transfer rates, 75–84, 168t
  inverted region, 80–82
  molecular shape, 76
  self-exchange, 172–173
  single-sphere model, 77
  temperature dependence, 76
  three-step mechanisms, 82–84
Reorganization parameter, free-energy surfaces, 67–68
Resistivity, as a function of temperature, 269
Resistivity data, rapid thermal annealing, 356–357
Resonance
  H_{eff} expression, 45
  McConnell's approach, 51
  nonresonant bridge levels, 51
  two-level and bridged systems, 44
Resonance Raman enhancement
  Franck–Condon factor, 261
  magnitude, 262
Resonant bridge coupling, model, 52–53
Ridge-line crossing, friction, 139–140

Ring-current effects
  electron delocalization, 414
  enhanced diamagnetic susceptibilities, 414
Rock-salt structure
  MgO, 288
  with BO_2 planes, 317
Room-temperature lattice parameters vs. quench temperatures, $La_2CuO_{4.05}$, 316f
Ruthenium-modified cytochrome c
  characterization studies, 95
  directional electron transfer, 91–100
  intramolecular electron transfer and reduction potential, 95t
  oxidation and reduction from the same inorganic modifier, 97
  rate of reduction, 96
  reducing agent, 96
  relative position of heme and ruthenium sites, 94f
  reorganization energy, 96–97
Ruthenium(II) polypyridine cytochrome c derivatives, electron-transfer kinetics, 181–193
Ruthenium reagents, ligands, 96f
Rutile, geometry, 331

S

Saddle-point crossing, friction, 139–140
Salt formation, admixture with fluorine, 394
Scavenger proteins
  represented by cytochrome c, 6f
  surfaces, 13–14
Screening of anion from anion, carbon sheets, 398
Second-order rate constant, 242f
Self-exchange rate, ligand atoms, 190
Self-exchange rate constant, cytochromes, 162, 173
Semiconductor
  crystals, examples, 8t
  energy bands, 270
  insulator, 270
  temperature, 269
Separated-dielectric-spheres model, permittivity, 203
Single chain, zeolite matrix, 436
Singlet electronic state
  electronic coupling between different ligands, 265
  energy gaps, 265
Site orbitals, density, 277
Small polaron, mobile electron and its local deformation, 298
Solution conformations, peptide-linked compounds, 114f
Solvent dynamics, 151–152
Solvent permittivity, 206

Solvent relaxation effects, dynamic extensions, 137–138
Solvent reorganization energy, free energy changes, 205
Space correlations, real vs. reciprocal, 280–282
Spacer structures
  chromophore–quencher compounds, 102–103
  rate of electron transfer, 102–103
Spectral congestion, 259
Spectroscopic results, polypyridyl complexes of Ru(II), 255–262
Spin-density wave state, periodic unequal densities, 278–279
Spin alignment, bulk ferromagnetism, 420
Spin direction, conduction electrons, 300
Spin polarization, total energy of a system, 279
Spinless conductivity, bipolarons, 435
Splitting
  crystalline-field, 301
  interatomic correlation, 301
Spontaneous magnetization, polycrystalline samples, 422
Stability, metallic state, 269
Stabilization energy
  charge-density wave state, 293
  orthorhombic structure, 332
Stereoselective oxidant, probe of mechanisms of electron transfer with metalloproteins, 238
Stereoselectivity
  analysis, 237
  chiral induction in reactions with metalloproteins, 244
  comparisons, 238
  detection, 243, 247
  optical activity, 248
  oxidation of horse cytochrome c(II), 249t
  source, 249
Steric considerations, 163–167
Steric factor
  evaluation, 166
  reorganizational energy, 173
Stretch alignment, directional anisotropy in preformed polymers, 437
Structural transformation, transition temperature, 330
Superconducting materials, mixed-valence compounds with polarizable oxide-bridging ligand, 230–231
Superconductivity
  bond angles, 317
  ceramic materials, 287–322
  compositional range, 293, 310–311
  Cooper pairs, 283
  copper oxide electron-superconductors, 285

INDEX

Superconductivity—*Continued*
  copper oxide hole-
    superconductors, 284
  critical temperature, 231, 351–368
  electronic instability, 282–285
  extended Hubbard models, 232
  Fermi surface nesting character, 282–285
  geometrical control, 323–347
  high-T_c, 287
  high-T_c copper oxide superconductors, 308–317
  interaction of occupied and unoccupied pair function, 284
  mechanism and structure, 323
  mixed-valence compounds of copper, 226
  narrow compositional range, 288
  orbital mixing among the band levels, 282–285
  perturbations causing orbital mixing, 285
  planar copper atoms, 339
  preparation of thin films, 351–368
  stabilization, 319
  temperature, 269
Superexchange mixing, nuclear configuration, 75
Superexchange spin–spin interaction, 302
Superlattice spots, long-range order, 282
Symmetric coordinate, 227
Symmetric trisubstituted complexes, 259
Symmetry, two-level systems, 36
Synchronous processes
  electron transfer, 131
  intramolecular electron transfer, 143

T

Temporal dynamics, 263
Tetrabutylammonium salts, electronic spectra, 415f
Tetracyanoethylene, 419–432
Tetragonal symmetry, CuO_6 octahedra, 311
Tetragonal T structure
  $La_{1.85}Sr_{0.15}CuO_4$, 310f
  Nd_2CuO_4, 318f
Thermal activation, higher energy singlet states, 266
Thermal equilibrium, temperature dependence of Raman spectrum, 262
Thermal processes, electron transfer, 131
Thermal rate constants, calculation from optical data, 208
Thermodynamic control, electron-transfer proteins, 5
Three-step mechanism, electron-transfer reaction, 82
Through-bridge kinetics, McConnell's approach, 51
Time delays, between pump and probe lasers, 262
Time-response function, two-level and bridged systems, 44

Tl–Ba–Ca–Cu–O film
  charge-transport characteristics, 363
  electrical resistivity data, 365f
  precursor, 359–362
  preparation methods, 362–363
  scanning electron micrograph, 364f
  stability, 359
  X-ray diffraction patterns, 362f
Total energy, one- and two-electron contributions, 231–233
Transient absorption, $Ru(bpy)_2(dcbpy$–cyt c) derivatives, 187–188
Transient Raman spectrum with two-color excitation, 256f
Transition energies, bridge-dependent quantities, 229
Transition metal(s)
  encapsulated in zirconium structures, 373
  group-three and group-four, 369–389
Transition-metal oxides, 300–308
Transition temperatures
  half-filled band, 295f
  like-atom interatomic resonance integral b, 307f
Translational motion, CO rebinding to myoglobin, 139
Translational symmetry, conduction bands, 290
Trap states, out of the conduction band, 300
Triplet electronic state, energy gaps, 265–266
Triplet–triplet energy transfer
  electronic damping factor, 147
  equivalence of electron transfer, 150–151
  rate, 150–151
Tunneling, spiro binuclear system, 200
Tunneling electron transfer, protein-bound redox sites, 103
Two-color spectra
  MLCT state, 263f
  one-color signal, 257f
  Raman spectra, 258f, 259f, 260f, 261f
Two-dimensional metal, closed Fermi surface, 274–275
Two-dimensional superconducting systems, 230–231
Two-electron calculations
  distortion of the CuO_2 planes, 342
  energy difference between high- and low-spin, 342
Two-level systems
  energy-transfer model, 29
  Fermi's golden rule, 38
  formulae for maximum rate constant, 39–41
  large-λ expansion, 36–38
  mid-range in λ, 38–41
  probabilities, 33
  small-λ expansion, 33–36

Two-level systems—*Continued*
   symmetric resonant, 34f, 35f
   symmetry relation, 36
   unit quantum yield, 33
Two-state models, nonresonant bridge
      limit, 61

U

Unstable conformer
   high-λ intermediate, 79
   low-λ intermediate, 78

V

Vibrational bands, assignment with respect to ligand parentage, 257
Vibrational coupling, dinuclear complex with solvent molecules, 199
Vibrational difference coordinate, 227
Vibrational modes, totally symmetric, 228
Vibrational sum coordinate, 227
Vibronic coupling, pressure, 199–200
Vibronic coupling problem, 142
Vibronic wells, Marcus potential energy diagram, 200
Volatility, vapor deposition, 352
Volume, effect on lattice enthalpy, 396–398
Volume of activation
   agreement of model and experimental system, 207–208
   distance between reacting species, 198
   intramolecular electron-transfer process, 207
   medium effects, 207
   solvent rearrangement, 201
   theoretical calculations, 200–202

W

Water, protonolytic reactant gas, 363
Wave vector
   Fermi surface, 274
   first Brillouin zone, 272, 273
   proportional to band filling, 273

Work term
   association constant, 167
   electrostatic considerations, 168–170

Y

Y–Ba–Cu–O film
   preferential orientation, 355–356
   scanning electron micrograph, 357f
   variable-temperature four-probe resistivity data, 356f
   X-ray diffraction patterns, 355f

Z

Zeolite(s)
   channel structures, 438
   description, 437–438
   large-pore, 433–449
   microporous host structures, 437–440
   structure types, 439t
Zeolite–pyrrole samples, 441
Zeolite Y
   maximum exchange level, 444
   rate of polymerization, 444
Zirconium cluster chlorides
   bond lengths, 385
   cluster dimensions, 385t, 386t, 387t
   D_{2d} symmetry, 384f
   enclosure of main-group elements, 373–374
   incorporation of iron within the cluster, 383
   interpenetrating networks, 374–375, 377f
   molecular orbitals, 372
   nonaqueous solutions, 381–388
   oxidized clusters, 383–384
   range of electronic stabilities, 381
   reducing agents, 383
   solution and precipitation, 381–383
Zirconium halides, contrasting behaviors, 373
^{3}Zn cyt c delayed emission, decay data, 151f

Copy editing and indexing: Colleen P. Stamm
Production: Donna Lucas
Acquisitions: Cheryl J. Shanks
Cover Design: Amy Meyer Phifer

Typeset by Techna Type, Inc., York, PA
Printed and bound by Maple Press, York, PA

The paper in this book meets the minimum requirements of American National Standard for Information Sciences—Permanence of Paper for Printed Library Materials, ANSI Z39.48—1984 ∞

Other ACS Books

Metal–DNA Chemistry
Edited by Thomas D. Tullius
ACS Symposium Series 402; 213 pp; ISBN 0-8412-1660-6

The Challenge of d and f Electrons
Edited by Dennis R. Salahub and Michael C. Zerner
ACS Symposium Series 394; 405 pp; ISBN 0-8412-1628-2

Silicon-Based Polymer Science: A Comprehensive Resource
Edited by John M. Zeigler and F. W. Gordon Fearon
Advances in Chemistry Series 224; 801 pp; ISBN 0-8412-1546-4

Chemical Structure Software for Personal Computers
Edited by Daniel E. Meyer, Wendy A. Warr, and Richard A. Love
ACS Professional Reference Book; 107 pp;
clothbound, ISBN 0-8412-1538-3; paperback, ISBN 0-8412-1539-1

Biotechnology and Materials Science: Chemistry for the Future
Edited by Mary L. Good
160 pp; clothbound, ISBN 0-8412-1472-7;
paperback, ISBN 0-8412-1473-5

Cancer: The Outlaw Cell, Second Edition
Edited by Richard E. LaFond
274 pp; clothbound, ISBN 0-8412-1419-0;
paperback, ISBN 0-8412-1420-4

The ACS Style Guide: A Manual for Authors and Editors
Edited by Janet S. Dodd
264 pp; clothbound, ISBN 0-8412-0917-0;
paperback, ISBN 0-8412-0943-X

For further information and a free catalog of ACS books, contact:
American Chemical Society
Distribution Office, Department 225
1155 16th Street, NW, Washington, DC 20036
Telephone 800-227-5558